MONOGRAPHS AND RESEARCH NOTES IN MATHEMATICS

Application of Fuzzy Logic to Social Choice Theory

John N. Mordeson

Creighton University
Omaha, Nebraska, USA

Davender S. Malik

Creighton University
Omaha, Nebraska, USA

Terry D. Clark

Creighton University
Omaha, Nebraska, USA

CRC Press
Taylor & Francis Group
Boca Raton   London   New York

CRC Press is an imprint of the
Taylor & Francis Group, an **informa** business

A CHAPMAN & HALL BOOK

First published 2015 by Chapman and Hall

Published 2019 by CRC Press
Taylor & Francis Group
6000 Broken Sound Parkway NW, Suite 300
Boca Raton, FL 33487-2742

© 2015 by Taylor & Francis Group, LLC
CRC Press is an imprint of Taylor & Francis Group, an Informa business

First issued in paperback 2019

No claim to original U.S. Government works

ISBN 13: 978-0-367-44583-6 (pbk)
ISBN 13: 978-1-4822-5098-5 (hbk)

Visit the Taylor & Francis Web site at
http://www.taylorandfrancis.com

and the CRC Press Web site at
http://www.crcpress.com

MONOGRAPHS AND RESEARCH NOTES IN MATHEMATICS

Series Editors

John A. Burns
Thomas J. Tucker
Miklos Bona
Michael Ruzhansky
Chi-Kwong Li

Published Titles

Application of Fuzzy Logic to Social Choice Theory, John N. Mordeson, Davender S. Malik and Terry D. Clark

Blow-up Patterns for Higher-Order: Nonlinear Parabolic, Hyperbolic Dispersion and Schrödinger Equations, Victor A. Galaktionov, Enzo L. Mitidieri, and Stanislav Pohozaev

Iterative Optimization in Inverse Problems, Charles L. Byrne

Modeling and Inverse Problems in the Presence of Uncertainty, H. T. Banks, Shuhua Hu, and W. Clayton Thompson

Set Theoretical Aspects of Real Analysis, Alexander B. Kharazishvili

Signal Processing: A Mathematical Approach, Second Edition, Charles L. Byrne

Sinusoids: Theory and Technological Applications, Prem K. Kythe

Special Integrals of Gradshetyn and Ryzhik: the Proofs – Volume I, Victor H. Moll

Forthcoming Titles

Actions and Invariants of Algebraic Groups, Second Edition, Walter Ferrer Santos and Alvaro Rittatore

Analytical Methods for Kolmogorov Equations, Second Edition, Luca Lorenzi

Complex Analysis: Conformal Inequalities and the Bierbach Conjecture, Prem K. Kythe
Cremona Groups and Icosahedron, Ivan Cheltsov and Constantin Shramov

Dictionary of Inequalities, Second Edition, Peter Bullen

Difference Equations: Theory, Applications and Advanced Topics, Third Edition, Ronald E. Mickens

Geometric Modeling and Mesh Generation from Scanned Images, Yongjie Zhang

Groups, Designs, and Linear Algebra, Donald L. Kreher

Handbook of the Tutte Polynomial, Joanna Anthony Ellis-Monaghan and Iain Moffat

Lineability: The Search for Linearity in Mathematics, Juan B. Seoane Sepulveda, Richard W. Aron, Luis Bernal-Gonzalez, and Daniel M. Pellegrinao

Line Integral Methods and Their Applications, Luigi Brugnano and Felice Iaverno

Microlocal Analysis on R^n and on NonCompact Manifolds, Sandro Coriasco

Monomial Algebra, Second Edition, Rafael Villarreal

Partial Differential Equations with Variable Exponents: Variational Methods and Quantitative Analysis, Vicentiu Radulescu

Forthcoming Titles (continued)

Practical Guide to Geometric Regulation for Distributed Parameter Systems,
Eugenio Aulisa and David S. Gilliam

Reconstructions from the Data of Integrals, Victor Palamodov

Special Integrals of Gradshetyn and Ryzhik: the Proofs – Volume II, Victor H. Moll

Stochastic Cauchy Problems in Infinite Dimensions: Generalized and Regularized Solutions, Irina V. Melnikova and Alexei Filinkov

Symmetry and Quantum Mechanics, Scott Corry

Contents

Preface

This book provides a comprehensive study of fuzzy social choice theory. Human thinking is marked by imprecision and vagueness. These are the very qualities that fuzzy logic seeks to capture. Thus, social choice theory suggests itself as a means for modeling the uncertainty and imprecision endemic in social life. Nonetheless, fuzzy logic has seen little application in the social sciences to include social choice theory. We attempt to partially fill this lacuna.

The main focus of Chapter 1 is the concept of a fuzzy maximal subset of a set of alternatives X. A fuzzy maximal subset gives the degree to which the elements of X are maximal with respect to a fuzzy relation on X and a fuzzy subset of X. This allows for alternative notions of maximality not allowed in the crisp case. The main result states that a fuzzy maximal subset is not the zero function if and only if the fuzzy preference relation involved is partially acyclic.

Chapter 2 deals with fuzzy choice functions. Classical revealed preference theory postulates a connection between choices and revealed preferences. If it is assumed that preferences of a set of individuals are not cyclic, then their collective choices are rationalizable. We examine this question in some detail for the fuzzy case. We determine conditions under which a fuzzy choice function is rationalizable. We also present results inspired by Georgescu [24]. Desai [16] characterized the rationality of fuzzy choice functions with reflexive, strongly connected, and quasi-transitive rationalization in terms of path independence and the fuzzy Condorcet property. We consider these results as well as those of Chaudhari and Desai [11] on various types of rationality.

Chapter 3 is concerned with the factorization of a fuzzy preference relation into the "union" of a strict fuzzy relation and an indifference operator, where union here means conorm. Such a factorization was motivated by Fono and Andjiga [12]. The factorization is useful in the examination of Arrowian type results. We show that there is an inclusion reversing correspondence between conorms and fuzzy strict preference relations in factorizations of fuzzy preference relations into their strict preference and indifference components.

In Chapter 4, we consider fuzzy non-Arrowian results. The first author felt that presenting this material before Arrowian type results would help emphasize the difference between the fuzzy case and the crisp case. We provide two types of fuzzy aggregation rules that satisfy certain fuzzy versions of

Arrowian conditions, but which are not dictatorial. We conclude the chapter by presenting the work of Duddy, Perote-Pena and Piggins [17] that demonstrates the central role max-∗ transitivity plays in fuzzy preference theory, and in particular Arrow's Theorem.

Chapter 5 is one of the main chapters of the book. Arrow's Theorem is one of the most important discoveries in social theory. It states that if preferences of individuals are aggregated under certain reasonable conditions: universal admissibility, transitivity, unanimity, and independence of irrelevant alternatives, then the aggregation method must be dictatorial. We consider versions of Arrow's conditions and show that they lead to fuzzy versions of Arrow's Theorem, and versions involving representation rules, oligarchies, and veto players. In the fuzzy situation, there are many types of transitivity, independence of irrelevant alternatives, and strict preference relations. We examine how various combinations of these concepts yield impossibility theorems. In particular, we consider the work of Dutta [11], where he demonstrates that a fuzzy aggregation rule is oligarchic rather than dictatorial when a rather strong definition of transitivity is used. However, dictatorship holds if positive responsiveness is assumed. We also present the ideas introduced by Banerjee [4]. These include the notion that for exact preference relations and fuzzy social preference relations, it is possible to distinguish between different degrees of power of a dictator.

One solution to the Arrowian Impossibility Theorem is to place restrictions on preferences. Among the more promising approaches is that of Black's Median Voter Theorem, which restricts preferences to single-peaked profiles. In Chapter 6, we show that the Median Voter Theorem holds for fuzzy preferences. We also demonstrate that Black's Median Voter Theorem holds for fuzzy single-dimensional spatial models even though there are significant differences between conventional models and their fuzzy counterparts. In crisp spatial models, preferences in space are most often represented by a single ideal point whereas in the fuzzy case, a political actor's ideal policy position encompasses a range of alternatives.

In Chapter 7, we consider the questions of how unambiguous or exact choices are generated by fuzzy preferences and whether exact choices induced by fuzzy preferences satisfy certain plausible rationality relations. The results are based on those of Barrett, Pattanaik, and Salles [9]. Several alternative rules for generating exact choice sets from fuzzy weak preference relations are presented. To what extent exact choice sets generated by these alternative rules satisfy certain fairly weak rationality conditions is also examined. We also consider Banerjee's results [5] concerning the situation where the domain of the choice function consists of all crisp finite subsets of the universal set of alternatives, but where the choice is fuzzy in the sense that the decision maker can state the degree of his choice of an alternative.

We extend some known Arrowian results involving fuzzy set theory to results involving intuitionistic fuzzy sets in Chapter 8. We show how the use of an involutive fuzzy complement can be used to obtain results. Nana and Fono

[24] introduced the concepts of positive transitivity and negative transitivity of a strict component of a fuzzy relation. They allow for the establishment of Arrowian results for crisp and fuzzy preferences.

In social decision-making contexts, a manipulator often has incentive to misrepresent his/her preference. The Gibbard–Satterthwaite theorem states that a social choice function over three or more alternatives that does not provide incentive to individuals to misrepresent their sincere preferences must be dictatorial. In Chapter 9, we extend their result to the case of fuzzy weak preference relations. The results in this chapter are based on those of Abdelaziz, Figueria, and Meddeb [1] and also Duddy, Perote-Pena, and Piggins [13].

In Chapter 10, we present Georgescu's degree of similarity of two fuzzy choice functions [19]. It induces a similarity relation on the set of all choice functions defined on a collection of fuzzy choice functions. The degree of similarity of two fuzzy choice functions allows for the evaluation of how rational a fuzzy choice function is. Thus a ranking of a collection of fuzzy choice functions with respect to their rationality can be obtained. We also consider Georgescu's Arrow index of a fuzzy choice function [22]. The Arrow index provides a degree to which a fuzzy choice function satisfies the fuzzy Arrow axiom. We also present a result that provides a degree to which a fuzzy choice function is fully rational.

John Mordeson dedicates the book to his grandchildren, John, Michael, Matthew, Marc, Jack, Jessica, David, Josh, Emily, Emma, Jenna, Elizabeth, and Isabelle. Davender Malik dedicates the book to his wife Sadhana Malik and daughter Shelly Malik. Terry Clark thanks John Mordeson, mathematics, Creighton University and Mark Wierman, computer science, Creighton University for eight years of intellectually stimulating and amazingly productive collaborative work.

Acknowledgments

We are indebted to the political science students in the mathematics colloquium, especially Crysta Price. We thank Dr. George Haddix and his late wife Sally for their generous endowments to the Department of Mathematics at Creighton University. We thank the journal *New Mathematics and Natural Computation* for allowing us to reuse some of our work.

John N. Mordeson
Davender S. Malik
Terry D. Clark

Chapter 1

Fuzzy Maximal Subsets

One paradigmatic change in science and mathematics this century concerns the concept of uncertainty. It was once thought science should strive for certainty and that uncertainty was regarded as unscientific. An alternate view now believes that uncertainty is essential to science. The publication of Lotfi A. Zadeh's seminal paper, [23], introducing the concept of fuzzy set theory is an important work in the modern concept of uncertainty. Zadeh's paper challenges both Aristotelian two-valued logic and probability theory as being the sole agent for uncertainty. Given a universal set X, a fuzzy subset μ of X is a function of X into the closed interval $[0, 1]$. For an element x in X, the statement that x is in μ is not necessarily true or false, but may be true to some degree. This degree is given by $\mu(x)$.

It is stated by Zadeh in [23] that in the realm of soft sciences, sciences in which imprecision, uncertainty, incompleteness of information and partiality of truth lie at the center rather than the periphery, crisp-set-based mathematics is not adequate as a modeling language. It is the inadequacy of crisp-set-based mathematics as a modeling language that underlies the paucity of solid theories in soft sciences. See also [4]. To overcome these inadequacies, we make use of fuzzy set theory.

By Zadeh's argument [23], we would expect to see considerable interest in fuzzy logic in the social sciences. However, we do not. Why is that so? One reason relates to the division in these disciplines between empirical and theoretical research. The two are often disconnected, with theoretical findings either being ignored by empiricists or those same conclusions having no reasonable empirical referent. This is particularly evident in political science and is the major impetus behind the National Science Foundation's (NSF) initiative to bring the two into more meaningful dialog [9]. Labeled the Empirical Implications of Theoretical Models (EITM) program, the effort has arguably yet to have made any notable impact in the discipline. Thus explicitly formal approaches, to which fuzzy logic might contribute, have a minimal impact on the empirical work that dominates the discipline's journals.

Further exacerbating the problem is that fuzzy logic has not been able to overcome what has become the paradigmatic hold of probability theory on empirical studies in the discipline. Claudio Cioffi-Rivilla [3] was the first to argue that fuzzy mathematics could have important applications in political science. Charles Ragin [11] subsequently proposed a fuzzy logic method for testing necessary and sufficient conditions as well as identifying multiple paths of causality between a set of conditions as well as identifying multiple paths of causality between a set of conditions and an outcome. Pennings [10] used Ragin's method in a cross-national analysis of executive accountability in parliamentary systems; Arfi [1], Koenig-Archibugi [7], and Sanjian [12-17] used fuzzy logic to test hypotheses of decision-making; and Tabor [21] and Seitz [19] employed fuzzy expert systems in computer simulations of regional rivalries.

Despite these promising efforts, the property ranking issue has proven an insurmountable obstacle to the wider application of fuzzy logic in empirical studies in the social sciences. See [20, 22]. Fuzzy logic approaches for establishing correlations between variables require that the data be measured on comparable scales, otherwise there is no way to overcome the charge that the conclusions reached are little more than consequences of how the data are measured. In short, fuzzy logic methods are substantially more sensitive to data measurement issues than are the conventional statistical methods that dominate empirical research in the social sciences.

Fuzzy logic applications in formal-theoretical approaches do not suffer from these problems. This is particularly the case for theoretical work in social choice, a research agenda that has arguably been the most dynamic formal-theoretical agenda in the social sciences. This book is a purposeful effort to lay out how fuzzy logic can contribute to this agenda. In the rest of this chapter, we consider some basic concepts in fuzzy social choice theory. In subsequent chapters, we explore more nuanced issues and applications.

1.1   Fuzzy Set Theory

We begin with a short review of some basic concepts from fuzzy set theory.

Let X be a nonempty set. If A and B are subsets of X, we let $A\backslash B$ denote the set $\{x \in X \mid x \in A, x \notin B\}$. We use the symbol $\cup$ for the union of two sets and $\cap$ for the intersection. A **binary relation** on X is a subset R of the set of ordered pairs $X \times X$. Let R be a binary relation on X. Then R is called **reflexive** if $(x,x) \in R$ for all $x \in X$, **transitive** if for all $x,y,z \in X$, $(x,y),(y,z) \in R$ implies $(x,z) \in R$ and **complete** if for all $x,y \in X$, either $(x,y) \in R$ or $(y,x) \in R$. R is called **symmetric** if for all $x,y \in X, (x,y) \in R$ implies $(y,x) \in R$. R is called **antisymmetric** if for all distinct $x,y \in X, (x,y) \in R$ implies $(y,x) \notin R$ and **irreflexive** if $(x,x) \notin R$ for all $x \in X$. We let $\mathcal{P}(X)$ denote the power set of X, i.e., the set of all subsets of X. We let $\Delta X = \{(x,x) \mid x \in X\}$. If $(x,y) \in R$, we sometimes write xRy. Given R, we define a binary relation P on X in terms of R by for all $x, y \in X$, $(x, y) \in P$ if and only if $(x, y) \in R$, $(y, x) \notin R$. We call P the **strict binary relation associated with** R. If f is a function from a set X into a set Y and A is a subset of X, we use the notation $f|_A$ to denote the restriction of f to A.

A fuzzy subset of X is a function of X into the closed interval $[0, 1]$. We use the notation $\mu : X \rightarrow [0, 1]$ to denote a fuzzy subset μ of X. We let $\mathcal{FP}(X)$ denote the fuzzy power set of X, i.e., the set of all fuzzy subsets of X. If $\mu \in \mathcal{FP}(X)$, we define the **support** of μ, written $\text{Supp}(\mu)$, to be $\{x \in X \mid \mu(x) > 0\}$. The **cosupport** of μ, written $\text{Cosupp}(\mu)$, is defined to be the set $\{x \in X \mid \mu(x) < 1\}$. Let $\mathcal{P}^*(X) = \mathcal{P}(X)\backslash\{\emptyset\}$ and $\mathcal{FP}^*(X) = \{\mu \in \mathcal{FP}(X) \mid \text{Supp}(\mu) \neq \emptyset\}$. If S is a subset of X, we let 1_S denote the **characteristic function** of S in X. That is, 1_S is the function of X into $\{0, 1\}$ such that $1_S(x) = 1$ if $x \in S$ and $1_S(x) = 0$ otherwise. Let $t \in [0, 1]$ and $x \in X$. Define the fuzzy subset $t_{\{x\}}$ of X by $t_{\{x\}}(x) = t$ and $t_{\{x\}}(y) = 0$ if $y \neq x$. Then $t_{\{x\}}$ is called a **fuzzy singleton**. Let $\mu, \nu \in \mathcal{FP}(X)$. Then we write $\mu \subseteq \nu$ if $\mu(x) \leq \nu(x)$ $\forall x \in X$ and we write $\mu \subset \nu$ if $\mu \subseteq \nu$ and there exists $x \in X$ such that $\mu(x) < \nu(x)$. Let $t \in [0, 1]$. We let $\mu_t = \{x \in X \mid \mu(x) \geq t\}$. Then μ_t is called the t-**cut** or the t-**level set** of μ. We let θ denote the fuzzy subset of X that maps all elements of X to 0. We use the notation $\wedge$ to denote the minimum or infimum of a set of real numbers and $\vee$ to denote the maximum or supremum of a set of real numbers. Let $\mathcal{R}(X)$ denote the set of all relations on X. A fuzzy subset of $X \times X$ is called a **fuzzy binary relation** on X or simply a **fuzzy relation** on X. Let $\mathcal{FR}(X)$ denote the set of all fuzzy relations on X. We let $\mathcal{FR}^*(X) = \mathcal{FP}^*(X \times X)$.

We next define the notion of a t-norm, a concept central to our study. The notion of a t-norm first appeared in the context of probabilistic metric spaces Schweizer and Sklar [6]. They were then used as a natural interpretation of the conjunction in semantics of mathematical fuzzy logics, Hajek [4].

Definition 1.1.1 *Let $*$ be a fuzzy binary operation on $[0, 1]$. Then $*$ is called a t-**norm** if the following conditions hold $\forall a, b, c \in [0, 1]$:*
(1) $*(a, 1) = a$ (boundary condition);
(2) $b \leq c$ implies $*(a, b) \leq *(a, c)$ (monotonicity);
(3) $*(a, b) = *(b, a)$ (commutativity);
(4) $*(a, *(b, c)) = *(*(a, b), c)$ (associativity).

Example 1.1.2 *Some basic t-norms are defined as follows. Let $a, b \in [0, 1]$.*
   Standard intersection: *$*(a, b) = a \wedge b$;*
   Algebraic product: *$*(a, b) = ab$;*
   Bounded difference: *$*(a, b) = 0 \vee (a + b - 1)$;*
   Drastic intersection: $*(a, b) = \begin{cases} a & \text{if } b = 1, \\ b & \text{if } a = 1, \\ 0 & \text{otherwise.} \end{cases}$

The bounded difference t-norm is also known as the Lukasiewicz t-norm.

Let $*$ be a t-norm and let $a, b \in [0, 1]$. Then a and b are called **zero divisors** with respect to $*$ if $a \neq 0 \neq b$ and $a * b = 0$.

Let $*$ denote a t-norm on $[0, 1]$. Let μ and ν be fuzzy subsets of X. Define the fuzzy subset $\mu \cap \nu$ of X by $\forall x \in X, (\mu \cap \nu)(x) = \mu(x) * \nu(x)$. (Even though the t-norm $*$ can vary, we use the notation $\cap$ since the context of its use will be clear.)

Definition 1.1.3 *Let $\rho \in \mathcal{FR}^*(X)$. Define the fuzzy relation ι on X by $\forall x, y \in X, \iota(x, y) = \rho(x, y) \wedge \rho(y, x)$. Then ι is called the* **indifference operator** *with respect to ρ.*

A fuzzy binary relation is used to indicate the degree to which each alternative is preferred to every other alternative. The indifference operator indicates the degree to which a player is indifferent with respect to a pair of alternatives.

Definition 1.1.4 *Let $\rho \in \mathcal{FR}^*(X)$ and let $*$ be a t-norm.*
   (1) If $\rho(x, x) = 1$ for all $x \in X$, then ρ is called **reflexive.**
   (2) If $\rho(x, y) = \rho(y, x)$ for all $x, y \in X$, then ρ is called **symmetric.**
   (3) If $\rho(x, y) > 0$ implies $\rho(y, x) = 0$ for all $x, y \in X$, then ρ is called **asymmetric.**
   (4) If $\rho(x, y) * \rho(y, z) \leq \rho(x, z)$ for all $x, y, z \in X$, then ρ is called **max-$*$ transitive.**
   (5) If $\rho(x, y) > 0$ or $\rho(y, x) > 0$ for all $x, y \in X$, then ρ is called **complete.**
   (6) If $\rho(x, y) + \rho(y, x) \geq 1$ for all $x, y \in X$, then ρ is called **strongly complete;**
   (7) If $\rho(x, y) \vee \rho(y, x) = 1$ for all $x, y \in X$, then ρ is called **strongly connected.**

Completeness for a fuzzy relation says that for any two alternatives, the degree of preference of at least one over the other must be positive. This may seem implausible, but an entirely different theory would be developed without the assumption of completeness.

A binary relation R on X is called **acyclic** if for all $x_1, \ldots, x_n \in X$, $x_1 P x_2$, $x_2 P x_3$, $\ldots$, $x_{n-1} P x_n$ implies $x_1 R x_n$.

For $\rho \in \mathcal{FR}^*(X)$, we often associate with ρ an asymmetric fuzzy relation π.

Definition 1.1.5 *Let $\rho \in \mathcal{FR}^*(X)$ and let $*$ be a t-norm. Let π be an asymmetric fuzzy relation associated with ρ.*
   (1) If $\pi(x_1, x_2) * \pi(x_2, x_3) * \ldots * \pi(x_{n-1}, x_n) \leq \rho(x_1, x_n)$ for all $x_1, \ldots, x_n \in X$, then ρ is called **acyclic** *(with respect to $*$)*
   (2) If $\pi(x_1, x_2) * \pi(x_2, x_3) * \ldots * \pi(x_{n-1}, x_n) > 0$ implies $\rho(x_1, x_n) > 0$ for all $x_1, \ldots, x_n \in X$, then ρ is called **partially acyclic** *(with respect to $*$).*

Various types of asymmetric preference relations are used in fuzzy social choice. In the following definition, we list some of the ones that are used the most.

Definition 1.1.6 *Let $\rho \in \mathcal{FR}(X)$. Define the fuzzy binary relations $\pi_{(i)}, i = 0, 1, 2, 3$ as follows:* $\forall x, y \in X$,

(1) $\pi_{(0)}(x, y) = \begin{cases} \rho(x, y) & \text{if } \rho(x, y) > \rho(y, x) = 0 \\ 0 & \text{otherwise.} \end{cases}$

(2) $\pi_{(1)}(x, y) = \begin{cases} \rho(x, y) & \text{if } \rho(x, y) > \rho(y, x) \\ 0 & \text{otherwise.} \end{cases}$

(3) $\pi_{(2)}(x, y) = 1 - \rho(y, x)$.

(4) $\pi_{(3)}(x, y) = 0 \vee (\rho(x, y) - \rho(y, x))$.

The strict preference $\pi_{(1)}$ requires only that the degree of strict preference of x to y is greater than its degree of preference of y to x. Then the degree of strict preference of x to y equals its preference. The size of the degree of preference of y to x is immaterial. This is in contrast to $\pi_{(3)}$ where the size of the degree of preference of y to x is taken into account.

Definition 1.1.7 *Let $\rho \in \mathcal{FR}(X)$ and let π be an asymmetric fuzzy preference relation associated with ρ. Then*
*(1) π is called **simple** if $\forall x, y \in X$, $\rho(x, y) = \rho(y, x)$ implies $\pi(x, y) = \pi(y, x)$;*
*(2) π is called **regular** if $\forall x, y \in X$, $\rho(x, y) > \rho(y, x)$ if and only if $\pi(x, y) > 0$.*

If a strict preference relation π associated with ρ is simple, then $\rho(x, y) = \rho(y, x)$ implies $\pi(x, y) = 0 = \pi(y, x)$ for all $x, y \in X$. The strict preference relations $\pi_{(0)}$, $\pi_{(1)}$, and $\pi_{(3)}$ are simple. However, $\pi_{(2)}$ in general is not. The strict preference relations $\pi_{(1)}$ and $\pi_{(3)}$ are in fact regular.

1.2 Fuzzy Maximal Sets

Let X be a finite set and R be a relation on X. In the crisp case, a necessary condition for the maximal set $M(R, S) = \{x \in S \mid \forall y \in S, xRy\}$ to be nonempty for all subsets S of X is that R be reflexive and complete. Transitivity of R together with completeness and reflexivity are sufficient for $M(R, S)$ to be nonempty for all subsets S of X. However, transitivity is not necessary for $M(R, S)$ to be nonempty.

The maximal set comprises the subset of alternatives that are at least weakly preferred to all others. If there is more than one alternative in the set, then the player is indifferent between them. An important finding of social choice theory is that any choice made by a collective will not necessarily be contained in the collective's maximal set.

Definition 1.2.1 *Let $\mathcal{C}$ be a function from $X \times X$ into $[0,1]$. Define $M :$ $\mathcal{FR}^*(X) \times \mathcal{FP}^*(X) \to \mathcal{FP}^*(X)$ by $\forall(\rho,\mu) \in \mathcal{FR}^*(X) \times \mathcal{FP}^*(X)$,*

$$M(\rho,\mu)(x) = \mu(x) * (\circledast\{\vee\{t \in [0,1] \mid \mathcal{C}(x,w) * t \le \rho(x,w)\} \mid w \in Supp(\mu)\})$$

*for all $x \in X$, where $\circledast$ is a t-norm. Then $M(\rho,\mu)$ is called a **maximal fuzzy subset** with respect to $\mathcal{C}$ associated with (ρ,μ).*

The point of the t-norm $\circledast$ different from $*$ is that in [5], $\circledast$ is $\wedge$ and consequently we want our results to correspond to those in [5], but yet have the freedom of letting $\circledast = *$. If $\mathcal{C}(x,w) = \mu(w) * \rho(w,x)$ $\forall x,w \in X$ and $\circledast = \wedge$, then M is the maximal fuzzy subset defined in [5]. We denote this maximal fuzzy subset by M_G, but allow $\circledast$ to replace $\wedge$. Whenever we use M_G, we assume $\circledast$ has no zero divisors. If $\mathcal{C}(x,w) = 1$ $\forall x,w \in X$ and $* = \wedge = \circledast$, then M is the maximal fuzzy subset defined in [8]. We denote this maximal fuzzy subset by M_M.

The definition of a fuzzy maximal subset associated with (ρ,μ) gives the degree to which the elements of X are maximal with respect to ρ and μ. The degree of maximality of an element x in X can neither be larger than the degree that it is a member of μ nor larger than the smallest degree for which x is preferred to those y which have positive membership in μ.

The above definition allows for alternative notions of a fuzzy maximal set that are not available in crisp logic. The fuzzy subset μ provides a measure of the degree to which each element in the set of alternatives is in the set of most preferred alternatives by each individual. The fuzzy relation ρ is a measure of the degree to which some alternative is preferred to another. Since π is asymmetric, it provides the degree to which x is strictly preferred to y. The fuzzy relation ρ addresses the symmetrical relationship between x and y. As the following example demonstrates, a political actor may prefer x to y at some relatively high level, nevertheless the choice for y is not unacceptable.

Example 1.2.2 *Let $X = \{x,y,z\}$. Let $\mu \in \mathcal{FP}^*(X)$ be such that $\mu(x) = 0, \mu(y) = 1/4,$ and $\mu(z) = 1/2$. Let $\rho \in \mathcal{FR}^*(X)$ be such that $\rho(w,w) = 1$ $\forall w \in X$, $\rho(x,y) = 1/4, \rho(y,z) = 1/8, \rho(x,z) = 3/4, \rho(\_,\_) = 0$ otherwise. Then since $\mu(x) = 0$, it is immediate that $M_G(\rho,\mu)(x) = 0$. Now*

$$
\begin{aligned}
M_G(\rho,\mu)(y) &= \mu(y) * [(\vee\{t \in [0,1] \mid \mu(y) * \rho(y,y) * t \le \rho(y,y)\}) \\
&\quad \circledast(\vee\{t \in [0,1] \mid \mu(z) * \rho(z,y) * t \le \rho(y,z)\})] \\
&= (1/4) * (1 \circledast 1) \\
&= 1/4
\end{aligned}
$$

and

$$
\begin{aligned}
M_G(\rho,\mu)(z) &= \mu(z) * [(\vee\{t \in [0,1] \mid \mu(y) * \rho(y,z) * t \le \rho(z,y)\}) \\
&\quad \circledast(\vee\{t \in [0,1] \mid \mu(z) * \rho(z,z) * t \le \rho(z,z)\})] \\
&= (1/2) * (0 \circledast 1) \\
&= 0.
\end{aligned}
$$

In the next several results, we determine the relationship between fuzzy maximal subsets and crisp maximal subsets.

Proposition 1.2.3 *Let R be a relation on X and let S be a nonempty subset of X. Then*

$$M(1_R, 1_S)(x) = \begin{cases} 1 & \text{if } x \in S \text{ and } \forall w \in S, (x,w) \in R \\ 0 & \text{if } x \notin S \text{ or } \exists w \in X \text{ such that } w \in S, C(w,x) \neq 0, \\ & (x,w) \notin R. \end{cases}$$

Proof. We have that

$$\vee\{t \in [0,1] \mid C(x,w) * t \leq 1_R(x,w)\} = \begin{cases} 1 \text{ if } (x,w) \in R, \\ 1 \text{ if } C(x,w) = 0, \\ 0 \text{ if } w \in S, C(w,x) \neq 0, (x,w) \notin R. \end{cases}$$

■

In Proposition 1.2.3, we have that if $C(x,w) = 1_S(x) * \rho(w,x)$, then $M_G(\rho,\mu)(x) = 1$ if $x \in S$ and $\forall w \in S, [(x,w) \in R$ or $w \notin S$ or $(w,x) \notin R]$.

Example 1.2.4 *Let $X = \{x,y,z\}$ and $R = \Delta X \cup \{(y,z),(z,x)\}$. Let $S = \{x,y\}$. Then since $1_R(y,x) = 0$ and $1_S(z) = 0$,*

$$\begin{aligned} M_G(1_R, 1_S)(x) &= 1_S(x) * [(\vee\{t \in [0,1] \mid 1_S(x) * 1_R(x,x) * t \leq 1_R(x,x)\}) \circledast \\ (\vee\{t &\in [0,1] \mid 1_S(y) * 1_R(y,x) * t \leq 1_R(x,y)\})] \\ &= 1 * (1 \circledast 1) = 1. \end{aligned}$$

Note that R is not complete. However, $x \in S$, but $(x,v) \notin R$ $\forall v \in S$ since $(x,y) \notin R$. That is, Proposition 1.2.5 below doesn't hold. Thus this definition of $M_G(\rho,\mu)$ doesn't contain the crisp case.

Proposition 1.2.5 *Suppose $*$ has no zero divisors. Let R be a relation on X and let S be a nonempty subset of X. Suppose that $\forall x,w \in X, 1_R(x,w) = 0$ implies $C(x,w) \neq 0$. Then*

$$M(1_R, 1_S)(x) = \begin{cases} 1 & \text{if } x \in S \text{ and } (x,w) \in R \text{ } \forall w \in S, \\ 0 & \text{otherwise.} \end{cases}$$

Proof. Clearly, $M(1_R, 1_S)(x) = 1$ if $x \in S$ and $(x,w) \in R$ $\forall w \in S$. Suppose $x \notin S$ or $(x,w) \notin R$ for some $w \in S$. If $x \notin S$, then clearly $M(1_R, 1_S)(x) = 0$. Suppose $(x,w) \notin R$ for some $w \in S$. Then $1_R(x,w) = 0$ and so $C(x,w) \neq 0$. In this case, $t = 0$ is the only possible value such that $C(x,w) * t = 0$ since $*$ has no zero divisors. Thus $M(1_R, 1_S)(x) = 0$. ■

If R is complete and $C(x,w) = 1_S(x) * 1_R(w,x)$, then $C(x,w) \neq 0$ if $1_R(x,w) = 0$ and $x \in S$. By Proposition 1.2.5, $M_G(\rho,\mu)$ contains the crisp case if R is complete. We state this formally in the Corollary 1.2.6.

Corollary 1.2.6 *Let R be a relation on X and let S be a nonempty subset of X. Suppose R is complete. Then*

$$M(1_R, 1_S)(x) = \begin{cases} 1 & \text{if } x \in S \text{ and } (x, w) \in R \ \forall w \in S, \\ 0 & \text{otherwise.} \end{cases}$$

Proposition 1.2.7 *Let $(\rho, \mu) \in \mathcal{FR}^*(X) \times \mathcal{FP}^*(X)$ and $x \in X$. Then*

$$M_M(\rho, \mu)(x) = \mu(x) \wedge (\wedge\{\rho(x, w) \mid w \in Supp(\mu)\}).$$

Proof. Since X is finite, we have that $M_M(\rho, \mu)(x) = \vee\{t \in [0, 1] \mid \mu(x) \wedge \rho(x, w) \geq t \ \forall w \in Supp(\mu)\} = \wedge\{\mu(x) \wedge \rho(x, w) \mid w \in Supp(\mu)\} = \mu(x) \wedge (\wedge\{\rho(x, y) \mid w \in Supp(\mu)\})$. ∎

Proposition 1.2.8 *Let $\circledast = \wedge$. Then $\forall(\rho, \mu) \in \mathcal{FR}^*(X) \times \mathcal{FP}^*(X), M_M(\rho, \mu) \subseteq M_G(\rho, \mu)$.*

Proof. Let $(\rho, \mu) \ \mathcal{FR}^*(X) \times \mathcal{FP}^*(X)$ and $x \in X$. Let $w_0 \in X$ be such that $\rho(x, w_0) = \wedge\{\rho(x, w) \mid w \in Supp(\mu)\}$. Then $M_M(\rho, \mu)(x) = \mu(x) \wedge w_0$. Now $M_G(\rho, \mu)(x) = \mu(x) * (\wedge\{\vee\{t \in [0, 1] \mid \mu(w) * \rho(w, x) * t \leq \rho(x, w)\} \mid w \in Supp(\mu)\})$

$$= \begin{cases} \mu(x) & \text{if } \mu(w) * \rho(w, x) \leq \rho(x, w) \ \forall w \in Supp(\mu), \\ \mu(x) * (\wedge\{\rho(x, w) \mid \mu(w) * \rho(w, x) > \rho(x, w), w \in Supp(\mu)\} & \text{otherwise.} \end{cases}$$

Let $w_1 = \wedge\{\rho(x, w) \mid \mu(w) * \rho(w, x) > \rho(x, w), w \in Supp(\mu)\}$ if $\exists w \in Supp(\mu)$ such that $\mu(w) * \rho(w, x) > \rho(x, w)$. Since $w_0 \leq w_1$, the desired result holds.

∎

Proposition 1.2.9 *Let $(\rho, \mu) \in \mathcal{FR}^*(X) \times \mathcal{FP}^*(X)$ and $x \in X$. Then*

$$M(\rho, \mu)(x) = \begin{cases} \mu(x) & \text{if } \mathcal{C}(x, w) \leq \rho(x, w) \ \forall w \in Supp(\mu), \\ \mu(x) * t_x & \text{otherwise,} \end{cases}$$

where $t_x = \vee\{t \in [0, 1] \mid \mathcal{C}(x, w) * t \leq \rho(x, w), w \in Supp(\mu)\}$ if $\exists w \in Supp(\mu)$ such that $\mathcal{C}(x, w) > \rho(x, w)$.

In Proposition 1.2.9, $t_x = \wedge\{\rho(x, w) \mid \mathcal{C}(x, w) > \rho(x, w), w \in Supp(\mu)\}$ if $* = \wedge$ and $t_x = \wedge\{\rho(x, w)/\mathcal{C}(x, w) \mid \mathcal{C}(x, w) > \rho(x, w), w \in Supp(\mu)\}$ if $*$ is the product t-norm.

The concept of a maximal set is of considerable importance to spatial models. If the maximal set is empty, then every alternative can be defeated by at least one other alternative. When this situation occurs, spatial models can not predict the outcome. The crisp characterization demonstrates that a maximal set is assured when R is reflexive, complete, and acyclic [2]. The next result gives a corresponding result for maximal fuzzy subsets.

Theorem 1.2.10 *Let* $\pi = \pi_{(0)}$. *Suppose* $*$ *has no zero divisors. Let* $\rho \in \mathcal{FR}^*(X)$ *be reflexive and complete. Then* $M(\rho, \mu) \neq \theta$ *for all* $\mu \in \mathcal{FP}^*(X)$ *if and only if* ρ *is partially acyclic.*

Proof. Suppose ρ is partially acyclic. Let $\mu \in \mathcal{FP}^*(X)$. Then $\text{Supp}(\mu) \neq \emptyset$. Let $x_1 \in \text{Supp}(\mu)$. If $\rho(x_1, w) > 0 \ \forall w \in \text{Supp}(\mu)$, then by Proposition 1.2.9, $M(\rho, \mu)(x_1) > 0$ and so $M(\rho, \mu) \neq \theta$. Suppose there exists $x_2 \in \text{Supp}(\mu)\backslash\{x_1\}$ such that $\rho(x_1, x_2) = 0$. Then $\rho(x_2, x_1) > 0$ since ρ is complete. Thus $\pi(x_2, x_1) > 0$. Suppose there exists $x_1, \dots, x_k \in \text{Supp}(\mu)$ such that $x_i \in \text{Supp}(\mu)\backslash\{x_1, \dots, x_{i-1}\}$ and $\pi(x_i, x_{i-1}) > 0$ for $i = 2, \dots, k$. If $\rho(x_k, w) > 0$ for all $w \in \text{Supp}(\mu)$, then $M(\rho, \mu) \neq \theta$ as above. Suppose this is not the case and there exists $x_{k+1} \in \text{Supp}(\mu)\backslash\{x_1, \dots, x_k\}$ such that $\rho(x_k, x_{k+1}) = 0$. Then $\rho(x_{k+1}, x_k) > 0$ since ρ is complete. Hence by induction, either there exists $x \in \text{Supp}(\mu)$ such that $\rho(x, w) > 0$ for all $w \in \text{Supp}(\mu)$, in which case $M(\rho, \mu) \neq \theta$, or since $\text{Supp}(\mu)$ is finite, $\text{Supp}(\mu) = \{x_1, \dots, x_n\}$ is such that $\pi(x_i, x_{i-1}) > 0$ for $i = 2, \dots, n$. Since ρ is partially acyclic and ρ is reflexive, $\rho(x_n, x_i) > 0$ for $i = 1, \dots, n$. That is, $\rho(x_n, w) > 0$ for all $w \in \text{Supp}(\mu)$. Thus $M(\rho, \mu)(x_n) > 0$ and so $M(\rho, \mu) \neq \theta$.

Conversely, suppose $M(\rho, \mu) \neq \theta$ for all $\mu \in \mathcal{FP}^*(X)$. Suppose $x_1, \dots, x_n \in X$ are such that $\pi(x_1, x_2) > 0, \pi(x_2, x_3) > 0, \dots, \pi(x_{n-1}, x_n) > 0$. We must show $\rho(x_1, x_n) > 0$. Let $S = \{x_1, \dots, x_n\}$. Then $M(\rho, 1_S) \neq \theta$. Since $\pi(x_{i-1}, x_i) > 0$ and so $\rho(x_i, x_{i-1}) = 0$, $i = 2, \dots, n$, we have that $M(\rho, 1_S)(x_i) = 0, i = 2, \dots, n$. Thus it must be the case that $M(\rho, 1_S)(x_1) \neq 0$ and so $\rho(x_1, x_i) > 0$ for $i = 1, \dots, n$. Thus $\rho(x_1, x_n) > 0$ and so ρ is partially acyclic. ∎

Definition 1.2.11 *Let* $(\rho, \mu) \in \mathcal{FR}^*(X) \times \mathcal{FP}^*(X)$. *Define the fuzzy subset* nd *of* X *by* $\forall x \in X, nd(x) = 1 - \vee\{\pi(y, x) \mid y \in \text{Supp}(\mu)\}$, *where* π *is the strict preference relation associated with* ρ. *Then* nd *is called the* **degree of nondominance** *of* x *with respect to* (ρ, μ). *Let* $ND(\rho, \mu) = \{x \in X \mid nd(x) = 1\}$.

Clearly $x \in ND(\rho, \mu)$ if and only if $\pi(y, x) = 0 \ \forall y \in X$.

Definition 1.2.12 *Let* $(\rho, \mu) \in \mathcal{FR}^*(X) \times \mathcal{FP}^*(X)$. *Suppose* ρ *is regular. Define the fuzzy subset* $M_d(\rho, \mu)$ *of* X *as follows:* $\forall x \in X$,

$$M_d(\rho, \mu)(x) = \begin{cases} M(\rho, \mu)(x) & \text{if } x \in ND(\rho, \mu), \\ 0 & \text{otherwise.} \end{cases}$$

Definition 1.2.13 *Let* $\rho \in \mathcal{FR}^*(X)$. *If* $\pi(x_1, x_2) * \pi(x_2, x_3) * \dots * \pi(x_{n-1}, x_n) > 0$ *implies not* $\pi(x_n, x_1) > 0$ *for all* $x_1, \dots, x_n \in X$, *then* ρ *is called* **strictly acyclic**.

Let $\pi = \pi_{(0)}$. Then not $\pi(x_n, x_1) > 0 \Leftrightarrow$ not $(\rho(x_n, x_1) > 0$ and $\rho(x_1, x_n) = 0) \Leftrightarrow \rho(x_n, x_1) = 0$ or $\rho(x_1, x_n) > 0 \Leftrightarrow \rho(x_1, x_n) > 0$, where $\rho(x_n, x_1) = 0 \Rightarrow \rho(x_1, x_n) > 0$ assuming ρ is complete. That is, for complete ρ the definitions of partial acyclicity and strict acyclicity are equivalent for $\pi = \pi_{(0)}$.

Now let ρ be a fuzzy preference relation on X and define π by $\forall x, y \in X, \pi(x, y) = 1$ if $\rho(x, y) = 1, \rho(y, x) = 0$; and $\pi(x, y) = 0$ otherwise. Then π is asymmetric. Suppose $\rho(x_1, x_n) = 0$ and $\rho(x_n, x_1) = 1/2$. Then $\pi(x_1, x_n) = 0 = \pi(x_n, x_1)$, Thus $\pi(x_n, x_1) = 0 \not\Rightarrow \rho(x_1, x_n) > 0$. Note also if ρ is a fuzzy preference relation such that $\rho(x_1, x_n) = 1/2$ and $\rho(x_n, x_1) = 3/4$, then $\pi(x_n, x_1) > 0$ for many types of strict preference relations associated with ρ. This discussion shows that strict acyclicity does not imply partial acyclicity and partial acyclicity does not imply strict acyclicity.

Theorem 1.2.14 *Suppose $*$ has no zero divisors. Let $\rho \in \mathcal{FR}^*(X)$ be reflexive and complete. Suppose ρ is regular. Then ρ is strictly acyclic if and only if $M_d(\rho, \mu) \neq \theta \; \forall \mu \in \mathcal{FP}^*(X)$.*

Proof. Suppose ρ is strictly acyclic. Let $\mu \in \mathcal{FP}^*(X)$. Then $\text{Supp}(\mu) \neq \emptyset$. Let $x_1 \in \text{Supp}(\mu)$. If $\rho(x_1, w) \geq \rho(w, x_1) \; \forall w \in \text{Supp}(\mu)$, then $\rho(x_1, w) > 0 \; \forall w \in \text{Supp}(\mu)$. Thus $M_d(\rho, \mu)(x_1) > 0$ and so $M_d(\rho, \mu) \neq \theta$. Suppose there exists $x_2 \in \text{Supp}(\mu)\backslash\{x_1\}$ such that $\rho(x_2, x_1) > \rho(x_1, x_2)$ Then $\pi(x_2, x_1) > 0$. Suppose there exists $x_1, \ldots, x_k \in \text{Supp}(\mu)$ such that $x_i \in \text{Supp}(\mu)\backslash\{x_1, \ldots, x_{i-1}\}$ and $\rho(x_i, x_{i-1}) > \rho(x_{i-1}, x_i)$ and so $\pi(x_i, x_{i-1}) > 0$ for $i = 2, \ldots, k$. If $\rho(x_k, w) \geq \rho(w, x_k)$ for all $w \in \text{Supp}(\mu)$, then $M_d(\rho, \mu) \neq \theta$ as above. Suppose this is not the case and there exists $x_{k+1} \in \text{Supp}(\mu)\backslash\{x_1, \ldots, x_k\}$ such that $\rho(x_{k+1}, x_k) > \rho(x_k, x_{k+1})$. Then $\pi(x_{k+1}, x_k) > 0$. Hence by induction, either there exists $x \in \text{Supp}(\mu)$ such that $\rho(x, w) \geq \rho(w, x)$ for all $w \in \text{Supp}(\mu)$, in which case $M_d(\rho, \mu) \neq \theta$, or since $\text{Supp}(\mu)$ is finite, $\text{Supp}(\mu) = \{x_1, \ldots, x_n\}$ is such that $\pi(x_i, x_{i-1}) > 0$ for $i = 2, \ldots, n$. Since ρ is strictly acyclic and ρ is reflexive, not $\pi(x_i, x_n) > 0$ for $i = 1, \ldots, n$ Thus $\rho(x_n, x_i) \geq \rho(x_i, x_n)$ and so $\rho(x_n, x_i) > 0$ for $i = 1, \ldots, n$. That is, $\rho(x_n, w) \geq \rho(w, x_n)$ for all $w \in \text{Supp}(\mu)$. Thus $M_d(\rho, \mu)(x_n) > 0$ and so $M_d(\rho, \mu) \neq \theta$.

Suppose $M_d(\rho, \mu) \neq \theta \; \forall \mu \in \mathcal{FP}^*(X)$. Suppose $x_1, \ldots, x_n \in X$ are such that $\pi(x_1, x_2) > 0, \pi(x_2, x_3) > 0, \ldots, \pi(x_{n-1}, x_n) > 0$. We must show not $\pi(x_n, x_1) > 0$. Let $S = \{x_1, \ldots, x_n\}$. Then $M_d(\rho, 1_S) \neq \theta$. Since $\pi(x_{i-1}, x_i) > 0, M_d(\rho, 1_S)(x_i) = 0$ for $i = 2, \ldots, n$. Thus $M_d(\rho, 1_S)(x_1) > 0$. Hence $\rho(x_1, x_i) \geq \rho(x_i, x_1)$ for $i = 1, \ldots, n$. Thus not $\pi(x_n, x_1) > 0$. Hence ρ is strictly acyclic.

∎

The preceding result and the following corollary give corresponding results to Theorem 1.2.10 for regular fuzzy preference relations.

Corollary 1.2.15 *Suppose $*$ has no zero divisors. Let $\rho \in \mathcal{FR}^*(X)$ be reflexive and complete. Suppose ρ is regular. If ρ is strictly acyclic, then $M(\rho, \mu) \neq \theta \; \forall \mu \in \mathcal{FP}^*(X)$.*

Example 1.2.16 *For regular strict fuzzy preference relations, $M(\rho, \mu) \neq \theta$ $\forall \mu \in \mathcal{FP}^*(X) \not\Rightarrow \rho$ is strictly acyclic: Let $X = \{x, y, z\}$. Define $\rho \in \mathcal{FR}^*(X)$ by for all $w \in X, \rho(w, w) = 1$ and*

$$\rho(x, y) = \rho(y, z) = \rho(x, z) = 1/2,$$
$$\rho(y, x) = \rho(z, y) = 1/4, \ \rho(z, x) = 3/4.$$

Suppose ρ is regular. Then $\pi(x, y) > 0, \pi(y, z) > 0, \pi(z, x) > 0$ (since $\pi \neq \pi_{(0)}$). Thus ρ is not strictly acyclic. Let $\mu \in \mathcal{FP}^*(X)$. Then $M(\rho, \mu) \neq \theta$ since $\rho(u, v) > 0 \ \forall u, v \in X$. Note $M_d(\rho, \mu) = \theta$.

Proposition 1.2.17 *$M(\rho, \mu) = M_d(\rho, \mu) \ \forall (\rho, \mu) \in \mathcal{FR}^*(X) \times \mathcal{FP}^*(X)$ if and only if π is of type $\pi_{(0)}$.*

 Proof. Suppose π is of type $\pi_{(0)}$. Let $(\rho, \mu) \in \mathcal{FR}^*(X) \times \mathcal{FP}^*(X)$. Suppose $M(\rho, \mu)(x) > 0$ for $x \in \text{Supp}(\mu)$. Then $\rho(x, w) > 0 \ \forall w \in \text{Supp}(\mu)$. If $\rho(w, x) > \rho(x, w) > 0$ for some $w \in \text{Supp}(\mu)$, then since π is of type $\pi_{(0)}, \pi(x, w) = 0 = \pi(w, x)$. Thus $M_d(\rho, \mu)(x) > 0$. Hence $M_d(\rho, \mu)(x) = M(\rho, \mu)(x)$.
 Conversely, suppose $M(\rho, \mu) = M_d(\rho, \mu) \ \forall (\rho, \mu) \in \mathcal{FR}^*(X) \times \mathcal{FP}^*(X)$. Suppose π is not of type $\pi_{(0)}$. Then there exists $x, y \in X$ such that $\rho(y, x) > 0$ and $\pi(x, y) > 0$. Let $S = \{x, y\}$. Then $M_d(\rho, 1_S)(y) = 0$, but $M(\rho, 1_S)(y) > 0$ since $\rho(y, x) > 0$. Hence $M_d(\rho, 1_S) \neq M(\rho, 1_S)$. ∎

 We next consider strict preferences π of type $\pi_{(2)}$.

Theorem 1.2.18 *Suppose π is of type $\pi_{(2)}$. Suppose $*$ has no zero divisors. Let $\rho \in \mathcal{FR}^*(X)$ be reflexive and complete. If ρ is strictly acyclic, then $M(\rho, \mu) \neq \theta \ \forall \mu \in \mathcal{FP}^*(X)$.*

 Proof. Suppose ρ is partially acyclic. Let $\mu \in \mathcal{FP}^*(X)$. Then $\text{Supp}(\mu) \neq \emptyset$. Let $x_1 \in \text{Supp}(\mu)$. If $\rho(x_1, w) \geq \rho(w, x_1) \ \forall w \in \text{Supp}(\mu)$, then $\rho(x_1, w) > 0$ $\forall w \in \text{Supp}(\mu)$. Thus $M(\rho, \mu)(x_1) > 0$ and so $M(\rho, \mu) \neq \theta$. Suppose there exists $x_2 \in \text{Supp}(\mu) \backslash \{x_1\}$ such that $\rho(x_2, x_1) > \rho(x_1, x_2)$ Then $\pi(x_2, x_1) > 0$. Suppose there exists $x_1, \ldots, x_k \in \text{Supp}(\mu)$ such that $x_i \in \text{Supp}(\mu) \backslash \{x_1, \ldots, x_{i-1}\}$ and $\rho(x_i, x_{i-1}) > \rho(x_{i-1}, x_i)$ and so $\pi(x_i, x_{i-1}) > 0$ for $i = 2, \ldots, k$. If $\rho(x_k, w) \geq \rho(w, x_k)$ for all $w \in \text{Supp}(\mu)$, then $M(\rho, \mu) \neq \theta$ as above. Suppose this is not the case and there exists $x_{k+1} \in \text{Supp}(\mu) \backslash \{x_1, \ldots, x_k\}$ such that $\rho(x_{k+1}, x_k) > \rho(x_k, x_{k+1})$. Then $\pi(x_{k+1}, x_k) > 0$. Hence by induction, either there exists $x \in \text{Supp}(\mu)$ such that $\rho(x, w) \geq \rho(w, x)$ for all $w \in \text{Supp}(\mu)$, in which case $M(\rho, \mu) \neq \theta$, or since $\text{Supp}(\mu)$ is finite, $\text{Supp}(\mu) = \{x_1, \ldots, x_n\}$ is such that $\pi(x_i, x_{i-1}) > 0$ for $i = 2, \ldots, n$. Since ρ is strictly acyclic and ρ is reflexive, not $\pi(x_i, x_n) > 0$ for $i = 1, \ldots, n$ Thus $\rho(x_n, x_i) = 1$ for $i = 1, \ldots, n$. That is, $\rho(x_n, w) > 0$ for all $w \in \text{Supp}(\mu)$. Thus $M(\rho, \mu)(x_n) > 0$ and so $M(\rho, \mu) \neq \theta$. ∎

Example 1.2.19 *Consider ρ of Example 1.2.16. Then $M(\rho, \mu) \neq \theta \ \forall \mu \in \mathcal{FP}^*(X)$. Clearly ρ is not strictly acyclic when π is of type $\pi_{(2)}$ since $\pi(u, v) > 0 \ \forall u, v \in X$ such that $u \neq v$.*

Definition 1.2.20 *Let $(\rho, \mu) \in \mathcal{FR}^*(X) \times \mathcal{FP}^*(X)$. Define the fuzzy subset $M_2(\rho, \mu)$ of X by $\forall x \in X$,*

$$M_2(\rho, \mu)(x) = \begin{cases} M(\rho, \mu)(x) & \text{if } \rho(x, w) = 1 \ \forall w \in Supp(\mu), \\ 0 & \text{otherwise.} \end{cases}$$

Proposition 1.2.21 *Suppose π is of type $\pi_{(2)}$. Suppose $*$ has no zero divisors. Let $\rho \in \mathcal{FR}^*(X)$ be reflexive and complete. If $M_2(\rho, \mu) \neq \theta \ \forall \mu \in \mathcal{FP}^*(X)$, then ρ is strictly acyclic.*

Proof. Suppose $M_2(\rho, \mu) \neq \theta \ \forall \mu \in \mathcal{FP}^*(X)$. Suppose $x_1, \ldots, x_n \in X$ are such that $\pi(x_1, x_2) > 0, \pi(x_2, x_3) > 0, \ldots, \pi(x_{n-1}, x_n) > 0$. We must show not $\pi(x_n, x_1) > 0$. Let $S = \{x_1, \ldots, x_n\}$. Then $M_2(\rho, 1_S) \neq \theta$. Since $\pi(x_{i-1}, x_i) > 0, \rho(x_i, x_{i-1}) < 1$ and so $M_2(\rho, 1_S)(x_i) = 0$ for $i = 2, \ldots, n$. Thus $M_2(\rho, 1_S)(x_1) > 0$. Hence $\rho(x_1, x_i) = 1$ for $i = 1, \ldots, n$. Thus not $\pi(x_n, x_1) > 0$. Hence ρ is strictly acyclic. ∎

Suppose we change the definition of strictly acyclic when π is of type $\pi_{(2)}$ to the following:

Let $\rho \in \mathcal{FR}^*(X)$. If $\pi(x_1, x_2) * \pi(x_2, x_3) * \ldots * \pi(x_{n-1}, x_n) > 0$ implies not $\rho(x_n, x_1) > \rho(x_1, x_n)$ for all $x_1, \ldots, x_n \in X$.

Then Theorem 1.2.18 holds for this definition with the same proof except for the conclusion: Since ρ is strictly acyclic and ρ is reflexive, not $\rho(x_i, x_n) > \rho(x_n, x_i)$ for $i = 1, \ldots, n$ Thus $\rho(x_n, x_i) > 0$ for $i = 1, \ldots, n$. Hence $M(\rho, \mu)(x_n) > 0$ and so $M(\rho, \mu) \neq \theta$.

Example 1.2.22 *Consider ρ of Example 1.2.19. Then $M(\rho, \mu) \neq \theta \ \forall \mu \in \mathcal{FP}^*(X)$. Clearly ρ is not acyclic when π is of type $\pi_{(2)}$ and with the altered definition of acyclic given immediately above since $\pi(x, y) > 0, \pi(y, z) > 0$ and $\rho(z, x) > \rho(x, z)$.*

The following result gives a necessary and sufficient condition for an element to belong to a level set of a fuzzy maximal subset.

Proposition 1.2.23 *Let $(\rho, \mu) \in \mathcal{FR}^*(X) \times \mathcal{FP}^*(X)$. Let $* = \wedge = \circledast$. Then $\forall x \in X, \forall t \in [0, 1], x \in M(\rho, \mu)_t$ if and only if $x \in \mu_t$ and $\forall w \in X, \mathcal{C}(x, w) \wedge t \leq \rho(x, w)$.*

Proof. We have that

$$
\begin{aligned}
x \;&\in\; M(\rho,\mu)_t \Leftrightarrow M(\rho,\mu)(x) \geq t \\
&\Leftrightarrow\; \mu(x) \wedge (\wedge_{w \in X}(\vee\{r \in [0,1] \mid \mathcal{C}(x,w) \wedge r \leq \rho(x,w)\})) \geq t \\
&\Leftrightarrow\; x \in \mu_t \text{ and } \wedge_{w \in X} (\vee\{r \in [0,1] \mid \mathcal{C}(x,w) \wedge r \leq \rho(x,w)\})) \geq t \\
&\Leftrightarrow\; x \in \mu_t \text{ and } \vee \{r \in [0,1] \mid \mathcal{C}(x,w) \wedge r \leq \rho(x,w)\}) \geq t \; \forall w \in X \\
&\Leftrightarrow\; x \in \mu_t \text{ and } \mathcal{C}(x,w) \wedge t \leq \rho(x,w) \; \forall w \in X.
\end{aligned}
$$

∎

Proposition 1.2.24 *Let S be a nonempty subset of X and let ρ be a fuzzy relation on X. Let $x \in X$ and $S_x = \{w \in X \mid \mathcal{C}(x,w) > \rho(x,w)\}$ and $t_{x,w} = \vee\{t \in [0,1] \mid \mathcal{C}(x,w) * t \leq \rho(x,w)\}$, where $w \in S_x$. Then $\forall x \in X$,*

$$
M(\rho,1_S)(x) = \begin{cases} 1 & \text{if } x \in S \text{ and } S_x = \emptyset, \\ \circledast\{t_{x,w} \mid w \in S_x\} & \text{if } x \in S \text{ and } S_x \neq \emptyset, \\ 0 & \text{if } x \notin S. \end{cases}
$$

Proof. Let $x \in S$. Then

$$
\begin{aligned}
M(\rho,1_S(x)) \;&=\; 1_S(x) * (\circledast_{w \in X}(\vee\{t \in [0,1] \mid \mathcal{C}(x,w) * t \leq \rho(x,w)\})) \\
&=\; \circledast_{w \in X}(\vee\{t \in [0,1] \mid \mathcal{C}(x,w) * t \leq \rho(x,w)\})
\end{aligned}
$$

from which the desired result follows easily. ∎

We next provide some examples involving fuzzy maximal subsets for various t-norms.

Example 1.2.25 *Let $X = \{x,y,z\}$. Define the fuzzy relation ρ on X as follows:*

$$
\begin{aligned}
\rho(x,x) \;&=\; \rho(y,y) = \rho(z,z) = 1, \\
\rho(x,y) \;&=\; 3/4, \rho(y,z) = 1/2, \rho(x,z) = 1/4, \\
\rho(y,x) \;&=\; \rho(z,y) = \rho(z,x) = 0.
\end{aligned}
$$

Let $* = \wedge = \circledast$. Then $M_G(\rho,1_X)(x) = 1, M_G(\rho,1_X)(y) = 0, M_G(\rho,1_X)(z) = 0$ by Proposition 1.2.24. In this example, $M_G(\rho,1_X) = 1_T$, where $T = \{x\}$.

Example 1.2.26 *Now let $* = \circledast = $ product. Then*

$$
\begin{aligned}
M_G(\rho,1_X)(x) \;&=\; *_{w \in X}\{\vee\{t \in [0,1] \mid \rho(w,x) * t \leq \rho(x,w)\}\} \\
&=\; 1 * (\vee\{t \in [0,1] \mid \rho(y,x) * t \leq \rho(x,y)\}) * \\
&\quad\;\, (\vee\{t \in [0,1] \mid \rho(z,x) * t \leq \rho(x,z)\}) \\
&=\; (\vee\{t \in [0,1] \mid 0 * t \leq 3/4\}) * \\
&\quad\;\, (\vee\{t \in [0,1] \mid 0 * t \leq 1/4\}) \\
&=\; 1,
\end{aligned}
$$

$$M_G(\rho, 1_X)(y) = (\vee\{t \in [0,1] \mid \rho(x,y) * t \le \rho(y,x)\})*$$
$$1 * (\vee\{t \in [0,1] \mid \rho(z,y) * t \le \rho(y,z)\})$$
$$= (\vee\{t \in [0,1] \mid 3/4 * t \le 0\})*$$
$$1 * (\vee\{t \in [0,1] \mid 0 * t \le 1/2\})$$
$$= 0,$$

$$M_G(\rho, 1_X)(z) = (\vee\{t \in [0,1] \mid \rho(x,z) * t \le \rho(z,x)\})*$$
$$1 * (\vee\{t \in [0,1] \mid \rho(y,z) * t \le \rho(z,y)\})$$
$$= (\vee\{t \in [0,1] \mid 1/4 * t \le 0\})*$$
$$(\vee\{t \in [0,1] \mid 1/2 * t \le 0\}) * 1$$
$$= 0.$$

Let $*$ and $\circledast$ denote the Lukasiewicaz t-norm. Then

$$M_G(\rho, 1_X)(x) = 1,$$

$$M_G(\rho, 1_X)(y) = (\vee\{t \in [0,1] \mid 3/4 * t \le 0\})*$$
$$1 * (\vee\{t \in [0,1] \mid 0 * t \le 1/2\})$$
$$= 1/4 * 1 * 1/2$$
$$= 0,$$

$$M_G(\rho, 1_X)(z) = (\vee\{t \in [0,1] \mid 1/4 * t \le 0\})*$$
$$(\vee\{t \in [0,1] \mid 1/2 * t \le 0\}) * 1$$
$$= 3/4 * 1/2 * 1$$
$$= 1/4.$$

Example 1.2.27 *Let $X = \{x, y, z\}$. Define the fuzzy relation ρ on X as follows:*

$$\rho(x, x) = \rho(y,y) = \rho(z,z) = 1,$$
$$\rho(x, y) = 3/4, \rho(y,z) = 1/2, \rho(x,z) = 1/4,$$
$$\rho(y, x) = 7/8, \rho(z,y) = \rho(z,x) = 0.$$

Let $* = \wedge = \circledast$. Then $M_G(\rho, 1_X)(x) = 3/4, M_G(\rho, 1_X)(y) = 1, M_G(\rho, 1_X)(z) = 0$. In this example, $M_G(\rho, 1_X) \ne 1_T$ for any subset T of X.

Let $*$ and $\circledast$ denote product. Then

$$M_G(\rho, 1_X)(x) = *_{w \in X}\{\vee\{t \in [0,1] \mid \rho(w,x) * t \le \rho(x,w)\}\}$$
$$= 1 * (\vee\{t \in [0,1] \mid \rho(y,x) * t \le \rho(x,y)\})*$$
$$(\vee\{t \in [0,1] \mid \rho(z,x) * t \le \rho(x,z)\})$$
$$= (\vee\{t \in [0,1] \mid 7/8 * t \le 3/4\})*$$
$$(\vee\{t \in [0,1] \mid 0 * t \le 1/4\})$$
$$= 6/7,$$

$$M_G(\rho, 1_X)(y) = (\vee\{t \in [0,1] \mid \rho(x,y) * t \le \rho(y,x)\}) * 1$$
$$*(\vee\{t \in [0,1] \mid \rho(z,y) * t \le \rho(y,z)\})$$
$$= (\vee\{t \in [0,1] \mid 3/4 * t \le 7/8\}) * 1$$
$$*(\vee\{t \in [0,1] \mid 0 * t \le 1/2\})$$
$$= 1,$$

$$
\begin{aligned}
M_G(\rho, 1_X)(z) &= (\vee\{t \in [0,1] \mid \rho(x,z) * t \le \rho(z,x)\}) * 1 \\
&\quad * (\vee\{t \in [0,1] \mid \rho(y,z) * t \le \rho(z,y)\}) \\
&= (\vee\{t \in [0,1] \mid 1/4 * t \le 0\}) * \\
&\quad (\vee\{t \in [0,1] \mid 1/2 * t \le 0\}) * 1 \\
&= 0.
\end{aligned}
$$

Let $*$ and $\circledast$ denote the Lukasiewicaz t-norm. Then

$$
\begin{aligned}
M_G(\rho, 1_X)(x) &= (\vee\{t \in [0,1] \mid 7/8 * t \le 3/4\}) * \\
&\quad (\vee\{t \in [0,1] \mid 0 * t \le 1/4\}) \\
&= 7/8 * 1 \\
&= 7/8,
\end{aligned}
$$

$$
\begin{aligned}
M_G(\rho, 1_X)(y) &= (\vee\{t \in [0,1] \mid 3/4 * t \le 7/8\}) * 1 \\
&\quad * (\vee\{t \in [0,1] \mid 0 * t \le 1/2\}) \\
&= 1 * 1 * 1 \\
&= 1,
\end{aligned}
$$

$$
\begin{aligned}
M_G(\rho, 1_X)(z) &= (\vee\{t \in [0,1] \mid 1/4 * t \le 0\}) * \\
&\quad (\vee\{t \in [0,1] \mid 1/2 * t \le 0\}) * 1 \\
&= 3/4 * 1/2 * 1 \\
&= 1/4.
\end{aligned}
$$

Example 1.2.28 *Let $X = \{x,y,z\}$. Define the fuzzy relation ρ on X as follows:*

$$
\begin{aligned}
\rho(x,x) &= \rho(y,y) = \rho(z,z) = 1, \\
\rho(x,y) &= 3/4, \rho(y,z) = 1/2, \rho(z,x) = 1/4, \\
\rho(y,x) &= \rho(z,y) = \rho(x,z) = 0.
\end{aligned}
$$

For any t-norm $*$ without zero divisors, $M_G(\rho, 1_X)(x) = (\vee\{t \in [0,1] \mid 0 * t \le \rho(x,y)\}) \circledast (\vee\{t \in [0,1] \mid 1/4 * t \le 0\}) = 3/4 \circledast 0 = 0$. Similarly, $M_G(\rho, 1_X)(y) = 0, M(\rho, 1_X)(z) = 0$. In this example, $M_G(\rho, 1_X) = 1_T$, where $T = \emptyset$. In this example, we have a cycle.

We next consider when $M(\rho, 1_S)$ is a characteristic function even though ρ may not be. That is, we are interested in determining when the fuzzy maximal subset is crisp given that the set S is crisp and when the preference relation is fuzzy, i.e., when the decision maker states this degree of preference.

Proposition 1.2.29 *Let $\rho \in \mathcal{FR}^*(X)$ and let S be a subset of X. Suppose $*$ is a t-norm without zero divisors. Then $M(\rho, 1_S) = 1_T$ for some $T \subseteq S$ if and only if $\forall x, y \in S, \rho(x,y) * C(y,x) > 0$ implies $\rho(x,y) \ge C(y,x)$.*

Proof. Suppose $M(\rho, 1_S) = 1_T$ for some $T \subseteq S$. Let $y \in S$. $\forall w \in S_y$, let $t_{y,w} = \vee\{t \in [0,1] \mid C(w,y) * t \le \rho(y,w)\}$. Then $M(\rho, 1_S)(y) = \circledast\{t_{y,w} \mid w \in S_y\}$, where $S_y = \{w \in X \mid C(w,y) > \rho(y,w)\}$ and if $S_y \ne \emptyset$. Suppose $S_y \ne \emptyset$.

Then it follows that $0 < \wedge\{\rho(y,w) \mid w \in S_y\} < 1$. However, this contradicts the hypothesis. Hence $S_y = \emptyset$. Thus $\mathcal{C}(w,y) \leq \rho(y,w) \; \forall w \in S, t_{y,w} = 1$ $\forall w \in S$.

Conversely, suppose that $\forall x, y \in S, \rho(x,y) * \mathcal{C}(y,x) > 0$ implies $\rho(x,y) \geq \mathcal{C}(y,x)$. Let $T = \{x \in S \mid \not\exists w \in S, \mathcal{C}(w,x) > \rho(x,w)\}$. Suppose $x \in S \backslash T$. Then $\exists w \in S$ such that $\mathcal{C}(w,x) > \rho(x,w)$. By our hypothesis, $\rho(x,w) = 0$. Thus $M(\rho, 1_S)(x) = 0$. Hence $M(\rho, 1_S) = 1_T$. ∎

We next consider the relationship between fuzzy maximal subsets and level sets.

Proposition 1.2.30 Let $\rho \in \mathcal{FR}^*(X)$. Suppose that there exists $t \in (0,1]$ such that ρ_t is reflexive, complete, and acyclic. Then $\forall \mu \in \mathcal{FP}^*(X)$ such that $\mu_t \neq \emptyset, \{x \in X \mid x \in \mu_t, (x,y) \in \rho_t \; \forall y \in \text{Supp}(\mu)\} \neq \emptyset$.

Proof. By [22], $\{x \in X \mid x \in \text{Supp}(\mu), (x,y) \in \rho_t \; \forall y \in \text{Supp}(\mu)\} \neq \emptyset$. The desired result follows since $\mu_t \subseteq \text{Supp}(\mu)$. ∎

Corollary 1.2.31 Let $\rho \in \mathcal{FR}^*(X)$. Suppose that there exists $t \in (0,1]$ such that ρ_t is reflexive, complete, and acyclic. Then $\forall \mu \in \mathcal{FP}^*(X)$ such that $\mu_t \neq \emptyset, M_M(\rho, \mu) \neq \theta$.

Proposition 1.2.32 Let $(\rho, \mu) \in \mathcal{FR}^*(X) \times \mathcal{FP}^*(X)$. Then $\forall x \in X, \forall t \in [0,1], x \in M_M(\rho, \mu)_t$ if and only if $x \in \mu_t$ and $(x,y) \in \rho_t \; \forall y \in \text{Supp}(\mu)$.

Proof. We have that $x \in M_M(\rho, \mu)_t \Leftrightarrow M_M(\rho, \mu)(x) \geq t \Leftrightarrow \vee\{r \in [0,1] \mid \mu(x) \geq r$ and $\rho(x,y) \geq r \; \forall y \in \text{Supp}(\mu)\} \geq t \Leftrightarrow \exists r \in [0,1]$ such that $r \geq t$ and $\mu(x) \geq r, \rho(x,y) \geq r \; \forall y \in \text{Supp}(\mu) \Leftrightarrow \exists r \in [0,1]$ such that $r \geq t$ and $x \in \mu_r, (x,y) \in \rho_r \; \forall y \in \text{Supp}(\mu) \Leftrightarrow x \in \mu_t$ and $(x,y) \in \rho_t \; \forall y \in \text{Supp}(\mu)$. ∎

Corollary 1.2.33 Let $(\rho, \mu) \in \mathcal{FR}^*(X) \times \mathcal{FP}^*(X)$. If $t \in (0, s_*]$, where $s_* = \vee\{s \in \text{Im}(\mu) \mid \mu_s = \text{Supp}(\mu)\}$, then $M_M(\rho, \mu)_t = M_M(\rho_t, \text{Supp}(\mu))$.

Proof. $M_M(\rho, \mu)_t = \{x \in \mu_t \mid (x,y) \in \rho_t \; \forall y \in \text{Supp}(\mu)\} = M_M(\rho_t, \text{Supp}(\mu))$ since $\mu_t = \text{Supp}(\mu)$. ∎

Proposition 1.2.34 Let $\rho \in \mathcal{FR}^*(X)$. Then $\forall S \in \mathcal{FP}^*(X) \exists T \subseteq S$ such that $M_M(\rho, 1_S) = 1_T$ if and only if $\not\exists x, y \in X$ (not necessarily distinct) such that $0 < \rho(x,x) \wedge \rho(x,y) < 1$.

Proof. Suppose there exists $x, y \in X$ such that $0 < \rho(x,x) \wedge \rho(x,y) < 1$. Let $S = \{x,y\}$. Then $M_M(\rho, 1_S)(x) = \rho(x,x) \wedge \rho(x,y)$. Thus $\not\exists T \subseteq S$ such that $M_M(\rho, 1_S) = 1_T$.

Conversely, suppose $\not\exists x, y \in X$ such that $0 < \rho(x,x) \wedge \rho(x,y) < 1$. Let $S \in \mathcal{FP}^*(X)$. Then $\forall x \in S, \rho(x,x) \wedge \rho(x,y) = 0$ or $\rho(x,x) \wedge \rho(x,y) = 1$ for all $y \in S$. Hence $\forall x \in S$, $\vee\{t \in [0,1] \mid x \in (1_S)_t, (x,y) \in \rho_t \,\forall y \in (1_S)_t\} = 0$ or 1, i.e., $M_M(\rho, 1_S)(x) = 0$ or 1. Thus the desired conclusion holds. ∎

Let $X = \{x,y\}$. Define $\rho : X \times X \rightarrow [0,1]$ as follows: $\rho(x,x) = 0, \rho(x,y) = 1/2, \rho(y,y) = \rho(y,x) = 1$. Then $M(\rho, 1_X)(x) = 0$ and $M(\rho, 1_X)(y) = 1$. Thus $M(\rho, 1_X) = 1_T$, where $T = \{y\}$ even though $0 < \rho(x,y) < 1$. Also, $M_M(\rho, 1_{\{y\}}) = 1_{\{y\}}$ and $M_M(\rho, 1_{\{x\}}) = 1_\emptyset$. If we require in Proposition 1.2.34 that $T \neq \emptyset$, then we get the following result.

Proposition 1.2.35 *Let $\rho \in \mathcal{FR}^*(X)$. Then $\forall S \in \mathcal{P}^*(X) \, \exists T \subseteq S, T \neq \emptyset$, such that $M_M(\rho, 1_S) = 1_T$ if and only if $\nexists x, y \in X$ (not necessarily distinct) such that $0 < \rho(x,y) < 1$.*

Proof. The result follows from Proposition 1.2.34 and the fact that if $T \neq \emptyset$, then $\rho(x,x) > 0 \,\forall x \in X$ else if $\rho(x,x) = 0$ for some $x \in X$, $M_M(\rho, 1_{\{x\}}) = 1_T$, where $T = \emptyset$. Note also that the assumption $\nexists x, y \in X$ (not necessarily distinct) such that $0 < \rho(x,y) < 1$ implies $\rho(x,x) > 0 \,\forall x \in X$. ∎

1.3 Exercises

1. Let X be a finite set and a R relation on X. Prove that a necessary condition for the maximal set $M(R,S)$ to be nonempty for all subsets S of X is that R be reflexive and complete. Prove also that transitivity of R together with completeness and reflexivity are sufficient for $M(R,S)$ to be nonempty for all subsets S of X.

2. Show by example that the transitivity of R is not necessary for $M(R,S)$ to be nonempty for all S.

3. Let R be reflexive and complete. Prove that $M(R,S) \neq \emptyset$ for all subsets S of X if and only if R is acyclic.

1.4 References

1. B. Arfi, Fuzzy decision making in politics: A linguistic fuzzy set approach (LFSA), *Political Analysis*, 13 (2005) 23–56.

2. D. Austen-Smith and J. S. Banks, *Positive Political Theory I: Collective Preference*, University of Michigan Press 2000.

3. C. A. Cioffi-Revilla, Fuzzy sets and models of international relations, *American Journal of Political Science*, 25 (1981) 129–159.

4. T. D. Clark, J. M. Larson, J. N. Mordeson, J. D. Potter, M. J. Wierman, *Applying Fuzzy Mathematics to Formal Models in Comparative Politics*, Springer-Verlag Berlin Heidelberg, Studies in Fuzziness and Soft Computing 225, 2008.

5. Georgescu, *Fuzzy Choice Functions: A Revealed Preference Approach*, Studies in Fuzziness and Soft Computing, Springer-Verlag Berlin Heidelberg 2007.

6. P. Hajek, *Metamathematics of Fuzzy Logic*, Kluwer, Dordrecht, 1998.

7. M. Koenig-Archibugi, Explaining government preferences for institutional change in EU foreign and security policy, *International Organizations*, 58 (2004) 137–174.

8. J. N. Mordeson, K. R. Bhutani, and T. D. Clark, The rationality of fuzzy choice functions, *New Mathematics and Natural Computation*, 4 (2008) 309–327.

9. R. B. Morton, *Methods and models: A guide to the empirical analysis of formal models in political science*, Cambridge, Cambridge University Press, 1999.

10. P. Pennings, Beyond dichotomous explanations: Explaining constitutional control of the executive with fuzzy set-sets, *European Journal of Political Research*, 42 (2003) 541–567.

11. C. C. Ragin, *Fuzzy-set social choice*, Chicago, University of Chicago Press, 2000.

12. G. S. Sanjian, Fuzzy set theory and US arms transfers: Modeling the decision-making process, *American Journal of Political Science*, 32 (1988) 1018–1046.

13. G. S. Sanjian, Great power arms transfers: Modeling the decision-making processes of hegemonic, industrial, and restrictive exporters, *International Studies Quarterly*, 35 (1991) 173–193.

14. G. S. Sanjian, A fuzzy set model of NATO decision-making: The case of short-range nuclear forces in Europe, *Journal of Peace Research*, 29 (1992) 271–285.

15. G. S. Sanjian, Cold War imperatives and quarrelsome clients: Modeling US and USSR arms transfers to India and Pakistan, *The Journal of Conflict Resolution*, 42 (1998) 97–127.

16. G. S. Sanjian, Promoting stability or instability? Arms transfers and regional rivalries, 1950-1991, *International Studies Quarterly*, 43 (1999) 641–670.

17. G. S. Sanjian, Arms and arguments: Modeling the effects of weapons transfers on subsystem relationships, *Political Research Quarterly*, 54 (2001) 285–309.

18. B Schweizer and A. Sklar, *Probabilistic Metric Spaces*, North-Holland, New York, 1983.

19. S. T. Seitz, Apollo's oracle: Strategizing for peace, *Syntese*, 100 (1994) 461–495.

20. M. Smithson and J. Verkuilen, *Fuzzy set theory: Applications in the social sciences*, Thousand Oaks, Sage Publications, 2006.

21. C. S. Taber, POLI: An expert system model of US foreign policy belief systems, *American Political Science Review*, 86 (1992) 888–904.

22. J. V. Verkuilen, Assigning membership in a fuzzy set analysis, *Sociological Methods and Research*, 33 (2005) 462–496.

23. L. A. Zadeh, Fuzzy sets, *Information and Control*, 8 (1965) 338–353.

Chapter 2

Fuzzy Choice Functions

Classical revealed preference theory postulates a connection between choices and revealed preferences. Political scientists have employed this theory in a wide variety of approaches and issues. For instance, comparativists predict parliamentary vote outcomes based on the preferences for legislators revealed in party manifestos. Social choice theorists initially expected on the basis of revealed preference theory that collective choice was based on collective preferences. In the previous chapter, we demonstrated that a maximal set may not exist under all conditions. This conclusion has led scholars to question the relationship between revealed social preferences and social choice. In particular, they have asked whether the choices made by collective actors are consistent with their collective social preference. In other words, if we assume that the preferences of a set of individuals are not cyclic, we would like to know if their collective choices are rationalizable. Formally, do their collective choices comprise a proper subset of the maximal set when one exists? The question is not a trivial one. For example, scholars often estimate the preferences of legislators based on roll-call votes [39]. If we do not assume that legislators' roll-call votes are related to their preferences, then we must estimate their preferences by some other means.

Considerable work has been done on the rationalizability of fuzzy social preference and fuzzy choice functions [5, 6, 7, 19, 20, 33]. Orlovsky first considered human preferences as fuzzy relations for drawing conclusions in decision making problems [37]. Banerjee [4] introduced the concept of a fuzzy choice function whose domain is crisp and range is fuzzy. Banerjee also introduced three congruence axioms in order to characterize rationality of choice functions. Wang in [50] later proved that these axioms are not independent. (We examine these results in Chapter 7.) Georgescu [21] considers the fuzzy choice functions defined on non-empty subsets of non-zero fuzzy subsets of the set of alternatives. Georgescu's work is a natural extension of [4]. In subsequent papers [21, 22, 23, 25] and the monograph [24], fuzzy forms of revealed preference axioms and congruence axioms are developed. In [25], Georgescu introduced

21

a fuzzy Arrow axiom, a weak fuzzy congruence axiom, and a strong fuzzy congruence axiom.

2.1 Basic Properties

The first two sections of this chapter address sufficient conditions for a partially acyclic fuzzy choice function to be rationalizable. We find that certain fuzzy choice functions that satisfy conditions α and β are rationalizable. Furthermore, any fuzzy choice function that satisfies these two conditions satisfies Arrow and WARP

Formal models have found increasing application in comparative politics in recent years. Spatial models have been particularly popular. The individual and collective preferences of humans are inherently vague and ambiguous no matter how they are aggregated. The conventional crisp approach forces upon spatial models a degree of precision in human thinking that is not realistic. Moreover, as was demonstrated in [14], a crisp approach requires a larger number of highly restrictive assumptions in order to generate stable predictions.

In the absence of a maximal set, spatial models predict cycling. Cycling is a condition under which every alternative can be majority defeated by at least one other alternative. Austen-Smith and Banks [2, pp. 6-21] determine under what conditions observable political outcomes (i.e., the choices made by political actors) are consistent with the preferences of players. If the choice function used to select an alternative is rationalizable, then choices should be a proper subset of a maximal set, when one exists.

Unless otherwise stated, we assume in this chapter that strict preferences are of type $\pi_{(0)}$.

Definition 2.1.1 Let $C : \mathcal{FP}^*(X) \to \mathcal{FP}^*(X)$ be such that $C(\mu) \subseteq \mu \; \forall \mu \in \mathcal{FP}^*(X)$. Then C is called a **fuzzy choice function** on X. Define $C^* : \mathcal{P}^*(X) \to \mathcal{P}^*(X)$ by $\forall S \in \mathcal{P}^*(X), C^*(S) = Supp(C(1_S))$.

A fuzzy choice function C assigns to every element in X the degree to which it is chosen with respect to every fuzzy subset μ of X. The degree to which it is selected cannot be larger than the degree for which it is a member of μ.

Proposition 2.1.2 Let C be a fuzzy choice function on X. Then C^* is a choice function on X.

Proof. Let $S \in \mathcal{P}^*(X)$. Let $x \in C^*(S)$. Then $x \in Supp(C(1_S))$. Now $C(1_S) \subseteq 1_S$ and so $Supp(C(1_S)) \subseteq Supp(1_S) = S$. Thus $x \in S$. Hence $C^*(S) \subseteq S$. Thus C^* is a choice function on X. ∎

Suppose C is a fuzzy choice function on X and $\mu \in \mathcal{FP}^*(X)$. Then it follows that $C(\mu)_t \subseteq \mu_t \ \forall t \in \mathrm{Im}(\mu)$. The condition (not assumed here) that $\forall S \in \mathcal{P}^*(X), C(1_S) = 1_T$ for some nonempty $T \subseteq S$ assures that C maps characteristic functions onto characteristic functions and so C can be considered to be a choice function on X. Consequently, if this condition is assumed, then we can abuse the notation and write $C(\mu_t) \subseteq \mu_t \ \forall t \in \mathrm{Im}(\mu) \ \forall \mu \in \mathcal{FP}^*(X)$.

Lemma 2.1.3 *Suppose $\pi, \rho \in \mathcal{FR}^*(X)$. Then π is the strict preference relation associated with ρ if and only if $\mathrm{Supp}(\pi)$ is the strict preference relation associated with $\mathrm{Supp}(\rho)$ and $\pi = \rho$ on $\mathrm{Supp}(\pi)$.*

Proof. π is the strict preference relation associated with $\rho \Leftrightarrow (\forall x, y \in X,$ $\pi(x, y) = \rho(x, y)$ if $\rho(x, y) > 0$ and $\rho(y, x) = 0$; $\pi(x, y) = 0$ otherwise) $\Leftrightarrow$ $\forall x, y \in X, (x, y) \in \mathrm{Supp}(\pi)$ if $(x, y) \in \mathrm{Supp}(\rho)$ and $(y, x) \notin \mathrm{Supp}(\rho)$ and $\pi = \rho$ on $\mathrm{Supp}(\pi) \Leftrightarrow \mathrm{Supp}(\pi)$ is the strict preference relation associated with $\mathrm{Supp}(\rho)$ and $\pi = \rho$ on $\mathrm{Supp}(\pi)$. ∎

If the preferences of players are known, reflexivity, completeness, and acyclicity are easily determined. However, they are not always known. For a variety of reasons ranging from electoral strategy to national security, politicians and political actors often hide or misrepresent their preferences. Therefore, scholars must deduce their preferences from their observed choices. Among the conventional approaches is to estimate the preferences of parties and presidents based on roll-call votes and the executive veto. In so doing, individuals assume that observable choices comprise a subset of the maximal set. That is, the choice function used by political actors is rationalizable, by which we mean that observable political outcomes (choices over a set of alternatives) reflect preference.

Assuming that a fuzzy maximal set has nonempty support, we next consider the conditions under which fuzzy choice functions are rationalizable.

Definition 2.1.4 *Let C be a fuzzy choice function on X. Let $\rho \in \mathcal{FR}^*(X)$. Then C is called **rationalizable** with respect to ρ if $C(\mu) = M(\rho, \mu)$ for all $\mu \in \mathcal{FP}^*(X)$.*

Example 2.1.5 *Let $X = \{x, y, z\}$. Define $\rho : X \times X \to [0, 1]$ as follows:*

$$\rho(x, x) = \rho(y, y) = \rho(z, z) = 1, \rho(x, y) = \rho(x, z) = 1/2,$$
$$\rho(y, z) = \rho(z, y) = 1/4, \rho(y, x) = \rho(z, x) = 0.$$

Clearly ρ is reflexive and complete. Now $\forall u, v \in X, \pi(u, v) = 1/2$ if and only if $u = x$ and $v = y$ or z, and $\pi(u, v) = 0$ otherwise. Thus it follows that ρ is acyclic. Define $C : \mathcal{FP}^*(X) \to \mathcal{FP}^*(X)$ by $\forall \mu \in \mathcal{FP}^*(X), C(\mu)(w) =$

$M_M(\rho, \mu)(w) \; \forall w \in X$. Then ρ rationalizes C by Theorem 1.2.10. By Proposition 1.2.9, we have the following table.

S	$\{x\}$	$\{y\}$	$\{z\}$	$\{x,y\}$	$\{x,z\}$	$\{y,z\}$	$\{x,y,z\}$
$C(1_S)(x)$ $= M(\rho, 1_S)(x)$	1	0	0	1/2	1/2	0	1/2
$C(1_S)(y)$ $= M(\rho, 1_S)(y)$	0	1	0	0	0	1/4	0
$C(1_S)(z)$ $= M(\rho, 1_S)(z)$	0	0	1	0	0	1/4	0

It follows that $C^*(\{x,y\}) = C^*(\{x,z\}) = C^*(\{x,y,z\}) = \{x\}$ and $C^*(\{y, z\}) = \{y,z\}$. Define $\mu : X \to [0,1]$ as follows: $\mu(x) = 1, \mu(y) = \mu(z) = 1/4$. Then $C(\mu)(x) = 1, C(\mu)(y) = C(\mu)(z) = 0$. Hence $C^*(\mu_1) = C^*(\{x\}) = \{x\} = C(\mu)_1$ and $C^*(\mu_{1/4}) = C^*(\{x,y,z\}) = \{x\} = C(\mu)_{1/4}$. Note that $C(1_{\{x,z\}})(x) = 1/2$ and so $C(1_{\{x,z\}}) \neq 1_T$ for some $T \subseteq \{x,z\}$. Thus it is not the case that $\forall S \in \mathcal{P}^*(X), C(1_S) = 1_T$ for some $T \subseteq S$.

Define $\nu : X \to [0,1]$ by $\nu(x) = \nu(y) = \nu(z) = 1/8$. Since $\nu_t = \emptyset$ for $t \in (1/8, 1], \nu_{1/8} = X$, and $(y,x), (z,x) \notin \rho_t \; \forall t \in (0,1]$, it follows that $C(\nu)(x) = 1/8$, $C(\nu)(y) = 0$, and $C(\nu)(z) = 0$.

Example 2.1.5 shows that it is not always the case C restricted to $\mathcal{P}^*(X)$ can be considered a choice function on X. In fact, Proposition 1.2.35 shows that this can only be the case if ρ is crisp. Nevertheless, if this condition is assumed, a theory involving fuzzy subsets of X is still possible.

We now proceed with a formal definition of the degree to which a fuzzy choice function will select an outcome in the fuzzy maximal set, i.e., whether a fuzzy choice function is rationalizable.

Proposition 2.1.6 *Let $\rho \in \mathcal{FR}^*(X)$ be such that ρ is complete, partially acyclic, and $(x,x) \in \mathrm{Supp}(\rho) \; \forall x \in X$. Suppose $*$ has no zero divisors. Define $C : \mathcal{FP}^*(X) \to \mathcal{FP}^*(X)$ by $\forall \mu \in \mathcal{FP}^*(X), C(\mu)(x) = M(\rho, \mu)(x) \; \forall x \in X$. Then ρ rationalizes C.*

Proof. Let $x \in X$. Then

$$M(\rho, \mu)(x) > 0$$
$$\Leftrightarrow \quad \mu(x) * (\circledast_{w \in X}(\vee\{t \in [0,1] \mid C(w,x) * t \leq \rho(x,w)\})) > 0$$
$$\Leftrightarrow \quad \mu(x) > 0 \text{ and } \vee \{t \in [0,1] \mid C(w,x) * t \leq \rho(x,w)\})) > 0 \; \forall w \in X$$
$$\Leftrightarrow \quad \mu(x) > 0 \text{ and } \forall w \in X, \exists t_w \in (0,1] \text{ such that } C(w,x) * t_w \leq \rho(x,w).$$

Let $t_* = \wedge(\{\rho(x,y) \mid (x,y) \in X \times X, \rho(x,y) > 0\})$. Then $t_* > 0$ since X is finite. Since ρ is complete, we have that $\forall x, y \in X$ either $\rho(x,y) \geq t_*$ or $\rho(y,x) \geq t_*$. Let $x_1, \ldots, x_n \in X$ and $t \in (0, t_*]$. Let π be the strict preference relation associated with ρ. Since ρ is acyclic, it follows that $(x_1, x_2), (x_2, x_3), \ldots, (x_{n-1}, x_n) \in \pi_t = \mathrm{Supp}(\pi)$ implies $(x_1, x_n) \in \rho_t = \mathrm{Supp}(\rho)$. Since $\mathrm{Supp}(\pi)$ is the strict preference relation associated with $\text{Supp}(\rho)$ by Lemma 2.1.3, it follows that $\text{Supp}(\rho) = \rho_t$ is acyclic for all $t \in (0, t_*]$. Since $(x, x) \in \rho_t$ $\forall t \in [0, t_*]$ and $\forall x \in X$, ρ_t is reflexive $\forall t \in [0, t_*]$. Let $s^* = \vee\{\mu(x) \mid x \in X\}$. Then it follows that $\{x \in X \mid x \in \mu_t \text{ and } (x, y) \in \rho_t \ \forall y \in \mu_t\} \neq \emptyset$ if $t \in [0, t_* \wedge s^*]$ by [2, Theorem 1.1, p. 5]. Thus $\{x \in X \mid x \in \mu_t \text{ and } (x, y) \in \rho_t \ \forall y \in \text{Supp}(\mu)\} \neq \emptyset$ if $t \in [0, t_* \wedge s_*]$, where $s_* = \wedge\{\mu(x) \mid x \in \text{Supp}(\mu)\}$. Thus $\{x \in \mu_t \mid \rho(x, w) \geq t \ \forall w \in \text{Supp}(\mu)\} \neq \emptyset$ $\forall t \in (0, t_* \wedge s_*]$. Hence $\exists x \in \text{Supp}(\mu)$ such that $\forall w \in X, \rho(x, w) \geq t_w$, where $t_w \in (0, t_* \wedge s_*]$. (Actually, t_w can be taken independent of w.) Thus $\rho(x, w) \geq C(w, x) * t \ \forall w \in X$, where $t = \wedge\{t_w \mid w \in X\}$. Since X is finite, $t > 0$. Thus $\exists x \in X$ such that $M(\rho, \mu)(x) > 0$. ∎

Example 2.1.7 *Let $X = \{x, y, z\}$. Suppose $*$ has no zero divisors. Define the fuzzy relation ρ on X as follows:*

$$\rho(x, x) = \rho(y, y) = \rho(z, z) = 1, \rho(x, y) = 1/2, \rho(x, z) = 1/2,$$
$$\rho(y, z) = \rho(z, y) = 1/4, \rho(y, x) = \rho(z, x) = 0.$$

Clearly, ρ is reflexive and complete. Now $\forall u, v \in X$, $\pi(u, v) = 1/2$ if $u = x$ and $v = y$ or z, and $\pi(u, v) = 0$ otherwise. Thus it follows that ρ is acyclic. Define $C : \mathcal{FP}^*(X) \to \mathcal{FP}^*(X)$ by $\forall \mu \in \mathcal{FP}^*(X), C(\mu)(w) = M(\rho, \mu)(w)$ $\forall w \in X$. Then ρ rationalizes C by Proposition 2.1.6. By Proposition 1.2.9, we have the following table:

S	$\{x\}$	$\{y\}$	$\{z\}$	$\{x,y\}$	$\{x,z\}$	$\{y,z\}$	$\{x,y,z\}$
$C(1_S)(x)$ $= M_G(\rho, 1_S)(x)$	1	0	0	1	1	0	1
$C(1_S)(y)$ $= M_G(\rho, 1_S)(y)$	0	1	0	0	0	1	0
$C(1_S)(z)$ $= M_G(\rho, 1_S)(z)$	0	0	1	0	0	1	0

For $*$ the Lukasiewicaz t-norm, $M_G(\rho, 1_{\{x,y\}})(y) = 1/2$ since $1/2 * t = 0$ for $t = 1/2$. Recall that the Lukasiewicaz t-norm has zero divisors.

Definition 2.1.8 *Let C be a fuzzy choice function on X. We say that C is* **crisply rationalized** *if there exists $R \in \mathcal{R}(X)$ such that $C(\mu) = M(1_R, \mu)$ $\forall \mu \in \mathcal{FP}^*(X)$.*

Proposition 2.1.9 *Let C be a fuzzy choice function on X such that $\forall S \in \mathcal{P}^*(X)$, $C(1_S) = 1_T$ for some $T \subseteq S$. Then C is crisply rationalized if and only if C is crisply rationalized by R_C, where R_C is defined by $\forall x, y \in X$, $x R_C y$ if and only if $x \in C^*(\{x, y\})$.*

Proof. If C is crisply rationalized by R_C, then C is crisply rationalized. Conversely, suppose C is crisply rationalized by R. Then since $\forall S \in \mathcal{P}^*(X)$,

$C(1_S) = 1_T$ for some $T \subseteq S$, $M(1_R, 1_S) = C(1_S) = 1_T$. Hence $C^*(S) =$ Supp$(C(1_S) = $Supp$(1_T) = T$. Thus

$$
\begin{aligned}
&\quad x \in M(R, S) \\
&\Leftrightarrow \quad x \in S \text{ and } xRy \ \forall y \in S \Leftrightarrow 1_S(x) = 1 \\
&\quad \text{and } 1_R(x, y) = 1 \ \forall y \text{ such that } 1_S(y) = 1 \\
&\Leftrightarrow \quad 1_S(x) * (\circledast\{\vee\{t \in [0, 1] \mid \mathcal{C}(y, x) * t \leq 1_R (x, y)\} \mid y \in X\} = 1 \\
&\Leftrightarrow \quad M(1_R, 1_S)(x) = 1 \\
&\Leftrightarrow \quad C(1_S)(x) = 1 \Leftrightarrow 1_T(x) = 1 \Leftrightarrow x \in T \\
&\Leftrightarrow \quad x \in \text{ Supp}(C(1_S)) \Leftrightarrow x \in C^*(S).
\end{aligned}
$$

Hence $C^*(S) = M(R, S)$, i.e., R rationalizes C^* and in fact $R = R_C$ by [2, Lemma 1.1, p. 8]. Let $\mu \in \mathcal{FP}^*(X)$. Then $\forall x \in X$,

$$
C(\mu)(x) = M(1_R, \mu)(x) = M(1_{R_C}, \mu)(x).
$$

That is, R_C rationalizes C. ∎

The following result states that if there exists a fuzzy binary relation ρ on X such that the level sets μ_t of every fuzzy subset μ of X is rationalized by ρ_t for every t, then ρ must be crisp and in fact ρ is the base relation for C^*.

Proposition 2.1.10 *Let C be a fuzzy choice function on X. If there exists $\rho \in \mathcal{FR}(X)$ such that $\forall \mu \in \mathcal{FP}^*(X)$ and $\forall t \in \text{Im}(\mu), C^*(\mu_t) = M(\rho_t, \mu_t)$,then there exists a relation R on X such that $\rho = 1_R$.*

Proof. Let $S \in \mathcal{P}^*(X)$ and let $t \in (0, 1]$. Define $\mu : X \to [0, 1]$ by $\mu(x) = t$ if $x \in S$ and $\mu(x) = 0$ otherwise. Then $C^*(\mu_t) = M(\rho_t, \mu_t)$ by hypothesis. Since $\mu_t = S$ and S is arbitrary, ρ_t rationalizes C^*. Hence $\rho_t = R_C$. Since t is arbitrary, $\rho_t = R_C \ \forall t \in (0, 1]$. Thus it follows that $\rho = 1_R$ for some relation R on X. (In fact, $R = R_C$.) ∎

2.2   Consistency Conditions

In this section, we consider the effect of placing restrictions on the behavior of choice functions. We consider the fuzzification of conditions α, γ, β,path independence, Arrow axiom, and WARP. These conditions place restrictions on the behavior of choice functions. The fuzzification follows without difficulty because of our choice of the strict fuzzy preference relation to associate with fuzzy preference relations, namely strict fuzzy preference relations of type $\pi_{(0)}$. We use the notation M for crisp maximal sets at times as well as for the fuzzy maximal set defined in Definition 1.2.1.

Rational behavior of consumers is an important concern in welfare economics. To study the rationality of consumers, Samuelson in [42] introduced the theory of revealed preferences through preference relations associated with

demand functions. Since then Uzawa [49] and Arrow [1] have developed revealed preference theory in an abstract framework. The work of Uzawa and
Arrow was continued mainly by Richter [40], Sen [42, 44] and Suzumura [47,
48]. The work of Uzawa, Arrow and Sen was based on the assumption that the
domain of choice functions under consideration contains all non-empty finite
subsets of the set of alternatives of the universal set, whereas Richter [40] and
Suzumura [47] defined choice functions on an arbitrary non-empty class of
non-empty subsets of the universal set and studied rationality of choice functions by means of properties as the revealed preference axioms, the congruence
axioms and the consistency conditions.

Example 2.2.1 *Consider the crisp example,* $X = \{x, y, z\}, S = \{y, z\},$ *and*
$R = \Delta X \cup \{(x, y), (y, z), (x, z)\}.$ *Then* $M(R, S) = \{w \in S \mid (w, v) \in R$ *for all*
$v \in S\} = \{y\}.$

Recall that $S_y = \{w \in X \mid \rho(w, y) > \rho(y, w)\}.$

Proposition 2.2.2 *Let* $\rho \in \mathcal{FR}^*(X)$ *and let* S *be a subset of* $X.$ *Suppose* $*$
is a t-norm without zero divisors. Then $M_G(\rho, 1_S) = 1_T$ *for some* $T \subseteq S$ *if
and only if* $\forall x, y \in S, \rho(x, y) * \rho(y, x) > 0$ *implies* $\rho(x, y) = \rho(y, x).$

Proof. Suppose $M_G(\rho, 1_S) = 1_T$ for some $T \subseteq S.$ Let $x, y \in S.$ Suppose
$\rho(x, y) * \rho(y, x) > 0.$ Then $\rho(x, y) > 0$ and $\rho(y, x) > 0.$ Suppose $\rho(x, y) >$
$\rho(y, x) > 0.$ $\forall w \in S_y,$ let $t_{y,w} = \vee\{t \in [0, 1] \mid \rho(w, y) * t \leq \rho(y, w)\}.$ Then
$M_G(\rho, 1_S)(y) = \circledast\{t_{y,w} \mid w \in S_y\}.$ Since $S_y \neq \emptyset$ and $\rho(y, x) > 0$ it follows
that $0 < \circledast\{\rho(y, w) \mid w \in S_y\} < 1.$ However, this contradicts the hypothesis.
Thus $\rho(x, y) = \rho(y, x).$
Conversely, suppose that $\forall x, y \in S, \rho(x, y) * \rho(y, x) > 0$ implies $\rho(x, y) =$
$\rho(y, x).$ Let $T = \{x \in S \mid \nexists w \in S, \rho(w, x) > \rho(x, w)\}.$ Suppose $x \in S \setminus T.$ Then
$\exists w \in S$ such that $\rho(w, x) > \rho(x, w).$ By our hypothesis, $\rho(x, w) = 0.$ Thus
$M_G(\rho, 1_S)(x) = 0.$ Hence $M_G(\rho, 1_S) = 1_T.$ ∎

The next result says that if a fuzzy choice function C when restricted to
crisp sets S yields a crisp subset of S and if C is rationalizable with respect
to ρ in $\mathcal{FR}(X),$ then ρ is also crisp.

Proposition 2.2.3 *Let* C *be a fuzzy choice function on* X *such that* $\forall S \in$
$\mathcal{P}^*(X), C(1_S) = 1_T$ *for some nonempty* $T \subseteq S.$ *Suppose that there exists*
$\rho \in \mathcal{FR}(X)$ *such that* $C(1_S) = M_G(\rho, 1_S) \ \forall S \in \mathcal{P}^*(X).$ *Then there exists a
relation* R *on* X *such that* $\rho = 1_R.$

Proof. Let $S \in \mathcal{P}^*(X)$ and let $t \in (0, 1].$ Then $x \in M_G(\rho, 1_S)_t \Leftrightarrow \wedge\{\vee\{u \in$
$[0, 1] \mid \rho(w, x) \wedge u \leq \rho(x, w)\} \mid w \in S\} \geq t \Leftrightarrow \rho(x, w) \geq \rho(w, x) \wedge t \ \forall w \in S$
and $x \in M_G(\rho_t, S) \Leftrightarrow M_G(1_{\rho_t}, 1_S)(x) > 0 \Leftrightarrow \wedge\{\vee\{u \in [0, 1] \mid 1_{\rho_t}(w, x) \wedge$
$u \leq 1_{\rho_t}(x, w)\} \mid w \in S\} > 0 \Leftrightarrow \forall w \in S, [(w, x) \in \rho_t \Rightarrow (x, w) \in \rho_t$ or
$(w, x) \notin \rho_t].$ Thus $M_G(\rho, 1_S)_t = M_G(\rho_t, S).$ Hence $C^*(S) = \text{Supp}(C(1_S)) =$

$\mathrm{Supp}(1_T) = T = (1_T)_t = C(1_S)_t = M_G(\rho, 1_S)_t = M_G(\rho_t, (1_S)_t) = M_G(\rho_t, S)$.
Since S is arbitrary, ρ_t rationalizes C^* $\forall t \in (0,1]$. Thus $\rho_t = R_C$ $\forall t \in (0,1]$.

∎

Definition 2.2.4 *Let C be a fuzzy choice function on X. Define $\rho_C : X \times X \to [0,1]$ by $\forall x, y \in X$, $\rho_C(x,y) = C(1_{\{x,y\}})(x)$.*

Proposition 2.2.5 *Let C be a choice function on X and let $\rho \in \mathcal{FR}(X)$. Suppose ρ rationalizes C. Then $\forall x, y \in X$,*

$$\rho_C(x,y) = \begin{cases} 1 & \text{if } \rho(y,x) \leq \rho(x,y), \\ t_{x,y} & \text{otherwise,} \end{cases}$$

Furthermore, ρ_C rationalizes C for characteristic functions.

Proof. Let $x, y \in X$. Then

$$\begin{aligned}
\rho_C(x,y) &= C(1_{\{x,y\}})(x) \\
&= 1_{\{x,y\}}(x)* \\
&\quad (\circledast_{w \in X}(\vee\{t \in [0,1] \mid 1_{\{x,y\}}(w) * \rho(w,x) * t \leq \rho(x,w))) \\
&= (\vee\{t \in [0,1] \mid \rho(x,x) * t \leq \rho(x,x)\}) \\
&\quad \circledast (\vee\{t \in [0,1] \mid \rho(y,x) * t \leq \rho(x,y)\}) \\
&= \vee\{t \in [0,1] \mid \rho(y,x) * t \leq \rho(x,y)\}
\end{aligned}$$

from which the desired result is immediate. We now show ρ_C rationalizes C for characteristic functions. It suffices to compare $\rho_C(w,x) * t \leq \rho_C(x,w)$ and $\rho(w,x) * t \leq \rho(x,w)$. Now $\rho_C(w,x) * t \leq \rho_C(x,w)$ implies $t = 1$ is the largest possible t if $\rho_C(w,x) \leq \rho_C(x,w)$ and $\rho_C(w,x) * t \leq \rho_C(x,w)$ implies $t_{x,w}$ is the largest possible t if $\rho_C(w,x) \geq \rho_C(x,w)$. Also, $\rho(w,x) * t \leq \rho(x,w)$ implies $t = 1$ is the largest possible t if $\rho(w,x) \leq \rho(x,w)$ and $\rho(w,x) * t \leq \rho(x,w)$ implies $t_{x,w}$ is the largest possible t if $\rho(w,x) \geq \rho(x,w)$. The desired result now follows since $\rho_C(x,w) = t_{x,w}$ if $\rho(w,x) > \rho(x,w)$ and $\rho(x,w) > \rho(w,x) \Leftrightarrow \rho_C(x,w) > \rho_C(w,x)$. ∎

Corollary 2.2.6 *Let C be a fuzzy choice function on X and let $\rho \in \mathcal{FR}(X)$. Suppose ρ rationalizes C. Then $\forall x \in X, \rho_C(x,x) = 1$.*

Recall that in the crisp case, condition α requires that if an alternative x is chosen from a set T and S is a subset of T containing x, then x should still be chosen.

Definition 2.2.7 *Let C be a fuzzy choice function on X. Then C is said to satisfy condition α if $\forall \mu, \nu \in \mathcal{FP}^*(X), \mu \subseteq \nu$ implies $C(\nu) \cap \mu \subseteq C(\mu)$. C is said to satisfy condition α for characteristic functions if $\forall S, T \in \mathcal{P}^*(X), S \subseteq T$ implies $C(1_T) \cap 1_S \subseteq C(1_S)$.*

Condition α says that the degree of membership of an element in $C(\mu)$ is at least as large as the minimum of the degrees of membership of the elements in $C(\nu)$ and μ.

Lemma 2.2.8 *Let C be a fuzzy choice function on X. Suppose $*$ has no zero divisors. Let $x_1, \ldots, x_n \in X$. Suppose C satisfies condition α. Then $\pi_C(x_1, x_2) * \pi_C(x_2, x_3) * \ldots * \pi_C(x_{n-1}, x_n) > 0$ implies $\rho_C(x_1, x_2) > 0$.*

Proof. Suppose $C(1_{\{x_1,\ldots,x_n\}})(x_i) > 0$ for some $i = 2, \ldots, n$. Then since C satisfies condition α,

$$
\begin{aligned}
0 &< C(1_{\{x_1,\ldots,x_n\}})(x_i) * 1_{\{x_{i-1},x_i\}}(x_i) = (C(1_{\{x_1,\ldots,x_n\}}) \cap 1_{\{x_{i-1},x_i\}})(x_i) \\
&\le C(1_{\{x_{i-1},x_i\}})(x_i) = \rho_C(x_i, x_{i-1}),
\end{aligned}
$$

which contradicts $\pi_C(x_{i-1}, x_i) > 0$. Thus $C(1_{\{x_1,\ldots,x_n\}})(x_i) = 0$ for $i = 2, \ldots, n$. Hence $C(1_{\{x_1,\ldots,x_n\}})(x_1) > 0$. Thus

$$
\begin{aligned}
0 &< C(1_{\{x_1,\ldots,x_n\}})(x_1) * 1_{\{x_1,x_n\}}(x_1) = C(1_{\{x_1,\ldots,x_n\}}) \cap 1_{\{x_1,x_n\}})(x_1) \\
&\le C(1_{\{x_1,x_n\}})(x_1) = \rho_C(x_1, x_n).
\end{aligned}
$$

∎

Proposition 2.2.9 *Suppose C is a fuzzy choice function on X that satisfies condition α. Suppose $*$ has no zero divisors. If ρ_C is crisp, then ρ_C is acyclic.*

Proof. Let $x_1, \ldots, x_n \in X$. Suppose that $\pi_C(x_1, x_2) * \pi_C(x_2, x_3) * \ldots * \pi_C(x_{n-1}, x_n) > 0$. Then $\rho_C(x_1, x_n) > 0$ by Lemma 2.2.8 and so $\rho_C(x_1, x_n) = 1$. Thus

$$
\pi_C(x_1, x_2) * \pi_C(x_2, x_3) * \ldots * \pi_C(x_{n-1}, x_C) \le \rho_C(x_1, x_n).
$$

∎

In the crisp case, condition γ requires that if an alternative x was chosen from a set S and also from a set T, then it should be chosen from the set $S \cup T$. In the next definition, we fuzzify this condition.

Definition 2.2.10 *Let C be a fuzzy choice function on X. Then C is said to satisfy **condition** γ if $\forall \mu, \nu \in FP^*(X)$, $C(\mu) \cap C(\nu) \subseteq C(\mu \cup \nu)$. C is said to satisfy **condition** γ **for characteristic functions** if $\forall S, T \in \mathcal{P}^*(X)$, $C(1_S) \cap C(1_T) \subseteq C(1_S \cup 1_T)$.*

Condition γ says that the degree of membership of any alternative x in $C(\mu \cup \nu)$ should not be strictly less than the smallest degree of membership of x in $C(\mu)$ and $C(\nu)$.

Proposition 2.2.11 *Suppose C is a fuzzy choice function on X. If ρ_C is partially acyclic and C satisfies conditions α and γ for characteristic functions, then ρ_C rationalizes C for characteristic functions, i.e., $C(1_S) = M_G(\rho_C, 1_S)$ $\forall S \in \mathcal{P}^*(X)$. Conversely, if ρ_C rationalizes C for characteristic functions, then C satisfies condition α and satisfies condition γ if $* = \wedge = \circledast$.*

Proof. Suppose that ρ_C is partially acyclic and C satisfies conditions α and γ for characteristic functions. Clearly ρ_C is complete and reflexive. Since ρ_C is acyclic, $\exists x \in X$ such that $M_G(\rho, \mu)(x) > 0 \ \forall \mu \in \mathcal{FP}^*(X)$ by the proof of Proposition 2.1.6. By condition α, $1_{\{x,y\}} \cap C(1_S) \subseteq C(1_{\{x,y\}})$, where $x, y \in S$. By the definition of ρ_C, $\rho_C(x, y) = C(1_{\{x,y\}})(x) \geq 1 * C(1_S)(x) = C(1_S)(x)$. Now

$$
\begin{aligned}
M_G(\rho_C, 1_S)(x) &= 1_S(x)* \\
&\quad (\circledast_{w \in X}(\vee\{t \in [0,1] \mid 1_S(w) * \rho_C(w,x) * t \leq \rho_C(x,w)\})) \\
&= \circledast_{w \in X}(\vee\{t \in [0,1] \mid 1_S(w) * \rho_C(w,x) * t \leq \rho_C(x,w)\})) \\
&= \circledast_{w \in S}(\vee\{t \in [0,1] \mid \rho_C(w,x) * t \leq \rho_C(x,w)\})) \\
&\geq C(1_S)(x),
\end{aligned}
$$

where the latter inequality holds since the largest t for which $\rho_C(w,x) * t \leq \rho_C(x,w)$ is 1 if $\rho_C(w,x) < \rho_C(x,w)$ and is $\geq \rho_C(x,w)$ (which is $\geq C(1_S)(x)$) if $\rho_C(w,x) > \rho_C(x,w)$. Thus $C(1_S) \subseteq M_G(\rho_C, 1_S)$. Suppose $S = \{x\}$. Then it follows easily that $C(1_{\{x\}}) = 1_{\{x\}} = M_G(\rho_C, 1_{\{x\}})$. Now suppose $S = \{x, y_1, \ldots, y_n\}$. If $n = 1$, then

$$
\begin{aligned}
&C(1_{\{x,y_1\}})(x) \\
=\ & \rho_C(x, y_1) \geq \ \vee\{t \in [0,1] \mid \rho_C(y_1, x) * t \leq \rho_C(x, y_1)\} \\
=\ & \circledast_{w \in \{x,y_1\}}(\vee\{t \in [0,1] \mid \rho_C(w,x) * t \leq \rho_C(x,w)\}) \\
=\ & 1_{\{x,y_1\}}(x) * (\circledast_{w \in X}(\vee\{t \in [0,1] \mid 1_{\{x,y_1\}}(w) * \rho_C(w,x) * t \leq \rho_C(x,w)\})) \\
=\ & M_G(\rho_C, 1_S)(x).
\end{aligned}
$$

Hence $C(1_S) \supseteq M_G(\rho_C, 1_S)$. Suppose $n \geq 1$ and $C(1_{\{x,y_1\}}) * \cdots * C1_{\{x,y_k\}}) \subseteq C(1_{\{x,y_1,\ldots,y_k\}})$ for $1 \leq k < n$, the induction hypothesis. By condition γ, $C(1_{\{x,y_1,\ldots,y_k\}}) * C(1_{\{x,y_{k+1}\}}) \subseteq C(1_{\{x,y_1,\ldots,y_{k+1}\}})$. Thus it follows by induction that $*_{y \in S} C(1_{\{x,y\}}) \subseteq C(1_S)$. Thus $C(1_S) \supseteq M_G(\rho_C, 1_S)$ and so $C(1_S) = M_G(\rho_C, 1_S)$.

Conversely, suppose ρ_C rationalizes C for characteristic functions. Let $x \in X$. Let $S, T \in \mathcal{P}^*(X)$ be such that $S \subseteq T$. If $(C(1_T) \cap 1_S)(x) = 0$, then $(C(1_T) \cap 1_S)(x) \leq C(1_S)(x)$. Suppose that $(C(1_T) \cap 1_S)(x) > 0$. Then $x \in S$ and $(C(1_T) \cap 1_S)(x) = C(1_T)(x)$. Now

$$
\begin{aligned}
0\ <\ & C(1_T)(x) \\
=\ & M_G(\rho_C, 1_T)(x) \\
=\ & 1_T(x)* \\
& (\circledast_{w \in X}(\vee\{t \in [0,1] \mid 1_T(w) * \rho_C(w,x) * t \leq \rho_C(x,w)\})) \\
\leq\ & 1 * (\circledast_{w \in X}(\vee\{t \in [0,1] \mid 1_S(w) * \rho_C(w,x) * t \leq \rho_C(x,w)\})) \\
=\ & M_G(\rho_C, 1_S)(x) \\
=\ & C(1_S)(x),
\end{aligned}
$$

where the inequality holds since $x \in S \subseteq T$. Thus $C(1_T) \cap 1_S \subseteq C(1_S)$. Hence condition α holds for characteristic functions. Let $S, T \in \mathcal{P}^*(X)$ and $x \in X$. Suppose that $0 = (C(1_S) \cap C(1_T))(x)$. Then $(C(1_S) \cap C(1_T))(x) \leq C(1_{S \cup T})(x)$. Suppose that $(C(1_S) \cap C(1_T))(x) > 0$. Since $S, T \subseteq S \cup T$, it follows that $M_G(\rho_C, 1_S) \cap M_G(\rho_C, 1_T) \supseteq M_G(\rho_C, 1_{S \cup T})$ (since $* = \wedge = \circledast) = M_G(\rho_C, 1_S \cup 1_T)$. Thus $C(1_S) \cap C(1_T) \subseteq C(1_S \cup 1_T)$. Hence condition γ holds for characteristic functions. ∎

Example 2.2.12 *Let* $X = \{x, y\}$. *Let* C *be the choice function on* X *such that*

$$C(1_{\{x\}})(x) = 1, C(1_{\{x\}})(y) = 0, C(1_{\{y\}})(x) = 0, C(1_{\{y\}})(y) = 1,$$
$$C(1_{\{x,y\}})(x) = 1/2, C(1_{\{x,y\}})(y) = 0.$$

Then

$$\rho_C(x, x) = \rho_C(y, y) = 1, \rho_C(x, y) = 1/2, \rho_C(y, x) = 0.$$

It follows easily that $C(1_S) = M_G(\rho_C, 1_S)$ $\forall S = \{x\}$ *or* $\{y\}$, *but* $C(1_X)(x) = 1/2 < 1 = M_G(\rho_C, 1_X)(x)$. *Thus* $C(1_X) \subset M_G(\rho_C, 1_X)$. *Hence* ρ_C *does not rationalize* C. *Clearly,* ρ_C *is reflexive, complete, and acyclic.*

Path independence in the crisp case allows for dividing the set of alternatives into smaller subsets, say S and T, then choosing a subset from $S \cup T$ and making the final decision from this subset.

Definition 2.2.13 *Let* C *be a fuzzy choice function on* X. *Then* C *is said to satisfy* **path independence** *(PI) if* $\forall \mu, \nu \in \mathcal{FP}^*(X), C(\mu \cup \nu) = C(C(\mu) \cup C(\nu))$.

Definition 2.2.13 is a fuzzification of the crisp case which says that choices from $\mu \cup \nu$ are the same as those arrived at by first choosing from μ and from ν, and then choosing among these chosen alternatives.

Path independence yields the following result: An alternative $x \in \text{Supp}(C(\mu \cup \nu))$ if and only if $x \in \text{Supp}(C(C(\mu) \cup C(\nu)))$. Since $x \in \text{Supp}(C(\mu \cup \nu))$ implies $x \in \text{Supp}(\mu \cup \nu)$ and also $x \in \text{Supp}(C(C(\mu) \cup C(\nu)))$ implies $x \in \text{Supp}(C(\mu) \cup C(\nu))$ similar comments concerning decentralization as those in [2, p. 13] can be made under certain conditions.

There are several ways to define set difference, $\nu \backslash \mu$, of two fuzzy subsets of μ and ν of X. For example, consider the following definition.

Definition 2.2.14 *Let* $\mu, \nu \in \mathcal{FP}(X)$.
(1) *Define* $\nu \backslash \mu$ *by* $\forall x \in X$,

$$(\nu \backslash \mu)(x) = \begin{cases} \nu(x) & \text{if } \mu(x) = 0, \\ 0 & \text{if } \mu(x) > 0. \end{cases}$$

(2) *Define* $\nu \backslash \mu$ *by* $\forall x \in X$,

$$(\nu \backslash \mu)(x) = \begin{cases} \nu(x) & \text{if } \mu(x) < \nu(x), \\ 0 & \text{if } \mu(x) \geq \nu(x). \end{cases}$$

(3) *Define* $\nu \backslash \mu$ *by* $\forall x \in X$,

$$(\nu \backslash \mu)(x) = \begin{cases} \nu(x) - \mu(x) & \text{if } \mu(x) < \nu(x), \\ 0 & \text{if } \mu(x) \geq \nu(x). \end{cases}$$

Let $\backslash$ be a set difference of $\mathcal{FP}(X)$. Let $\mathcal{S}(\backslash) = \{(\mu, \nu) \mid \mu, \nu \in \mathcal{FP}(X), \mu \cup (\nu \backslash \mu) = \mu \cup \nu$ and $\mu \cap (\nu \backslash \mu) = 1_\emptyset\}$. We note that if $\backslash$ is defined as in Definition 2.2.14(1), then $\mu \cap (\nu \backslash \mu) = 1_\emptyset$ and $\mu \cup (\nu \backslash \mu) = \mu \cup \nu$ for those μ and ν such that $\text{Supp}(\mu) \cap \text{Supp}(\nu) \neq \emptyset$ and $\mu = \nu$ on $\text{Supp}(\mu) \cap \text{Supp}(\nu)$.

Proposition 2.2.15 *Let C be a fuzzy choice function on X. Suppose C satisfies PI. Then $\forall (\mu, \nu) \in \mathcal{S}(\backslash)$ such that $\mu \subseteq \nu, C(\nu) \cap \mu \subseteq C(\mu)$, i.e., C satisfies condition α for $\mathcal{S}(\backslash)$.*

Proof. Let $x \in X$. Then

$$\begin{aligned} (C(\nu) \cap \mu)(x) &= C(\nu)(x) * \mu(x) = C(\mu \cup \nu)(x) * \mu(x) \\ &= C(\mu \cup (\nu \backslash \mu))(x) * \mu(x) = (C(C(\mu) \cup C(\nu \backslash \mu))(x)) * \mu(x) \\ &\leq (C(\mu) \cup C(\nu \backslash \mu))(x)) * \mu(x) \\ &= C(\mu)(x) * \mu(x)) \vee (C(\nu \backslash \mu)(x) * \mu(x)) \\ &\leq (C(\mu)(x) * \mu(x)) \vee ((\nu \backslash \mu)(x) * \mu(x)) = C(\mu)(x) \vee 0 \\ &= C(\mu)(x). \end{aligned}$$

Thus $C(\nu) \cap \mu \subseteq C(\mu)$. ∎

The notion of quasi-transitivity will play an important role in our consideration of dictatorships and oligarchies later in the book. We use it here in the study of rationalization.

Definition 2.2.16 *Let $\rho \in \mathcal{FR}^*(X)$. Then*
   *(1) ρ is said to be **quasi-transitive** if $\forall x, y, z \in X$, $\pi(x, y) * \pi(y, z) \leq \pi(x, z)$;*
   *(2) ρ is said to be **partially quasi-transitive** if $\forall x, y, z \in X, \pi(x, y) * \pi(y, z) > 0$ implies $\pi(x, z) > 0$.*

Proposition 2.2.17 *Let $\rho \in \mathcal{FR}^*(X)$. If ρ_t is quasi-transitive $\forall t \in [0, 1]$, then ρ is quasi-transitive. Conversely, if $* = \wedge$ and ρ is quasi-transitive, then ρ_t is quasi-transitive $\forall t \in [0, 1]$*

Proof. Let $x, y, z \in X$. Suppose ρ_t is quasi-transitive $\forall t \in [0, 1]$. Let $\rho(x, y) * \rho(y, z) = t$. Then $(x, y), (y, z) \in \rho_t$ and so $(x, z) \in \rho_t$. Thus $\rho(x, z) \geq \rho(x, y) * \rho(y, z)$. Hence ρ is quasi-transitive. Conversely, suppose $* = \wedge$ and ρ is quasi-transitive. Let $t \in [0, 1]$. Then $(x, y), (y, z) \in \rho_t \Leftrightarrow \rho(x, y) \geq t, \rho(y, z) \geq t \Rightarrow \rho(x, z) \geq \rho(x, y) \wedge \rho(y, z) \geq t$. Hence $(x, z) \in \rho_t$ and so ρ_t is quasi-transitive. ∎

Definition 2.2.18 *Let C be a fuzzy choice function on X. Then C is said to satisfy* **condition** ω *if $\forall x, y \in X, C(1_{\{x,y\}}) = t_{\{x\}}$ for some $t \in (0,1]$ if and only if $\forall s \in [0,1], C(t_{\{x\}} \cup s_{\{y\}}) = t'_{\{x\}}$ for some $t' \in (0,1]$.*

Proposition 2.2.19 *Let C be a fuzzy choice function on X. Suppose C satisfies condition ω. If C satisfied PI, then ρ_C is partially quasi-transitive.*

Proof. Let $x, y, z \in X$. Suppose $\pi_C(x,y) * \pi_C(y,z) > 0$. Then $\pi_C(x,y) > 0$ and $\pi_C(y,z) > 0$ and so $C(1_{\{x,y\}})(x) = t' > 0, C(1_{\{x,y\}})(y) = 0$ and $C(1_{\{y,z\}})(y) = s' > 0, C(1_{\{y,z\}})(z) = 0$. Now $C(1_{\{x,y,z\}}) = C(C(1_{\{x\}}) \cup C(1_{\{y,z\}}) = C(t_{\{x\}} \cup s'_{\{y\}}) = t''_{\{x\}}$. Also, $C(1_{\{x,y,z\}}) = C(C(1_{\{x,y\}} \cup C(1_{\{z\}})) = C(t'_{\{x\}} \cup r_{\{z\}})$. Thus $C(t'_{\{x\}} \cup r_{\{z\}}) = t''_{\{x\}}$. Hence $C(1_{\{x,z\}})(z) = 0$ and so $\rho_C(z,x) = 0$. Now $\rho_c(x,z) > 0$ since $1_\emptyset \neq C(1_{\{x,z\}}) \subseteq 1_{\{x,z\}}$. Thus $\pi_C(x,z) > 0$. $\blacksquare$

Example 2.2.20 *Let $X = \{x,y,z\}$. Suppose C is a fuzzy choice function on X such that $C(1_{\{x\}}) = t^*_{\{x\}}, C(1_{\{y\}})(y) = s^*_{\{y\}}$, and $C(1_{\{z\}})(z) = r^*_{\{z\}}$ for some $t^*, s^*, r^* \in (0,1]$. Suppose also that*

$$
\begin{aligned}
C(t_{\{x\}} \cup s_{\{y\}}) &= (t^*t)_{\{x\}} \ \forall t, s \in (0,1], \\
C(t_{\{x\}} \cup r_{\{z\}}) &= (t^*t)_{\{x\}} \ \forall t, r \in (0,1], \\
C(s_{\{y\}} \cup r_{\{z\}}) &= (t^*s)_{\{y\}} \ \forall s, r \in (0,1], \\
C(t_{\{x\}} \cup s_{\{y\}} \cup r_{\{z\}}) &= (t^*t)_{\{x\}} \ \forall t, s, r \in (0,1].
\end{aligned}
$$

Then

$$
\begin{aligned}
C(C1_{\{x\}}) \cup C(1_{\{y,z\}})) &= C(t^*_{\{x\}} \cup t^*_{\{y\}}) = (t^*t^*)_{\{x\}}, \\
C(C1_{\{y\}}) \cup C(1_{\{x,z\}})) &= C(t^*_{\{y\}} \cup t^*_{\{x\}}) = (t^*t^*)_{\{x\}}, \\
C(C1_{\{z\}}) \cup C(1_{\{x,y\}})) &= C(t^*_{\{z\}} \cup t^*_{\{x\}}) = (t^*t^*)_{\{x\}}.
\end{aligned}
$$

Hence it follows that C satisfies PI for characteristic functions. Since

$$
\begin{aligned}
C(1_{\{x,y\}} &\subseteq 1_{\{x\}} \ and \ C(t_{\{x\}} \cup s_{\{y\}}) \subseteq 1_{\{x\}}, \\
C(1_{\{x,z\}} &\subseteq 1_{\{x\}} \ and \ C(t_{\{x\}} \cup r_{\{z\}}) \subseteq 1_{\{x\}}, \\
C(1_{\{y,z\}} &\subseteq 1_{\{y\}} \ and \ C(s_{\{y\}} \cup r_{\{z\}}) \subseteq 1_{\{x\}},
\end{aligned}
$$

C satisfies condition ω.

Proposition 2.2.21 *Let C be a fuzzy choice function on X. If ρ_C is partially acyclic and C satisfies PI, condition ω, and condition γ for characteristic functions, then C is rationalized for characteristic functions by a partially quasi-transitive fuzzy relation on X. Conversely, if C is rationalized for characteristic functions by a quasi-transitive fuzzy relation on X, then C satisfies PI and conditions α and γ for characteristic functions.*

Proof. Suppose ρ_C is partially acyclic and C satisfies PI, condition ω, and condition γ for characteristic functions. Since C satisfies PI, C satisfies condition α for $S(\backslash)$ by Proposition 2.2.15. Hence C satisfies condition α for characteristic functions. By Proposition 2.2.11, ρ_C rationalizes C for characteristic functions. By Proposition 2.2.19, ρ_C is partially quasi-transitive.

Conversely, suppose C is rationalized for characteristic functions by a quasi-transitive fuzzy relation on X. By Proposition 2.2.11, C satisfies conditions α and γ for characteristic functions. Now

$$
\begin{aligned}
C(C(1_S) \cup C(1_T)) &= M_G(\rho_C, C(1_S) \cup C(1_T)) \\
&= M_G(\rho_C, M(\rho_C, 1_S) \cup M_G(\rho_C, 1_T)).
\end{aligned}
$$

Since $C(1_S \cup 1_T) = M_G(\rho_C, 1_{S \cup T})$, we must show that

$$
M_G(\rho_C, M_G(\rho_C, 1_S) \cup M_G(\rho_C, 1_T)) = M_G(\rho_C, 1_{S \cup T}).
$$

By [36, Corollary 2.15, p. 314], Proposition 2.2.11, and [2, Theorem 1.3, p. 14], we have $\forall t \in [0,1]$ (and writing ρ for ρ_C) that

$$
\begin{aligned}
M_G(\rho_C, 1_{S \cup T})_t &= M_G(\rho_t, S \cup T) = M_G(\rho_t, M(\rho_t, S) \cup M_G(\rho_t, T)) \\
&= M_G(\rho_t, M_G(\rho_t, (1_S)_t) \cup M_G(\rho_t, (1_T)_t) \\
&= M_G(\rho_t, M_G(\rho_t, (1_S)_t) \cup M_G(\rho_t, (1_T)_t) \\
&= M_G(\rho_t, (M_G(\rho, 1_S) \cup M_G(\rho, 1_T))_t) \\
&= M_G(\rho, M_G(\rho, S) \cup M_G(\rho, T))_t
\end{aligned}
$$

Thus $M_G(\rho, 1_{S \cup T}) = M_G(\rho, M_G(\rho, 1_S) \cup M_G(\rho, 1_T))$. ∎

Definition 2.2.22 *Let $\rho \in \mathcal{FR}(X)$. Then ρ is called* **partially transitive** *if $\forall x, y, z \in X$, $\rho(x, y) \wedge \rho(y, z) > 0$ implies $\rho(x, z) > 0$.*

In the crisp case, condition β requires that if an alternative is chosen from a set S and is still chosen from a larger set T, then all previously chosen alternatives from S must be chosen from T.

Definition 2.2.23 *Let C be a fuzzy choice function on X. Then C is said to satisfy* **condition** β *if for all $\mu, \nu \in \mathcal{FP}^*(X), \mu \subseteq \nu$ and $C(\mu) \cap C(\nu) \neq \theta$ imply $C(\mu) \subseteq C(\nu)$.*

Condition β says that if ν "extends" μ and the degree of membership of an alternative in both $C(\mu)$ and $C(\nu)$ is positive, then the degree of membership of every alternative in $C(\nu)$ is as large or larger than its degree of membership in $C(\mu)$.

Proposition 2.2.24 *Let C be a fuzzy choice function on X. Suppose C satisfies conditions α and β. Then the following properties hold.*

(1) C satisfies condition γ for all $(\mu, \nu) \in \mathcal{S}_C$, where $\mathcal{S}_C = \{(\mu, \nu) \mid \mu, \nu \in \mathcal{FP}^*(X), \mu \cap (\mu \cup \nu) \neq 1_\emptyset\}$.

(2) ρ_C is partially transitive.

Proof. (1) Let $\mu, \nu \in \mathcal{FP}^*(X)$. Suppose $C(\mu) \cap C(\nu) = 1_\emptyset$. Then condition γ clearly holds. Suppose $C(\mu) \cap C(\nu) \neq 1_\emptyset$. By condition α, $C(\mu \cup \nu) \cap \mu \subseteq C(\mu)$. Let $(\mu, \nu) \in \mathcal{S}_C$. Then $1_\emptyset \neq C(\mu \cup \nu) \cap \mu = C(\mu \cup \nu) \cap C(\mu)$. Hence by condition β, $C(\mu) \subseteq C(\mu \cup \nu)$. Thus $C(\mu) \cap C(\nu) \subseteq C(\mu \cup \nu)$.

(2) Let $x, y, z \in X$ be such that $\rho_C(x, y) > 0, \rho_C(y, z) > 0$. By condition α, $C(1_{\{x,y,z\}}) \cap 1_{\{x,z\}} \subseteq C(1_{\{x,z\}})$. Thus (i) if $C(1_{\{x,y,z\}})(x) > 0$, then $C(1_{\{x,z\}})(x) > 0$ by condition α. (ii) Suppose $C(1_{\{x,y,z\}})(y) > 0$. Then $C(1_{\{x,y\}})(y) > 0$ by condition α. Hence by condition β, $C(1_{\{x,y\}}) \subseteq C(1_{\{x,y,z\}})$. Since $C(1_{\{x,y\}})(x) = \rho_C(x, y) > 0, C(1_{\{x,y,z\}})(x) > 0$. Thus by (i), $C(1_{\{y,z\}})(x) > 0$. Suppose $C(1_{\{x,y,z\}})(z) > 0$. Then by condition α, $C(1_{\{y,z\}})(z) > 0$. By condition β, $C(1_{\{y,z\}}) \subseteq C(1_{\{x,y,z\}})$ and so $C(1_{\{x,y,z\}})(y) > 0$ since $C(1_{\{y,z\}})(y) = \rho_C(y, z) > 0$. Thus by (ii), $C(1_{\{x,z\}})(x) > 0$. Since $C(1_{\{x,y,z\}}) \neq 1_\emptyset$, we have in all cases that $\rho_C(x, z) = C(1_{\{x,z\}})(x) > 0$. ∎

Definition 2.2.25 *Let C be a fuzzy choice function in X. Then C is called **resolute** if $|Supp(C(\mu))| = 1 \ \forall \mu \in \mathcal{FP}^*(X)$.*

Let $\mathcal{S}_C$ be as defined in Proposition 2.2.24.

Theorem 2.2.26 *Let C be a fuzzy choice function on X. If C is rationalized by a partially transitive fuzzy relation, then $Supp(C(1_S) \subseteq Supp(C(1_T))$ $\forall S, T \in \mathcal{P}^*(X)$ such that $S \subseteq T$. Conversely, if C satisfies conditions α and β, then is C rationalizable for characteristic functions in $\mathcal{S}_C$.*

Proof. Suppose C satisfies conditions α and β. Then C satisfies condition γ for $\mathcal{S}_C$ by Proposition 2.2.24. Since condition α holds for C for characteristic functions, C is rationalizable for characteristic functions in $\mathcal{S}_C$ by Proposition 2.2.11.

Suppose C is rationalizable by a partially transitive fuzzy relation. Then C satisfies conditions α and γ for characteristic functions by Proposition 2.2.11. Let $S, T \in \mathcal{P}^*(X)$ be such that $S \subseteq T$. Suppose $C(1_S) \cap C(1_T) \neq 1_\emptyset$. Then $(C(1_S) \cap C(1_T))(x) > 0$ for some $x \in X$. Suppose $C(1_S)(y) > 0$. Then $\rho_C(y, x) > 0$ since ρ_C rationalizes C. Now $C(1_T)(x) > 0$ and $\rho_C(y, z) > 0 \forall z \in T$. Thus $\rho_C(y, z) > 0 \ \forall z \in T$ since C is rationalized by a partially transitive relation. Hence $C(1_T)(y) > 0$. Thus $Supp(C(1_S) \subseteq Supp(C(1_T))$.

∎

We note that if $S, T \in \mathcal{P}^*(X)$ are such that $S \subseteq T$ and if C is rationalized for characteristic functions, then $C(1_T)|_S \subseteq C(1_S)$ on S. Hence if $C(1_S) \subseteq C(1_T)$, then $C(1_T)|_S = C(1_S)$ on S.

Corollary 2.2.27 *Let C be a fuzzy choice function on X that is resolute. If C satisfies condition α, then C is rationalizable for characteristic functions in $\mathcal{S}_C$.*

Proof. Since C is resolute, C satisfies condition β. ∎

The next two conditions each identify choice functions admitting transitive rationalizations. In the crisp case, the Arrow axiom says that for $S \subseteq T$ that if there are alternatives in S that are chosen from T, then exactly these alternatives must be chosen from S, i.e., $S \cap C(T) = \emptyset$ or $S \cap C(T) = C(S)$.

Definition 2.2.28 *Let C be a fuzzy choice function on X. Then C is said to satisfy the **Arrow axiom** if $\forall \mu, \nu \in \mathcal{FP}^*(X)$ such that $\mu \subseteq \nu, \mu \cap C(\nu) = 1_\emptyset$ or $\mu \cap C(\nu) = C(\mu)$.*

In the crisp case, condition WARP says that if an alternative x is chosen from S and alternative y isn't and y is chosen from T, then x is not a member of T.

Definition 2.2.29 *Let C be a fuzzy choice function on X. Then C is said to satisfy the **weak axiom of revealed preference** (WARP) if $\forall \mu, \nu \in \mathcal{FP}^*(X), x \in \mathrm{Supp}(\mu), y \in \mathrm{Supp}(\mu) \backslash \mathrm{Supp}(C(\mu))$ and $y \in \mathrm{Supp}(C(\nu))$ imply $x \notin \mathrm{Supp}(\nu)$.*

Theorem 2.2.30 *Let C be a fuzzy choice function on X. Then $(1) \Leftrightarrow (2) \Rightarrow (3)$, where*
(1) C satisfies conditions α and β;
(2) C satisfies Arrow;
(3) C satisfies WARP.

Proof. $(1) \Rightarrow (2)$: Let $\mu, \nu \in \mathcal{FP}^*(X)$ be such that $\mu \subseteq \nu$. Then $\mu \cap C(\nu) \subseteq C(\mu)$ by condition α. Suppose $\mu \cap C(\nu) \neq 1_\emptyset$. Then $C(\mu) \neq 1_\emptyset$. In fact, $(\mu \cap C(\nu))(x) > 0$ implies $C(\mu)(x) > 0$ and so $(C(\mu) \cap C(\nu))(x) > 0$. Thus $C(\mu) \cap C(\nu) \neq 1_\emptyset$. Hence by condition $\beta, C(\mu) \subseteq C(\nu)$. Since $C(\mu) \subseteq \mu, C(\mu) \subseteq \mu \cap C(\nu)$. Thus $\mu \cap C(\nu) = C(\mu)$.

$(2) \Rightarrow (1)$: Let $\mu, \nu \in \mathcal{FP}^*(X)$ be such that $\mu \subseteq \nu$. Suppose that $\mu \cap C(\nu) \neq 1_\emptyset$. Then $C(\nu) \supseteq \mu \cap (\nu) = C(\mu)$ by Arrow. Hence condition β holds. That condition α holds is immediate.

$(2) \Rightarrow (3)$: Let $x \in \mathrm{Supp}(C(\mu)), y \in \mathrm{Supp}(\mu) \backslash \mathrm{Supp}(C(\mu)), y \in \mathrm{Supp}(\nu)$. Let $\sigma = \mu \cap \nu$. Now $\mathrm{Supp}(\sigma) = \mathrm{Supp}(\mu) \cap \mathrm{Supp}(\nu) \neq \emptyset$ since $y \in \mathrm{Supp}(\mu)$ and $y \in \mathrm{Supp}(C(\nu)) \subseteq \mathrm{Supp}(\nu)$. In fact, $y \in \mathrm{Supp}(\sigma)$. Since $y \in \mathrm{Supp}(C(\nu)), y \in \mathrm{Supp}(\sigma) \cap \mathrm{Supp}(C(\nu)) = \mathrm{Supp}(\sigma \cap C(\nu))$. Thus $\mathrm{Supp}(\sigma \cap (\nu)) \neq \emptyset$ and so $\sigma \cap C(\nu) \neq 1_\emptyset$. Now $\sigma \subseteq \nu$ and so by Arrow, $\sigma \cap C(\nu) = C(\mu)$. Hence $y \in \mathrm{Supp}(C(\sigma))$. But $y \notin \mathrm{Supp}(\sigma \cap C(\mu))$ and so $\sigma \cap C(\mu) \neq C(\sigma)$. Thus since $\sigma \subseteq \mu$, we have by Arrow that $\sigma \cap C(\mu) = 1_\emptyset$. Hence $x \in \mathrm{Supp}(C(\mu))$ implies $x \notin \mathrm{Supp}(\sigma)$ which in turn implies $x \notin \mathrm{Supp}(\nu)$ since $x \in \mathrm{Supp}(\mu)$ and $\mathrm{Supp}(\sigma) = \mathrm{Supp}(\mu) \cap \mathrm{Supp}(\nu)$. ∎

2.3 M-Rationality and G-Rationality

In this section, we consider the situation in which choices are deduced from vague preferences. We follow Georgescu [24]. We consider M-Rationality and G-Rationality for choice functions, where M fuzzifies the notion of a ρ-maximal element and G fuzzifies the notion of a ρ-greatest element. Let C be a fuzzy choice function. Rationality means that C is recoverable from a fuzzy preference relation and for rationality which is normal C is recoverable from the fuzzy revealed preference relation ρ associated with C.

Let $*$ be a continuous t-norm. Let $a \to b = \vee\{t \in [0,1] \mid a * t \leq b\}$. The operation $\to$ is called the **residuum** associated with $*$, [24].

Lemma 2.3.1 *For any $a, b, c \in [0,1]$, the following properties hold:*
(1) $a * b \leq c \Leftrightarrow a \leq b \to c$;
(2) $a * (a \to b) = a \wedge b$.

Proof. (1) follows immediately from the definition of $b \to c$.
(2) Since $a * (a \wedge b) \leq b, a \to b \geq a \wedge b$. Also, $a * (a \to b) \leq b$ since $*$ is continuous and clearly, $a * (a \to b) \leq a$. ∎

Definition 2.3.2 *Let $\rho \in \mathcal{FR}(X)$. Then we call ρ a **fuzzy weak preference relation** (**FWPR**) if ρ is reflexive and complete.*

Let $\mathcal{B}$ be a nonempty family of nonzero fuzzy subsets of X.

Definition 2.3.3 *Let $\mu \in \mathcal{FP}(X)$ and ρ be a FWPR on X. Define the fuzzy subsets $M(\mu, \rho)$ and $G(\mu, \rho)$ of X as follows:$\forall x \in X$,*

$$M(\mu, \rho)(x) = \mu(x) * \wedge\{\mu(y) * \rho(y, x) \to \rho(x, y) \mid y \in X\},$$
$$G(\mu, \rho)(x) = \mu(x) * \wedge(\mu(y) \to \rho(x, y) \mid y \in X\}.$$

Proposition 2.3.4 *Let $\mu \in \mathcal{FP}(X)$ and ρ be a FWPR on X. Then*

$$M(\mu, \rho)(x) = \mu(x) * \wedge\{\vee\{t \in [0,1] \mid \mu(y) * \rho(y, x) * t \leq \rho(x, y)\} \mid y \in X\},$$
$$G(\mu, \rho)(x) = \mu(x) * \wedge\{\vee\{t \in [0,1] \mid \mu(y) * t \leq \rho(x, y)\} \mid y \in X\}.$$

Proposition 2.3.5 *Let $\mu \in \mathcal{FP}(X)$ and ρ be a FWPR on X. Let $* = \wedge$. Then*

$$G(\mu, \rho)(x) = \begin{cases} \mu(x) * \wedge\{\rho(x, y) \mid \mu(y) > \rho(x, y), y \in Supp(\mu)\} \\ \qquad\qquad if\ \exists y \in Supp(\mu)\ such\ that\ \mu(y) > \rho(x, y), \\ \mu(x) \qquad if\ \mu(y) \leq \rho(x, y)\ \forall y \in Supp(\mu). \end{cases}$$

Proposition 2.3.6 *Let S be a subset of X and ρ be a FWPR on X. Then*

$$M(1_S, \rho)(x) = \begin{cases} \wedge\{\rho(y,x) \to \rho(x,y) \mid y \in S\} & \text{if } x \in S, \\ 0 & \text{if } x \notin S. \end{cases}$$

and

$$G(1_S, \rho)(x) = \begin{cases} \wedge\{\rho(x,y) \mid y \in S\} & \text{if } x \in S, \\ 0 & \text{if } x \notin S. \end{cases}$$

Recall that our definition of a fuzzy maximal set is $M(\rho, \mu)(x) = \mu(x) * (\circledast\{\vee\{t \in [0,1] \mid \mathcal{C}(y,x)*t \le \rho(x,y)\} \mid y \in X\})$. We obtain M above by letting $\mathcal{C}(y,x) = \mu(y) * \rho(y,x)$ and we obtain G above by letting $\mathcal{C}(y,x) = \mu(y)$. Note that M above is M_G. We obtain M_M by letting $\mathcal{C}(y,x) = 1$.

The negation operator associated with a continuous t-norm $*$ is defined by

$$\neg a = a \to 0 = \vee\{t \in [0,1] \mid a * t = 0\}.$$

Recall the negation associated with Lukasiewicz, Godel, and product t-norms:

Lukasiewicz: $\neg a = 1 - a$;

Godel and product t-norms: $\neg a = \begin{cases} 1 \text{ if } a = 0. \\ 0 \text{ if } a > 0. \end{cases}$

A fuzzy subset μ of a set X is called **normal** if $\exists x \in X$ such that $\mu(x) = 1$.

Let $\mathcal{B}$ denote a set of non-zero fuzzy subsets of X. We often assume the following two conditions, [24, p. 91].

(H1) Every $\mu \in \mathcal{B}$ and $C(\mu)$ are normal fuzzy subsets of X.

(H2) $\mathcal{B}$ includes all fuzzy subsets $1_{\{x_1,\ldots,x_n\}}$ for all $x_1,\ldots,x_n \in X, n \ge 1$.

Definition 2.3.7 *Let $C : \mathcal{B} \to \mathcal{FP}(X)$ be a fuzzy choice function on $(X,\mathcal{B})$. Define the following revealed preference relations ρ, π, ι on X by $\forall x, y \in X$,*
(1) $\rho(x,y) = \vee\{C(\mu)(x) * \mu(y) \mid \mu \in \mathcal{B}\}$;
(2) $\pi(x,y) = \rho(x,y) * \neg\rho(y,x)$;
(3) $\iota(x,y) = \rho(x,y) * \rho(y,x)$.

Suppose $*$ has no zero divisors in Definition 2.3.7(2). Then

$$\begin{aligned} \rho(x,y) * \neg\rho(y,x) &= \rho(x,y) * \vee\{t \in [0,1] \mid \rho(y,x) * t = 0\} \\ &= \begin{cases} \rho(x,y) * 1 \text{ if } \rho(y,x) = 0 \\ \rho(x,y) * 0 \text{ if } \rho(y,x) > 0 \end{cases} \end{aligned}$$

Thus $\pi(x,y) = \rho(x,y)$ if $\rho(y,x) = 0$ and $\pi(x,y) = 0$ if $\rho(y,x) > 0$. (Suppose ρ is complete. Then π here is our $\pi_{(0)}$ because if $\rho(y,x) = 0$, then $\rho(x,y) > 0$.)

Definition 2.3.8 *Assume hypothesis (H2) holds. Let $C : \mathcal{B} \to \mathcal{FP}(X)$ be a fuzzy choice function on $(X,\mathcal{B})$. Define the following revealed preference relations $\overline{\rho}, \overline{\pi}, \overline{\iota}$ on X by $\forall x, y \in X$,*

(1) $\overline{\rho}(x,y) = C(1_{\{x,y\}})(x)$;
(2) $\overline{\pi}(x,y) = \overline{\rho}(x,y) * \neg\overline{\rho}(y,x)$;
(3) $\overline{\iota}(x,y) = \overline{\rho}(x,y) * \overline{\rho}(y,x)$.

$\overline{\rho}(x,y)$ is the degree of truth of the statement "at least the alternative x is chosen from the set $\{x,y\}$."

Definition 2.3.9 *Let $C : \mathcal{B} \to \mathcal{FP}(X)$ be a fuzzy choice function on $(X,\mathcal{B})$. Define the following revealed preference relations $\widetilde{\rho}, \widetilde{\pi}, \widetilde{\iota}$ on X by $\forall x, y \in X$,*
   (1) $\widetilde{\pi}(x,y) = \vee\{(C(\mu)(x) * \mu(y)) * \neg C(\mu)(y) \mid \mu \in \mathcal{B}\}$;
   (2) $\widetilde{\rho}(x,y) = \neg\widetilde{\pi}(y,x)$;
   (3) $\widetilde{\iota}(x,y) = \widetilde{\rho}(x,y) * \widetilde{\rho}(y,x)$.

$\widetilde{\pi}(x,y)$ is the degree of truth of the statement "there exists a criterion μ with respect to which x is chosen and y is rejected."

Let $* = \wedge$. Since $\neg C(\mu)(y) = 1$ if $C(\mu)(y) = 0$ and $\neg C(\mu)(y) = 0$ if $C(\mu)(y) > 0$, we have that

$$\widetilde{\pi}(x,y) = \begin{cases} \vee\{C(\mu)(x) \wedge \mu(y) \mid \mu \in \mathcal{B}\} & \text{if } C(\mu)(y) = 0 \text{ for some } \mu \in \mathcal{B}, \\ 0 & \text{if } C(\mu)(y) > 0 \text{ for all } \mu \in \mathcal{B}. \end{cases}$$

Thus $\widetilde{\rho}(x,y) = 1$ if $C(\mu)(y) > 0$ for all $\mu \in \mathcal{B}$ and $\widetilde{\rho}(x,y) = 0$ if there exists $\mu \in \mathcal{B}$ such that $C(\mu)(y) = 0, \mu(y) > 0$, and $C(\mu)(x) > 0$.

Definition 2.3.10 *Let $C : \mathcal{B} \to \mathcal{FP}(X)$ be a fuzzy choice function on (X, B). Define the following revealed preference relations ρ_* and ρ_1 on X by $\forall \mu \in \mathcal{B}$, $\forall x, y \in X$,*
   (1) $\rho_*(x,y) = \wedge\{(\mu(x) * C(\mu)(y)) \to C(\mu)(x) \mid \mu \in \mathcal{B}\}$;
   (2) $\rho_1(x,y) = \wedge(\neg\mu(x) \vee C(\mu)(y) \vee \neg C(\mu)(x) \mid \mu \in \mathcal{B}\}$.

Proposition 2.3.11 [24] *Suppose $*$ is the Godel t-norm. Let ρ be a reflexive and max-$*$ transitive fuzzy preference relation on a finite set X. Then $G(\_,\rho)$ and $M(\_,\rho)$ are fuzzy choice functions on $(X,\mathcal{B})$.*

Proof. Suppose that $G(\_,\rho)$ is not a fuzzy choice function. Then there exists $\mu \in \mathcal{B}$ such that $G(\mu,\rho)(x) = 0$ for all $x \in X$. Let $x_1 \in X$ be such that $\mu(x_1) \neq 0$. Then

$$\mu(x_1) \wedge (\wedge\{\mu(y) \to \rho(x_1,y) \mid y \in X\}) = G(\mu,\rho)(x_1) = 0.$$

Since $\mu(x_1) \neq 0$ and X is finite, there exists $x_1' \in X$ such that

$$0 = \mu(x_1') \to \rho(x_1,x_1') = \begin{cases} 1 & \text{if } \mu(x_1') \leq \rho(x_1,x_1') \\ \rho(x_1,x_1') & \text{if } \mu(x_1') > \rho(x_1,x_1'). \end{cases}$$

Hence it follows that $\rho(x_1,x_1') = 0$ and $\mu(x_1') > 0$. Since ρ is reflexive, $x_1 \neq x_1'$. Let $y_1 = x_1$ and $y_2 = x_1'$. By applying the same argument to y_2, there exists

$y_3 \in X$ such that $\mu(y_3) > 0$ and $\rho(y_2, y_3) = 0$. Continuing this process, we obtain a sequence $y_1, y_2, \ldots, y_k, \ldots$ such that $\mu(y_k) > 0$ and $\rho(y_k, y_{k+1}) = 0$, but $y_{k+1} = y_j$ some $j < k+1$ since X is finite. Since ρ is max-* transitive, $0 = \rho(y_k, y_{k+1}) = \rho(y_k, y_j) \geq \rho(y_k, \rho_{k-1}) \wedge \ldots \wedge \rho(y_{j+1}, y_j) > 0$, a contradiction. Thus no such x exists and so $G(\_, \rho)$ is a fuzzy choice function. Since $G(\_, \rho) \subseteq M(\_, \rho)$, it follows that $M(\_, \rho)$ is a fuzzy choice function. $\blacksquare$

An interesting paper by Martinetti, de Baets, Diaz, and Montes [35 35] concerning when $G(\_, \rho)$ is a fuzzy choice function has recently been published. We present some of its results in the exercises at the end of this chapter.

Definition 2.3.12 [24] *A fuzzy choice function C on $(X, \mathcal{B})$ is called G-**rational** if there exists a fuzzy preference relation ρ on X such that $C(\mu) = G(\mu, \rho)$ for all $\mu \in \mathcal{B}$. A fuzzy choice function C on $(X, \mathcal{B})$ is called M-**rational** if there exists a fuzzy preference relation ρ on X such that $C(\mu) = M(\mu, \rho)$ for all $\mu \in \mathcal{B}$.*

Definition 2.3.13 [24] *Let C be a fuzzy choice function on $(X, \mathcal{B})$ and let ρ be the associated revealed preference relation (Definition 2.3.7). Define the functions $G^* : \mathcal{B} \to \mathcal{FP}(X)$ and $M^* : \mathcal{B} \to \mathcal{FP}(X)$ by $\forall \mu \in \mathcal{B}, \forall x \in X$,*

$$G^*(\mu)(x) = G(\mu, \rho)(x) = \mu(x) * \wedge\{\mu(y) \to \rho(x, y) \mid y \in X\}$$

and

$$M^*(\mu)(x) = M(\mu, \rho)(x) = \mu(x) * \wedge\{\mu(y) * \rho(y, x) \to \rho(x, y) \mid y \in X\}.$$

$$I(\mu, \nu) = \wedge\{\mu(x) \to \nu(x) \mid x \in X\}.$$

The **transitive closure** of a fuzzy binary relation ρ on a set X is the smallest transitive fuzzy binary relation on X containing ρ.

Let ρ^{tc} denote the transitive closure of ρ and π^* the transitive closure of $\underset{\sim}{\pi}$.

Definition 2.3.14 [24] *A fuzzy choice function C is called G-**normal** if $C(\mu) = G^*(\mu)$ for all $\mu \in \mathcal{B}$. A fuzzy choice function C is called M-**normal** if $C(\mu) = M^*(\mu)$ for all $\mu \in \mathcal{B}$.*

Social choice theory most often assumes a binary preference ordering is complete, reflexive, and transitive. A particular case is consumer theory. If preferences satisfy these axioms, then they can be characterized by utility functions. Samuelson developed a theory of revealed preference in [40] based on utility functions. He provided the following concepts.

Directly revealed preferred. An alternative x is directly revealed preferred to a different alternative y if y was available when x was chosen.

Weak axiom of revealed preference (WARP): If alternative x is directly revealed preferred to y, then y cannot be directly revealed preferred to x.

Revealed preferred. The revealed preferred relation is the transitive closure of the directly revealed preferred relation.

Strong axiom of revealed preference (SARP): If alternative x is revealed preferred to y, then y will never be revealed preferred to x.

Generalized axiom of revealed preference (GARP): If an alternative x is revealed preferred to y, then y is never strictly revealed preferred to x.

We next list axioms of fuzzy revealed preference. We lay the foundation for their further study later in the chapter and in Chapter 7.

$WAFRP$ **weak axiom of fuzzy revealed preference**: For all $x, y \in X$, $\widetilde{\pi}(x, y) \leq \neg \rho(y, x)$, i.e., $\widetilde{\pi}(x, y) \leq 1$ if $\rho(y, x) = 0$ and $\widetilde{\pi}(x, y) = 0$ if $\rho(y, x) > 0$.

$SAFRP$ **strong axiom of fuzzy revealed preference**: For all $x, y \in X$, $\pi^*(x, y) \leq \neg \rho(y, x)$.

The next axioms are congruence axioms for fuzzy choice functions.

$WFCA$ **weak fuzzy congruence axiom**: For all $\mu \in \mathcal{B}$ and for all $x, y \in X$, $\rho(x, y) * C(\mu)(y) * \mu(x) \leq C(\mu)(x)$.

$SFCA$ **strong fuzzy congruence axiom**: For all $\mu \in B$ and for all $x, y \in X$, $\rho^{tc}(x, y) * C(\mu)(y) * \mu(x) \leq C(\mu)(x)$.

We now consider fuzzy versions $F\alpha$ and $F\beta$ of conditions α and β, respectively. The conditions $F\alpha'$ and $F\beta'$ are weak forms of $F\alpha$ and $F\beta$, respectively.

$F\alpha : I(\mu, \nu) * \mu(x) * C(\nu)(x) \leq C(\mu)(x)$.

$F\alpha' : \forall \mu, \nu \in \mathcal{FP}(X)$, $\forall x \in X$, if $\mu(x) \leq \nu(x)$, then $\mu(x) * C(\nu)(x) \leq C(\mu)(x)$. $(\forall \mu, \nu \in \mathcal{FP}(X)$, if $\mu \subseteq \nu$, then $\mu \cap C(\nu) \subseteq C(\mu)$.)

$F\beta : I(\mu, \nu) * C(\mu)(x) * C(\mu)(y) \leq C(\nu)(x) \longleftrightarrow C(\nu)(y)$.

$F\beta' : \forall \mu, \nu \in \mathcal{FP}(X)$, $\forall x \in X$, if $\mu(x) \leq \nu(x)$, then $C(\mu)(x) * C(\mu)(y) \leq C(\nu)(x) \longleftrightarrow C(\nu)(y)$.

The following conditions are also fuzzy versions of known consistency conditions. We list the corresponding crisp consistency conditions first for the convenience of the reader.

δ : For $S, T \in \mathcal{P}^*(X)$ and $\forall x, y \in C(S)$, $x \neq y$ if $S \subseteq T$ then $C(T) \neq \{x\}$.

$\alpha 2$: For all $S \in \mathcal{P}^*(X)$ and $\forall x \in S$, if $x \in C(S)$, then $x \in C(\{x, y\})$ for all $y \in S$.

$\gamma 2$: For all $S \in \mathcal{P}^*(X)$ and for all $x \in S$, if $x \in C(\{x, y\})$ for all $y \in S$, then $x \in C(S)$.

$\beta(+)$: For all $S, T \in \mathcal{P}^*(X)$ such that $S \subseteq T$ and for all $x \in C(S)$ and $y \in S$, if $y \in C(T)$, then $x \in C(T)$.

$F\delta : \forall S, T \in \mathcal{P}^*(X), \forall x, y \in X, I(1_S, 1_T) \leq (C(1_S)(x) \wedge C(1_S)(y) \rightarrow \neg(C(1_T)(x) \wedge (\wedge\{\neg C(1_T)(z) \mid z \in X, z \neq x\}.$

$F\alpha 2 : \forall \mu \in \mathcal{B}$ and $\forall x, y \in X, C(\mu)(x) \wedge \mu(y) \leq C(1_{\{x,y\}})(x).$

$F\gamma 2 : \forall \mu \in \mathcal{B}$ and $\forall x \in X, \mu(x) \wedge (\wedge\{\mu(y) \rightarrow C(1_{\{x,y\}})(x)) \mid y \in X\} \leq C(\mu)(x).$

$F\beta(+) : \forall \mu, \nu \in B$ and $\forall x, y \in X, I(\mu, \nu) \wedge C(\mu)(x) \wedge \mu(y) \leq C(\nu)(y) \rightarrow C(\nu)(x).$

The choice theory developed by Uzawa [49], Arrow [1], and Sen [43-46] was based on many of the consistency conditions just listed. Ritcher [40], Hansoon [29] and Suzumura [47] extended these results.

For any choice function C, we consider the following statements:

(i) ρ is a regular preference relation and C is G-normal;

(ii) $\overline{\rho}$ is a regular preference relation and C is G-normal;

(iii) C satisfies $WFCA$;

(iv) C satisfies $SFCA$:

(v) C satisfies $WAFRP$;

(vi) C satisfies $SAFRP$;

(vii) $\rho = \overline{\rho}$;

$(viii)$ $\overline{\rho} = \widetilde{\rho}$ and C is G-normal.

Theorem 2.3.15 (*Georgescu* [24]) *Let C be a fuzzy choice function then the following properties hold.*

(1) *Conditions $(i), (ii)$ are equivalent.*

(2) *The implication $(i) \Rightarrow (iii)$ holds; if $* = \wedge$, then the implication $(iii) \Rightarrow (i)$ holds;*

(3) *Conditions (iii) and (iv) are equivalent;*

(4) *If $*$ is the Lukasiewicz t-norm, then conditions $(iii), (v), (vi), (vii)$ are equivalent;*

(5) *The implication $(viii) \Rightarrow (iii)$ holds.*

Let C be a fuzzy choice function. Let $\widehat{C}$ denote the fuzzy choice function $G^* = G(\_, \rho)$, [22, p. 109]. Recall ρ is the associated revealed preference relation of C.

Proposition 2.3.16 (*Georgescu* [24]) *If C is a G-normal fuzzy choice function, then C satisfies $F\alpha$.*

Proof. Since $C = \widehat{C}$, it suffices to show for all $\mu, \nu \in \mathcal{B}$ and $\forall x \in X$ that the following inequality holds

$$I(\mu, \nu) * \mu(x) * \widehat{C}(\nu)(x) \leq \widehat{C}(\mu)(x).$$

By Lemma 2.3.1(2), it follows for all $u \in X$ that

$$\mu(u) * (\mu(u) \rightarrow \nu(u)) * (\nu(u) \rightarrow \rho(x,u))$$
$$= \mu(u) \wedge \nu(u) * (\nu(u) \rightarrow \rho(x,u))$$
$$= \mu(u) \wedge \nu(u) \wedge \rho(x,u) \le \rho(x,u).$$

By Lemma 2.3.1(1), with $a = (\mu(u) \rightarrow \nu(u)) * (\nu(u) \rightarrow \rho(x,u)), b = \mu(u), c = \rho(x,u))$, we have

$$(\mu(u) \rightarrow \nu(u)) * (\nu(u) \rightarrow \rho(x,u))$$
$$\le \mu(u) \rightarrow \rho(x,u).$$

Thus

$$I(\mu, \nu) * \mu(x) * \widehat{C}(\nu)(x)$$
$$= \wedge\{\mu(u) \rightarrow \nu(u) \mid u \in X\} * \mu(x) * \nu(x) * \wedge\{\nu(u) \rightarrow \rho(x,u) \mid u \in X\}$$
$$\le \mu(x) * \wedge\{(\mu(u) \rightarrow \nu(u)) * (\nu(u) \rightarrow \rho(x,u))\} \mid u \in X\}$$
$$\le \mu(x) * \wedge\{\mu(u) \rightarrow \rho(x,u))\} \mid u \in X\}$$
$$= \widehat{C}(\mu)(x).$$

∎

Let $\mu = 1_S$, where S is a nonempty subset of X. Let C be a G^*-normal fuzzy choice function with respect to ρ. Then $\forall x \in X$,

$$C(1_S)(x) = G^*(1_S, \rho)$$
$$= 1_S(x) \wedge (\wedge\{1_S(y) \rightarrow \rho(x,y) \mid y \in X\})$$
$$= \begin{cases} \wedge\{\rho(x,y) \mid y \in S\} \text{ if } x \in S, \\ 0 \text{ otherwise.} \end{cases}$$

Hence if C is G^*-normal, then $C(1_S) = M_M(\rho, 1_S)$. Thus ρ rationalizes C with respect to characteristic functions.

Corollary 2.3.17 *Suppose C is G^*-normal. Then C satisfies condition $F\alpha$ for characteristic functions.*

Proof. By the argument immediately preceding this corollary, we have for $x \in S$ that $C(1_S)(x) = M(\rho, 1_S)(x) = \wedge\{\rho(x,y) \mid y \in S\} \ge \wedge\{\rho(x,y) \mid y \in T\} = M(\rho, 1_T)(x) = C(1_T)(x)$. Suppose $x \in T \backslash S$. Then $1_S(x) \wedge C(1_T)(x) = 0 \le C(1_S)(x)$. ∎

Proposition 2.3.18 (*Georgescu* [24]) *If a fuzzy choice function C satisfies $WFCA$, then conditions $F\alpha$ and $F\beta$ hold.*

Proof. Since C is $WFCA$, C is G-normal by Theorem 2.3.15. Hence by Proposition 2.3.16, condition $F\alpha$ holds. Suppose that condition $F\beta$ does not hold. Then there exists $\mu, \nu \in \mathcal{B}$ and $x, y \in X$ such that

$$I(\mu, \nu) \wedge C(\mu)(x) \wedge C(\mu)(y) \nleq C(\nu)(x) \leftrightarrow C(\nu)(y) =$$
$$(C(\nu)(x) \rightarrow C(\nu)(y)) \wedge (C(\nu)(y) \rightarrow C(\nu)(x)).$$

Thus either

$$I(\mu, \nu) \wedge C(\mu)(x) \wedge C(\mu)(y) \nleq C(\nu)(x) \rightarrow C(\nu)(y)$$

or

$$I(\mu, \nu) \wedge C(\mu)(x \wedge C(\mu)(y) \nleq C(\nu)(y) \rightarrow C(\nu)(x).$$

Assume the first case. Then by Lemma 2.3.1(1),

$$(a)\ I(\mu, \nu) \wedge C(\mu)(x) \wedge C(\mu)(y) \wedge C(\nu)(x) \nleq C(\nu)(y).$$

Now

$$I(\mu, \nu) \wedge C(\mu)(x) \wedge C(\mu)(y) \wedge C(\nu)(x) \leq C(\mu)(y) \vee \mu(x) \wedge C(\nu)(x)$$
$$= C(\mu)(y) \wedge \mu(x) \wedge C(\nu)(x) \wedge \mu(y)$$

since $C(\mu)(y) \wedge \mu(y) = C(\mu)(y)$. Since $C(\mu)(y) \wedge \mu(x) \leq \rho(y, x)$, we have that

$$(b)\ I(\mu, \nu) \wedge C(\mu)(x) \wedge C(\mu)(y) \wedge C(\nu)(x) \leq \rho(y, x) \wedge C(\nu)(x) \wedge \mu(y).$$

By (a) and (b), it follows that $\rho(y, x) \wedge C(\nu)(x) \wedge \mu(y) \nleq C(\nu)(y)$. Hence $WFCA$ does not hold.

A similar argument can be used if the second case holds. ∎

Proposition 2.3.19 (*Georgescu* [24]) *If a fuzzy choice function C satisfies $F\alpha$ and $F\beta$, then $WFCA$ holds.*

Proof. Let $\mu \in \mathcal{B}$ and $x, y \in X$. Since $I(1_{\{x,y\}}, \mu) = \mu(x) \wedge \mu(y)$, it follows that

$$\mu(x) \wedge C(\mu)(y) \wedge \rho(x, y)$$
$$= \mu(x) \wedge \mu(y) \wedge C(\mu)(y) \wedge \rho(x, y)$$
$$= I(1_{\{x,y\}}, \mu) \wedge \mu(x) \wedge \mu(y) \wedge C(\mu)(y) \wedge \rho(x, y)$$
$$= I(1_{\{x,y\}}, \mu) \wedge 1_{\{x,y\}}(y) \wedge C(\mu)(y) \wedge \mu(x) \wedge \mu(y) \wedge \rho(x, y)$$
$$\leq C(1_{\{x,y\}})(y) \wedge \mu(x) \wedge \mu(y) \wedge C(\mu)(y) \wedge \rho(x, y)$$

since $I(1_{\{x,y\}}, \mu) \wedge 1_{\{x,y\}}(y) \wedge C(\mu)(y) \leq C(1_{\{x,y\}})(y)$ by condition $F\alpha$. Replacing $\rho(x, y)$ with its expression in Definition 2.3.7, we obtain

$$\mu(x) \wedge C(\mu)(y) \wedge \rho(x, y)$$
$$\leq C(1_{\{x,y\}})(y) \wedge \mu(x) \wedge \mu(y) \wedge C(\mu)(y) \wedge (\vee\{C(\eta)(x) \wedge \eta(y) \mid \eta \in \mathcal{B}\})$$
$$= \vee\{C(1_{\{x,y\}})(y) \wedge \mu(x) \wedge \mu(y) \wedge C(\mu)(y) \wedge C(\eta)(x) \wedge \eta(y) \mid \eta \in \mathcal{B}\}.$$

By condition $F\alpha$,

$$
\begin{aligned}
C(\eta)(x) \wedge \eta(y) &= C(\eta)(x) \wedge \eta(x) \wedge \eta(y) \\
&= I(1_{\{x,y\}}, \eta) \wedge 1_{\{x,y\}}(x) \wedge C(\eta)(x) \\
&\leq C(1_{\{x,y\}})(x).
\end{aligned}
$$

Thus by $F\beta$ and Lemma 2.3.1(2),

$$
\begin{aligned}
&C(1_{\{x,y\}})(y) \wedge \mu(x) \wedge \mu(y) \wedge C(\mu)(y) \wedge C(\eta)(x) \wedge \eta(y) \\
\leq\ &C(1_{\{x,y\}})(y) \wedge \mu(x) \wedge \mu(y) \wedge C(\mu)(y) \wedge C(1_{\{x,y\}})(x) \\
=\ &I(1_{\{x,y\}}, \mu) \wedge C(1_{\{x,y\}})(x) \wedge C(1_{\{x,y\}})(y) \wedge C(\mu)(y) \\
\leq\ &[C(\mu)(x) \leftrightarrow C(\mu)(y)] \wedge C(\mu)(y) \\
\leq\ &C(\mu)(y) \wedge [C(\mu)(y) \rightarrow C(\mu)(x)] \\
=\ &C(\mu)(y) \wedge C(\mu)(x) \\
\leq\ &C(\mu)(x).
\end{aligned}
$$

Since these inequalities hold for all $\eta \in \mathcal{B}, \mu(x) \wedge C(\mu)(y) \wedge \rho(x,y) \leq C(\mu)(x)$. Thus $WFCA$ holds for C. ∎

Theorem 2.3.20 *For a fuzzy choice function $C, WFCA$ holds if and only if $F\alpha$ and $F\beta$ hold*

Proof. The result holds from Propositions 2.3.18 and 2.3.19. ∎

A fuzzy preference relation ρ is said to be **strongly total** if $\forall x, y \in X, x \neq y$, either $\rho(x,y) = 1$ or $\rho(y,x) = 1$.

Definition 2.3.21 *Let $C : \mathcal{B} \rightarrow \mathcal{FP}(X)$ be a fuzzy choice function. Then C is said to be **full rational** if $C = G(\_, \rho)$ for some reflexive, max-* transitive, and strongly total fuzzy preference relation ρ.*

Definition 2.3.22 *Let C be a fuzzy choice function on a fuzzy choice space $(X, \mathcal{B})$. Then C is said to satisfy the fuzzy Arrow axiom **FAA** if for all $\mu_1, \mu_2 \in \mathcal{B}$ and for all $x \in X$,*

$$
I(\mu_1, \mu_2) * \mu_1(x) * C(\mu_2)(x) \leq E(\mu_1 \cap C(\mu_2), C(\mu_1)).
$$

Theorem 2.3.23 *(Georgescu [24]) If $C : \mathcal{B} \rightarrow \mathcal{FP}(X)$ is a fuzzy choice function, then the following statements are equivalent:*
(1) C is full rational;
(2) C satisfies FAA.

2.4 Full Rationality of Fuzzy Choice Functions on Base Domain

We now examine full rationality of fuzzy choice functions on base domain. The results are based on those of Desai and Chaudhari [17]. The concept of rational choice as an optimizing choice is binary in nature in the view of Bossert, Surmont, and Suzumura [9]. Consequently, they discussed classical choice functions defined on a domain that contained all singleton and all two-element subsets of the universal set of alternatives. They subsequently characterize full rationality of choice functions in terms of the T-congruence axiom. The choice of functions may or may not contain subsets of three or more elements of the universal set of alternatives in the approach adopted in [9]. In contrast, fuzzy choice functions are defined on the set domain consisting of all characteristic functions of all single and two element subsets of the universal set in [10, 17, 12]. Moreover, the effects of introducing various congruence and revealed preference axioms on the rationality of fuzzy choice functions are examined. Chaudhari and Desai [10] subsequently introduce the notion of a fuzzy T-congruence axiom to characterize full rationality of fuzzy choice functions.

In this section, we present the concept of a weak form of fuzzy T-congruence axiom, which was introduced in [17]. We characterize the full rationality of fuzzy choice functions in terms of this axiom and the fuzzy Chernoff axiom. We show that this form is insufficient to imply full rationality of fuzzy choice functions defined on base domain. However, the weak fuzzy T-congruence axiom together with the fuzzy Chernoff axiom characterizes full rationality. We also show that the fuzzy Arrow axiom alone does not imply full rationality of fuzzy choice functions defined on a base domain and also that G-rational fuzzy choice functions with transitive rationalizations characterize their full rationality if the fuzzy Chernoff axiom holds.

Recall that the operation $\wedge$ is a continuous t-norm called the **Gödel t-norm**. The fuzzy implication operator associated with $\wedge$ is the binary operation $\longrightarrow$ on $[0,1]$ defined as follows: $\forall a, b \in [0,1]$,

$$a \rightarrow b = \begin{cases} 1 & \text{if } a \le b \\ b & \text{if } a > b \end{cases}$$

Lemma 2.4.1 [21, 24, 31, 32] *If $\longrightarrow$ is the fuzzy implication associated with the Gödel t-norm $\wedge$ on $[0,1]$, then*
  (1) $a \wedge b \le c \Longleftrightarrow a \le b \longrightarrow c$;
  (2) $a \le b \Longleftrightarrow a \longrightarrow b = 1$;
  (3) $1 \longrightarrow a = a$.

Recall that a fuzzy subset μ of X is called normal if $\mu(x) = 1$ for some $x \in X$ and is called non-zero if $\mu(x) > 0$ for some $x \in X$. For $\mu, \nu \in \mathcal{F}(X)$, we denote the degree of subsethood of μ in ν by $I(\mu, \nu)$ and the degree of equality of μ and ν by $E(\mu, \nu)$, where $I(\mu, \nu) = \wedge\{\mu(z) \longrightarrow \nu(z) \mid z \in X\}$ and $E(\mu, \nu) = \wedge\{\mu(z) \leftrightarrow \nu(z) \mid z \in X\}$. Clearly, $\mu \subseteq \nu$ if and only if $I(\mu, \nu) = 1$ and $\mu = \nu$ if and only if $E(\mu, \nu) = 1$.

A fuzzy binary relation on X is also called **fuzzy preference relation** on X. Recall that a fuzzy binary relation ρ is said to be strongly connected if $\rho(x, y) = 1$ or $\rho(y, x) = 1$ for all $x, y \in X$. We next recall the definitions of fuzzy choice function and fuzzy revealed preference relation for base domains and few basic results discussed in [10].

Let X be a non-empty set of alternatives and $\mathcal{B}$ be a non-empty family of non-zero fuzzy subsets of X which contain characteristic functions of all singletons and two-element subsets of X. Then $\mathcal{B}$ is called the **base domain**. A fuzzy choice function on $(X, \mathcal{B})$ is a function $C : \mathcal{B} \longrightarrow \mathcal{FP}(X)$ such that for all $\mu \in \mathcal{B}$, $C(\mu) \neq \theta$ and $C(\mu) \subseteq \mu$.

In this approach the domain, $\mathcal{B}$, of the fuzzy choice functions do not necessarily contain characteristic functions of the set having three or more elements. In classical social choice theory, it is assumed that every choice set is non-empty. Consequently, we assume in this section as was done in [17] that for every $\mu \in \mathcal{B}, C(\mu)$ is a normal fuzzy subset of X.

Let $C : \mathcal{B} \longrightarrow \mathcal{FP}(X)$ be a fuzzy choice function on $(X, \mathcal{B})$. The **fuzzy revealed preference relation** ρ^* of C is defined as follows: $\forall \mu \in \mathcal{B}, \forall x, y \in X$,

$$\rho^*(x, y) = \vee\{C(\mu)(x) \wedge \mu(y) \mid \mu \in \mathcal{B}\}.$$

Define the fuzzy relation $\overline{\rho}$ on X as follows: $\forall x, y \in X$,

$$\overline{\rho}(x, y) = C(1_{\{x,y\}})(x).$$

Clearly, $\overline{\rho} \subseteq \rho^*$.

Lemma 2.4.2 *Let C be a fuzzy choice function on $(X, \mathcal{B})$. Then ρ^* and $\overline{\rho}$ both are reflexive and strongly connected.*

Proof. Since $\mathcal{B}$ is the base domain and the fuzzy choice function C is normal, we have for every $x, y \in X, 1_{\{x\}}, 1_{\{y\}}, 1_{\{x,y\}} \in \mathcal{B}$. Also, $C(1_{\{x\}})(x) = 1$ and $C(1_{\{x,y\}})(x) = 1$ or $C(1_{\{x,y\}})(y) = 1$. Thus ρ^* and $\overline{\rho}$ are reflexive and strongly connected. ∎

Let ρ be a fuzzy preference relation on X and $\mathcal{B}$ a nonempty family of nonzero fuzzy subsets of X. For all $\mu \in \mathcal{B}$, define the fuzzy set $G(\mu, \rho)$ of X is as follows: $\forall x \in X$,

$$G(\mu, \rho)(x) = \mu(x) \wedge (\wedge\{\mu(y) \longrightarrow \rho(x, y) \mid y \in X\}).$$

Then $G(\mu, \rho)(x)$ represents the greatest degree of an alternative x in relation to the fuzzy set μ.

Definition 2.4.3 [10, 22] *Let C be a fuzzy choice function. Then C is said to be G-**rational**, if there exists a fuzzy preference relation ρ on X such that $C(\mu) = G(\mu, \rho)$ for all $\mu \in \mathcal{B}$. The fuzzy preference relation ρ for which the fuzzy choice function C is G-rational is called a **rationalization** for C. The fuzzy choice function C is called **full rational** if C is G-rational with reflexive, strongly connected, and max-min transitive rationalization.*

Theorem 2.4.4 [10] *If a fuzzy choice function C defined on base domain is G-rational with rationalization ρ, then $\rho^* \subseteq \rho$.*

The following is the fuzzy T-congruence axiom given in [10].

Fuzzy T-congruence axiom: For all $x, y, z \in X$ and $\mu \in \mathcal{B}$,

$$\rho^*(x, y) \wedge \rho^*(y, z) \wedge C(\mu)(z) \wedge \mu(x) \leq C(\mu)(x)$$

Theorem 2.4.5 [10] *A fuzzy choice function C defined on base domain is full rational if and only if C satisfies the fuzzy T-congruence axiom.*

Fuzzy Chernoff axiom: [24] For all $\mu_1, \mu_2 \in \mathcal{B}$ and $x \in X$,

$$I(\mu_1, \mu_2) \wedge \mu_1(x) \wedge C(\mu_2)(x) \leq I(\mu_1 \cap C(\mu_2), C(\mu_1)).$$

The following theorem was presented by Georgescu in [22] for general domains. As shown in [17], the result also holds for fuzzy choice functions defined on a base domain.

Theorem 2.4.6 (*Desai and Chaudhari* [17]) *If a fuzzy choice function C defined on base domain satisfies the fuzzy Chernoff axiom, then $\rho^* = \overline{\rho}$.*

Proof. Let $x, y \in X$ and $\mu \in \mathcal{B}$. Then by definition of I, we have $I(1_{\{x,y\}}, \mu) = \mu(x) \wedge \mu(y)$. For all $x, y \in X$, we have

$$
\begin{aligned}
\rho^*(x, y) &= \vee\{C(\mu)(x) \wedge \mu(y) \mid \mu \in \mathcal{B}\} \\
&= \vee\{\mu(x) \wedge \mu(y) \wedge 1_{\{x,y\}}(x) \wedge C(\mu)(x) \wedge C(\mu)(x) \mid \mu \in \mathcal{B}\} \\
&= \vee\{I(1_{\{x,y\}}, \mu) \wedge 1_{\{x,y\}}(x) \wedge C(\mu)(x) \wedge C(\mu)(x) \mid \mu \in \mathcal{B}\} \\
&\leq \vee\{I(1_{\{x,y\}} \cap C(\mu), C(1_{\{x,y\}})) \wedge C(\mu)(x) \mid \mu \in \mathcal{B}\} \\
&\leq \vee\{C(\mu)(x) \wedge \big(C(\mu)(x) \longrightarrow C(1_{\{x,y\}})(x)\big) \mid \mu \in \mathcal{B}\} \\
&= \vee\{C(\mu)(x) \wedge C(1_{\{x,y\}})(x) \mid \mu \in \mathcal{B}\} \\
&\leq \vee\{C(1_{\{x,y\}})(x) \mid \mu \in \mathcal{B}\} \\
&= \overline{\rho}(x, y)
\end{aligned}
$$

Thus $\overline{\rho} \supseteq \rho^*$. Since clearly $\overline{\rho} \subseteq \rho^*$, we have $\rho^* = \overline{\rho}$. ∎

In Theorem 2.4.5, we have that the fuzzy choice function C defined on the base domain is full rational if and only if it satisfies the fuzzy T-congruence axiom. In the remainder of this section, we present the work of Desai and

Chaudhari [17]. We show that the weak form of fuzzy T-congruence axiom does not imply the full rationality of the fuzzy choice function.

Weak fuzzy T-congruence axiom: For any $x, y, z \in X$ and $\mu \in \mathcal{B}$,

$$\overline{\rho}(x, y) \wedge \overline{\rho}(y, z) \wedge C(\mu)(z) \wedge \mu(x) \leq C(\mu)(x).$$

The weak fuzzy T-congruence axiom and the fuzzy T-congruence axiom coincide if the domain of fuzzy choice function contains only characteristic functions of single and two-element subsets of the universal set.

Lemma 2.4.7 *If a fuzzy choice function C defined on a base domain satisfies the weak fuzzy T-congruence axiom, then $\overline{\rho}$ is max-min transitive.*

Proof. Since the fuzzy choice function C is defined on base domain $\mathcal{B}$ and C is normal, we have $1_{\{x,z\}} \in \mathcal{B}$ for all $x, z \in X$ and also $C(1_{\{x,z\}})(x) = 1$ or $C(1_{\{x,z\}})(z) = 1$. Suppose $C(1_{\{x,z\}})(x) = 1$. Then $\overline{\rho}(x, z) = 1$. Thus $\overline{\rho}$ is max-min transitive. Suppose $C(1_{\{x,z\}})(z) = 1$. Then

$$\begin{aligned}
\overline{\rho}(x, y) \wedge \overline{\rho}(y, z) &= \overline{\rho}(x, y) \wedge \overline{\rho}(y, z) \wedge 1_{\{x,z\}}(x) \wedge C(1_{\{x,z\}})(z) \\
&\leq C(1_{\{x,z\}})(x) \\
&= \overline{\rho}(x, z).
\end{aligned}$$

Hence $\overline{\rho}$ is max-min transitive. ∎

Clearly, the fuzzy T-congruence axiom implies the weak fuzzy T-congruence axiom. The following example shows the converse does not hold.

Example 2.4.8 *Let $X = \{x, y, z, w\}$ and $\mathcal{B} = \{1_{\{x\}}, 1_{\{y\}}, 1_{\{z\}}, 1_{\{w\}}, 1_{\{x,y\}},$ $1_{\{x,z\}}, 1_{\{x,w\}}, 1_{\{y,z\}}, 1_{\{y,w\}}, 1_{\{z,w\}}, 1_X\}$. Let r, s be such that $0 < r < s < 1$. Define a fuzzy choice function C on $\mathcal{B}$ as follows: $C(1_{\{v\}})(v) = 1$ for all $v \in X$, $C(1_{\{x,y\}})(x) = 1$, $C(1_{\{x,y\}})(y) = r$, $C(1_{\{x,z\}})(x) = 1$, $C(1_{\{x,z\}})(z) = r$, $C(1_{\{y,z\}})(y) = 1$, $C(1_{\{y,z\}})(z) = r$, $C(1_{\{y,w\}})(y) = r$, $C(1_{\{y,w\}})(w) = 1$, $C(1_{\{z,w\}})(z) = r$, $C(1_{\{z,w\}})(w) = 1$, $C(1_{\{x,w\}})(x) = 1$, $C(1_{\{x,w\}})(w) = s$, $C(1_X)(x) = 1$, $C(1_X)(y) = s$, $C(1_X)(z) = r$ and $C(1_X)(w) = s$. Then the fuzzy revealed preference relation ρ^* and $\overline{\rho}$ are as follows:*

$$\rho^* = \begin{array}{c@{\;}c} & \begin{array}{cccc} x & y & z & w \end{array} \\ \begin{array}{c} x \\ y \\ z \\ w \end{array} & \left[\begin{array}{cccc} 1 & 1 & 1 & 1 \\ s & 1 & 1 & s \\ r & r & 1 & r \\ s & 1 & 1 & 1 \end{array}\right] \end{array} \quad \text{and } \overline{\rho} = \begin{array}{c@{\;}c} & \begin{array}{cccc} x & y & z & w \end{array} \\ \begin{array}{c} x \\ y \\ z \\ w \end{array} & \left[\begin{array}{cccc} 1 & 1 & 1 & 1 \\ r & 1 & 1 & r \\ r & r & 1 & r \\ s & 1 & 1 & 1 \end{array}\right] \end{array}$$

Clearly, C satisfies the weak fuzzy T-congruence axiom, but for $1_{\{y,w\}}$, we have $\rho^*(y, x) \wedge \rho^*(x, w) \wedge C(1_{y,w})(w) \wedge 1_{\{y,w\}}(y) = s$ and $C(1_{\{y,w\}})(y) = r$. Hence C does not satisfy the fuzzy T-congruence axiom.

Example 2.4.8 also shows the weak fuzzy T-congruence axiom does not imply the full rationality of the fuzzy choice function by Theorem 2.4.5.

In the following theorem, we characterize the full rationality of a fuzzy choice function by combining the weak fuzzy T-congruence axiom and the fuzzy Chernoff axiom. First we show that the weak fuzzy T-congruence axiom and the fuzzy Chernoff axioms are independent.

In Example 2.4.8, the fuzzy choice function C satisfies the weak fuzzy T-congruence axiom, but for $1_{\{x,y\}}$ and 1_X we have, $I(1_{\{x,y\}}, 1_X) \wedge 1_{\{x,y\}}(y) \wedge C(1_X)(y) = s$ and $I(1_{\{x,y\}} \cap C(1_X), C(1_{\{x,y\}})) = r$. Hence C does not satisfy fuzzy the Chernoff axiom.

Example 2.4.9 *Let* $X = \{x, y, z\}$ *and* $\mathcal{B} = \{1_{\{x\}}, 1_{\{y\}}, 1_{\{z\}}, 1_{\{x,y\}}, 1_{\{x,z\}}, 1_{\{y,z\}}\}$. *Let* r, s *be such that* $0 < r < s < 1$. *Define a fuzzy choice function* C *on* $\mathcal{B}$ *as follows:* $C(1_{\{v\}})(v) = 1$ *for all* $v \in X$, $C(1_{\{x,y\}})(x) = 1, C(1_{\{x,y\}})(y) = r, C(1_{\{x,z\}})(x) = r, C(1_{\{x,z\}})(z) = 1, C(1_{\{y,z\}})(y) = 1$, *and* $C(1_{\{y,z\}})(z) = s$. *Then* $\overline{\rho}$ *is given by*

$$
\overline{\rho} \;=\; \begin{array}{c} \\ x \\ y \\ z \end{array}
\begin{array}{c} x \quad y \quad z \end{array}
\left[\begin{array}{ccc} 1 & 1 & r \\ r & 1 & 1 \\ 1 & s & 1 \end{array} \right]
$$

Clearly, C *satisfies the fuzzy Chernoff axiom. However, for* $x, y \in X$, *we have* $\overline{\rho}(x, y) \wedge \overline{\rho}(y, z) \wedge C(1_{\{x,z\}})(z) \wedge 1_{\{x,z\}}(x) = 1$ *and* $C(1_{\{x,z\}})(x) = r$. *Thus* C *does not satisfy the weak fuzzy T-congruence axiom.*

Theorem 2.4.10 (*Desai and Chaudhari* [17]) *Let* ρ *be a fuzzy preference relation on* X. *Let* C *be a fuzzy choice function defined on a base domain such that* $C(\mu) = G(\mu, \rho)$ *for all* $\mu \in \mathcal{B}$. *Then* ρ *is max-min transitive if and only if* C *satisfies the weak fuzzy T-congruence axiom.*

Proof. Let ρ be a fuzzy preference relation on X such that $C(\mu) = G(\mu, \rho)$ for all $\mu \in \mathcal{B}$. Then ρ is reflexive and for all $x, y \in X$, it follows that $C(1_{\{x,y\}})(x) = \rho(x, y)$. For all $x, y, z \in X$ and $\mu \in \mathcal{B}$, we have

$$
\begin{aligned}
&\overline{\rho}(x, y) \wedge \overline{\rho}(y, z) \wedge \mu(x) \wedge C(\mu)(z) \\
&= C(1_{\{x,y\}})(x) \wedge C(1_{\{y,z\}})(y) \wedge \mu(x) \wedge C(\mu)(z) \\
&= \rho(x, y) \wedge \rho(y, z) \wedge \mu(x) \wedge C(\mu)(z) \\
&\leq \rho(x, z) \wedge S(x) \wedge C(\mu)(z),
\end{aligned}
$$

where the inequality holds by the max-min transitivity of ρ. Thus

$$
\overline{\rho}(x, y) \wedge \overline{\rho}(y, z) \wedge \mu(x) \wedge C(\mu)(z) \leq \rho(x, z) \wedge \mu(x) \wedge C(\mu)(z) \qquad (2.1)
$$

By the max-min transitivity of ρ and Theorem 2.4.4, we have $\rho(x, z) \wedge C(\mu)(z) \wedge \mu(t) \leq \rho(x, t)$ for all $t \in X$. Therefore, by Lemma 2.4.1(2), we have $\rho(x,z) \wedge C(\mu)(z) \leq \mu(t) \longrightarrow \rho(x,t)$ for all $t \in X$. Hence

$$\rho(x,z) \wedge C(\mu)(z) \wedge \mu(x) \leq \mu(x) \wedge (\wedge\{\mu(t) \longrightarrow \rho(x,t) \mid t \in X\})$$
$$= G(\mu,\rho)(x)$$
$$= C(\mu)(x).$$

Thus by (2.1) we have that C satisfies the weak fuzzy T-congruence axiom.

Conversely, suppose that C satisfies the weak fuzzy T-congruence axiom. Then $1_{\{x,z\}} \in \mathcal{B}$ for all $x, z \in X$. Since C is normal, we have $C(1_{\{x,z\}})(x) = 1$ or $C(1_{\{x,z\}})(z) = 1$. If $C(1_{\{x,z\}})(x) = 1$, then by Lemma 2.4.1(4), $C(1_{\{x,z\}})(x) \leq 1_{\{x,z\}}(x) \longrightarrow \rho(x,z)$ implies $C(1_{\{x,z\}})(x) \leq \rho(x,z)$ and hence $\rho(x,z) = 1$ and ρ is clearly max-min transitive. Suppose $C(1_{\{x,z\}})(z) = 1$. We also have $\overline{\rho}(x,y) = C(1_{\{x,y\}})(x) = \rho(x,x) \wedge \rho(x,y) = \rho(x,y)$ for all $y \in X$. Thus

$$\rho(x,y) \wedge \rho(y,z) = \overline{\rho}(x,y) \wedge \overline{\rho}(y,z)$$
$$= \overline{\rho}(x,y) \wedge \overline{\rho}(y,z) \wedge C(1_{\{x,z\}})(z) \wedge 1_{\{x,z\}}(x)$$
$$\leq C(1_{\{x,z\}})(x)$$
$$= \overline{\rho}(x,z)$$
$$= \rho(x,z),$$

where the inequality holds by the weak T-congruence axiom. Hence ρ is max-min transitive. ∎

The following is a characterization theorem for full rationality of fuzzy choice functions.

Theorem 2.4.11 (*Desai and Chaudhari* [17]) *A fuzzy choice function C defined on the base domain $\mathcal{B}$ is full rational if and only if it satisfies the weak fuzzy T-congruence axiom and the fuzzy Chernoff axiom.*

Proof. Suppose C satisfies the weak fuzzy T-congruence axiom and the fuzzy Chernoff axiom. Then by Theorems 2.4.5 and 2.4.6, C is full rational.

Conversely, suppose C is full rational with rationalization ρ. Then by Theorem 2.4.5 and Lemma 2.4.2(1), C satisfies the weak fuzzy T-congruence axiom. Let $\mu, \nu \in \mathcal{B}$ and $x \in X$. By Lemma 2.4.1(2), we have for all $y, z \in X$ that

$$I(\mu,\nu) \wedge \mu(x) \wedge C(\nu)(x) \wedge \mu(y) \wedge C(\nu)(y) \wedge \mu(z)$$
$$\leq \mu(z) \wedge I(\mu,\nu) \wedge C(\nu)(y)$$
$$\leq \mu(z) \wedge [\mu(z) \longrightarrow \nu(z)] \wedge C(\nu)(y)$$
$$= \mu(z) \wedge \nu(z) \wedge C(\nu)(y)$$
$$\leq \nu(z) \wedge C(\nu)(y)$$
$$= \nu(z) \wedge G(\nu,\rho)(y)$$
$$\leq \nu(z) \wedge [\nu(z) \longrightarrow \rho(y,z)]$$
$$= \nu(z) \wedge \rho(y,z)$$
$$\leq \rho(y,z).$$

By Lemma 2.4.1(1), the above inequality reduces to

$$I(\mu, \nu) \wedge \mu(x) \wedge C(\nu)(x) \wedge \mu(y) \wedge C(\nu)(y) \leq \mu(z) \longrightarrow \rho(y, z) \text{ for all } z \in X.$$

Hence

$$
\begin{aligned}
&I(\mu, \nu) \wedge \mu(x) \wedge C(\nu)(x) \wedge \mu(y) \wedge C(\nu)(y) \\
&\leq \mu(y) \wedge (\wedge\{\mu(z) \longrightarrow \rho(y, z) \mid z \in X\}) \\
&= C(\mu)(y).
\end{aligned}
$$

By Lemma 2.4.1(1), we have

$$I(\mu, \nu) \wedge \mu(x) \wedge C(\nu)(x) \leq \mu(y) \wedge C(\nu)(y) \longrightarrow C(\mu)(y) \text{ for all } y \in X$$

Thus $I(\mu, \nu) \wedge \mu(x) \wedge C(\nu)(x) \leq I(\mu \cap C(\nu), C(\mu))$. ∎

In [25], Georgescu establishes that the fuzzy Arrow axiom and the property of full rationality of fuzzy choice functions defined on general domain are equivalent, [25, Theorem 4.5]. As remarked in [17], "if this result is true for fuzzy choice functions defined on the base domain, then the full rationality of fuzzy choice functions implies FCA and the above theorem will be trivial. But the following discussion rules out the possibility."

Fuzzy Arrow axiom (FAA): [21] For all $\mu_1, \mu_2 \in \mathcal{B}$ and $x \in X$,

$$I(\mu_1, \mu_2) \wedge \mu_1(x) \wedge C(\mu_2)(x) \leq E(\mu_1 \cap C(\mu_2), C(\mu_1)).$$

In Example 2.4.8, if $C(1_X)(y) = s$ is replaced by $C(1_X)(y) = r$, then C is full rational with rationalization $\overline{\rho}$. For $1_{\{x,w\}}$ and 1_X, we have $I(1_{\{x,w\}}, X) \wedge 1_{\{x,w\}}(x) \wedge C(1_X)(x) = 1$ and

$$E(C(1_{\{x,w\}}), 1_{\{x,w\}} \cap C(X)) \leq 1_{\{x,w\}}(w) \wedge C(1_X)(w) \longrightarrow C(1_{\{x,w\}})(w) = s.$$

Hence the fuzzy Arrow axiom does not hold on the base domain. Hence full rationality does not imply the FAA on the base domain.

The following result follows from Theorems 2.4.10 and 2.4.11.

Theorem 2.4.12 *Let C be a fuzzy choice function defined on a base domain. Then C is full rational if and only if it satisfies the fuzzy Chernoff axiom and is G-rational with transitive rationalization.*

The following result follows easily from Theorems 2.4.5, 2.4.11, and 2.4.12.

Theorem 2.4.13 *Let C be any fuzzy choice function defined on the base domain. Then the following statements are equivalent*

(1) *C is full rational.*
(2) *C satisfies fuzzy T-congruence axiom.*
(3) *C satisfies the weak fuzzy T-congruence axiom and fuzzy Chernoff axiom.*
(4) *C satisfies the fuzzy Chernoff axiom and C is G-rational with transitive rationalization.*

Chaudhari and Desai [10, 11] have studied full and acyclic rationality, G-rationality, G-normality by introducing various congruence axioms and revealed preference axioms.

2.5   Quasi-Transitive Rationality of Fuzzy Choice Functions

In this section, we follow the lead of Desai [16] in a discussion of the rationality of fuzzy choice functions. He characterized the rationality of fuzzy choice functions with reflexive, strongly connected and quasi-transitive rationalization in terms of the path independence property and the fuzzy Condorcet property. The definition of quasi-transitivity in Definition 2.2.16 uses a t-norm $*$. In this section, we let $* = \wedge$.

In order to study the rationality of a consumer, Samuelson [42] introduced the concept of the revealed preference theory in terms of a preference relation associated with a demand function. Uzawa [49] states that the behavior of a consumer is rational if he has a definite preference over all conceivable commodity bundles and he chooses those commodity bundles that are optimum with respect to his preference subject to the budgetary constraints. The problem of rationality appears in many other disciplines. Consequently, the problem of rationality has been examined in a more general framework, see Georgescu-Rogen [28 28], Uzawa [49], Arrow [1], Sen [43, 44], Richter [40], Hansson [29], and Suzumura [47, 48]. They introduced notions such as the revealed preference axioms, the consistency conditions, and the congruence axioms.

De Baets and Fodor [19] and Fodor [20] studied the case of vague preferences and exact choices. Barrett, Pattanaik, and Salles [6, 7] studied the case when crisp choice functions are generated by the fuzzy preference relations, see Section 2.4. In [4 4], Banerjee studied the revealed preference theory in the context of the fuzzy set theory. In his approach, the domain of a choice function is made only of crisp sets of alternatives and the range is made of fuzzy sets of alternatives, Section 7.2. Georgescu [21, 22] examined fuzzy choice functions whose domain and range are both fuzzy sets of alternatives. She studied the rationality of fuzzy choice functions and various fuzzy revealed preference axioms, fuzzy congruence axioms and fuzzy consistency conditions. In the previous section, rationality of fuzzy choice functions defined on the domain that contains characteristic functions of all single and two-element subsets of the universal set was examined.

We recall the definition of the residuum operation $\longrightarrow$ on $[0, 1]$, i.e.,

$$a \to b = \vee\{t \in [0,1] \mid a \wedge t \leq b\} = \begin{cases} 1 & \text{if } a \leq b \\ b & \text{if } a > b \end{cases}$$

for all $a, b \in [0, 1]$. The **biresiduum operation** $\longleftrightarrow$ on $[0, 1]$ is defined by

$$a \leftrightarrow b = (a \rightarrow b) \wedge (b \rightarrow a),$$

for all $a, b \in [0, 1]$. The corresponding negation operation $\neg$ has the following form

$$\neg a = a \rightarrow 0 = \begin{cases} 1 & \text{if } a = 0 \\ 0 & \text{if } a > 0 \end{cases}$$

for all $a, b \in [0, 1]$.

Lemma 2.5.1 [8, 30, 31] *For any $a, b, c \in [0, 1]$ the following properties hold:*
(1) $(a \longrightarrow b) \wedge (b \longrightarrow c) \leq a \longrightarrow c$;
(2) $a \leq b \Longrightarrow c \longrightarrow a \leq c \longrightarrow b$ *and* $b \longrightarrow c \leq a \longrightarrow c$;
(3) $a \wedge \neg a = 0$;
(4) $a \leq \neg b \Longleftrightarrow a \wedge b = 0$

Let ρ be a fuzzy relation on X. In this section, the asymmetric part of ρ is the fuzzy relation π on X defined by $\pi(x, y) = \rho(x, y) \wedge \neg\rho(y, x)$ for all $x, y \in X$. We think of $\pi(x, y)$ as showing the degree to which x is strictly preferred to y. Recall that under this definition, π is of type $\pi_{(0)}$.

Let X be a non-empty set of alternatives and $\mathcal{B}$ is a non-empty family of non-zero fuzzy subsets of X. Then the pair $(X, \mathcal{B})$ is called fuzzy choice space. In this section, a fuzzy choice function on the fuzzy choice space $(X, \mathcal{B})$ is a function $C : \mathcal{B} \longrightarrow \mathcal{FP}(X)$ such that for each $\mu \in \mathcal{B}$, $C(\mu) \neq \theta$ and $C(\mu) \subseteq \mu$.

In the language of fuzzy consumers, X is called the set of bundles and $\mathcal{B}$ is called the family of fuzzy budgets. For any fuzzy budget μ the real numbers $\mu(x)$ and $C(\mu)(x)$ denote the availability degree of the bundle x in the fuzzy budget μ and the degree to which the bundle x is chosen from the fuzzy budget μ respectively.

Georgescu extended the results of Uzawa-Arrow-Sen theory using the following conditions presented in Section 2.3:
(H1) Every $\mu \in \mathcal{B}$ and $C(\mu)$ are normal fuzzy subsets of X,
(H2) $\mathcal{B}$ includes all fuzzy sets 1_S, where $S \subseteq X$.

In this section, we also consider these conditions.

Let $(X, \mathcal{B})$ be a fuzzy choice space and ρ be a fuzzy relation on X. For any $\mu \in \mathcal{B}$ the fuzzy subsets $M(\mu, \rho)$ and $G(\mu, \rho)$ of X are defined as in Definition 2.3.3 with $*$ replaced by $\wedge$. Thus we have

$$M(\mu, \rho)(x) = \mu(x) \wedge (\wedge\{\mu(z) \wedge \rho(z, x) \longrightarrow \rho(x, z) \mid z \in X\})$$
$$G(\mu, \rho)(x) = \mu(x) \wedge (\wedge\{\mu(z) \longrightarrow \rho(x, z) \mid z \in X\}).$$

We note that $G(\mu, \rho) \subseteq M(\mu, \rho)$ for all $\mu \in \mathcal{B}$ and $G(\mu, \rho) = M(\mu, \rho)$ if ρ is strongly connected. Note also that $G(\mu, \rho^*) \subseteq C(\mu)$ for all $\mu \in \mathcal{B}$.

Definition 2.5.2 [21, 23, 25] *Let* $C : \mathcal{B} \longrightarrow \mathcal{F}(X)$ *be a fuzzy choice function. Then* C *is said to be*

(1) *G-**rational** if there exists a fuzzy preference relation* ρ *on* X *such that* $C(\mu) = G(\mu, \rho)$ *for all* $\mu \in \mathcal{B}$; *M-**rational** if there exists a fuzzy preference relation* ρ *on* X *such that* $C(\mu) = M(\mu, \rho))$ *for all* $\mu \in \mathcal{B}$;

(2) *G-**normal** if* $C(\mu) = G(\mu, \rho)$ *for all* $\mu \in \mathcal{B}$; *M-**normal** if* $C(\mu) = M(\mu, \rho)$ *for all* $\mu \in \mathcal{B}$;

(3) *full **rational** if there exists a reflexive, strongly connected and max-min transitive fuzzy preference relation* ρ *on* X *such that* $C(\mu) = G(\mu, \rho)$ *for all* $\mu \in \mathcal{B}$.

If a fuzzy choice function C is G-rational with rationalization ρ, then $\rho^* \subseteq \rho$.

The fuzzy path independence property and the fuzzy Condorcet property are needed to characterize quasi-transitive rationality of fuzzy choice functions.

We first discuss the Condorcet property briefly. Approval voting allows a voter to vote for as many candidates as he wishes in a multicandidate race. Condorcet argued that the winner should be the candidate who is preferred by a simple majority of voters to each of the other candidates in pairwise contests, assuming such a candidate exists. In the following definition, assume μ is crisp. If z belongs to μ and x is chosen over z, then x is chosen with respect to μ. Recall that a fuzzy choice function $C : \mathcal{B} \longrightarrow \mathcal{F}(X)$ is said to satisfy fuzzy path independence if $C(\mu \cup \nu) = C(C(\mu) \cup C(\nu))$ for all $\mu, \nu \in \mathcal{B}$.

Definition 2.5.3 *Let* $C : \mathcal{B} \longrightarrow \mathcal{F}(X)$ *be a fuzzy choice function. Then* C *is said to satisfy the **fuzzy Condorcet property** if for all* $\mu \in B$ *and* $x \in X$,

$$\mu(x) \wedge (\vee\{\mu(z) \longrightarrow \overline{\rho}(x, z) \mid z \in X\}) \le C(\mu)(x).$$

The characterization of a choice function whose rationality is reflexive, strongly connected, and max-min transitive has previously been determined. In this section, transitivity is weakened to quasi-transitivity.

Definition 2.5.4 *A fuzzy choice function* C *on* $(X, \mathcal{B})$ *is said to be Q-**rational** if there exists a reflexive, strongly connected and quasi-transitive fuzzy preference relation* ρ *on* X *such that* $C(\mu) = G(\mu, \rho)$ *for all* $\mu \in \mathcal{B}$.

Since every max-min transitive fuzzy preference relation is quasi-transitive, every full rational fuzzy choice function is Q-rational.

Lemma 2.5.5 *Let* C *be a fuzzy choice function. If* C *satisfies the fuzzy path independence property, then* $\overline{\rho}$ *is quasi-transitive.*

Proof. We shall prove that for all $x, y, z \in X$,

$$\overline{\pi}(x, y) \wedge \overline{\pi}(y, z) \le \overline{\pi}(x, z)). \tag{2.2}$$

Let $x, y, z \in X$. If $\overline{\rho}(y, x) > 0$ or $\overline{\rho}(z, y) > 0$, then $\neg\overline{\rho}(y, x) = 0$ and $\neg\overline{\rho}(z, y) = 0$. In this case, inequality (2.2) holds. Suppose $\overline{\rho}(y, x) = 0$ and $\overline{\rho}(z, y) = 0$. Since $\overline{\rho}$ is strongly connected, $\overline{\rho}(x, y) = 1$ and $\overline{\rho}(y, z) = 1$. By the definition of $\overline{\rho}$, it follows that $C(1_{\{x,y\}}) = 1_{\{x\}}$ and $C(1_{\{y,z\}}) = 1_{\{y\}}$. Thus by fuzzy path independence property, we have that

$$
\begin{aligned}
\overline{\pi}(x, y) \wedge \overline{\pi}(y, z) &\leq C(1_{\{x,y\}})(x) \\
&= C(C(1_{\{x\}}) \cup C(1_{\{y,z\}}))(x) \\
&\leq C(1_{\{x\}} \cup 1_{\{y,z\}})(x) \\
&= C(1_{\{x,y,z\}})(x) \\
&= C(1_{\{x,y\}} \cup 1_{\{z\}})(x) \\
&\leq C(C(1_{\{x,y\}}) \cup C(1_{\{z\}}))(x) \\
&= C(1_{\{x\}} \cup 1_{\{z\}})(x) \\
&= C(1_{\{x,z\}})(x) \\
&= \overline{\rho}(x, z).
\end{aligned}
$$

Hence

$$
\overline{\pi}(x, y) \wedge \overline{\pi}(y, z) \leq \overline{\rho}(x, z) \tag{2.3}
$$

Now

$$
\begin{aligned}
\overline{\pi}(x, y) \wedge \overline{\pi}(y, z) \wedge \overline{\rho}(z, x) &\leq \overline{\rho}(z, x) \\
&= C(1_{\{x,z\}})(z) \\
&= C(C(1_{\{x,y\}}) \cup C(1_{\{z\}}))(z) \\
&\leq C(1_{\{x,y\}} \cup 1_{\{z\}})(z) \\
&= C(1_{\{x,y,z\}})(z) \\
&= C(1_{\{x,y\}} \cup 1_{\{y,z\}})(z) \\
&\leq C(C(1_{\{x,y\}}) \cup C(1_{\{y,z\}}))(z) \\
&= C(1_{\{x\}} \cup 1_{\{y\}})(z) \\
&= C(1_{\{x,y\}})(z) \\
&= 0.
\end{aligned}
$$

By Lemma 2.5.1(8), it follows that

$$
\overline{\pi}(x, y) \wedge \overline{\pi}(y, z) \leq \neg\overline{\rho}(z, x). \tag{2.4}
$$

By the idempotent property of the Gödel t-norm and inequalities (2.3) and (2.4), inequality (2.2) holds. ∎

Lemma 2.5.6 *Let C be a fuzzy choice function. Then C is G-rational if and only if it satisfies the fuzzy Condorcet property.*

Proof. Let C satisfy the fuzzy Condorcet property. Then $G(\mu, \overline{\rho}) \subseteq C(\mu)$ for all $\mu \in \mathcal{B}$. Since $\overline{\rho} \subseteq \rho^*$, by Lemma 2.5.1(3), it follows that

$$\mu(y) \longrightarrow \rho^*(y, x) \leq \mu(y) \longrightarrow \overline{\rho}(y, x)$$

for all $\mu \in \mathcal{B}$ and $x, y \in X$. Thus

$$\mu(x) \wedge (\wedge\{\mu(y) \longrightarrow \rho^*(y, x) \mid y \in X\}) \leq \mu(x) \wedge (\wedge\{\mu(y) \longrightarrow \overline{\rho}(y, x) \mid y \in X\}).$$

Hence $G(\mu, \rho^*) \subseteq G(\mu, \overline{\rho})$ for all $\mu \in \mathcal{B}$. However, $C(\mu) \subseteq G(\mu, \rho^*)$ for all $\mu \in \mathcal{B}$. Thus $C(\mu) \subseteq G(\mu, \overline{\rho})$. Hence $C(\mu) = G(\mu, \overline{\rho})$ for all $\mu \in \mathcal{B}$. Therefore, C is G-rational.

Conversely, suppose $C(\mu) = G(\mu, \rho)$ for some fuzzy preference relation ρ on X. Then for all $\mu \in \mathcal{B}$ and $x \in X$, we have by Lemmas 2.3.1(2) and 2.5.5 that

$$\begin{aligned}
G(\mu, \overline{\rho})(x) \wedge \mu(z) &\leq \mu(z) \wedge [\mu(z) \longrightarrow \overline{\rho}(x, z)] \\
&= \mu(z) \wedge \overline{\rho}(x, z) \\
&\leq \overline{\rho}(x, z) \\
&\leq \rho^*(x, z) \\
&\leq \rho(x, z).
\end{aligned}$$

By Lemma 2.3.1(2), it follows that

$$G(\mu, \overline{\rho})(x) \leq \mu(z) \longrightarrow \rho(x, z) \tag{2.5}$$

for all $z \in X$. Thus

$$\begin{aligned}
G(\mu, \overline{\rho})(x) &\leq \mu(x) \wedge (\wedge\{\mu(z) \longrightarrow \rho(x, z) \mid z \in X\}) \\
&= G(\mu, \rho)(x) \\
&= C(\mu)(x).
\end{aligned}$$

Hence C satisfies the fuzzy Condorcet property. $\blacksquare$

We now show that the fuzzy independence property and the fuzzy Condorcet property are independent.

Example 2.5.7 Let $X = \{x, y, z\}$ and $\mathcal{B} = \{1_{\{x\}}, 1_{\{y\}}, 1_{\{z\}}, 1_{\{x,y\}}, 1_{\{x,z\}}, 1_{\{y,z\}}, 1_X\}$. Define a fuzzy choice function C on $\mathcal{B}$ as follows: $C(1_{\{v\}})(v) = 1$ for all $v \in X$; $C(1_{\{x,y\}})(x) = 1$; $C(1_{\{x,y\}})(y) = 1$; $C(1_{\{x,z\}})(x) = 1$; $C(1_{\{x,z\}})(z) = 1$; $C(1_{\{y,z\}})(y) = 1$; $C(1_{\{y,z\}})(z) = 1$; $C(1_X)(x) = 0.5$; $C(1_X)(y) = 1$ and $C(1_X)(z) = 1$. Then $\overline{\rho}(u, v) = 1$ for all $u, v \in X$. Hence C satisfies the fuzzy path independence property. However, for $1_X \in \mathcal{B}$, we have $1_X(x) \wedge (\wedge\{1_X(z) \longrightarrow \overline{\rho}(x, z) \mid z \in X\}) = 1$ and $C(1_X)(x) = 0.5$. Thus C does not satisfy the fuzzy Condorcet property.

Example 2.5.8 *Let* $X = \{x, y, z\}$ *and* $\mathcal{B} = \{1_{\{x\}}, 1_{\{y\}}, 1_{\{z\}}, 1_{\{x,y\}}, 1_{\{x,z\}}, 1_{\{y,z\}}, 1_X\}$. *Define a fuzzy choice function* C *on* $\mathcal{B}$ *as follows:* $C(1_{\{v\}})(v) = 1$ *for all* $v \in X$, $C(1_{\{x,y\}})(x) = 1$, $C(1_{\{x,y\}})(y) = 0$, $C(1_{\{x,z\}})(x) = 1$, $C(1_{\{x,z\}})(z) = 1$, $C(1_{\{y,z\}})(y) = 1$, $C(1_{\{y,z\}})(z) = 0$, $C(1_X)(x) = 1$, $C(1_X)(y) = 0.5$ *and* $C(1_X)(z) = 0$. *Then the fuzzy revealed preference relation* ρ^* *is given as follows:* $\rho^*(y, x) = 0 = \rho^*(z, y)$ *and* $\rho^*(u, v) = 1$ *otherwise. Hence* C *satisfies the fuzzy Condorcet property. However, for* $1_{\{x,y\}}, 1_{\{x,z\}} \in \mathcal{B}$, *we have* $C(1_{\{x,y\}} \cup 1_{\{y,z\}})(y) = C(1_X)(y) = 0.5$ *and* $C(C(1_{\{x,y\}}) \cup C(1_{\{y,z\}}))(y) = C(1_{\{x,y\}})(y) = 0$. *Thus* C *does not satisfy the fuzzy path independence property.*

The following is a characterization theorem for quasi-transitive rationality of fuzzy choice functions.

Theorem 2.5.9 (*Desai* [16]) *Let* C *be a fuzzy choice function. Then* C *is* Q-*rational if and only if* C *satisfies the fuzzy path independence property and the fuzzy Condorcet property.*

Proof. Suppose that C satisfies the fuzzy path independence property and the fuzzy Condorcet property. Then by Lemmas 2.5.5 and 2.5.6, C is Q-rational with rationalization $\overline{\rho}$.

Conversely, suppose that C is Q-rational with rationalization ρ. To prove C satisfies the fuzzy path independence property, first we prove the following inequality. Let $\mu, \nu \in \mathcal{B}$ and $x \in X$. Then for all $y, z \in X$, we have by Lemma 2.3.1(2) that

$$
\begin{aligned}
I(\mu, \nu) \wedge \mu(y) \wedge C(\nu)(y) \wedge \mu(z) &\leq \mu(z) \wedge [\mu(z) \longrightarrow \nu(z)] \wedge C(\nu)(y) \\
&= \mu(z) \wedge \nu(z) \wedge C(\nu)(y) \\
&\leq \nu(z) \wedge C(\nu)(y) \\
&= \nu(z) \wedge G(\nu, \rho)(y) \\
&\leq \nu(z) \wedge [\nu(z) \longrightarrow \rho(y, z)] \\
&= \nu(z) \wedge \rho(y, z) \\
&\leq \rho(y, z).
\end{aligned}
$$

Thus by Lemma 2.3.1(2), the above inequality reduces to

$$
I(\mu, \nu) \wedge \mu(y) \wedge C(\nu)(y) \leq \mu(z) \longrightarrow \rho(t, z) \text{ for all } z \in X.
$$

Hence

$$
\begin{aligned}
I(\mu, \nu) \wedge \mu(y) \wedge C(\nu)(y) &\leq \mu(y) \wedge (\wedge\{\mu(z) \longrightarrow \rho(t, z) \mid z \in X\}) \\
&= C(\mu)(y).
\end{aligned}
$$

By Lemma 2.3.1(2), it follows that

$$
I(\mu, \nu) \leq \mu(y) \wedge C(\nu)(y) \longrightarrow C(\mu)(y) \text{ for all } y \in X.
$$

Thus
$$I(\mu, \nu) \le I(\mu \cap C(\nu), C(\mu)). \tag{2.6}$$

Since $\mu \subseteq \mu \cup \nu$ and $\nu \subseteq \mu \cup \nu$, we have that $I(\mu, \mu \cup \nu) = 1$ and $I(\nu, \mu \cup \nu) = 1$. Hence by the above inequality, it follows that

$$
\begin{aligned}
C(\mu \cup \nu)(x) &\le I(\mu, \mu \cup \nu) \\
&\le I(\mu \cap C(\mu \cup \nu), C(\mu)) \\
&\le \mu(x) \wedge C(\mu \cup \nu)(x) \longrightarrow C(\mu)(x).
\end{aligned}
$$

By Lemma 2.3.1(2) we have that

$$C(\mu \cup \nu)(x) \wedge \mu(x) \wedge C(\mu \cup \nu)(x) \le C(\mu)(x).$$

By Lemma 2.3.1(2), the above inequality reduces to

$$C(\mu \cup \nu)(x) \le \mu(x) \longrightarrow C(\mu)(x).$$

By the definition of $\longrightarrow$, $a > b$, implies $(a \longrightarrow b) = b$. Since $C(\mu)(x) \le \mu(x)$, we have $\mu(x) \longrightarrow C(\mu)(x) = C(\mu)(x)$. Thus $C(\mu \cup \nu)(x) \le C(\mu)(x)$. Similarly we have that $C(\mu \cup \nu)(x) \le C(\nu)(x)$ for all $x \in X$. Consequently, $C(\mu \cup \nu) \subseteq C(\mu) \cup C(\nu)$. Since $C(\mu) \cup C(\nu) \subseteq \mu \cup \nu$, it follows by inequality (2.6) and above inclusion that

$$
\begin{aligned}
C(\mu \cup \nu)(x) &\le I(C(\mu) \cup C(\nu), \mu \cup \nu) \\
&\le I((C(\mu) \cup C(\nu)) \cap C(\mu \cup \nu), C(C(\mu) \cup C(\nu))) \\
&\le (C(\mu) \cup C(\nu))(x) \wedge C(\mu \cup \nu)(x) \longrightarrow C(C(\mu) \cup C(\nu))(x).
\end{aligned}
$$

for all $x \in X$. By Lemma 2.3.1(1), we have

$$C(\mu \cup \nu)(x) \wedge (C(\mu) \cup C(\nu))(x) \wedge C(\mu \cup \nu)(x) \le C(C(\mu) \cup C(\nu))(x).$$

Since $C(\mu \cup \nu) \subseteq C(\mu) \cup C(\nu)$, we have that

$$C(\mu \cup \nu)(x) \le C(C(\mu) \cup C(\nu))(x) \text{ for all } x \in X.$$

Therefore,
$$C(\mu \cup \nu) \subseteq C(C(\mu) \cup C(\nu)). \tag{2.7}$$

Now for any $\mu, \nu \in \mathcal{B}$ and $x \in X$, we have

$$
\begin{aligned}
C(C(\mu) \cup C(\nu))(x) &= G(C(\mu) \cup C(\nu), \rho)(x) \\
&\le (C(\mu) \cup C(\nu))(z) \longrightarrow \rho(x, z).
\end{aligned}
$$

Since $C(\mu) \cup C(\nu) \subseteq \mu \cup \nu$, we have by Lemma 2.5.1(2) that

$$(C(\mu) \cup C(\nu))(x) \longrightarrow \rho(x, z) \le (\mu \cup \nu)(x) \longrightarrow \rho(x, z).$$

Therefore the above inequality becomes

$$C(C(\mu) \cup C(\nu))(x) \le (\mu \cup \nu)(z) \longrightarrow \rho(x, z)$$

for all $z \in X$. Therefore

$$C(C(\mu) \cup C(\nu)) \subseteq C(\mu \cup \nu). \tag{2.8}$$

Hence inequality (2.7) and (2.8) prove that C satisfies the fuzzy path independence property. To prove the fuzzy Condorcet property, consider $\mu \in \mathcal{B}$ and $x \in X$. Since C is Q-rational, we have $\rho^* \subseteq \rho$. Also, $\bar{\rho} \subseteq \rho$ and $\bar{\rho} \subseteq \rho^*$. Thus $\bar{\rho} \subseteq \rho$. Hence by Lemma 2.5.1(2), we have

$$\mu(z) \longrightarrow \bar{\rho}(x, z) \le \mu(z) \longrightarrow \rho(x, z)$$

for all $z \in X$. Thus

$$\wedge\{\mu(z) \longrightarrow \bar{\rho}(x, z) \mid z \in X\} \le \wedge\{\mu(z) \longrightarrow \rho(x, z) \mid z \in X\}.$$

Hence

$$\mu(x) \wedge (\wedge\{\mu(z) \longrightarrow \bar{\rho}(x, z) \mid z \in X\}) \le \mu(x) \wedge (\wedge\{\mu(z) \longrightarrow \rho(x, z) \mid z \in X\})$$
$$= G(\mu, \rho)(x)$$
$$= C(\mu)(x).$$

Consequently, C satisfies the fuzzy Condorcet property. ∎

Georgescu [21, 22, 23, 25] has studied fuzzy choice theory in a very general manner. She extended the results of the Uzawa–Arrow–Sen theory and the Richter–Hansson–Suzumura theory in the context of fuzzy set theory. The results of the Uzawa–Arrow–Sen theory were extended in the context of fuzzy set theory under the hypotheses (H1) and (H2). The results of the Richter–Hansson–Suzumura were extended without assuming (H1) and (H2).

2.6 Full Rationality and Congruence Axioms of Fuzzy Choice Functions

We present two axioms, the fuzzy direct revelation axiom (FDRA) and the fuzzy transitive closure coherence axiom (FTCCA). Connections between full rationality, G-rationality, G-normality, the fuzzy congruence axiom, and the weak fuzzy congruence axiom are also presented. This section is based on Chaudhari and Desai [11].

Let ρ be a fuzzy binary relation on X. Recall that ρ is called strongly connected if $\forall x, y \in X, \rho(x, y) = 1$ or $\rho(y, x) = 1$. The max-* transitive closure ρ^{tc} of ρ is defined as follows: $\forall x, y \in X$,

$$\rho^{tc}(x, y) = \rho(x, y) \vee$$
$$\{\vee\{\vee\{\rho(x, z_1) * \rho(z_1, z_2) * \ldots * \rho(z_k, y) \mid z_1, z_2, \ldots, z_k \in X\}$$
$$\mid k \in \mathbb{N}\}.$$

Let $\mathcal{B}$ denote a non-empty family of non-zero fuzzy subsets of X. A fuzzy choice function is a function $C : \mathcal{B} \to \mathcal{FP}^*(X)$ such that $C(\mu) \subseteq \mu$ for all $\mu \in \mathcal{B}$. Let π denote the strict fuzzy preference relation associated with ρ. Recall that π is defined by $\forall x, y \in X, \pi(x, y) = \rho(x, y) * \neg \rho(y, x)$ and that if $*$ has no zero divisors, then π is of type $\pi_{(0)}$. If $\pi^{tc}(x, y) \leq \neg \pi(y, x)$ for all $x, y \in X$, then ρ is said to be **acyclic**. Clearly ρ is max-$*$ transitive is equivalent to $\rho \circ \rho \subseteq \rho$. Also ρ^{tc} is the smallest max-$*$ transitive relation containing ρ. If $\rho_1 \subseteq \rho_2$, then $\rho_1^{tc} \subseteq \rho_2^{tc}$ for any two fuzzy binary relations ρ_1 and ρ_2 on X.

Definition 2.6.1 *Let $C : \mathcal{B} \to \mathcal{FP}(X)$ be a fuzzy choice function on $(X, \mathcal{B})$. Define the fuzzy binary operations on X as follows: $\forall x, y \in X$,*

$$\rho^*(x, y) = \vee \{C(\mu)(x) * \mu(y) \mid \mu \in \mathcal{B}\},$$
$$\pi^*(x, y) = \rho^*(x, y) * \neg \rho^*(y, x),$$
$$\iota^*(x, y) = \rho^*(x, y) * \rho^*(y, x).$$

*Then ρ^* is called the **fuzzy revealed preference relation** generated by C, π^* is called the **strict fuzzy revealed preference relation** generated by C, and ι^* is called the **indifference fuzzy revealed preference relation** generated by C.*

Definition 2.6.2 *Let $C : \mathcal{B} \to \mathcal{FP}(X)$ be a fuzzy choice function on $(X, \mathcal{B})$. C is said to satisfy **weak fuzzy congruence axiom** (**WFCA**) if $\forall \mu \in \mathcal{B}$ and $\forall x, y \in X, \rho(x, y) * C(\mu)(y) * \mu(x) \leq C(\mu)(x)$.*

Proposition 2.6.3 [24] *Let C be a fuzzy choice function. If C satisfies WFCA, then ρ^* is max-$*$ transitive, reflexive, and strongly connected on X.*

Proposition 2.6.4 [24] *Let C be a fuzzy choice function. If C satisfies WFCA, then condition $F\alpha$ holds and $\forall x \in X$,*

$$\mu(x) * (\wedge \{\mu(z) \to \rho^*(x, y) \mid z \in X\}) \leq C(\mu)(x).$$

Definition 2.6.5 *Let C be a fuzzy choice function. Then C is said to satisfy the **fuzzy transitive closure coherence axiom** (**FTCCA**) if $\forall \mu \in \mathcal{B}$ and $\forall x \in X$,*

$$\mu(x) * (\wedge \{\mu(z) \to (\rho^*)^{tc}(x, y) \mid z \in X\}) \leq C(\mu)(x),$$

where $(\rho^*)^{tc}$ is the transitive closure of ρ^*.

Definition 2.6.6 *Let C be a fuzzy choice function. Then C is said to satisfy the **fuzzy congruence axiom** (**FCA**) if $\forall \mu, \nu \in \mathcal{B}$ and $\forall x, y \in X$,*

$$(\rho^*)^{tc}(x, y) * C(\mu)(y) * \mu(x) \leq C(\mu)(x).$$

Theorem 2.6.7 (*Chaudhari and Desai* [11]) *Let C be a fuzzy choice function. If C is G-rational with max-*transitive rationalization, then C satisfies WFCA.*

Proof. Let $\mu \in B$ and $x, y \in X$. Then

$$
\begin{aligned}
C(\mu)(y) &= G(\mu, \rho^*)(y) = \mu(y) * \wedge\{\mu(z) \to \rho^*(y, z) \mid z \in X\} \\
&\leq \mu(z) \to \rho^*(y, z) \; \forall z \in X.
\end{aligned}
$$

Thus $C(\mu)(y) * \mu(z) \leq \rho^*(y, z)$ by Lemma 2.3.1(1). Hence $\rho(x, y) * C(\mu)(y) * \mu(z) \leq \rho^*(x, y) * \rho^*(y, z) \leq \rho^*(x, z)$ for all $z \in X$. By Lemma 2.3.1(1), we have

$$
\rho^*(x, y) * C(\mu)(y) \leq \mu(z) \to \rho^*(x, z).
$$

Hence $\rho^*(x, y) * C(\mu)(y) \leq \{\mu(z) \to \rho^*(x, z) \mid z \in X\}$. Thus

$$
\begin{aligned}
\rho^*(x, y) * C(\mu)(y) * \mu(x) &\leq \mu(x) * \wedge\{\mu(z) \to \rho^*(x, z) \mid z \in X\} \\
&= G(\mu, \rho^*)(x).
\end{aligned}
$$

Hence $\rho^*(x, y) * C(\mu)(y) * \mu(x) \leq C(\mu)(x)$ since C is G-rational. Thus the desired result follows by Lemma 2.5.6. ∎

Definition 2.6.8 *Let $C : B \to \mathcal{FP}(X)$ be a fuzzy choice function on (X, B). C is said to satisfy **fuzzy direct relation axiom** (**FDRA**) if $\forall \mu \in B$ and $\forall x, y \in X$, $\rho(x, y) * C(\mu)(y) * \mu(x) \leq C(\mu)(x)$.*

We are now able to establish connections between the above axioms and connections between G-normal and FDRA, FTCCA, WFCA, and FCA.

Theorem 2.6.9 (*Chaudhari and Desai* [11]) *Let C be a fuzzy choice function. If C satisfies WFCA, then C satisfies FDRA.*

Proof. Let $\mu \in B$ and $x \in X$. Since C is normal, there exists $y \in X$ such that $C(\mu)(y) = 1$. Hence $\wedge\{\mu(z) \to \rho^*(x, z) \mid z \in X\} \leq \mu(y) \to \rho^*(x, y)$ by Lemma 2.4.1(3), we have that $\wedge\{\mu(z) \to \rho^*(x, z) \mid z \in X\} \leq \rho^*(x, y)$. Thus

$$
\mu(x) * \wedge\{\mu(z) \to \rho^*(x, z) \mid z \in X\} \leq \rho^*(x, y) * C(\mu)(y) * \mu(x).
$$

Since C satisfies WFCA, we have that

$$
\rho^*(x, y) * C(\mu)(y) * \mu(x) \leq C(\mu)(x).
$$

Therefore, $\mu(x) * \wedge\{\mu(z) \to \rho^*(x, z) \mid z \in X\} \leq C(\mu)(x)$. ∎

Lemma 2.6.10 *Let C be a fuzzy choice function. If C satisfies FTCCA, then C is G-normal.*

Proof. Since $\rho^* \subseteq (\rho^*)^{tc}$, $\mu(y) \to \rho^*(x, y) \leq \mu(y) \to (\rho^*)^{tc}(x, y)$ for all $\mu \in \mathcal{B}$ and $y \in X$. Thus for all $x \in X$,

$$\mu(x) * (\wedge\{\mu(y) \to \rho^*(x, y) \mid y \in X\})$$
$$\leq \mu(x) * (\wedge\{\mu(y) \to (\rho^*)^{tc}(x, y) \mid y \in X\})$$
$$\leq C(\mu)(x).$$

Hence $G(\mu, \rho^*)(x) \leq C(\mu)(x)$ for all $\mu \in \mathcal{B}$ and $x \in X$. Thus $G(\mu, \rho^*) \subseteq C(\mu)$. Hence C is G-normal. ∎

Lemma 2.6.11 *Let C be a fuzzy choice function and ρ be a fuzzy preference relation on X. If ρ is a max-* transitive G-rationalization of C, then $(\rho^*)^{tc} \subseteq \rho$.*

Proof. Now

$$(\rho^*)^{tc}(x, y)$$
$$= \rho^*(x, y) \vee (\vee\{\vee\{\rho^*(x, z_1) * \ldots * \rho^*(z_k, y) \mid z_i \in X, i = 1, \ldots, k\} \mid k \in \mathbb{N}\}).$$

Thus by Lemma 2.5.6,

$$(\rho^*)^{tc}(x, y)$$
$$= \rho(x, y) \vee (\vee\{\vee\{\rho(x, z_1) * \ldots * \rho(z_k, y) \mid z_i \in X, i = 1, \ldots, k\} \mid k \in \mathbb{N}\})$$
$$\leq \rho(x, y) \vee (\vee\{\vee\{\rho(x, y) \mid z_i \in X, i = 1, \ldots, k\} \mid k \in \mathbb{N}\}) = \rho(x, y).$$

Hence $(\rho^*)^{tc} \subseteq \rho$. ∎

Theorem 2.6.12 (*Chaudhari and Desai* [11]) *Let C be a fuzzy choice function. Then C is G-rational with max-* transitive rationalization if and only if C satisfies FTCCA.*

Proof. Suppose that C is G-rational with max-* transitive rationalization. Then by Lemma 2.6.11, $(\rho^*)^{tc} \subseteq \rho$, where ρ is a fuzzy preference relation on X. By Lemma 2.5.1(6), we have for all $y \in X$ that $\mu(y) \to (\rho^*)^{tc}(x, y) \leq \mu(y) \to \rho(x, y)$. Thus $\wedge\{\mu(y) \to (\rho^*)^{tc}(x, y) \mid y \in X\} \leq \wedge\{\mu(y) \to \rho(x, y) \mid y \in X\}$. Hence

$$\mu(x) * (\wedge\{\mu(y) \to (\rho^*)^{tc}(x, y) \mid y \in X\})$$
$$\leq \mu(x) * (\wedge\{\mu(y) \to \rho(x, y) \mid y \in X\})$$
$$= G(\mu, \rho)(x)$$
$$= C(\mu)(x)$$

since C is full rational.

Conversely, suppose C satisfies FTCCA. Let $\rho = (\rho^*)^{tc}$. Then ρ is max-* transitive. Let $\mu \in \mathcal{B}$ and $x \in X$. By the definition of $G(\mu, \rho)$, it follows that

$$G(\mu, \rho)(x) = \mu(x) * (\wedge\{\mu(y) \to \rho(x, y) \mid y \in X\})$$
$$= \mu(x) * (\wedge\{\mu(y) \to (\rho^*)^{tc}(x, y) \mid y \in X\})$$
$$\leq C(\mu)(x).$$

Thus $G(\mu, \rho) \subseteq C(\mu)$ and the desired result follows. ∎

Theorem 2.6.13 *Let C be a fuzzy choice function. If C satisfies WFCA and FTCCA, then C is full rational.*

Proof. By Proposition 2.6.3, ρ^* is reflexive, strongly complete, and transitive. Since C satisfies FTCCA, C is G-rational by Theorem 2.6.12. Thus C is full rational. ∎

Theorem 2.6.14 (*Chaudhari and Desai* [11]) *Let C be a fuzzy choice function. If C satisfies FDRA and ρ^* is max-$*$ transitive, then C satisfies WFCA.*

Proof. Let $\mu \in \mathcal{B}$ and $x, y \in X$. Then for all $z \in X$,

$$\rho^*(x, y) * C(\mu)(y) * \mu(z) \leq \rho^*(x, y) * \rho^*(y, z) \leq \rho^*(x, z).$$

Thus $\rho^*(x, y) * C(\mu)(y) \leq \mu(z) \rightarrow \rho^*(x, z)$. By Lemma 2.3.1(1),

$$\rho^*(x, y) * C(\mu)(y) * \mu(x) \leq \mu(x) * (\wedge\{\mu(z) \rightarrow \rho^*(x, z) \mid z \in X\})$$

and so the desired result holds. ∎

Theorem 2.6.15 *Let C be a fuzzy choice function. If C satisfies FDRA and ρ^* is max-$*$ transitive, then C satisfies conditions $F\alpha$ and $F\beta$.*

Proof. The proof follows from Theorem 2.6.14 and Proposition 2.6.4. ∎

Theorem 2.6.16 *Let C be a fuzzy choice function. Then C is G-rational if and only if C satisfies FDRA.*

Proof. Suppose that C is G-rational with rationalization ρ. By Lemma 2.5.6, $\rho^* \subseteq \rho$. Thus $\mu(y) \rightarrow \rho^*(x, y) \leq \mu(y) \rightarrow \rho(x, y)$ for all $y \in X$. Thus

$$\begin{aligned} \mu(y) * (\wedge\{\mu(y) \;\;\rightarrow\;\; \rho^*(x, y) \mid y \in X\}) &\leq \mu(x) * (\wedge\{\mu(y) \rightarrow \rho(x, y) \mid y \in X\}) \\ &= G(\mu, \rho^*)(x) = C(\mu)(x). \end{aligned}$$

Conversely, suppose C satisfies FDRA. Then $\mu(y) * (\wedge\{\mu(y) \rightarrow \rho^*(x, y) \mid y \in X\}) \leq C(\mu)(x)$ and so $G(\mu, \rho^*)(x) \leq C(\mu)(x)$ for all $\mu \in B$ and $x \in X$. ∎

Theorem 2.6.17 (*Chaudhari and Desai* [11]) *Let C be a fuzzy choice function. Then C satisfies FTCCA if and only if C satisfies FCA.*

Proof. Suppose C satisfies FTCCA. Let $\mu \in \mathcal{B}$ and $x, y, z \in X$. By the definition of ρ^*, $\rho^*(y, z) \geq C(\mu)(y) * \mu(z)$. Thus for all $z \in X$,

$$
\begin{aligned}
(\rho^*)^{tc}(x, y) * C(\mu)(y) * \mu(z) &\leq (\rho^*)^{tc}(x, y) * \rho^*(y, z) \\
&\leq (\rho^*)^{tc}(x, y) * (\rho^*)^{tc}(y, z) \\
&\leq (\rho^*)^{tc}(x, z).
\end{aligned}
$$

Hence $(\rho^*)^{tc}(x, y) * C(\mu)(y) \leq \mu(z) \to (\rho^*)^{tc}(x, z)$ by Lemma 2.5.1(3). Thus $(\rho^*)^{tc}(x, y) * C(\mu)(y) \leq \wedge\{\mu(z) \to (\rho^*)^{tc}(x, z) \mid z \in X\}$. Hence

$$
\begin{aligned}
(\rho^*)^{tc}(x, y) * C(\mu)(y) * \mu(x) &\leq \mu(x) * (\wedge\{\mu(z) \to (\rho^*)^{tc}(x, z) \mid z \in X\}) \\
&\leq C(\mu)(x)
\end{aligned}
$$

since C satisfies FTCCA.

Conversely, suppose C satisfies FCA. Let $\mu \in \mathcal{B}$ and $x \in X$. Since C is normal, there exists $y \in X$ such that $C(\mu)(y) = 1$. Thus $\mu(y) = 1$. Now $\wedge\{\mu(z) \to (\rho^*)^{tc}(x, z) \mid z \in X\} \leq \mu(y) \to (\rho^*)^{tc}(x, y) \leq (\rho^*)^{tc}(x, y)$. Hence

$$
\mu(x) * \wedge(\{\mu(z) \to (\rho^*)^{tc}(x, z) \mid z \in X\}) \leq (\rho^*)^{tc}(x, t) * C(\mu)(t) * \mu(x) \leq C(\mu)(x)
$$

since C satisfies FCA. $\blacksquare$

In this section, we introduced the notions of FDRA and FTCCA given in [1]. They are related to each other in the sense that a fuzzy choice function satisfying FDRA with max-$*$ transitive rationalization satisfies FTCCA and conversely. Also, if a fuzzy choice function satisfies WFCA and FTCCA, then it is full rational.

2.7 Exercises

Let ρ be a fuzzy preference relation on X. Let $tr(\rho)$ denote the transitive closure of ρ. Define the consistent closure of ρ, written $\widehat{\rho}$, by for all $x, y \in X$, $\widehat{\rho}(x, y) = \rho(x, y) \vee (\widehat{\rho}(x, y) * \rho(y, x))$. Let $\pi, \widetilde{\pi}$, and ρ^* be defined as in Chapter 2. Define the fuzzy relation ι on X by $\iota(x, y) = \rho(x, y) * \rho(y, x)$ for all $x, y \in X$. Let WFCA, SFCA, FAA, WAFRP and FDRA, FTCCA, FCCCA, FICA be defined as in Chapter 2.

1. [13] Let C be a fuzzy choice function. Prove that FTCCA $\Rightarrow$ FICA $\Rightarrow$ FCCCA $\Rightarrow$ FDRA .

2. [13] Let C be a fuzzy choice function. Prove that the following statements hold:

   (*i*) FTCCA $\Leftrightarrow$ SFCA;

   (*ii*) FICA $\Rightarrow$ WFCA;

   (*iii*) WFCA $\Rightarrow$ FDRA;

   (*iv*) FICA $\Rightarrow$ WAFRP;

   (*v*) SFCA $\Rightarrow$ FICA.

3. [13] Let C be a fuzzy choice function such that $\mathcal{B}$ is closed under intersection. Prove that the fuzzy Arrow axiom and the weak fuzzy congruence axiom are equivalent for $* = \wedge$.

4. Let $*$ be a continuous t-norm without zero divisors. Let ρ be a complete and $*$-transitive fuzzy preference relation on a finite set X and $(X, \mathcal{B})$ a fuzzy choice space. Prove that $G(\_, \rho)$ is a fuzzy choice function on $(X, \mathcal{B})$.

   Let $*$ be a t-norm. Let complete and strongly complete be defined as in this book.

5. [35] Let ρ be fuzzy preference relation on X. If ρ is acyclic, prove that it is $*$-acyclic for any t-norm that is left continuous or that has no zero divisors.

6. [35] Let ρ be a fuzzy preference relation on a finite set X. Prove that $\vee\{\wedge\{\rho(x, y) \mid y \in X\} \mid x \in X\} > 0$ if either of the following two conditions hold:

   (1) ρ is complete and $*$-acyclic and $*$ has no zero divisors.

   (2) ρ is strongly complete and $*$-acyclic.

7. [35] If a fuzzy preference relation ρ on a finite set X is acyclic, prove that $\vee\{\wedge\{\rho(y, x) \rightarrow \rho(x, y) \mid y \in X\} \mid x \in X\} > 0$.

8. [35] Let ρ be a fuzzy preference relation on a finite set X. Prove that $G(\_, \rho)$ is a fuzzy choice function if either of the following conditions hold:

   (1) ρ is complete and $*$-acyclic, where $*$ is continuous and without zero divisors.

   (2) ρ is strongly complete and $*$-acyclic, where $*$ is continuous.

2.8   References

1. K. J. Arrow, Rational choice functions and ordering, *Economica*, 26(1959) 121–127.

2. D. Austen-Smith and J. S. Banks, *Positive Political Theory I: Collective Preference*, The University of Michigan Press 2000.

3. D. Austen-Smith and J. S. Banks, *Positive Political Theory II: Strategy and Structure*, The University of Michigan Press 2005.

4. A. Banerjee, Fuzzy choice functions, revealed preference and rationality, *Fuzzy Sets and Systems*, 70 (1995) 31–43.

5. C. R. P. Barrett, K. Pattanaik and M. Salles, On the structure of fuzzy social welfare functions, *Fuzzy Sets and Systems,* 19 (1986) 1–11.

6. C. R. Barrett, P. K. Pattanaik and M. Salles, On choosing rationally when preferences are fuzzy, *Fuzzy Sets and Systems,* 34 (1990) 197–212.

7. C. R. Barrett, P. K. Pattanaik and M. Salles, Rationality and aggregation of preferences in an ordinal fuzzy framework, *Fuzzy Sets and Systems,* 49 (1992) 9–13.

8. R. Bělohlávek, *Fuzzy relational systems. Foundations and principles,* (Kluwer, 2002).

9. W. Bossert, Y. Srumont, and K. Suzumura, Rationality of choice functions on general domains without full rationality, *Social Choice and Welfare,* 27 (2006) 435–458.

10. S. R. Chaudhari and S. S. Desai, Transitive and acyclic rationality of fuzzy choice functions, *International Journal of Mathematical Sciences and Engineering Applications,* 1 (2010) 209–224.

11. S. R. Chaudhari and S. S. Desai, On full rationality and congruence axioms of fuzzy choice functions. *International Journal of Computational and Applied Mathematics,* 5 (2010) 313–324.

12. S. R. Chaudhari and S. S. Desai, Congruence axioms and rationality of fuzzy choice functions, *International Journal of Applied Mathematics,* 5 (2010) 843–856.

13. S. R. Chaudhari and S. S. Desai, On interrelations between fuzzy congruence axioms, *New Mathematics and Natural Computation,* 8 (2012) 297-310.

14. T. D. Clark, J. M. Larson, J. N. Mordeson, and M. J. Wierman, *Applying Fuzzy Mathematics to Formal Modeling in Comparative Politics,* (Springer-Verlag, Berlin 2008).

15. T. D. Clark, J. M. Larson, J. N. Mordeson, and M. J. Wierman, Extension of the portfolio allocation model to surplus majority governments: A fuzzy approach, *Public Choice,* 134 (2008) 179–199.

16. S. S. Desai, On quasi-transitive rational fuzzy choice functions, *New Mathematics and Natural Computation,* 10 (2014) 91–102.

17. S. S. Desai and S. R. Chaudhari, On the full rationality of fuzzy choice functions on base domains, *New Mathematics and Natural Computation,* 8 (2012) 183–193.

18. S. S. Desai and S. R. Chaudhari, On interrelations between fuzzy congruence axioms through indicators, *Annals of Fuzzy Mathematics and Informatics,* to appear.

19. B. De Baets and J. Fodor, Twenty years of fuzzy preference relations (1978–1997), *Belgian Journal of Operations Research, Statistics and Computer Science,* 37 (1997) 61–82.

20. J. Fodor and M. Roubens, *Fuzzy Preference Modelling and Multicriteria Decision Support,* (Kluwer, 1994).

21. I. Georgescu, On the axioms of revealed preference in fuzzy consumer theory, *Journal of Systems Science and Systems Engineering,* 13 (2004) 279–296.

22. I. Georgescu, Consistency conditions in fuzzy consumer theory, *Fundamenta Informaticae,* 61 (2004) 223–245.

23. I. Georgescu, Revealed preference, congruence and rationality: A fuzzy approach, *Fundamenta Informaticae,* 65 (2005) 307–328.

24. I. Georgescu, *Fuzzy Choice Functions: A Revealed Preference Approach,* Springer-Berlin, 2007.

25. I. Georgescu, Arrow's axiom and full rationality for fuzzy choice functions *Social Choice and Welfare,* 28 (2007) 303–319.

26. I. Georgescu, Acyclic rationality indicators of fuzzy choice functions *Fuzzy Sets and Systems,* 160 (2009) 2673–2685.

27. I. Georgescu, Similarity of fuzzy choice functions, *Fuzzy Sets and Systems,* 158 (2007) 1314–1326.

28. I. Georgescu and N. Roegen, Choice and revealed preference *Southern Economic Journal,* 21 (1954) 119–130.

29. B. Hansson, Choice structures and preference relations *Synthese,* 18 (1968) 443–458.

30. J. Kacprzyk, M. Fedrizzi, and H. Nurmi, Group decision making and consensus under fuzzy preferences and fuzzy majority, *Fuzzy Sets and Systems* 49 (1992) 21–31.

31. K. P. Klement, R. Mesiar and E. Pap, *Triangular Norms,* (Kluwer, 2000).

32. G. J. Klir and B. Yuan, *Fuzzy Sets and Fuzzy Logic: Theory and Applications,* Prentice Hall, New York, 1995.

33. P. Kulshreshtha and B. Shekar, Interrelationship among fuzzy prefer-
    ence based choice functions and significance of rationality conditions: A
    taxonomic and intuitive perspective, *Fuzzy Sets and Systems*, 109 (2000)
    429–445.

34. E. Laver (ed), *Estimating the Policy Position of Political Actors*, Lon-
    don: Routledge, 2001.

35. D. Martinetti, B, de Baets, S. Diaz, and S. Montes, On the role of
    acyclicity in the study of rationality of fuzzy choice functions, *Fuzzy
    Sets and Systems*, 239 (2014) 35–50.

36. J. N. Mordeson, K. R. Bhutani and T. D. Clark, The rationality of fuzzy
    choice function, *New Mathematics and Natural Computation*, 4 (2008)
    309–327.

37. S. A. Orlovsky, Decision-making with a fuzzy preference relation, *Fuzzy
    Sets and Systems*, 1 (1978) 155–167.

38. C. R. Plott, Path Independence, Rationality, and Social Choice, *Econo-
    metrica*, 41 (1973) 1075–1091.

39. K. T. Poole, *Spatial Models of Parliamentary Voting*, Cambridge: Cam-
    bridge University Press 2005.

40. M. Richter, Revealed preference theory, *Econometrica*, 34 (1966) 635–
    645.

41. P. A. Samuelson, Consumption theory in terms of revealed preference,
    Econometrica, 15 (1948) 243–253.

42. P. A. Samuelson, A note on the pure theory of consumers behavior,
    Economica, 5 (1938) 61–71.

43. A. K. Sen, Quasi-transitivity, rational choice and collective decisions,
    Review of Economic Studies, 36 (1969) 381–393.

44. A. K. Sen, Choice functions and revealed preference, *Review of Economic
    Studies,* 38 (1971) 307–312.

45. A. K. Sen, Social choice theory: A re-examination, *Econometrica,* 45
    (1977) 53–89.

46. A. K. Sen, Internal consistency of choice, *Econometrica* ,61 (1993) 495–
    521.

47. K. Suzumura, Rational choice and revealed preference, *Review of Eco-
    nomic Studies,* 43 (1976) 149–159.

48. K. Suzumura, *Rational Choice, Collective Decisions and Social Welfare,*
    Cambridge University Press, 1983.

49. H. Uzawa, A note on preference and axioms of choice, *Annals of the Institute of Statistical Mathematics*, 8 (1956) 35–40.

50. X. Wang, A note on congruence conditions of fuzzy choice functions, *Fuzzy Sets and Systems*, 145 (2004) 355–358.

Chapter 3

Factorization of Fuzzy Preference Relations

We show that there is an inclusion reversing correspondence between conorms and fuzzy strict preference relations in factorizations of fuzzy preference relations into their strict preference and indifference components. We also associate various conorms with asymmetric components in a factorization of a fuzzy preference relation.

3.1 Basic Definitions and Results

The factorization $R = P \cup I$ of a preference relation R into a strict preference relation P and an indifference relation I is unique in the crisp case. In the fuzzy setting, there are several factorizations of a fuzzy weak preference relation that generalize the crisp case. In Fono and Andjiga [12], classical factorization of a fuzzy relation into a symmetric component (indifference) and an asymmetric component (regular fuzzy strict preference) were generalized. The results were used to obtain fuzzy versions of Gibbard's oligarchy theorem and Arrow's impossibility theorem. Previous fuzzy versions of Gibbard'a oligarchy theorem and Arrow's impossibility theorem were established in Dutta [11] and Richardson [24]. This factorization is important in the examination of Arrowian type results. In Banerjee [3], it was shown using factorization that the fuzzy analog of the general possibility theorem is valid under relatively weak transitivity restrictions. Other works can be found in [4, 5, 8, 9, 17, 18, 19, 26].

Proposition 3.1.1 *Let* $\rho, \pi, \iota \in \mathcal{FR}(X)$. *Suppose that* (*i*) $Supp(\rho) = Supp(\pi) \cup Supp(\iota)$, (*ii*) π *is asymmetric, and* (*iii*) ι *is symmetric. Then the following properties hold.*

(1) $\forall x, y \in X$, $\rho(x, y) > 0$ and $\rho(y, x) = 0$ if and only if $\pi(x, y) > 0$ and $\iota(x, y) = \iota(y, x) = 0$.

(2) $\forall x, y \in X$, $\rho(x, y) > 0$ and $\rho(y, x) > 0$ if and only if $\iota(x, y) = \iota(y, x) > 0$.

Proof. (1) Suppose $\rho(x, y) > 0$ and $\rho(y, x) = 0$. Then $(x, y) \in \mathrm{Supp}(\pi) \cup \mathrm{Supp}(\iota)$ and $(y, x) \notin \mathrm{Supp}(\pi) \cup \mathrm{Supp}(\iota)$. Thus $\iota(x, y) = \iota(y, x) = 0$ and so $\pi(x, y) > 0$. Conversely, suppose $\pi(x, y) > 0$ and $\iota(x, y) = \iota(y, x) = 0$. Then $(x, y) \in \mathrm{Supp}(\rho)$ by (i) and since $\pi(y, x) = 0$, $\rho(y, x) = 0$ by (i).

(2) Suppose $\rho(x, y) > 0$ and $\rho(y, x) > 0$. Then $(x, y), (y, x) \in \mathrm{Supp}(\pi) \cup \mathrm{Supp}(\iota)$. Since not both $\pi(x, y) > 0$ and $\pi(y, x) > 0$, either (x, y) or (y, x) is in $\mathrm{Supp}(\iota)$ and so both $(x, y), (y, x) \in \mathrm{Supp}(\iota)$. Conversely, suppose $\iota(x, y) = \iota(y, x) > 0$. Then by (i), $(x, y), (y, x) \in \mathrm{Supp}(\rho)$. ∎

Since $\mathrm{Supp}(\pi \cup \iota) = \mathrm{Supp}(\pi) \cup \mathrm{Supp}(\iota)$, it follows that $\rho = \pi \cup \iota$ implies $\mathrm{Supp}(\rho) = \mathrm{Supp}(\pi) \cup \mathrm{Supp}(\iota)$. Hence Propositions 3.1.1 and 3.1.2 hold for factorization results to follow.

Proposition 3.1.2 *Let $\rho, \pi, \iota \in \mathcal{FR}(X)$. Suppose that (i) $\mathrm{Supp}(\rho) = \mathrm{Supp}(\pi) \cup \mathrm{Supp}(\iota)$, (ii) π is asymmetric, and (iii) ι is symmetric. Then $(1) \Leftrightarrow (2) \Rightarrow (3)$, where*

(1) $\mathrm{Supp}(\pi) \cap \mathrm{Supp}(\iota) = \emptyset$;

(2) $\forall x, y \in X$, $\pi(x, y) > 0$ implies $\rho(x, y) > 0$ and $\rho(y, x) = 0$;

(3) $\forall x, y \in X$, $\rho(x, y) = \rho(y, x)$ implies $\pi(x, y) = \pi(y, x) = 0$.

Proof. $(1) \Rightarrow (2)$: Let $x, y \in X$. Suppose $\pi(x, y) > 0$. Since $\mathrm{Supp}(\pi) \cap \mathrm{Supp}(\iota) = \emptyset$, $\iota(x, y) = \iota(y, x) = 0$. Suppose $\rho(x, y) > 0$ and $\rho(y, x) > 0$. Then by (i), $\pi(y, x) > 0$, but this is impossible since $\pi(x, y) > 0$. Thus $\pi(y, x) = 0$ and so $\iota(x, y) = \iota(y, x) > 0$ by (i), but this is also impossible. Hence not both $\rho(x, y) > 0$ and $\rho(y, x) > 0$. By (i), $\rho(x, y) > 0$ since $\pi(x, y) > 0$ Thus $\rho(x, y) > 0$ and $\rho(y, x) = 0$.

$(2) \Rightarrow (1)$: Suppose $\mathrm{Supp}(\pi) \cap \mathrm{Supp}(\iota) \neq \emptyset$. Let $(x, y) \in \mathrm{Supp}(\pi) \cap \mathrm{Supp}(\iota)$. Then $\pi(x, y) > 0$ and $\iota(y, x) = \iota(x, y) > 0$. By (2), $\rho(x, y) > 0$ and $\rho(y, x) = 0$. However, $\rho(y, x) = 0$ and $\iota(y, x) > 0$ is impossible by (i). Thus $\mathrm{Supp}(\pi) \cap \mathrm{Supp}(\iota) = \emptyset$.

$(2) \Rightarrow (3)$: Let $x, y \in X$. Suppose $\rho(x, y) = \rho(y, x)$. If either $\pi(x, y) > 0$ or $\pi(y, x) > 0$, then by (2) $\rho(x, y) \neq \rho(y, x)$. Hence $\pi(x, y) = \pi(y, x) = 0$. ∎

Note: $[\iota(x, y) = \iota(y, x) > 0 \Rightarrow \pi(x, y) = \pi(y, x) = 0] \Leftrightarrow \mathrm{Supp}(\pi) \cap \mathrm{Supp}(\iota) = \emptyset$.

Let $\rho \in \mathcal{FR}(X)$ and π be an asymmetric fuzzy binary preference relation associated with ρ. Recall that π is called simple if $\forall x, y \in X$, $\rho(x, y) = \rho(y, x)$ implies $\pi(x, y) = \pi(y, x)$.

As previously mentioned, the first appearance of t-norms appeared in the context of probabilistic metric spaces Schweizer and Sklar [25]. This was followed by their use as an interpretation of the conjunction in the semantics of fuzzy mathematical logics Hajek [14]. The notion of a t-conorm is a natural companion to the concept of a t-norm in that a t-conorm is the interpretation of the disjunction in fuzzy logic. These operators are applied in fuzzy control to formulate assumptions of rules as the fuzzy intersection of fuzzy subsets. There are many ways to model these connectives. The combining of these connectives in logical statements, called data fusion (combining of evidence), is crucial in building expert systems. The types of connectives used are application dependent. For example, connectives used in data fusion are in medical science and geophysics. We next present the definition of a t-conorm.

Definition 3.1.3 *Let* $\cup : [0,1] \times [0,1] \to [0,1]$. *Then* $\cup$ *is called a* t-**conorm** *or simply a* **conorm** *if* $\forall a, b, c \in [0,1]$,
  (1) $\cup(a,0) = a$ *(boundary condition)*;
  (2) $b \le c$ *implies* $\cup(a,b) \le \cup(a,c)$ *(monotonicity)*;
  (3) $\cup(a,b) = \cup(b,a)$ *(commutativity)*;
  (4) $\cup(a, \cup(b,c)) = \cup(\cup(a,b),c)$ *(associativity)*.

Let $\rho, \pi, \iota \in \mathcal{FR}(X)$. Suppose $\cup$ is a conorm. When we write, $\rho = \pi \cup \iota$, we mean $\forall x, y \in X$, $\rho(x,y) = \pi(x,y) \cup \iota(x,y)$.

Proposition 3.1.4 *Let* $\rho, \pi, \iota \in \mathcal{FR}(X)$. *Suppose* $\cup$ *is a conorm. Suppose that* $(i) \rho = \pi \cup \iota$, (ii) π *is asymmetric, and* (iii) ι *is symmetric. Then* $\forall x, y \in X$, $\iota(x,y) = \rho(x,y) \wedge \rho(y,x)$.

Proof. Let $x, y \in X$. Since $\rho = \pi \cup \iota$, $\rho(x,y) \ge \iota(x,y)$ and $\rho(y,x) \ge \iota(y,x)$ by the union axioms. Since ι is symmetric, $\iota(x,y) = \iota(y,x) \le \rho(x,y) \wedge \rho(y,x)$. Since π is asymmetric, either $\pi(x,y) = 0$ or $\pi(y,x) = 0$, say $\pi(y,x) = 0$. Then $\rho(y,x) = \iota(y,x)$. Thus $\iota(x,y) = \rho(x,y) \wedge \rho(y,x)$. ∎

Proposition 3.1.5 *Let* $\rho, \pi, \iota \in \mathcal{FR}(X)$. *Suppose* $\cup$ *is a conorm. Suppose that* (i) $\rho = \pi \cup \iota$, (ii) π *is asymmetric, and* (iii) ι *is symmetric. If* π *is simple, then* $\forall x, y \in X$, $\rho(y,x) \le \rho(x,y)$ *if and only if* $\pi(y,x) = 0$.

Proof. Suppose $\rho(y,x) \le \rho(x,y)$, but $\pi(y,x) > 0$. Then since π is asymmetric, $\pi(x,y) = 0$. Since $\pi(y,x) > \pi(x,y)$ and $\iota(y,x) = \iota(x,y)$, $\rho(y,x) \ge \rho(x,y)$. Hence $\rho(y,x) = \rho(x,y)$ and so $\pi(y,x) = \pi(x,y)$ since π is simple, a contradiction. Thus $\pi(y,x) = 0$. Conversely, suppose $\pi(y,x) = 0$. Then $\rho(y,x) = \iota(y,x) = \rho(x,y) \wedge \rho(y,x)$ and so $\rho(y,x) \le \rho(x,y)$. ∎

Propositions 3.1.4 and 3.1.5 provide intuitively appealing results for π and ι when $\rho = \pi \cup \iota$.

Definition 3.1.6 *Define the conorm* $\cup_2$ *on* $[0,1]$ *by* $\forall a, b \in [0,1]$, $a \cup_2 b = 1 \wedge (a+b)$ *for all* $a, b \in [0,1]$.

The following factorization was given by Orlovsky [22] and characterized by Richardson [24].

Proposition 3.1.7 *Let* $\rho, \pi, \iota \in \mathcal{FR}(X)$. *Suppose that* (i) $\rho = \pi \cup_2 \iota$, (ii) π *is asymmetric, and* (iii) ι *is symmetric. Then* $\forall x, y \in X$, $\pi(x, y) + \iota(x, y) \leq 1 \Leftrightarrow \pi(x, y) = 0 \vee (\rho(x, y) - \rho(y, x))$.

Proof. Let $x, y \in X$. Suppose $\pi(x, y) + \iota(x, y) \leq 1$. Then $\rho(x, y) = (\pi \cup_2 \iota)(x, y) = 1 \wedge (\pi(x, y) + \iota(x, y))$ by (i). By hypothesis, $\rho(x, y) = \pi(x, y) + \iota(x, y)$ and so

$$
\begin{aligned}
\pi(x, y) &= \rho(x, y) - \iota(x, y) = \rho(x, y) - \rho(x, y) \wedge \rho(y, x) \\
&= (\rho(x, y) - \rho(x, y)) \vee (\rho(x, y) - \rho(y, x)) \\
&= 0 \vee (\rho(x, y) - \rho(y, x)).
\end{aligned}
$$

Conversely, suppose $\pi(x, y) = 0 \vee (\rho(x, y) - \rho(y, x))$. Then by Proposition 3.1.4,

$$
\begin{aligned}
&\pi(x, y) + \iota(x, y) \\
= {}& 0 \vee (\rho(x, y) - \rho(y, x)) + \rho(x, y) \wedge \rho(y, x) \\
= {}& \begin{cases} \rho(x, y) - \rho(y, x) + \rho(y, x) = \rho(x, y) \text{ if } \rho(x, y) > \rho(y, x) \\ 0 + \rho(x, y) = \rho(x, y) \text{ if } \rho(x, y) \leq \rho(y, x) \end{cases} \\
\leq {}& 1.
\end{aligned}
$$

∎

The next factorization was given by Ovchinnikov [23] and was characterized by Dutta [11]. We replaced the assumption of connectedness with $\cup = \max$.

Proposition 3.1.8 *Suppose* $\cup = \max$. *Let* $\rho, \pi, \iota \in \mathcal{FR}(X)$. *Suppose that* (i) $\rho = \pi \cup \iota$, (ii) π *is asymmetric, and* (iii) ι *is symmetric. Then* $\forall x, y \in X$,

$$
\begin{aligned}
&[\rho(x, y) = \rho(y, x) \Rightarrow \pi(x, y) = \pi(y, x)] \\
\Leftrightarrow \pi(x, y) = {}& \begin{cases} \rho(x, y) & \text{if } \rho(x, y) > \rho(y, x), \\ 0 & \text{otherwise.} \end{cases}
\end{aligned}
$$

Proof. Let $x, y \in X$. Suppose $\rho(x, y) = \rho(y, x) \Rightarrow \pi(x, y) = \pi(y, x)$. By (i), $\rho(x, y) \geq \pi(x, y)$. Suppose $\rho(x, y) > \rho(y, x)$. Assume $\rho(x, y) > \pi(x, y)$. Then since $\cup = \max$, $\rho(x, y) = \iota(x, y) = \iota(y, x) = \rho(y, x)$, where the last equality holds by Proposition 3.1.4. However, this is a contradiction. Thus $\rho(x, y) = \pi(x, y)$. Now suppose $\rho(x, y) \leq \rho(y, x)$. If $\rho(x, y) = \rho(y, x)$, then $\pi(x, y) = \pi(y, x)$ and so $\pi(x, y) = \pi(y, x) = 0$ by (iii). If $\rho(x, y) < \rho(y, x)$, then it has just been shown that $\pi(y, x) = \rho(y, x) > 0$. Hence $\pi(x, y) = 0$ by (iii).
Conversely, suppose

$$
\pi(x, y) = \begin{cases} \rho(x, y) \text{ if } \rho(x, y) > \rho(y, x), \\ 0 \text{ otherwise.} \end{cases}
$$

Suppose $\rho(x, y) = \rho(y, x)$. Then $\pi(x, y) = \pi(y, x) = 0$ by the hypothesis. ∎

Let $\rho \in \mathcal{FR}(X)$. Recall that ρ is strongly complete if $\forall x, y \in X$, $\rho(x, y) + \rho(y, x) \geq 1$.

Proposition 3.1.9 *Let* $\cup = max$. *Let* $\rho, \pi, \iota \in \mathcal{FR}(X)$. *Suppose that* (i) $\rho = \pi \cup \iota$, (ii) π *is asymmetric, and* (iii) ι *is symmetric. If* ρ *is strongly complete and* $Supp(\pi) \cap Supp(\iota) = \emptyset$, *then*
  (1) $\forall x, y \in X$, $\pi(x, y) > \iota(x, y) \Rightarrow \rho(x, y), \rho(y, x) \in \{0, 1\}$;
  (2) $\forall x, y \in X$, $\iota(x, y) > \pi(x, y) \Rightarrow \rho(x, y) = \rho(y, x) = \iota(x, y) = \iota(y, x)$;
  (3) $\forall x, y \in X$, $\pi(x, y) = \iota(x, y) \Rightarrow \rho(x, y), \rho(y, x) \in \{0, 1\}$.

Proof. Let $x, y \in X$. Then $\rho(x, y) = \pi(x, y) \vee \iota(x, y)$ and $\rho(y, x) = \pi(y, x) \vee \iota(y, x)$.
  (1) Suppose $\pi(x, y) > \iota(x, y)$. Then $\rho(x, y) = \pi(x, y)$. Thus $\iota(x, y) = 0$ by hypothesis. Hence $\iota(y, x) = 0$. By (ii), $\pi(y, x) = 0$. Thus $\rho(y, x) = 0$. Since ρ is strongly complete, $\rho(x, y) + \rho(y, x) \geq 1$. Hence $\rho(x, y) = 1$. Thus $\rho(x, y), \rho(y, x) \in \{0, 1\}$.
  (2) Suppose $\iota(x, y) > \pi(x, y)$. Then by hypothesis, $\pi(x, y) = 0$. Also, $\iota(y, x) = \iota(x, y) > 0$ and so $\pi(y, x) = 0$ by hypothesis. Hence $\rho(x, y) = \iota(x, y) = \iota(y, x) = \rho(y, x)$.
  (3) Finally, suppose $\iota(x, y) = \pi(x, y)$. Then $\rho(x, y) = \pi(x, y) = \iota(x, y) = 0$, where the last equality holds since $Supp(\pi) \cap Supp(\iota) = \emptyset$. Since ρ is strongly complete, $\rho(y, x) = 1$. Thus $\rho(x, y), \rho(y, x) \in \{0, 1\}$. ∎

In the following example, we illustrate the previous proposition.

Example 3.1.10 *Let* $X = \{x, y, z\}$. *Let* ρ *be the fuzzy relation on* X *defined as follows:* $\rho(x, x) = \rho(y, y) = \rho(z, z) = 1, \rho(x, y) = 1, \rho(y, x) = 0, \rho(x, z) = \rho(z, x) = \rho(y, z) = \rho(y, z) = 1/2$. *Let* ι *and* π *be as in Propositions 3.1.4 and 3.1.9, respectively. Then* $\pi(x, y) = 1$ *and* $\pi(u, v) = 0$ $\forall (u, v) \in X \times X \backslash \{(x, y)\}$ *and* $\iota(x, y) = \iota(y, x) = 0, \iota(u, v) = 1/2$ $\forall (u, v) \in X \times X \backslash \{(x, y), (y, x)\}$. *It is easily seen that* ρ *is strongly complete and that* $Supp(\pi) \cap Supp(\iota) = \emptyset$.

Banerjee proved the next factorization result [3].

Proposition 3.1.11 *Let* $\rho, \pi, \iota \in \mathcal{FR}(X)$. *Suppose that* (i) $\rho = \pi \cup_2 \iota$, (ii) π *is asymmetric, and* (iii) ι *is symmetric. If* (iv) $\forall x, y \in X$, $\rho(x, y) < 1$ *implies* $\pi(y, x) > 0$ *and* (v) $\pi(x, y) + i(x, y) \leq 1$, *then* $\forall x, y \in X$, $\pi(x, y) = 1 - \rho(y, x)$ *and* $\iota(x, y) = \rho(x, y) \wedge \rho(y, x)$.

Proof. Let $x, y \in X$. It follows that $\cup_2$ satisfies the union axioms. Hence $i(x, y) = \rho(x, y) \wedge \rho(y, x)$ by Proposition 3.1.4. By (i) and (v), it follows that $\rho(x, y) = \pi(x, y) + \iota(x, y)$. Suppose $\rho(y, x) = 1$. Then $\rho(x, y) \leq \rho(y, x)$. Thus by Proposition 3.1.5, $\pi(x, y) = 0$. Hence $\rho(x, y) = \iota(x, y)$. Thus $\pi(x, y) = 1 - \rho(y, x)$. Suppose $\rho(y, x) < 1$. Then by (iv), $\pi(x, y) > 0$. Thus $\pi(y, x) = 0$. By (iv), $\rho(x, y) = 1$. Thus $\rho(x, y) > \rho(y, x)$ and so $\iota(x, y) = \rho(y, x)$. Hence

$1 = \rho(x,y) = \pi(x,y) + \iota(x,y) = \pi(x,y) + \rho(y,x)$. Thus $\pi(x,y) = 1 - \rho(y,x)$.

∎

3.2   Quasi-Subtraction

Due to the nature of the results to follow, we often use the notation $\oplus$ for a t-norm rather than $\cup$. Let $\rho \in \mathcal{FR}(X)$. Recall that ρ is called a fuzzy weak preference relation (FWPR) if ρ is reflexive and complete.

We call $\rho = \pi \oplus \iota$, where π is an asymmetric fuzzy relation and ι is a symmetric fuzzy relation, a **factorization** of ρ. We will see that under very mild assumptions, a factorization of ρ into $\pi \oplus \iota$ is such that $\iota(x,y) = \rho(x,y) \wedge \rho(y,x)$ and that π will be of a known type depending on the definition of $\oplus$. Consequently, a better understanding of ρ is determined.

Let $\oplus$ be a continuous t-conorm. The **quasi-subtraction** of $\oplus$, denoted $\ominus$, is the function from $[0,1]^2$ into $[0,1]$ defined by $\forall a, b \in [0,1]$,

$$a \ominus b = \wedge\{t \in [0,1] \mid a \oplus t \geq b\}.$$

The following result is crucial to the factorization results.

Proposition 3.2.1 (*Fono and Andjiga* [12]) *Let ρ be a FWPR. Let $x, y \in X$. Suppose $\rho(x,y) > \rho(y,x)$. Then the following statements hold.*
(1) *The equation*

$$\rho(x,y) = \rho(y,x) \oplus t, \ t \in [0,1] \tag{3.1}$$

has a solution for t. The real number $\rho(y,x) \ominus \rho(x,y)$ is the smallest solution.
(2) *If $\oplus = \max$ or $\oplus$ is a strict t-conorm, then $\rho(y,x) \ominus \rho(x,y)$ is the unique solution.*

Proof. (1) Let $g : [0,1] \to [0,1]$ be such that g is continuous, $g(0) = \rho(y,x) \oplus 0 = \rho(y,x)$ and $g(1) = \rho(y,x) \oplus 1 = 1$. Then there exists $t_0 \in [\rho(y,x), 1]$ such that $g(t_0) = \rho(x,y)$ by the Intermediate Value Theorem. Hence Eq. (3.1) has a solution, namely t_0. Let $t^* = \wedge\{t \in [0,1] \mid \rho(y,x) \oplus t \geq \rho(x,y)\} = \rho(y,x) \ominus \rho(x,y)$. Then $\rho(y,x) \oplus t^* \geq \rho(x,y)$ and $\rho(y,x) \oplus t_0 = \rho(x,y)$, where t_0 is any solution to Eq. (3.1). Thus $t^* \leq t_0$. That is, $\rho(y,x) \ominus \rho(x,y)$ is the smallest solution.
(2) Suppose $\oplus = \max$. Then $\rho(y,x) \ominus \rho(x,y) = \wedge\{t \in [0,1] \mid \rho(y,x) \vee t \geq \rho(x,y)\} = \rho(x,y)$. Eq. (3.1) becomes $\rho(x,y) = \rho(y,x) \vee t$ from which the uniqueness clearly follows.
Suppose $\oplus$ is strict. Then there exists unique t_0 such that $\rho(y,x) \oplus t_0 = \rho(x,y)$. ∎

The following result is essentially due to Fono and Andjiga [12].

Proposition 3.2.2 *Let ρ be FWPR on X and let π and ι be fuzzy binary relations on X. Then statements (1) and (2) are equivalent.*

(1) $\forall x, y \in X$, $\rho(x, y) = \pi(x, y) \oplus \iota(x, y)$, π is asymmetric, ι is symmetric, and π is simple.

(2) $\forall x, y \in X$,

(i) $\iota(x, y) = \rho(x, y) \wedge \rho(y, x)$,

(ii) $\rho(x, y) \leq \rho(y, x) \Leftrightarrow \pi(x, y) = 0$,

(iii) $\rho(x, y) > \rho(y, x) \Rightarrow \pi(x, y)$ is a solution to Eq. (3.1).

Proof. Suppose (1) holds. Then statement (i) follows from Proposition 3.1.4 and statement (ii) follows from Proposition 3.1.5. For (iii), suppose $\rho(x, y) > \rho(y, x)$. Then $\iota(x, y) = \rho(y, x)$. By hypothesis, $\rho(x, y) = \pi(x, y) \oplus \rho(y, x)$. Thus $\pi(x, y)$ is a solution to Eq. (3.11).

Suppose (2) holds. By (i), ι is symmetric. By (ii), π is asymmetric. ($\pi(x, y) > 0 \Leftrightarrow \rho(x, y) > \rho(y, x) \Rightarrow \pi(y, x) = 0$.) Suppose $\rho(x, y) = \rho(y, x)$. Then $\pi(x, y) = 0 = \pi(y, x)$ by (ii). Thus π is simple.

Suppose $\rho(y, x) \geq \rho(x, y)$. Then $\pi(x, y) \oplus \iota(x, y) = 0 \oplus \iota(x, y) = 0 \oplus \rho(x, y) = \rho(x, y)$ by (i) and (ii).

Suppose $\rho(y, x) < \rho(x, y)$. Then $\pi(x, y)$ is a solution to Eq. (3.1). Hence $\rho(x, y) = \rho(y, x) \oplus \pi(x, y) = \iota(x, y) \oplus \pi(x, y)$. ∎

Corollary 3.2.3 *Let ρ be FWPR on X and let π and ι be fuzzy binary relations on X. Let $\oplus$ be a strict t-conorm or $\oplus = \max$. Then statements (1) and (2) are equivalent.*

(1) $\forall x, y \in X$, $\rho(x, y) = \pi(x, y) \oplus \iota(x, y)$, π is asymmetric, ι is symmetric, and π is simple.

(2) $\forall x, y \in X$, (i) $\iota(x, y) = \rho(x, y) \wedge \rho(y, x)$ and (ii) $\pi(x, y) = \rho(y, x) \ominus \rho(x, y)$.

Proof. By Proposition 3.2.2, it suffices to show that (2) of Proposition 3.2.2 is equivalent to (2) of Corollary 3.2.3. By the hypothesis concerning $\oplus$ and Proposition 3.2.1, $\rho(y, x) \ominus \rho(x, y)$ is the unique solution to Eq. (3.1). Assume (2) of Corollary 3.2.3 holds. Then $\pi(x, y) = \rho(y, x) \ominus \rho(x, y)$. Now $\pi(x, y) > 0$ if and only if $\rho(x, y) > \rho(y, x)$ by the definition of $\ominus$. Thus (2) of Proposition 3.2.2 holds. Assume (2) of Proposition 3.2.2 holds. Then $\pi(x, y)$ is a solution to Eq. (3.11). Hence $\pi(x, y) = \rho(y, x) \ominus \rho(x, y)$ since Eq. (3.1) has a unique solution. Thus (2) of Corollary 3.2.3 holds. ∎

Let $\cup_d : [0, 1]^2 \to [0, 1]$ be defined by $\forall a, b \in [0, 1]$,

$$\cup_d(a, b) = \begin{cases} a \text{ if } b = 0, \\ b \text{ if } a = 0, \\ 1 \text{ otherwise.} \end{cases}$$

Then $\cup_d$ is called the **drastic t-conorm**.

Let $\cup_A : [0,1]^2 \to [0,1]$ be defined by $\forall a, b \in [0,1]$, $\cup_A(a,b) = a + b - ab$. Then $\cup_A$ is a t-conorm called the **algebraic sum**.

Let $\cup_Y : [0,1]^2 \to [0,1]$ be defined by $\forall a, b \in [0,1]$, $\cup_Y(a,b) = 1 \wedge (a^\omega + b^\omega)^{1/\omega}$, where $\omega > 0$. Then $\cup_Y$ is a t-conorm in the Yager class.

In general, suppose $\rho(x,y) = \rho(y,x)$. Then $\rho(x,y) = \rho(x,y) \oplus t$ and $t = 0$ is a solution. Thus if $g(t) = \rho(x,y) \oplus t$ is strictly increasing in t, then $t = 0$ is the only solution to Eq. (3.11). Hence we would define $\pi(x,y) = 0$ in this case.

Proposition 3.2.4 *Let $\oplus = \cup_A$. Then the solution to Eq. (3.1) is given as follows:* $\forall x, y \in X$,

$$\pi(x,y) = \begin{cases} \frac{\rho(x,y) - \rho(y,x)}{1 - \rho(y,x)} & \text{if } \rho(x,y) > \rho(y,x), \\ 0 & \text{otherwise.} \end{cases}$$

We let $\pi = \pi_{(4)}$ in this case.

Proof. Let $x, y \in X$. Suppose $\rho(x,y) > \rho(y,x)$. Then $\rho(x,y) = \rho(y,x) \oplus t = \rho(y,x) + t - \rho(y,x)t$. Thus $\rho(x,y) - \rho(y,x) = t(1 - \rho(y,x))$. Hence the desired result now follows easily. ∎

Proposition 3.2.5 *Let $\oplus = \cup_Y$. Then the solution to Eq. (3.1) is given as follows:* $\forall x, y \in X$,

$$\pi(x,y) = \begin{cases} (\rho(x,y)^\omega - \rho(y,x)^\omega)^{1/\omega} & \text{if } \rho(x,y) > \rho(y,x), \\ 0 & \text{otherwise.} \end{cases}$$

We let $\pi = \pi_{(5)}$ in this case.

Proof. Let $x, y \in X$. Suppose $\rho(x,y) > \rho(y,x)$. Then $\rho(x,y) = \rho(y,x) \oplus t = 1 \wedge (\rho(y,x)^\omega + t^\omega)^{1/\omega}$. Suppose $(\rho(y,x)^\omega + t^\omega)^{1/\omega} \le 1$. Then $\rho(x,y) = (\rho(y,x)^\omega + t^\omega)^{1/\omega}$. Hence $\rho(x,y)^\omega = \rho(y,x)^\omega + t^\omega$ and so $t = (\rho(x,y)^\omega - \rho(y,x)^\omega)^{1/\omega}$. Thus the desired result follows easily. ∎

Proposition 3.2.6 *Let ρ be a FWPR. Let $\cup_d$ denote the drastic conorm. Then ρ has a factorization $\rho = \pi_{(0)} \cup_d \iota$ if and only if $\forall x, y \in X$, $x \ne y$, either $\rho(x,y) = \rho(y,x)$ or not both $\rho(x,y) > 0$, $\rho(y,x) > 0$.*

Proof. Suppose $\rho = \pi \cup_d \iota$. Let $x, y \in X$ be such that $x \ne y$. Since ρ is complete, either $\rho(x,y) > 0$ or $\rho(y,x) > 0$, say $\rho(x,y) > 0$. Suppose $\rho(y,x) > 0$. Then $\pi_{(0)}(x,y) = 0 = \pi_{(0)}(y,x)$. Thus $\rho(x,y) = \iota(x,y)$ and $\rho(y,x) = \iota(y,x)$. Hence $\rho(x,y) = \rho(y,x)$.

Conversely, suppose either $\rho(x,y) = \rho(y,x)$ or not both $\rho(x,y) > 0$ and $\rho(y,x) > 0$. Suppose $\rho(x,y) = \rho(y,x)$. Then $\pi_{(0)}(x,y) = 0 = \pi_{(0)}(y,x)$. Thus $0 \cup_d \iota(x,y) = \rho(x,y) = \rho(y,x)$. Hence $\rho = \pi_{(0)} \cup_d \iota$. Suppose not both $\rho(x,y) >$

0 and $\rho(y, x) > 0$. Since ρ is complete either $\rho(x, y) > 0$ or $\rho(y, x) > 0$, say $\rho(x, y) > 0$. Thus $\rho(y, x) = 0$ and so $\pi_{(0)}(x, y) = \rho(x, y)$ and $\iota(x, y) = 0$. Hence $\rho(x, y) = \pi_{(0)}(x, y) \cup_d \iota(x, y)$ and $\rho(y, x) = 0 = 0 \cup_d 0 = \pi_{(0)}(y, x) \cup_d \iota(y, x)$.

∎

We present the next result in this particular form because the notation is needed in later chapters.

Proposition 3.2.7 *Let ρ be a FWPR. Let π be a solution to Eq. (3.1). Then the following statements hold.*

(1) *If $\oplus = \cup_d$, then $\pi = \pi_{(0)}$.*
(2) *If $\oplus = \vee$, then $\pi = \pi_{(1)}$.*
(3) *If $\oplus = \cup_2$, then $\pi = \pi_{(3)}$.*
(4) *If $\oplus = \cup_2$, and (iv) and (v) of Proposition 3.1.11 hold, then $\pi = \pi_{(2)}$.*
(5) *If $\oplus = \cup_A$, then $\pi = \pi_{(4)}$.*
(6) *If $\oplus = \cup_Y$, then $\pi = \pi_{(5)}$.*

Proof. (1) Let $x, y \in X$. Suppose $\rho(y, x) < \rho(x, y)$. Suppose that $\rho(y, x) > 0$. Hence $\wedge\{t \in [0, 1] \mid \rho(y, x) \cup_d t \geq \rho(x, y)\}$ does not exist since $\forall t > 0$, $\rho(y, x) \cup_d t = 1$ and for $t = 0, \rho(y, x) \cup_d 0 = \rho(y, x) < \rho(x, y)$. Thus $\wedge\{t \in [0, 1] \mid \rho(y, x) \cup_d t \geq \rho(x, y)\}$ exists and equals $\rho(x, y)$ if and only if $\rho(y, x) = 0$. Hence $\pi = \pi_{(0)}$.

(2) For all $x, y \in X$, $\pi(x, y) = \wedge\{t \in [0, 1] \mid \rho(y, x) \vee t \geq \rho(x, y)\}$

$$= \begin{cases} \rho(x, y) \text{ if } \rho(x, y) > \rho(y, x), \\ 0 \text{ if } \rho(y, x) \geq \rho(x, y). \end{cases}$$

Hence $\pi = \pi_{(1)}$.

(3) For all $x, y \in X$,

$$\begin{aligned} \pi(x, y) &= \wedge\{t \in [0, 1] \mid \rho(y, x) \cup_2 t \geq \rho(x, y)\} \\ &= \wedge\{t \in [0, 1] \mid 1 \wedge (\rho(y, x) + t) \geq \rho(x, y)\} = t^*, \end{aligned}$$

where $t^* = 0 \vee (\rho(x, y) - \rho(y, x))$. Hence $\pi = \pi_{(3)}$.

(4) The desired result follows from Proposition 3.1.11.
(5) The desired result follows from Proposition 3.2.4.
(6) The desired result follows from Proposition 3.2.5. ∎

Let ρ be a FWPR and π an asymmetric fuzzy binary relation associated with ρ. Recall that π is said to be regular if $\forall x, y \in X$, $\pi(x, y) > 0$ if and only if $\rho(x, y) > \rho(y, x)$.

It follows easily that $\pi_{(1)}, \pi_{(3)}, \pi_{(4)}$, and $\pi_{(5)}$ are regular.

Let $\cup_n : [0, 1]^2 \rightarrow [0, 1]$ be defined by $\forall a, b \in [0, 1]$,

$$\cup_n(a, b) = \begin{cases} a \vee b \text{ if } a + b < 1, \\ 1 \text{ otherwise} \end{cases}.$$

Then $\cup_n$ is a t-conorm called the **nilpotent maximum conorm**.

Let $\cup_{H_2} : [0,1]^2 \to [0,1]$ be defined by $\forall a, b \in [0,1]$,

$$\cup_2(a,b) = (a+b)/(1+ab).$$

Then $\cup_{H_2}$ is a t-conorm called the **Einstein sum**.

Let $\cup_p : [0,1]^2 \to [0,1]$ be defined by $\forall a, b \in [0,1]$,

$$\cup_p(a,b) = 1 - 0 \vee ((1-a)^p + (1-b)^p - 1)^{\frac{1}{p}}, \quad p \neq 0.$$

Then $\cup_A$ is a t-conorm called the **Schweizer and Sklar conorm**.

Let $\cup_w : [0,1]^2 \to [0,1]$ be defined by $\forall a, b \in [0,1]$,

$$\cup_w(a,b) = 1 \wedge (a^w + b^w)^{\frac{1}{w}}, \quad w > 0.$$

Then $\cup_A$ is a t-conorm called the **Yager conorm**.

Let $\cup_s : [0,1]^2 \to [0,1]$ be defined by $\forall a, b \in [0,1]$,

$$\cup_s(a,b) = 1 - \log_s \left(1 + \frac{(s^{1-a}-1)(s^{1-b}-1)}{s-1}\right), \quad s > 0, \ s \neq 1.$$

Then $\cup_s$ is a t-conorm called the **Frank conorm**.

Proposition 3.2.8 *Let ρ be a FWPR. Let $\cup_n$ denote the nilpotent conorm. Then ρ has a factorization $\rho = \pi \cup_n \iota$ if and only if $\forall x, y \in X$, either (1) $\rho(x,y) = 1$ or $\rho(y,x) = 1$, or (2) $\rho(x,y) = \rho(y,x)$, or (3) $\rho(x,y) + \rho(y,x) < 1$ and $\rho(x,y) \neq \rho(y,x)$.*

Proof. Suppose not (1) or (2) or (3). Then there exists $x, y \in X$ such that $\rho(x,y) \neq \rho(y,x), \rho(x,y) < 1, \rho(y,x) < 1$, and $\rho(x,y) + \rho(y,x) \geq 1$. Suppose $\rho(x,y) > \rho(y,x)$. Then $\iota(x,y) = \rho(y,x)$. Suppose there exists π such that $\rho = \pi \cup_n \iota$. Then $\rho(x,y) = \pi(x,y) \cup_n \iota(x,y) \neq 1$ and so $\pi(x,y) = \rho(x,y)$. However, this is impossible since $\pi(x,y) \cup_n \iota(y,x) \geq 1$ and so $\rho(x,y) = 1$. Thus no such π exists.

Conversely, suppose $\rho(x,y) = \rho(y,x)$. Then let $\pi(x,y) = 0$ and we have $\rho(x,y) = 0 \cup_n \rho(x,y) = \pi(x,y) \cup_n \iota(x,y)$. Suppose $\rho(x,y) = 1$ or $\rho(y,x) = 1$ and $\rho(x,y) \neq \rho(y,x)$, say $\rho(x,y) = 1$. Then let $\pi(x,y) = 1$ and $\pi(y,x) = 0$. Then $\rho(x,y) = 1 = 1 \cup_n \rho(y,x) = \pi(x,y) \cup_n \iota(x,y)$. Suppose $\rho(x,y) + \rho(y,x) < 1$ and $\rho(x,y) \neq \rho(y,x)$, say $\rho(x,y) > \rho(y,x)$. Then let $\pi(x,y) = \rho(x,y)$ and so $\rho(x,y) = \rho(x,y) \cup_n \rho(y,x) = \pi(x,y) \cup_n \iota(x,y)$. We can let $\pi(y,x) \in [0, \rho(y,x)]$ here. ∎

We next determine results for π similar to those in Proposition 3.2.7 for other conorms.

Proposition 3.2.9 *Let ρ be a FWPR. Let π be a solution to Eq. (3.1). Then the following statements hold.*

(1) *If $\oplus = \cup_{H_2}$, then π is defined by $\forall x, y \in X$,*

$$\pi(x,y) = \begin{cases} \frac{\rho(x,y) - \rho(y,x)}{1 + \rho(x,y)(\rho(y,x))} & \text{if } \rho(x,y) > \rho(y,x), \\ 0 & \text{otherwise.} \end{cases}$$

(2) *If $\oplus = \cup_p$, then π is defined by $\forall x, y \in X$,*

$$\pi(x,y) = \begin{cases} 1 - [(1 - \rho(x,y))^p - (1 - \rho(y,x))^p + 1]^{\frac{1}{p}} & \text{if } \rho(x,y) > \rho(y,x), \\ 0 & \text{otherwise.} \end{cases}$$

(3) *If $\oplus = \cup_w$, then π is defined by $\forall x, y \in X$,*

$$\pi(x,y) = \begin{cases} (\rho(x,y)^w - \rho(y,x)^w)^{\frac{1}{w}} & \text{if } \rho(x,y) > \rho(y,x), \\ 0 & \text{otherwise.} \end{cases}$$

(4) *If $\oplus = \cup_s$, then π is defined by $\forall x, y \in X$,*

$$\pi(x,y) = 1 - \log_s \left[\frac{(s-1)(s^{1-\rho(x,y)} - 1)}{s^{1-\rho(y,x)} - 1} + 1 \right].$$

Proof. (1) Let $x, y \in X$. Define π as in (1). Suppose $\rho(x,y) > \rho(y,x)$. Then

$$\pi(x,y) \cup_{H_2} \iota(x,y)$$
$$= \left[\frac{\rho(x,y) - \rho(y,x)}{1 + \rho(x,y)\rho(y,x)} + \rho(y,x) \right] \Big/ \left[1 + \frac{\rho(x,y) - \rho(y,x)}{1 + \rho(x,y)\rho(y,x)} \rho(y,x) \right]$$
$$= \frac{\rho(x,y) - \rho(y,x) + \rho(y,x) - \rho(y,x)\rho(x,y)\rho(y,x)}{1 - \rho(y,x)\rho(x,y)}$$
$$\bullet \frac{1 - \rho(y,x)\rho(x,y)}{1 - \rho(y,x)\rho(x,y) - \rho(x,y)\rho(y,x) - \rho(y,x)\rho(y,x)}$$
$$= \frac{\rho(x,y)(1 - \rho(y,x)\rho(y,x))}{1 - \rho(y,x)\rho(y,x)} = \rho(x,y).$$

Suppose $\rho(x,y) \leq \rho(y,x)$. Then $\pi(x,y) \cup_{H_2} \iota(x,y) = 0 \cup_{H_2} \rho(x,y) = \rho(x,y)$.

(2) Let $x, y \in X$. Define π as in (2). Suppose $\rho(x,y) > \rho(y,x)$. Then

$$\pi(x,y) \cup_p \iota(x,y)$$
$$= 1 - [(1 - \pi(x,y))^p + (1 - \rho(y,x))^p - 1]^{\frac{1}{p}}$$
$$= 1 -$$
$$\{[1 - [1 - (1 - \rho(x,y))^p - (1 - \rho(y,x))^p + 1]^{\frac{1}{p}}]^p + (1 - \rho(y,x))^p - 1\}^{\frac{1}{p}}.$$

We show this latter expression is $\rho(x,y)$. We have $[(1 - \rho(x,y))^p - (1 - \rho(y,x))^p + 1]^{\frac{1}{p}} = 1 - (1 - [(1 - \rho(x,y))^p - (1 - \rho(y,x))^p + 1]^{\frac{1}{p}}$ and so

$[(1-\rho(x,y))^p-(1-\rho(y,x))^p+1] = \{1-(1-[(1-\rho(x,y))^p-(1-\rho(y,x))^p+1]^{\frac{1}{p}}\}^p$.
Hence $(1-\rho(x,y))^p = \{1-(1-[(1-\rho(x,y))^p-(1-\rho(y,x))^p+1]^{\frac{1}{p}}\}^p + (1-\rho(y,x))^p - 1$. Solving this latter equation for $\rho(x,y)$ yields the desired result.
Suppose $\rho(x,y) \leq \rho(y,x)$. Then $\pi(x,y) \cup_p \iota(x,y) = 0 \cup_p \rho(x,y) = \rho(x,y)$.

(3) Let $x,y \in X$. Define π as in (3). Suppose $\rho(x,y) > \rho(y,x)$. Then $\pi(x,y) \cup_\omega \iota(x,y)$

$$
\begin{aligned}
&= \{[(\rho(x,y)^\omega - \rho(y,x)^\omega)^{\frac{1}{\omega}}]^\omega + \rho(y,x)^\omega\}^{\frac{1}{\omega}} \\
&= \rho(x,y).
\end{aligned}
$$

Suppose $\rho(x,y) \leq \rho(y,x)$. Then $\pi(x,y) \cup_\omega \iota(x,y) = 0 \cup_\omega \rho(x,y) = \rho(x,y)$.

(4) Let $x,y \in X$. Define π as in (4). Suppose $\rho(x,y) > \rho(y,x)$. Then $\pi(x,y) \cup_s \iota(x,y)$

$$
= 1 - \log_s \left(1 + \frac{(s^{1-\pi(x,y)}-1)(s^{1-\rho(y,x)}-1)}{s-1}\right).
$$

Let $u = \pi(x,y) \cup_s \iota(x,y)$. Then $1 - u = \log_s \left(1 + \frac{(s^{1-\pi(x,y)}-1)(s^{1-\rho(y,x)}-1)}{s-1}\right)$
and so $s^{1-u} = 1 + \frac{(s^{1-\pi(x,y)}-1)(s^{1-\rho(y,x)}-1)}{s-1}$. Hence $s^{1-u} - 1 = \frac{(s^{1-\pi(x,y)}-1)(s^{1-\rho(y,x)}-1)}{s-1}$. Thus $(s-1)(s^{1-u}-1)/(s^{1-\rho(y,x)}-1) = s^{1-\pi(x,y)}-1$.
Hence $1 + (s-1)(s^{1-u}-1)/(s^{1-\rho(y,x)}-1) = s^{1-\pi(x,y)}$ and so $\pi(x,y) = 1 - \log_s \left(1 + \frac{(s-1)(s^{1-u}-1)}{s^{1-\rho(y,x)}-1}\right)$. Thus $u = \rho(x,y)$ by the definition of π. ∎

Henceforth all conorms are assumed to be continuous.

Proposition 3.2.10 *Let $\mathcal{C} = \{\cup \mid \cup$ is a continuous conorm$\}$ and let $\mathcal{A} = \{\rho \in \mathcal{FR}(X) \mid \rho$ is asymmetric$\}$. Then there exists a function f of $\mathcal{C} \times \mathcal{FR}(X)$ into $\mathcal{A}$ such that if $f(\cup, \rho) = f(\cup_1, \rho)$ $\forall \rho \in \mathcal{FR}(X)$, then $\cup = \cup_1$. Furthermore, $\mathrm{Im}(f) = \{\pi^\ominus \in \mathcal{A} \mid \exists (\cup, \rho) \in \mathcal{C} \times \mathcal{FR}(X), \rho = \pi^\ominus \cup \iota\}$.*

Proof. Let $(\cup, \rho) \in \mathcal{C} \times \mathcal{FR}(X)$ By Proposition 3.2.1, $\exists \pi^\ominus \in \mathcal{A}$ such that $\rho = \pi^\ominus \cup \iota$. Define $f : \mathcal{C} \times \mathcal{FR}(X) \to \mathcal{A}$ by $\forall (\cup, \rho) \in \mathcal{C} \times \mathcal{FR}(X), f((\cup, \rho)) = \pi^\ominus$, where $\rho = \pi^\ominus \cup \iota$. Clearly the domain of f is $\mathcal{C} \times \mathcal{FR}(X)$. Suppose $\cup = \cup_1$. Then $\forall x,y \in X, \pi^\ominus(x,y) = \wedge\{t \in [0,1] \mid t \cup \rho(y,x) \geq \rho(x,y)\} = \pi_1^\ominus(x,y)$, where the latter equality holds since $\cup = \cup_1$. Hence f is well defined.
Suppose that $f(\cup, \rho) = f(\cup_1, \rho)$ $\forall \rho \in \mathcal{FR}(X)$. Let $a,b \in [0,1]$. Let $x,y \in X$ and $\rho \in \mathcal{FR}(X)$ be such that $\rho(x,y) = a \cup b$ and $\rho(y,x) = b$. Let $t^* = \wedge\{t \in [0,1] \mid t \cup b \geq a \cup b\}$. Then $t^* = \wedge\{t \in [0,1] \mid t \cup b = a \cup b\}$. Hence $t^* \leq a$.
By hypothesis, $\rho(x,y) = \pi^\ominus(x,y) \cup \iota(x,y) = \pi^\ominus(x,y) \cup_1 \iota(x,y)$. Thus since $\pi^\ominus(x,y) = t^*, \rho(x,y) = a \cup b = t^* \cup b = t^* \cup_1 b \leq a \cup_1 b$, where the inequality holds since $t^* \leq a$. Hence $a \cup b \leq a \cup_1 b$. By symmetry, $a \cup_1 b \leq a \cup b$. Thus $a \cup b = a \cup_1 b$. ∎

Let ρ_1 and ρ_2 be fuzzy preference relations and let $\cup_1$ and $\cup_2$ be conorms. We now determine under which two factorizations of ρ, say $\rho = \pi_1 \cup_1 \iota =$

$\pi_2 \cup_2 \iota$ have the property that $\pi_1 \subseteq \pi_2$ for all such factorizations if and only if $\cup_1 \supseteq \cup_2$.

Proposition 3.2.11 *Let $\cup_1$ and $\cup_2 \in \mathcal{C}$. Let $\rho \in \mathcal{FR}(X)$. Suppose that $\rho = \pi_1 \cup_1 \iota = \pi_2 \cup_2 \iota$. Suppose further that either $\cup_1$ or $\cup_2$ is strict. If $\cup_1 \supseteq \cup_2$, then $\pi_1 \subseteq \pi_2$.*

Proof. Suppose that there exist $x, y \in [0, 1]$ such that $\pi_1(x, y) > \pi_2(x, y)$. Since $a \cup_1 b \geq a \cup_2 b$ for all $a, b \in [0, 1]$, we have

$$\pi_1(x, y) \cup_1 \iota(x, y) \geq \pi_1(x, y) \cup_2 \iota(x, y) \text{ and } \pi_2(x, y) \cup_1 \iota(x, y) \geq \pi_2(x, y) \cup_2 \iota(x, y).$$

Suppose $\cup_1$ is strict. Then $\pi_1(x, y) \cup_1 \iota(x, y) > \pi_2(x, y) \cup_1 \iota(x, y)$. Hence

$$
\begin{aligned}
\rho(x, y) &= \pi_1(x, y) \cup_1 \iota(x, y) > \pi_2(x, y) \cup_1 \iota(x, y) \\
&\geq \pi_2(x, y) \cup_2 \iota(x, y) = \rho(x, y)
\end{aligned}
$$

which is impossible. Thus no such x, y exist. Hence $\pi_1 \subseteq \pi_2$.

Suppose $\cup_2$ strict. Then $\pi_1(x, y) \cup_2 \iota(x, y) > \pi_2(x, y) \cup_2 \iota(x, y)$. Thus

$$
\begin{aligned}
\rho(x, y) &= \pi_2(x, y) \cup_2 \iota(x, y) < \pi_1(x, y) \cup_2 \iota(x, y) \\
&\leq \pi_1(x, y) \cup_1 \iota(x, y) = \rho(x, y)
\end{aligned}
$$

which is impossible. Hence no such x, y exist. Thus $\pi_1 \subseteq \pi_2$. ■

Proposition 3.2.12 *Let $\cup_1$ and $\cup_2 \in \mathcal{C}$. Suppose for all $\rho \in \mathcal{FR}(X)$ that $\rho = \pi_1^\ominus \cup_1 \iota = \pi_2^\ominus \cup_2 \iota$. If $\cup_1 \supseteq \cup_2$, then $\pi_1^\ominus \subseteq \pi_2^\ominus$ for all such factorizations.*

Proof. Suppose that $\cup_1 \supseteq \cup_2$. Suppose there exists $\rho \in \mathcal{FR}(X)$ and some $x.y \in X$ such that $\pi_1^\ominus(x, y) > \pi_2^\ominus(x, y)$. We have $\pi_2^\ominus(x, y) \cup_1 \iota(x, y) \geq \pi_2^\ominus(x, y) \cup_2 \iota(x, y)$. Also $\pi_1^\ominus(x, y) \cup_1 \iota(x, y) > \pi_2^\ominus(x, y) \cup_1 \iota(x, y)$, where the strict inequality holds by the minimality of $\pi_1^\ominus$. Hence $\rho(x, y) = \pi_1^\ominus(x, y) \cup_1 \iota(x, y) > \pi_2^\ominus(x, y) \cup_1 \iota(x, y) \geq \pi_2^\ominus(x, y) \cup_2 \iota(x, y) = \rho(x, y)$, which is a contradiction. Thus $\pi_1^\ominus \subseteq \pi_2^\ominus$. ■

Proposition 3.2.13 *Let $\cup_1$ and $\cup_2 \in \mathcal{C}$. Suppose for all $\rho \in \mathcal{FR}(X)$ that $\rho = \pi_1 \cup_1 \iota = \pi_2 \cup_2 \iota$. Suppose further that either $\cup_1$ or $\cup_2$ is strict. If $\pi_1 \subseteq \pi_2$ for all such factorizations, then $\cup_1 \supseteq \cup_2$.*

Proof. Suppose there exist $a, b \in [0, 1]$ such that $a \cup_1 b < a \cup_2 b$.

Suppose $\cup_1$ is strict. Let $\rho \in \mathcal{FR}(X)$ be such that there exist $x, y \in X$ such that $\rho(x, y) = a \cup_1 b$ and $\rho(y, x) = b$. Then $\rho(x, y) \geq \rho(y, x)$. Thus $\iota(x, y) = b$. Since $\cup_1$ is strict, $\pi_1(x, y) = a$. Thus $\rho(x, y) = \pi_1(x, y) \cup_1 \iota(x, y) = a \cup_1 b < a \cup_2 b \leq \pi_2(x, y) \cup_2 \iota(x, y) = \rho(x, y)$, a contradiction. Hence no such a, b exist. Thus $\cup_1 \supseteq \cup_2$.

Suppose $\cup_2$ is strict. Let $\rho \in \mathcal{FR}(X)$ be such that there exist $x, y \in X$ such that $\rho(x, y) = a \cup_2 b$ and $\rho(y, x) = b$. Then $\rho(x, y) \geq \rho(y, x)$. Thus $\iota(x, y) = b$. Since $\cup_2$ is strict, $\pi_2(x, y) = a$. Thus $\rho(x, y) = \pi_1(x, y) \cup_1 \iota(x, y) \leq a \cup_1 b < a \cup_2 b = \pi_2(x, y) \cup_2 \iota(x, y) = \rho(x, y)$, a contradiction. Hence no such a, b exist. Thus $\cup_1 \supseteq \cup_2$. ∎

Let $\cup$ be a conorm. Then $\cup$ is said to be **nearly strictly increasing** if $\forall b \in [0, 1]$, there exists $a_0 \in (0, 1]$ such that $\forall a \geq a_0$, $a \cup b = 1$ and g is strictly increasing on $[0, a_0]$.

Proposition 3.2.14 *Let $\cup_1$ and $\cup_2 \in \mathcal{C}$. Suppose for all $\rho \in \mathcal{FR}(X)$ that $\rho = \pi_1^{\ominus} \cup_1 \iota = \pi_2^{\ominus} \cup_2 \iota$. Suppose further that either $\cup_1$ or $\cup_2$ is nearly strict. If $\pi_1^{\ominus} \subseteq \pi_2^{\ominus}$ for all such factorizations, then $\cup_1 \supseteq \cup_2$.*

Proof. Suppose there exist $a, b \in [0, 1]$ such that $a \cup_1 b < a \cup_2 b$.

Suppose $\cup_1$ is nearly strict. Let $\rho \in \mathcal{FR}(X)$ be such that there exist $x, y \in X$ such that $\rho(x, y) = a \cup_1 b$ and $\rho(y, x) = b$. Then $\rho(x, y) \geq \rho(y, x)$. Thus $\iota(x, y) = b$. Suppose $a < a_0$, where $a_0 = \wedge\{t \in [0, 1] \mid t \cup_1 b = 1\}$. Since $\cup_1$ is nearly strict, $\pi_1^{\ominus}(x, y) = a$. Thus $\rho(x, y) = \pi_1^{\ominus}(x, y) \cup_1 \iota(x, y) = a \cup_1 b < a \cup_2 b \leq \pi_2^{\ominus}(x, y) \cup_2 \iota(x, y) = \rho(x, y)$, a contradiction. Hence no such a, b exist. Thus $\cup_1 \supseteq \cup_2$. Suppose $a \geq a_0$. Then $a \cup_1 b = 1$ and so $a \cup_1 b < a \cup_2 b$ is impossible. Once again no such a, b exist. Hence $\cup_1 \supseteq \cup_2$.

Suppose $\cup_2$ is nearly strict. Let $\rho \in \mathcal{FR}(X)$ be such that there exist $x, y \in X$ such that $\rho(x, y) = a \cup_2 b$ and $\rho(y, x) = b$. Then $\rho(x, y) \geq \rho(y, x)$. Thus $\iota(x, y) = b$. Suppose $a < a_0$, where $a_0 = \wedge\{t \in [0, 1] \mid t \cup_2 b = 1\}$. Since $\cup_2$ is nearly strict, $\pi_2^{\ominus}(x, y) = a$. Thus $\rho(x, y) = \pi_1^{\ominus}(x, y) \cup_1 \iota(x, y) \leq a \cup_1 b < a \cup_2 b = \pi_2^{\ominus}(x, y) \cup_2 \iota(x, y) = \rho(x, y)$, a contradiction. Hence no such a, b exist. Thus $\cup_1 \supseteq \cup_2$ Suppose $a \geq a_0$. Then $a \cup_2 b = 1$ and $\pi_2^{\ominus}(x, y) \leq a_0$. Hence $1 = \rho(x, y) = \pi_2^{\ominus}(x, y) \cup_2 b = a \cup_2 b > a \cup_1 b \geq \pi_2^{\ominus}(x, y) \cup_1 b \geq \pi_1^{\ominus}(x, y) \cup_1 b = \rho(x, y)$, a contradiction. Hence no such a, b exist. Thus $\cup_1 \supseteq \cup_2$. ∎

Proposition 3.2.15 *Let ρ be a FWPR. If π is regular, then π is simple.*

Proof. Let $x, y \in X$. Suppose $\rho(x, y) = \rho(y, x)$. Then since ρ is regular, $\pi(x, y) = 0 = \pi(y, x)$. ∎

Proposition 3.2.16 *Let ρ be a FWPR. If $\rho = \pi \cup \iota$ is a factorization of ρ and $\cup$ is strictly increasing, then π is regular.*

Proof. Let $x, y \in X$. Suppose $\rho(x, y) > \rho(y, x)$. Then $\iota(x, y) = \rho(y, x)$. Hence $\pi(x, y) > 0$ else $\rho(x, y) = 0 \cup \rho(y, x) = \rho(y, x)$ which is impossible. Conversely suppose $\pi(x, y) > 0$. Suppose $\rho(x, y) \leq \rho(y, x)$. Then $\iota(x, y) = \rho(x, y)$. Thus $\rho(x, y) = \pi(x, y) \cup \rho(x, y) > 0 \cup \rho(x, y) = \rho(x, y)$ which is impossible. Thus $\rho(x, y) > \rho(y, x)$. ∎

Proposition 3.2.17 *Let ρ is a FWPR. Suppose $\rho = \pi_\vee \vee \iota = \pi \cup \iota$ are factorizations of ρ, where $\vee$ denotes maximum. If π is regular, then $\pi_\vee \supseteq \pi$.*

Proof. Let $x, y \in X$. By Proposition 3.1.8,

$$\pi_\vee(x, y) = \begin{cases} \rho(x, y) \text{ if } \rho(x, y) > \rho(y, x), \\ 0 \text{ otherwise.} \end{cases}$$

Now $\rho(x, y) = \pi(x, y) \cup \iota(x, y) \geq \pi(x, y) \cup 0 = \pi(x, y)$. Since π is regular, the desired result is now immediate. ∎

3.3 Factorizations

The following list of t-conorms is taken from [16]. Many of the results are based on [15].

Conorms
Standard union: $a \oplus b = a \vee b$

Bounded sum: $a \oplus_2 b = 1 \wedge (a + b)$

Drastic union: $a \oplus_d b = \begin{cases} a \text{ if } b = 0 \\ b \text{ if } a = 0 \\ 1 \text{ otherwise} \end{cases}$

Algebraic sum: $a \oplus_A b = a + b - ab$

Yager: $a \oplus_Y b = 1 \wedge (a^w + b^w)^{1/w}$, $w > 0$

Nilpotent maximum: $a \oplus_n b = \begin{cases} a \vee b \text{ if } a + b < 1 \\ 1 \text{ otherwise} \end{cases}$

Einstein sum: $a \oplus_{H_2} b = (a + b)/(1 + ab)$

Schweizer and Sklar:

$$a \oplus_p b = 1 - 0 \vee ((1 - a)^p + (1 - b)^p - 1)^{1/p}, \ p \neq 0$$

Frank:

$$a \oplus_s b = 1 - \log_s \left(1 + \frac{(s^{1-a} - 1)(s^{1-b} - 1)}{s - 1}\right), \ s > 0, \ s \neq 1$$

Dombi: $a \oplus b = [1 + [(1/a - 1)^\lambda + (1/b - 1)^\lambda]^{-1/\lambda}]^{-1}$, $\lambda > 0$

Hamacher: $a \oplus b = [a + b + (r - 2)ab]/[r + (r - 1)ab$, $r > 0$

Schweizer and Sklar 2: $a \oplus b = [a^p + b^p - a^p b^p]^{1/p}$, $p > 0$

Schweizer and Sklar 3:

$$a \oplus b = 1 - e^{-(|\ln(1-a)|^p + |\ln(1-b)|^p)^{1/p}}, \ p > 0$$

Schweizer and Sklar 4:

$$a \oplus b = 1 - [(1-a)(1-b)]/[(1-a)^p + (1-b)^p - (1-a)^p(1-b)^p]^{1/p}, \ p > 0$$

Dubois and Prade:

$$a \oplus b = 1 - [(1-a)(1-b)/[(1-a) \vee (1-b) \vee \alpha], \ \alpha \in [0,1]$$

Weber:

$$a \oplus b = 1 \wedge \left(a + b - \frac{\lambda}{1-\lambda}ab \right), \ \lambda > -1$$

Yu: $a \oplus b = 1 \wedge (a + b + \lambda ab), \ \lambda > -1$

Proposition 3.3.1 [15] *Let* $\oplus = \vee$. *Let* $\rho \in \mathcal{FR}(X)$. *Suppose* $\rho = \pi \oplus \iota$. *Then* $\forall x, y \in X$,

$$\pi(x,y) = \begin{cases} \rho(x,y) & \text{if } \rho(x,y) > \rho(y,x) \\ 0 & \text{otherwise.} \end{cases}$$

Proof. The result follows from Proposition 3.1.8. ■

Proposition 3.3.2 *Let* $\oplus_2$ *be the t-conorm defined by* $\forall a, b \in [0,1]$, $a \oplus_2 b = 1 \wedge (a + b)$. *Let* $\rho \in \mathcal{FR}(X)$. *Suppose* $\rho = \pi \oplus_2 \iota$. *Then* $\forall x, y \in X$,

$$\pi^\ominus(x,y) = \begin{cases} 0 & \text{if } \rho(x,y) \leq \rho(y,x) \\ \rho(x,y) - \rho(y,x) & \text{if } \rho(x,y) > \rho(y,x). \end{cases}$$

Proof. If $\rho(x,y) \leq \rho(y,x)$, then $\iota(x,y) = \rho(x,y)$ and so $\rho(x,y) = 1 \wedge (\pi(x,y)+\rho(x,y))$. Thus $\pi(x,y) = 0$. Suppose $\rho(x,y) > \rho(y,x)$. Then $\rho(x,y) = 1 \wedge (\pi(x,y)+\rho(y,x))$. Suppose $\pi(x,y)+\rho(y,x) \geq 1$. Then $\pi(x,y) \geq 1-\rho(y,x)$ and $\rho(x,y) = 1$. Clearly, $\pi(x,y) = 1-\rho(y,x) = \rho(x,y)-\rho(y,x)$ is the minimal solution. Suppose $\pi(x,y)+\rho(y,x) \leq 1$. Then $\rho(x,y) = \pi(x,y)+\rho(y,x)$. Hence $\pi(x,y) = \rho(x,y) - \rho(y,x)$. ■

Proposition 3.3.3 [15] *Let* $\oplus = \oplus_d$. *Let* $\rho \in \mathcal{FR}(X)$. *Suppose* $\rho = \pi \oplus \iota$. *Then* $\forall x, y \in X$,

$$\pi(x,y) = \begin{cases} \rho(x,y) & \text{if } \rho(x,y) > \rho(y,x) = 0 \\ 0 & \text{otherwise.} \end{cases}$$

Proof. Proposition 3.2.7(1). ■

Theorem 3.3.4 [15] *Let* $\oplus = \oplus_A$. *Let* $\rho \in \mathcal{FR}(X)$. *Suppose* $\rho = \pi \oplus \iota$. *Then* $\forall x, y \in X$,

$$\pi(x,y) = \begin{cases} \frac{\rho(x,y)-\rho(y,x)}{1-\rho(y,x)} & \text{if } \rho(x,y) > \rho(y,x) \\ 0 & \text{otherwise.} \end{cases}$$

Proof. Proposition 3.2.7(5) and Proposition 3.2.4. ■

Proposition 3.3.5 [15] *Let* $\oplus = \oplus_Y$. *Let* $\rho \in \mathcal{FR}(X)$. *Suppose* $\rho = \pi \oplus \iota$. *Then* $\forall x, y \in X$,

$$\pi^\ominus(x,y) = \begin{cases} [\rho(x,y)^w - \rho(y,x)^w]^{1/w} & \text{if } \rho(x,y) > \rho(y,x) \\ 0 & \text{otherwise.} \end{cases}$$

Proof. Proposition 3.2.5 and 3.2.7(6). ■

Proposition 3.3.6 *Let* $\oplus = \oplus_n$. *Let* $\rho \in \mathcal{FR}(X)$. *Then* ρ *has a factorization* $\rho = \pi \oplus \iota$ *if and only if* $\forall x, y \in X$, *either* (1) $\rho(x,y) = 1$ *or* $\rho(y,x) = 1$ *or* (2) $\rho(x,y) = \rho(y,x)$ *or* (3) $\rho(x,y) + \rho(y,x) < 1$ *and* $\rho(x,y) \neq \rho(y,x)$.

Proof. Proposition 3.2.8. ■

Proposition 3.3.7 [15] *Let* $\oplus = \oplus_{H_2}$. *Let* $\rho \in \mathcal{FR}(X)$. *Suppose* $\rho = \pi \oplus \iota$. *Then* $\forall x, y \in X$,

$$\pi(x,y) = \begin{cases} \frac{\rho(x,y)-\rho(y,x)}{1-\rho(x,y)\rho(y,x)} & \text{if } \rho(x,y) > \rho(y,x) \\ 0 & \text{otherwise.} \end{cases}$$

Proof. Proposition 3.2.9(1). ■

Proposition 3.3.8 [15] *Let* $\oplus = \oplus_p$. *Let* $\rho \in \mathcal{FR}(X)$. *Suppose* $\rho = \pi \oplus \iota$. *Then* $\forall x, y \in X$,

$$\pi(x,y) = \begin{cases} 1 - [(1-\rho(x,y))^p - (1-\rho(y,x))^p + 1]^{1/p} & \text{if } \rho(x,y) > \rho(y,x) \\ 0 & \text{otherwise.} \end{cases}$$

Proof. Proposition 3.2.9(2). ■

Proposition 3.3.9 [15] *Let* $\oplus = \oplus_s$. *Let* $\rho \in \mathcal{FR}(X)$. *Suppose* $\rho = \pi \oplus \iota$. *Then* $\forall x, y \in X$,

$$\pi(x,y) = 1 - \log_s \left[\frac{(s-1)(s^{1-\rho(x,y)}-1)}{s^{1-\rho(y,x)}-1} + 1 \right].$$

Proof. The proof follows from Proposition 3.2.9(4). ■

Proposition 3.3.10 *Let* $\oplus$ *denote the Dubois and Prade conorm. Let* $\rho \in \mathcal{FR}(X)$. *Let* $x, y \in [0, 1]$. *Suppose* $\rho(x, y) > \rho(y, x)$ *and* $\rho(x, y) = \pi(x, y) \otimes \rho(y, x)$. *Then*

$$\pi(x, y) = \begin{cases} \rho(x, y) & \text{if } \alpha \leq 1 - \rho(y, x) \\ 1 - [\alpha(1 - \rho(x, y)]/[1 - \rho(y, x)] & \text{if } \alpha \geq 1 - \rho(y, x). \end{cases}$$

Proof. Let $t \in [0, 1]$ be such that $\rho(x, y) = t \oplus \rho(y, x)$. Then

$$\rho(x, y) = 1 - [(1 - t)(1 - \rho(y, x))]/[(1 - t) \vee (1 - \rho(y, x)) \vee \alpha]$$

(1) Suppose $1 - t = (1 - t) \vee (1 - \rho(y, x)) \vee \alpha$. Then $\rho(x, y) = 1 - (1 - \rho(y, x)) = \rho(y, x)$, a contradiction.

(2) Suppose $1 - \rho(y, x) = (1 - t) \vee (1 - \rho(y, x)) \vee \alpha$. Then $\rho(x, y) = 1 - (1 - t) = t$. (Note $1 - t = 1 - \rho(x, y) < 1 - \rho(y, x)$.)

(3) Suppose $\alpha = (1 - t) \vee (1 - \rho(y, x)) \vee \alpha$. Then $\rho(x, y) = 1 - [(1 - t)(1 - \rho(y, x)]/\alpha$. Thus $(1 - t)(1 - \rho(y, x)) = \alpha(1 - \rho(x, y))$. Hence $t = 1 - [\alpha(1 - \rho(x, y))/(1 - \rho(y, x))]$. (Note $(1 - \rho(x, y))/(1 - \rho(y, x)) \leq 1$ and so $1 - t = \alpha(1 - \rho(x, y))/(1 - \rho(y, x)) \leq \alpha$.) ∎

Theorem 3.3.11 *Let* $\oplus$ *be the Hamacher conorm. Then* $\forall x, y \in X$, $\forall \rho \in \mathcal{FR}(X)$ *such that* $\rho(x, y) > \rho(y, x)$, $\rho(x, y) = t \oplus \rho(y, x)$ *has a solution for* t *if and only if* $r = 1$. *If* $r = 1$, *then* $\pi(x, y) = (\rho(x, y) - \rho(y, x))/(1 - \rho(y, x))$.

Proof. Suppose $r = 1$. Then $\forall a, b \in [0, 1]$, $a \oplus b = a + b - ab$, i.e., $\oplus$ is the algebraic sum and the desired result is known. Conversely, suppose $r > 0$ and $\rho(x, y) = t \oplus \rho(y, x)$ has a solution $\forall \rho \in \mathcal{FR}(X)$ and $\forall x, y \in X$ such that $\rho(x, y) > \rho(y, x)$. Then

$$\rho(x, y) = [t + \rho(y, x) + (r - 2)t\rho(y, x)]/[r + (r - 1)t\rho(y, x)].$$

Thus

$$t = [\rho(y, x) - r\rho(x, y)]/[(r - 1)\rho(y, x)\rho(x, y) - 1 - (r - 2)\rho(y, x)]$$

and $0 \leq t \leq 1$. Hence either

(1) $0 \leq \rho(y, x) - r\rho(x, y) \leq (r - 1)\rho(y, x)\rho(x, y) - 1 - (r - 2)\rho(y, x)$

or

(2) $0 \geq \rho(y, x) - r\rho(x, y) \geq (r - 1)\rho(y, x)\rho(x, y) - 1 - (r - 2)\rho(y, x).$

For case (1), $r \leq \rho(y, x)/\rho(x, y)$. However, $\rho(y, x)/\rho(x, y)$ can be arbitrarily close to 0. Thus case (1) is impossible since $r > 0$ and r is fixed.

For case (2), $r \geq \rho(y, x)/\rho(x, y)$. Now $\rho(y, x)/\rho(x, y)$ can be made arbitrarily close to 1. Hence we can't have $r < 1$. Thus $r \geq 1$. Let $\rho(x, y) = 1$. Then from (2), $\rho(y, x) - r \geq (r - 1)\rho(y, x) - 1 - (r - 2)\rho(y, x)$. Hence $-r \geq -1$. Thus $r \leq 1$. Hence $r = 1$. ∎

Proposition 3.3.12 *Let* $\oplus$ *denote the Schweizer and Sklar* 2 *conorm. Let* $\rho \in \mathcal{FR}(X)$. *Let* $x, y \in X$. *Suppose* $\rho(x, y) > \rho(y, x)$ *and* $\rho(x, y) = \pi(x, y) \oplus \rho(y, x)$. *Then*

$$\pi(x, y) = \left[\frac{\rho(x, y)^p - \rho(y, x)^p}{1 - \rho(y, x)^p}\right]^{1/p}.$$

Proof. Suppose there exists $t \in [0, 1]$ such that $\rho(x, y) = t \oplus \rho(y, x)$. Then $\rho(x, y) = (t^p + \rho(y, x)^p - t^p \rho(y, x)^p)^{1/p}$. Thus $t^p = \frac{\rho(x,y)^p - \rho(y,x)^p}{1 - \rho(y,x)^p}$ and so

$$t = \left[\frac{\rho(x, y)^p - \rho(y, x)^p}{1 - \rho(y, x)^p}\right]^{1/p}.$$

The desired result follows since $0 \leq t \leq 1$. ∎

Proposition 3.3.13 *Let* $\oplus$ *denote the Schweizer and Sklar* 4 *conorm. Let* $\rho \in \mathcal{FR}(X)$. *Let* $x, y \in X$. *Suppose* $\rho(x, y) > \rho(y, x)$ *and* $\rho(x, y) = \pi(x, y) \oplus \rho(y, x)$. *Then*

$$\pi(x, y) = 1 - \left[\frac{(1 - \rho(x, y))^p(1 - \rho(y, x))^p}{(1 - \rho(y, x))^p + (1 - \rho(x, y))^p(1 - \rho(y, x))^p - (1 - \rho(x, y))^p}\right]^{1/p}.$$

Proof. Suppose there exists $t \in [0, 1]$ such that $\rho(x, y) = t \oplus \rho(y, x)$. Then

$$\rho(x, y) = 1 - \frac{(1 - t)(1 - \rho(y, x))}{[(1 - t)^p + (1 - \rho(y, x))^p - (1 - t)^p(1 - \rho(y, x))^p]^{1/p}}$$

and so

$$(1 - \rho(x, y))^p = \frac{(1 - t)^p(1 - \rho(y, x))^p}{(1 - t)^p + (1 - \rho(y, x))^p - (1 - t)^p(1 - \rho(y, x))^p},$$

i.e.,

$$\begin{aligned}
(1 - t)^p(1 - \rho(y, x))^p &= (1 - \rho(x, y))^p(1 - t)^p \\
&\quad + (1 - \rho(x, y))^p(1 - \rho(y, x))^p \\
&\quad - (1 - t)^p(1 - \rho(x, y))^p(1 - \rho(y, x))^p,
\end{aligned}$$

which implies that

$$\begin{aligned}
(1 - \rho(x, y))^p(1 - \rho(y, x))^p &= (1 - t^p)[(1 - \rho(y, x))^p + \\
&\quad (1 - \rho(x, y))^p(1 - \rho(y, x))^p \\
&\quad - (1 - \rho(x, y))^p].
\end{aligned}$$

Thus

$$(1 - t)^p = \frac{(1 - \rho(x, y))^p(1 - \rho(y, x))^p}{(1 - \rho(y, x))^p + (1 - \rho(x, y))^p(1 - \rho(y, x))^p - (1 - \rho(x, y))^p}.$$

Hence

$$t = 1 - \left[\frac{(1 - \rho(x,y))^p(1 - \rho(y,x))^p}{(1 - \rho(y,x))^p + (1 - \rho(x,y))^p(1 - \rho(y,x))^p - (1 - \rho(x,y))^p}\right]^{1/p}.$$

The desired result follows since $0 \le t \le 1$. ∎

Proposition 3.3.14 *Let* $\oplus$ *be the Dombi conorm. Then it is not the case that* $\forall x, y \in X$, $\forall \rho \in \mathcal{FR}(X)$ *such that* $\rho(x,y) > \rho(y,x)$, $\rho(x,y) = t \oplus \rho(y,x)$ *has a solution for* t.

Proof. Suppose $\rho(y,x) > 0$. Suppose there exists $t \in (0,1]$ such that
$$\rho(x,y) = t \oplus \rho(y,x). \text{ Then } \rho(x,y) = \left[1 + \left[\left(\tfrac{1}{t} - 1\right)^\lambda + \left(\tfrac{1}{\rho(y,x)} - 1\right)^\lambda\right]^{-1/\lambda}\right]^{-1}.$$
Thus

$$\frac{1}{\rho(x,y)} - 1 = \left[\left(\tfrac{1}{t} - 1\right)^\lambda + \left(\tfrac{1}{\rho(y,x)} - 1\right)^\lambda\right]^{-1/\lambda}$$

$$\left(\frac{1}{\rho(x,y)} - 1\right)^{-\lambda} = \left(\tfrac{1}{t} - 1\right)^\lambda + \left(\tfrac{1}{\rho(y,x)} - 1\right)^\lambda.$$

Hence

$$\left(\tfrac{1}{t} - 1\right)^\lambda = \left(\frac{1}{\rho(x,y)} - 1\right)^{-\lambda} - \left(\frac{1}{\rho(y,x)} - 1\right)^\lambda$$

$$\left(\tfrac{1}{t} - 1\right) = \left[\left(\frac{1}{\rho(x,y)} - 1\right)^{-\lambda} - \left(\frac{1}{\rho(y,x)} - 1\right)^\lambda\right]^{1/\lambda}.$$

Thus

$$t = \left[1 + \left[\left(\frac{1}{\rho(x,y)} - 1\right)^{-\lambda} - \left(\frac{1}{\rho(y,x)} - 1\right)^\lambda\right]^{1/\lambda}\right]^{-1}.$$

Now

$$\left[1 + \left[\left(\tfrac{1}{\rho(x,y)} - 1\right)^{-\lambda} - \left(\tfrac{1}{\rho(y,x)} - 1\right)^\lambda\right]^{1/\lambda}\right]^{-1} < 0$$

$$\Leftrightarrow \ 1 + \left[\left(\tfrac{1}{\rho(x,y)} - 1\right)^{-\lambda} - \left(\tfrac{1}{\rho(y,x)} - 1\right)^\lambda\right]^{1/\lambda} < 0.$$

This inequality can occur by taking $\rho(y,x)$ sufficiently close to 0. In this case, $t < 0$. ∎

Proposition 3.3.15 *Let* $\oplus$ *denote the Yu conorm. Let* $\rho \in \mathcal{FR}(X)$. *Let* $x, y \in X$. *Suppose* $\rho(x,y) > \rho(y,x)$ *and* $\rho(x,y) = \pi(x,y) \oplus \rho(y,x)$. *Then* $\forall x, y \in X$,

$$\pi(x,y) = \frac{\rho(x,y) - \rho(y,x)}{1 + \lambda\rho(y,x)} \, \text{if } \rho(x,y) < 1$$

and $\pi^{\oslash}(x,y) = 1$ *if* $\rho(x,y) = 1$.

Proof. If $\rho(x,y) = 1$, then $\rho(x,y) = 1 \oplus \rho(y,x)$. Suppose $\rho(x,y) < 1$. Suppose there exists $t \in [0,1]$ such that $\rho(x,y) = t \oplus \rho(y,x)$. Then $\rho(x,y) = t + \rho(y,x) + \lambda t\rho(y,x) = t(1 + \lambda\rho(y,x)) + \rho(y,x)$. Thus

$$t = \frac{\rho(x,y) - \rho(y,x)}{1 + \lambda\rho(y,x)}.$$

Suppose $\rho(y,x) > 0$. Since $\lambda > -1, \lambda > -1/\rho(y,x)$. Thus $1 + \lambda\rho(y,x) > 0$. Also, $\rho(x,y) - \rho(y,x) \le 1 - \rho(y,x)$ and so $\rho(x,y) - \rho(y,x) \le 1 + \lambda\rho(y,x)$. Thus $0 < t \le 1$ and so the desired result follows. ∎

Lemma 3.3.16 *Let* $\oplus$ *denote the Weber conorm. Let* $\rho \in \mathcal{FR}(X)$. *Let* $x, y \in X$. *Suppose* $\rho(x,y) > \rho(y,x)$ *and* $\rho(x,y) = \pi(x,y) \oplus \rho(y,x)$. *Then* $\forall x, y \in X$,

$$\pi(x,y) = \frac{\rho(x,y) - \rho(y,x)}{1 - \frac{\lambda}{1-\lambda}\rho(y,x)} \quad \text{if } -1 < \lambda \le 0$$

$$\text{or } (0 < \lambda < 1 \text{ and } \lambda \le \frac{1 - \rho(x,y) + \rho(y,x)}{1 + 2\rho(y,x) - \rho(x,y)}) \text{ or } (\lambda > 1).$$

Proof. Suppose there exists $t \in [0,1]$ such that $\rho(x,y) = t \oplus \rho(y,x)$. Then $\rho(x,y) = t + \rho(y,x) - \frac{\lambda}{1-\lambda}t\rho(y,x) = t\left(1 - \frac{\lambda}{1-\lambda}\rho(y,x)\right) + \rho(y,x)$. Thus

$$t = \frac{\rho(x,y) - \rho(y,x)}{1 - \frac{\lambda}{1-\lambda}\rho(y,x)}.$$

Suppose $-1 < \lambda < 0$. Then clearly $\frac{\lambda}{1-\lambda} < 0$. Thus $t > 0$. Also $1 - \frac{\lambda}{1-\lambda}\rho(y,x) \ge \rho(x,y) - \rho(y,x) > 0$. Hence $t < 1$.

Now suppose $0 < \lambda < 1$. Then since $1 - \lambda > 0$, $1 - \frac{\lambda}{1-\lambda}\rho(y,x) > 0 \Leftrightarrow 1 - \lambda - \lambda\rho(y,x) > 0 \Leftrightarrow \lambda < \frac{1}{1+\rho(y,x)}$ which holds since $\lambda < 1$. Hence $t > 0$. Now

$$
\begin{aligned}
t &\le 1 \Leftrightarrow 1 - \frac{\lambda}{1-\lambda}\rho(y,x) \ge \rho(x,y) - \rho(y,x) \\
&\Leftrightarrow 1 - \lambda - \lambda\rho(y,x) \ge (1-\lambda)(\rho(x,y) - \rho(y,x)) \\
&\Leftrightarrow 1 - \rho(x,y) + \rho(y,x) \ge \lambda + \lambda\rho(y,x) - \lambda\rho(x,y) + \lambda\rho(y,x) \\
&\Leftrightarrow 1 - \rho(x,y) + \rho(y,x) \ge \lambda(1 + 2\rho(y,x) - \rho(x,y)) \\
&\Leftrightarrow \lambda \le \frac{1 - \rho(x,y) + \rho(y,x)}{1 + 2\rho(y,x) - \rho(x,y)}.
\end{aligned}
$$

Now suppose $\lambda > 1$. Then as above $t \leq 1 \Leftrightarrow \lambda \geq \frac{1-\rho(x,y)+\rho(y,x)}{1+2\rho(y,x)-\rho(x,y)}$ since $1 - \lambda < 0$. However, $\lambda \geq \frac{1-\rho(x,y)+\rho(y,x)}{1+2\rho(y,x)-\rho(x,y)}$ holds since $\lambda > 1$. Thus $t \leq 1$. Now

$$t \quad \geq \quad 0 \Leftrightarrow 1 - \frac{\lambda}{1-\lambda}\rho(y,x) \geq 0 \Leftrightarrow 1 - \lambda - \lambda\rho(y,x) \leq 0$$

$$\Leftrightarrow \quad \lambda \geq \frac{1}{1+\rho(y,x)}.$$

However, $\lambda \geq \frac{1}{1+\rho(y,x)}$ since $\lambda > 1$. $\blacksquare$

Proposition 3.3.17 *Let $\oplus$ denote the Weber conorm. Let $x, y \in X$. Suppose $\forall \rho \in \mathcal{FR}(X)$, $\rho(x,y) > \rho(y,x)$ and $\rho(x,y) = \pi(x,y) \oplus \rho(y,x)$. Then*

$$\pi(x,y) = \frac{\rho(x,y) - \rho(y,x)}{1 - \frac{\lambda}{1-\lambda}\rho(y,x)} \quad \text{if } -1 < \lambda \leq 0 \text{ or } \lambda > 1.$$

Proof. Suppose $0 < \lambda < 1$. By Lemma 3.3.16, $\lambda \leq \frac{1-\rho(x,y)+\rho(y,x)}{1+2\rho(y,x)-\rho(x,y)}$. However, $\frac{1-\rho(x,y)+\rho(y,x)}{1+2\rho(y,x)-\rho(x,y)}$ can be made arbitrarily close to 1 by taking $\rho(y,x)$ sufficiently close to 0. This is impossible since λ is fixed. The desired result follows from Lemma 3.3.16. $\blacksquare$

Proposition 3.3.18 *Let $\oplus$ denote the Schweizer and Sklar 3 conorm. Let $\rho \in \mathcal{FR}(X)$. Let $x, y \in X$. Suppose $\rho(x,y) > \rho(y,x)$ and $\rho(x,y) = \pi(x,y) \oplus \rho(y,x)$. Then*

$$\pi(x,y) = \begin{cases} 1 & \text{if } \rho(x,y) = 1 \\ 1 - e^{-\left[\left(\ln\left(\frac{1}{1-\rho(x,y)}\right)\right)^p - |\ln(1-\rho(y,x)|^p\right]^{1/p}} & \text{if } \rho(x,y) < 1. \end{cases}$$

Proof. Suppose there exists $t \in [0,1)$ such that $\rho(x,y) = t \oplus \rho(y,x)$. Then

$$\rho(x,y) = 1 - e^{-(|\ln(1-t)|^p + |(1-\rho(y,x)|^p)^{1/p}}$$

$$1 - \rho(x,y) = e^{-(|\ln(1-t)|^p + |(1-\rho(y,x)|^p)^{1/p}}$$

$$\ln(1 - \rho(x,y)) = -(|\ln(1-t)|^p + |\ln(1-\rho(y,x)|^p)^{1/p}$$

$$\ln\left(\frac{1}{1-\rho(x,y)}\right) = (|\ln(1-t)|^p + |\ln(1-\rho(y,x)|^p)^{1/p}$$

$$\left(\ln\left(\frac{1}{1-\rho(x,y)}\right)\right)^p = |\ln(1-t)|^p + |\ln(1-\rho(y,x)|^p$$

$$|\ln(1-t)|^p = \left(\ln\left(\frac{1}{1-\rho(x,y)}\right)\right)^p - |\ln(1-\rho(y,x)|^p$$

$$|\ln(1-t)| = \left[\left(\ln\left(\frac{1}{1-\rho(x,y)}\right)\right)^p - |\ln(1-\rho(y,x)|^p\right]^{1/p}$$

$$\ln(1-t) = -\left[\left(\ln\left(\frac{1}{1-\rho(x,y)}\right)\right)^p - |\ln(1-\rho(y,x)|^p\right]^{1/p}$$

$$1 - t = e^{-\left[(\ln(\frac{1}{1-\rho(x,y)}))^p - |\ln(1-\rho(y,x)|^p\right]^{1/p}}$$

$$t = 1 - e^{-\left[(\ln(\frac{1}{1-\rho(x,y)}))^p - |\ln(1-\rho(y,x)|^p\right]^{1/p}}.$$

We must show that $t \geq 0$. We show that

$$\left[\left(\ln\left(\frac{1}{1-\rho(x,y)}\right)\right)^p - |\ln(1-\rho(y,x)|^p\right]^{1/p} > 0.$$

Now $\frac{1}{1-\rho(x,y)} > \frac{1}{1-\rho(y,x)}$ and so $\ln\left(\frac{1}{1-\rho(x,y)}\right) > \ln\left(\frac{1}{1-\rho(y,x)}\right)$. Consequently, $\ln\left(\frac{1}{1-\rho(x,y)}\right) > -\ln(1-\rho(y,x))$ and so $\ln\left(\frac{1}{1-\rho(x,y)}\right) > |\ln(1-\rho(y,x))|$. Hence $\left(\ln\left(\frac{1}{1-\rho(x,y)}\right)\right)^p > |\ln(1-\rho(y,x))|^p$ and $\left(\ln\left(\frac{1}{1-\rho(x,y)}\right)\right)^p - |\ln(1-\rho(y,x))|^p > 0$. (Note $e^{-a} \leq 1$ if $a > 0$.) ∎

3.4 Intuitionistic Fuzzy Relations

The proofs of many factorization results for an intuitionistic fuzzy binary relation $\langle \rho_\mu, \rho_\nu \rangle$ involve dual proofs, one for ρ_μ with respect to a t-conorm $\oplus$ and one for ρ_ν with respect to a t-norm $\otimes$. In this section, we show that one proof can be obtained from the other by considering $\oplus$ and $\otimes$ dual under an involutive fuzzy complement. We follow the lead in [2] in this and the next section.

Intuitionistic fuzzy sets were introduced by George Gargov. Krassimir Atanassov has been the leader in the field [1, 2]. An intuitionistic fuzzy set consists of a set X and two fuzzy subsets μ and ν of X such that $\mu(x) + \nu(x) \leq 1$.

The fuzzy subset μ can be thought of as providing the degree of membership of elements of X in a subset of X while ν can be thought of as providing the degree of nonmembership of elements of X in the subset. Other interpretations of μ and ν are possible, e.g., validity and nonvalidity, respectively. The fuzzy subset ω of X defined by $\omega(x) = 1 - \mu(x) - \nu(x)$ for all $x \in X$ is the uncertainty (indeterminacy) of x in the subset. These various interpretations of μ and ν open the door for a variety of interpretations of concepts in political science and other disciplines.

In Fono, Nana, Salles, and Gwet [13], a factorization of an intuitionistic fuzzy binary relation into a unique indifference component and a family of regular strict components was provided. This result generalizes a result in [10] with the (max, min) intuitionistic fuzzy t-conorm. In [13], a characterization of $\mathcal{C}$-transitivity for a continuous t-representable intuitionistic fuzzy t-conrm $\mathcal{C}$ was established. This enabled the authors to determine necessary and sufficient conditions on a $\mathcal{C}$-transitive intuitionistic fuzzy binary relation ρ under which a component of ρ satisfies pos-transitivity and negative transitivity.

In Cornelis, Deschrijver, and Kerre [6], some results on t-representable intuitionistic fuzzy t-norms (T, S) were established, where T is a fuzzy t-norm and S is a fuzzy t-conorm such that $T(a, b) = 1 - S(1 - a, 1 - b)$ for all $a, b \in [0, 1]$.

An **intuitionistic fuzzy binary relation** $\langle \rho_\mu, \rho_\nu \rangle$ on a set X is a pair of fuzzy binary relations ρ_μ and ρ_ν on X such that for all $x, y \in X$, $\rho_\mu(x, y) + \rho_\nu(x, y) \leq 1$. In [21], Nana and Fono obtained some intuitionistic versions of Arrow's impossibility theorem. They also provide an example of a nondictatorial intuitionistic fuzzy aggregation rule. They also establish an intuitionistic version of Gibbard's theorem. Many of their results rely on the factorization of intuitionistic fuzzy reference relations. The proofs of many factorization results for an intuitionistic fuzzy binary relation $\langle \rho_\mu, \rho_\nu \rangle$ involve dual proofs, one for ρ_μ with respect to a t-conorm $\oplus$ and one for ρ_ν with respect to a t-norm $\otimes$. We show that one proof can be obtained from the other by considering $\oplus$ and $\otimes$ dual under an involutive fuzzy complement. The key ideas are presented just prior to Theorem 3.5.8.

We present preliminary results needed for the rest of the chapter. We present basic properties concerning the factorization of intuitionistic fuzzy preference relations. In the final section, we present factorization results involving a wide variety of t-conorms and t-norms.

We say that a t-norm $\otimes$ is **strict** if $\forall a, b, d \in [0, 1]$, $b < d$ implies $\otimes(a, b) < \otimes(a, d)$. We say that a t-conorm is $\oplus$ **strict** if $\forall a, b, d \in [0, 1]$, $b < d$ implies $\oplus(a, b) < \oplus(a, d)$.

Definition 3.4.1 *Let $c : [0, 1] \to [0, 1]$. Then c is a called a* **fuzzy complement** *if the following two conditions hold:*
    (1) $c(0) = 1$ *and* $c(1) = 0$ *(boundary conditions).*

(2) $\forall a, b \in [0, 1]$, $a \leq b$ implies $c(a) \geq c(b)$ *(montonicity)*.
If $c(c(a)) = a$ for all $a \in [0, 1]$, then c is called **involutive**.

It is known that if $c : [0, 1] \to [0, 1]$ satisfies conditions (2) and is convolutive, then c is continuous and satisfies condition (1), [16]. It is also clear that if c is involutive, then c is one-to-one and thus strictly decreasing.

The following result is certainly known, at least for the standard complement. We present its proof for the sake of completeness.

Proposition 3.4.2 *Let c be an involutive fuzzy complement.*
(1) *Let $\oplus$ be a t-conorm. Define $\otimes : [0, 1]^2 \to [0, 1]$ by $\forall a, b \in [0, 1]$, $a \otimes b = c(c(a) \oplus c(b))$. Then $\otimes$ is a t-norm.*
(2) *Let $\otimes$ be a t-norm. Define $\oplus : [0, 1]^2 \to [0, 1]$ by $\forall a, b \in [0, 1]$, $a \oplus b = c(c(a) \otimes c(b))$. Then $\oplus$ is a t-conorm.*

Proof. (1) Let $a, b, d \in [0, 1]$. Then $1 \otimes b = c(c(1) \oplus c(b)) = c(0 \oplus c(b)) = c(c(b)) = b$. Also $a \otimes b = c(c(a) \oplus c(b)) = c(c(b) \oplus c(a)) = b \otimes a$. Now $b \leq d \Rightarrow c(a) \oplus c(b) \geq c(a) \oplus c(d) \Longrightarrow c(c(a) \oplus c(b)) \leq c(c(a) \oplus c(d)) \Rightarrow a \otimes b \leq a \otimes d$. We next show the associative law.

$$
\begin{aligned}
a \otimes (b \otimes d) &= a \otimes (c(c(b) \oplus c(d))) = c(c(a) \oplus (c(b) \oplus c(d))) \\
&= c((c(a) \oplus c(b)) \oplus c(d)) = c(c(a \otimes b) \oplus c(d)) \\
&= (a \otimes b) \otimes d.
\end{aligned}
$$

(2) The proof of (2) is similar to that of (1). ∎

Let c be an involutive fuzzy complement. Given $\oplus$. Define $\otimes$ in terms of $\oplus$ and c as above. Now define $\widehat{\oplus}$ in terms of $\otimes$ and c as above. Then $\widehat{\oplus} = \oplus$ since c is one-to-one. Similarly, given $\otimes$, then $\widehat{\otimes} = \otimes$, where $\oplus$ is defined in terms of $\otimes$ and c and then $\widehat{\otimes}$ is defined in terms of $\oplus$ and c.

Thus if $\otimes$ is determined from $\oplus$ by an involutive complement c, then $\oplus$ is determinable from $\otimes$ by c and vice versa. In either case, we call $\oplus$ and $\otimes$ **dual** under c.

Proposition 3.4.3 *Let c be an involutive fuzzy complement. Let $\oplus_i$ and $\otimes_i$ be dual t-conorms and t-norms under c, respectively, $i = 1, 2$. Then $\oplus_1 \subseteq \oplus_2$ if and only if $\otimes_1 \supseteq \otimes_2$.*

Proof. We have

$$
\begin{aligned}
\oplus_1 \leq \oplus_2 &\Leftrightarrow \forall a, b \in [0, 1], \ a \oplus_1 b \leq a \oplus_2 b \\
&\Leftrightarrow \forall a, b \in [0, 1], \ c(a) \oplus_1 c(b) \leq c(a) \oplus_2 c(b) \\
&\Leftrightarrow \forall a, b \in [0, 1], \ c(c(a) \oplus_1 c(b)) \geq c(c(a) \oplus_2 c(b)) \\
&\Leftrightarrow \forall a, b \in [0, 1], \ a \otimes_1 b \geq a \otimes_2 b.
\end{aligned}
$$

∎

Proposition 3.4.4 *Let c be a involutive fuzzy complement. Let $\oplus$ and $\otimes$ be dual t-conorms and t-norms under c, respectively. Then $\oplus$ is strict if and only if $\otimes$ is strict.*

Proof. Let $a, b, d \in [0, 1]$. Suppose $\oplus$ is strict. Then

$$b \ < \ d \Rightarrow a \oplus b < a \oplus d \Rightarrow c(a) \oplus c(b) > c(a) \oplus c(d)$$
$$\Rightarrow \ c(c(a) \oplus c(b)) < c(c(a) \oplus c(d)) \Rightarrow a \otimes b < a \otimes d.$$

Thus $\otimes$ is strict. That $\otimes$ is strict implies $\oplus$ is strict follows similarly. ∎

Proposition 3.4.5 *Let c be an involutive fuzzy complement. Let $\oplus$ and $\otimes$ be dual t-conorms and t-norms under c, respectively. Then $\oplus$ is continuous if and only if $\otimes$ is continuous.*

Proof. Suppose $\oplus$ is continuous. Let $c \times c$ be the function of $[0,1]^2$ into $[0,1]^2$ defined by $\forall(a,b) \in [0,1]^2$, $(c \times c)(a,b) = (c(a), c(b))$. Let $(a,b) \in [0,1]^2$. Then

$$
\begin{aligned}
(c \circ \oplus \circ (c \times c))(a,b) \ &= \ c(\oplus((c \times c)(a,b))) \\
&= \ c(\oplus(c(a), c(b))) \\
&= \ c(c(a) \oplus c(b)) \\
&= \ a \otimes b,
\end{aligned}
$$

where $\circ$ denotes composition of functions. Thus $\otimes = c \circ \oplus \circ c \times c$. Hence $\otimes$ is continuous since c, $\oplus$, and $c \times c$ are continuous. The converse follows in a similar manner. ∎

3.5 Intuitionistic Fuzzy Preference Relations and Their Factorization

The proofs of many factorization results for an intuitionistic fuzzy binary relation $\langle \rho_\mu, \rho_\nu \rangle$ involve dual proofs, one for ρ_μ with respect to a t-conorm and one for ρ_ν with respect to a t-norm $\otimes$. We show that one proof can be obtained from the other by considering $\oplus$ and $\otimes$ dual under an involutive fuzzy complement. In this section, we present some basic properties concerning the factorization of intuitionistic fuzzy preference relations.

Definition 3.5.1 *Let $\mathcal{T}$ denote the set of all continuous t-norms. The **quasi-division** of a t-norm $\otimes$, denoted $\oslash$, is the function $\oslash : [0,1]^2 \to [0,1]$ defined by $\forall a, b \in [0,1]$, $a \oslash b = \vee\{t \in [0,1] \mid a \otimes t \leq b\}$.*

Definition 3.5.2 *Let $\rho \in \mathcal{FR}(X)$. Then*
*(1) ρ is called **symmetric** if $\forall x, y \in X$, $\rho(x,y) = \rho(y,x)$;*
*(2) ρ is called c-**asymmetric** if $\forall x, y \in X$, $\rho(x,y) < 1$ implies $\rho(y,x) = 1$.*

For $\rho \in \mathcal{FR}(X)$, we often associate an c-asymmetric fuzzy relation π on X.

Definition 3.5.3 *Let $\rho \in \mathcal{FR}(X)$ and π be an c-asymmetric binary fuzzy relation associated with ρ. Then*
(1) π *is called **simple** if $\forall x, y \in X$, $\rho(x, y) = \rho(y, x)$ implies $\pi(x, y) = \pi(y, x)$;*
(2) π *is called c-**regular** if $\forall x, y \in X$, $\rho(x, y) < \rho(y, x)$ if and only if $\pi(x, y) < 1$.*

Let $\otimes$ and $\oplus$ be a t-norm and a t-conorm, respectively. Let $\rho, \pi, \iota \in \mathcal{FR}(X)$. By $\rho = \pi \otimes \iota$ or $\rho = \pi \oplus \iota$, we mean for all $x, y \in X$, $\rho(x, y) = \pi(x, y \otimes \iota(x, y))$ or $\rho(x, y) = \pi(x, y) \oplus \iota(x, y)$. In either case, we refer to $\rho = \pi \otimes \iota$ or $\rho = \pi \oplus \iota$ as a factorization of ρ.

Theorem 3.5.4 (*Fono, Nana, Salles, and Gwet* [13]) *Let $\otimes \in \mathcal{T}$ and $\rho \in \mathcal{FR}(X)$. Let $x, y \in X$. Suppose $\rho(x, y) < \rho(y, x)$. Then the following statement holds:*
(1) *The equation*

$$\rho(x, y) = t \otimes \rho(y, x), t \in [0, 1] \tag{3.2}$$

has a solution for t. The real number $\rho(y, x) \oslash \rho(x, y)$ is the largest solution to (3.2).
(2) *If $\otimes = min$ or if $\otimes$ is a strict t-norm, then $\rho(y, x) \oslash \rho(x, y)$ is the unique solution.*

Proof. (1) Let $g : [0, 1] \to [0, 1]$ be defined by $g(t) = t \otimes \rho(y, x)$ $\forall t \in [0, 1]$. Then $g(0) = 0$ and $g(1) = \rho(y, x)$. Since g is continuous, there exists $t_0 \in [0, 1]$ such that $\rho(x, y) = t_0 \otimes \rho(y, x)$ by the Intermediate Value Theorem. That is, Eq. (3.2) has a solution. Let

$$t^* = \vee \{t \in [0, 1] \mid t \otimes \rho(y, x) \le \rho(x, y)\}.$$

Then $t^* \otimes \rho(y, x) \le \rho(x, y)$ and $t_0 \otimes \rho(y, x) = \rho(x, y)$, where t_0 is any solution to Eq. (3.2). Thus $t^* \ge t_0$. Hence $\rho(y, x) \oslash \rho(x, y)$ is the largest solution.
(2) Suppose $\otimes = \wedge$. Then

$$\rho(y, x) \oslash \rho(x, y) = \vee \{t \in [0, 1] \mid t \otimes \rho(y, x) \le \rho(x, y)\} = \rho(x, y)$$

and so Eq. (3.2) becomes $\rho(x, y) = t \wedge \rho(y, x)$ and so $t = \rho(x, y)$.
Suppose $\otimes$ is strict. Then g is one-to-one. ∎

Proposition 3.5.5 *Let $\otimes \in \mathcal{T}$. Let $\rho, \pi, \iota \in \mathcal{FR}(X)$. Then statements (1) and (2) are equivalent.*
(1) $\forall x, y \in X$, $\rho(x, y) = \pi(x, y) \otimes \iota(x, y)$, π *is c-asymmetric, ι is symmetric, and π is simple.*

(2) $\forall x, y \in X$,

    (i) $\iota(x, y) = \rho(x, y) \vee \rho(y, x)$;

    (ii) $\rho(x, y) \leq \rho(y, x) \Leftrightarrow \pi(y, x) = 1$;

    (iii) $\rho(x, y) < \rho(y, x) \Rightarrow \pi(x, y)$ is a solution to (3.2).

Proof. Suppose (1) holds.

(i) Since π is c-asymmetric, either $\pi(x, y) = 1$ or $\pi(y, x) = 1$. Hence either $\rho(x, y) = \iota(x, y)$ or $\rho(y, x) = \iota(y, x)$. Say $\pi(x, y) = 1$. Then $\rho(x, y) = \iota(x, y) = \iota(y, x) \geq \rho(y, x)$, where the inequality holds since $\rho(y, x) = \pi(y, x) \otimes \iota(y, x)$ and $\pi(y, x) < 1$. Similarly, if $\pi(y, x) = 1$, then $\rho(y, x) = \iota(y, x) = \iota(x, y) \geq \rho(x, y)$. In either case, $\iota(x, y) = \rho(x, y) \vee \rho(y, x)$.

(ii) Suppose $\rho(x, y) \leq \rho(y, x)$. If $\rho(x, y) = \rho(y, x)$, then $\pi(x, y) = \pi(y, x) = 1$ since π is simple. Suppose $\rho(x, y) < \rho(y, x)$. Then $\pi(x, y) \otimes \iota(x, y) = \rho(x, y) < \rho(y, x) = \pi(y, x) \otimes \iota(y, x)$. Thus $\pi(x, y) \leq \pi(y, x)$. Since π is c-asymmetric, $\pi(y, x) = 1$. Conversely, suppose $\pi(y, x) = 1$. Then $\rho(y, x) = 1 \otimes \iota(y, x) = \iota(y, x) \geq \rho(x, y)$ by (i).

(iii) Suppose $\rho(x, y) < \rho(y, x)$. We're given $\rho(x, y) = \pi(x, y) \otimes \iota(x, y) = \pi(x, y) \otimes \rho(y, x)$. Hence $\pi(x, y)$ is a solution to Eq. (3.2).

Suppose (2) holds. By (i), ι is clearly symmetric. Suppose $\rho(x, y) = \rho(y, x)$. Then by (ii), $\pi(x, y) = 1 = \pi(y, x)$. Thus π is simple. Suppose $\pi(x, y) < 1$. Then $\rho(x, y) \neq \rho(y, x)$ since π is simple. Suppose $\rho(x, y) < \rho(y, x)$. Then $\pi(x, y)$ is a solution to (3.2) by (iii) and so $\rho(x, y) = \pi(x, y) \otimes \rho(y, x) = \pi(x, y) \otimes \iota(x, y)$. By (ii), $\pi(y, x) = 1$ since π is c-asymmetric. The case $\rho(x, y) \geq \rho(y, x)$ is not possible under the assumption $\pi(x, y) < 1$ by (ii). The case $\pi(y, x) < 1$ is similar. Suppose $\pi(x, y) = 1 = \pi(y, x)$. Then clearly π is c-asymmetric and $\rho(x, y) = \rho(y, x)$ by (ii) and $\rho(x, y) = 1 \otimes \iota(x, y)$ and $\rho(y, x) = 1 \otimes \iota(y, x)$. $\blacksquare$

The dual result to Proposition 3.5.5 is [15, Proposition 3.2, p. 25] or [12, Proposition 3, p. 378] and the dual result to Proposition 3.5.6 is [15, Proposition 2.6, p. 20] or [24, Proposition 1.2, p. 364].

Proposition 3.5.6 *Suppose $\otimes$ is a t-norm Suppose $\rho, \pi, \iota \in \mathcal{FR}(X)$. If (i) $\rho = \pi \otimes \iota$, (ii) π is c-asymmetric, and (iii) ι is symmetric, then $\iota(x, y) = \rho(x, y) \vee \rho(y, x)$.*

Let c be an involutive fuzzy complement and $\rho \in \mathcal{FR}(X)$. Define the fuzzy preference relation ρ^c on X by $\forall x, y \in X$, $\rho^c(x, y) = c(\rho(x, y))$.

Proposition 3.5.7 *Let c be a involutive fuzzy complement. Let $\oplus$ and $\otimes$ be dual t-conorms and t-norms under c, respectively.*

(1) *If $\rho = \pi \oplus \iota$, then $\rho^c = \pi^c \otimes \iota^c$;*

(2) *If $\rho = \pi \otimes \iota$, then $\rho^c = \pi^c \oplus \iota^c$.*

Proof. (1) Let $x, y \in X$. Then $\rho(x,y) = \pi(x,y) \oplus \iota(x,y) = c(c(\pi(x,y)) \otimes c(\iota(x,y)))$. Thus $c(\rho(x,y)) = c(\pi(x,y)) \otimes c(\iota(x,y))$. That is, $\rho^c(x,y) = \pi^c(x,y) \otimes \iota^c(x,y)$.

(2) The proof of (2) is similar to that of (1). ∎

Consider (1) of Proposition 3.5.7, where π is asymmetric and ι is symmetric. Then $\iota(x,y) = \rho(x,y) \wedge \rho(y,x)$ by [15, Proposition 2.6, p. 20] or [24, Proposition 1.2, p. 364] and so

$$\iota^c(x,y) = c(\rho(x,y) \wedge \rho(y,x)) = c(\rho(x,y)) \vee c(\rho(y,x)) = \rho^c(x,y) \vee \rho^c(y,x).$$

Hence we see that if $\rho^c = \pi^c \otimes \iota^c = \widehat{\pi} \otimes \widehat{\iota}$, where $\widehat{\pi}$ is c-asymmetric and $\widehat{\iota}$ is symmetric, then $\iota^c = \widehat{\iota}$ by Proposition 3.5.6. If $\otimes$ is strict, the solution to (3.2) is unique so it follows that $\widehat{\pi} = \pi^c$.

Similar comments can be made concerning (2).

Key idea: Let c be an involutive complement and let $\oplus$ and $\otimes$ be a t-conorm and t-norm, respectively, dual with respect to c. Let f be the function of $\mathcal{FR}(X)$ into $\mathcal{FR}(X)$ defined by $\forall \rho \in \mathcal{FR}(X)$, $f(\rho) = \rho^c$. Then f is one-to-one and onto. Hence when a result concerning ρ^c is determined from a result concerning ρ, then the result for ρ^c is a general result since f is one-to-one and onto. Thus if $\langle \rho_\mu, \rho_\nu \rangle$ is an intuitionistic fuzzy preference relation, a result for ρ_μ has an immediate corresponding result for ρ_ν.

Theorem 3.5.8 *Let c be an involutive fuzzy complement. Let $\oplus$ and $\otimes$ be dual t-conorms and t-norms under c, respectively. Let $\otimes_1, \otimes_2 \in \mathcal{C}(\mathcal{T})$. Let $\rho \in \mathcal{FR}(X)$. Suppose that $\rho = \pi_1 \otimes_1 \iota = \pi_2 \otimes_2 \iota$. Suppose further that either $\otimes_1$ or $\otimes_2$ is strict. Then $\otimes_1 \supseteq \otimes_2$ if and only if $\pi_1 \subseteq \pi_2$ for all such factorizations.*

Proof. 1. We know $\rho^c = \pi_1^c \oplus_1 \iota^c = \pi_2^c \oplus_2 \iota^c$. Also $\oplus_1 \subseteq \oplus_2$ if and only if $\otimes_1 \supseteq \otimes_2$ and in fact $\oplus_i$ is strict if and only if $\otimes_i$ is strict, $i = 1, 2$. From [14, Theorem 3.11, p. 32], we have that $\oplus_1 \subseteq \oplus_2$ if and only if $\pi_1^c \supseteq \pi_2^c$ and so $\otimes_1 \supseteq \otimes_2$ if and only if $\pi_1 \subseteq \pi_2$.

Proof. 2. Suppose $\otimes_1 \supseteq \otimes_2$. Suppose $\exists x, y \in X$ such that $\pi_1(x,y) > \pi_2(x,y)$. Now $\pi_1(x,y) \otimes_1 \iota(x,y) \geq \pi_1(x,y) \otimes_2 \iota(x,y)$ and $\pi_2(x,y) \otimes_1 \iota(x,y) \geq \pi_2(x,y) \otimes_2 \iota(x,y)$. Suppose $\otimes_1$ is strict. Then $\pi_1(x,y) \otimes_1 \iota(x,y) > \pi_2(x,y) \otimes_2 \iota(x,y)$. Hence $\rho(x,y) = \pi_1(x,y) \otimes_1 \iota(x,y) > \pi_2(x,y) \otimes_1 \iota(x,y) \geq \pi_2(x,y) \otimes_2 \iota(x,y) = \rho(x,y)$, a contradiction. Suppose $\otimes_2$ is strict. Then $\pi_1(x,y) \otimes_2 \iota(x,y) > \pi_2(x,y) \otimes_2 \iota(x,y)$. Thus $\rho(x,y) = \pi_2(x,y) \otimes_2 \iota(x,y) < \pi_1(x,y) \otimes_2 \iota(x,y) \leq \pi_1(x,y) \otimes_1 \iota(x,y) = \rho(x,y)$, a contradiction. Hence no such x, y exist. Thus $\pi_1 \subseteq \pi_2$.

Conversely, suppose $\pi_1 \subseteq \pi_2$. Suppose $\exists a, b \in [0,1]$ such that $a \otimes_1 b < a \otimes_2 b$. Suppose $\otimes_1$ is strict. Let $\rho \in \mathcal{FR}(X)$ be such that $\exists x, y \in X$ such that $\rho(x,y) = a \otimes_1 b$ and $\rho(y,x) = b$. Then $\rho(x,y) \leq \rho(y,x)$. Thus $\iota(x,y) = b$. Since $\rho(x,y) = \pi_1(x,y) \otimes_1 \iota(x,y)$ and $\otimes_1$ is strict, $\pi_1(x,y) = a$. Thus $\rho(x,y) = a \otimes_1 b < a \otimes_2 b \leq \pi_2(x,y) \otimes_2 \iota(x,y) = \rho(x,y)$, a contradiction. Suppose $\otimes_2$ is strict.

Let $\rho \in FR$ be such that $\exists x, y \in X$ such that $\rho(x, y) = a \otimes_2 b$ and $\rho(y, x) = b$. Then $\rho(x, y) \leq \rho(y, x)$. Thus $\iota(x, y) = b$. Since $\rho = \pi_2 \otimes_2 \iota$ and $\otimes_2$ is strict, $\pi_2(x, y) = a$. Thus $\rho(x, y) = \pi_1(x, y) \otimes_1 \iota(x, y) \leq a \otimes_1 b < a \otimes_2 b = \rho(x, y)$, a contradiction. Hence no such a and b exist. Thus $\otimes_1 \supseteq \otimes_2$. ∎

Proposition 3.5.9 *Let $\otimes \in \mathcal{T}$. Suppose $\otimes = \wedge$, (i) $\rho = \pi \otimes \iota$, (ii) π is c-asymmetric, (iii) ι is symmetric. Then π is simple if and only if $\forall x, y \in X$,*

$$\pi(x, y) = \begin{cases} \rho(x, y) & \text{if } \rho(x, y) < \rho(y, x) \\ 1 & \text{otherwise.} \end{cases}$$

Proof. Let $x, y \in X$. Assume π is simple. By (i), $\rho(x, y) \leq \pi(x, y)$. Suppose $\rho(x, y) < \rho(y, x)$. Assume $\rho(x, y) < \pi(x, y)$. Then $\rho(x, y) = \iota(x, y) = \iota(y, x) = \rho(y, x)$, where the last equality holds by Proposition 3.5.5. However, this is impossible. Thus $\rho(x, y) \geq \pi(x, y)$ and so $\rho(x, y) = \pi(x, y)$. Now suppose $\rho(x, y) \geq \rho(y, x)$. If $\rho(x, y) = \rho(y, x)$, then $\pi(x, y) = \pi(y, x)$ and so $\pi(x, y) = \pi(y, x) = 1$ since π is simple. If $\rho(x, y) > \rho(y, x)$, then it has just been shown that $\pi(y, x) = \rho(y, x) < 1$. Hence $\pi(x, y) = 1$ by Proposition 3.5.5.

Conversely, suppose $\pi(x, y) = \rho(x, y)$ if $\rho(x, y) < \rho(y, x)$ and 1 otherwise. Suppose $\rho(x, y) = \rho(y, x)$. Then $\pi(x, y) = \pi(y, x) = 1$ by the hypothesis. It follows that π is simple. ∎

Proposition 3.5.10 *Let $\rho \in \mathcal{FR}(X)$. Suppose $\rho = \pi^\ominus \oplus \iota$. Let c be an involutive fuzzy complement and let $\oplus$ and $\otimes$ be a dual t-conorm and a t-norm with respect to c, respectively. Then*
(1) $\pi^{\ominus c} = \pi^\oslash$, *where $\rho^c = \pi^\oslash \otimes \iota^c$.*
(2) $\pi^{\oslash c} = \pi^\ominus$, *where $\rho^c = \pi^\ominus \otimes \iota^c$.*

Proof. (1) Since $\rho = \pi^\ominus \oplus \iota, \rho^c = \pi^{\ominus c} \otimes \iota^c$. Suppose $\exists a \in [0, 1]$ and $x, y \in X$ such that $\rho^c = a \otimes \iota^c$, where $a > \pi^{\ominus c}(x, y)$. Then $\rho(x, y) = c(a \otimes \iota^c(x, y)) = c(c(c(a)) \otimes \iota^c(x, y)) = c(a) \oplus \iota(x, y)$. However, this contradicts the minimality of $\pi^\ominus$ since $c(a) < c(\pi^{\ominus c}(x, y)) = c(c(\pi^\ominus(x, y)) = \pi^\ominus(x, y)$.
(2) The proof for (2) is similar to that of (1). ∎

Theorem 3.5.11 *Let $\otimes_1, \otimes_2 \in \mathcal{C}(\mathcal{T})$. Let $\rho \in \mathcal{FR}(X)$. Suppose that $\rho = \pi_1^\oslash \otimes_1 \iota = \pi_2^\oslash \otimes_2 \iota$. Then $\pi_1^\oslash \subseteq \pi_2^\oslash$ for all such factorizations if and only if $\otimes_1 \supseteq \otimes_2$.*

Proof. 1. The proof follows from Proposition 3.5.10 and [15, Theorem 3.13, p. 33].

Proof. 2. Suppose $\pi_1^\oslash \subseteq \pi_2^\oslash$ for all such factorizations. Suppose there exists $a, b \in [0, 1]$ such that $a \otimes_1 b < a \otimes_2 b$. Let $\rho(x, y) = a \otimes_1 b$ and $\rho(y, x) = b$ for some $x, y \in X$. Then $\iota(x, y) = b$. Since $\pi_1^\oslash$ is the maximal solution, $\pi_1^\oslash(x, y) \geq a$ and thus $\pi_2^\oslash(x, y) \geq \pi_1^\oslash(x, y) \geq a$. Therefore, $\rho(x, y) = \pi_2^\oslash(x, y) \otimes_2 b \geq a \otimes_2 b > a \otimes_1 b = \rho(x, y)$, a contradiction. Hence no such a, b exist. Thus $\otimes_1 \supseteq \otimes_2$.

Conversely, suppose $\otimes_1 \supseteq \otimes_2$. Suppose there exists $\rho \in \mathcal{FR}(X)$ such that $\pi_1^\varnothing(x,y) > \pi_2^\varnothing(x,y)$ for some $x,y \in X$. Now $\pi_1^\varnothing(x,y) \otimes_1 \iota(x,y) \geq \pi_1^\varnothing(x,y) \otimes_2 \iota(x,y)$. Also $\pi_2^\varnothing(x,y) \otimes_2 \iota(x,y) < \pi_1^\varnothing(x,y) \otimes_2 \iota(x,y)$, where strict equality holds by the maximality of $\pi_2^\varnothing$. Thus $\rho(x,y) = \pi_1^\varnothing(x,y) \otimes_1 \iota(x,y) < \pi_1^\varnothing(x,y) \otimes_2 \iota(x,y) \leq \pi_1^\varnothing(x,y) \otimes_1 \iota(x,y) = \rho(x,y)$, a contradiction. Hence no such x,y exist. Thus $\pi_1^\varnothing \subseteq \pi_2^\varnothing$. ∎

Let T denote the set of all intuitionistic fuzzy preference relations $\langle \rho_\mu, \rho_\nu \rangle$ on X. Then T^n is the set of all such n-tuples.

Proposition 3.5.12 *Let c be an involutive fuzzy complement. Let $(\langle \rho_{1_\mu}, \rho_{1_\nu} \rangle,$ $\ldots, \langle \rho_{n_\mu}, \rho_{n_\nu} \rangle) \in T^n$ be such that $\rho_{i_\nu} = \rho_{i_\mu}^c$, $i = 1, \ldots, n$. Let $i \in N$. Let $x, y \in X$. Then*

$$[\pi_{i_\mu}(x,y) > 0 \Rightarrow \pi_\mu(x,y) > 0] \Leftrightarrow [\pi_{i_\mu}^c(x,y) < 1 \Rightarrow \pi_\mu^c(x,y) < 1].$$

Proof. Suppose $[\pi_{i_\mu}(x,y) > 0 \Rightarrow \pi_\mu(x,y) > 0]$. Suppose $\pi_{i_\mu}^c(x,y) < 1$. Then $\pi_{i_\mu}(x,y) = \pi_{i_\mu}^{cc}(x,y) > 0$. Hence $\pi_\mu(x,y) > 0$. Thus $\pi_\mu^c(x,y) < 1$. The remainder of the proof follows in a similar manner. ∎

Corollary 3.5.13 *Let c be an involutive fuzzy complement. Let $(\langle \rho_{1_\mu}, \rho_{1_\nu} \rangle,$ $\ldots, \langle \rho_{n_\mu}, \rho_{n_\nu} \rangle) \in T^n$ be such that $\rho_{i_\nu} = \rho_{i_\mu}^c$, $i = 1, \ldots, n$. If i is a dictator for ρ_μ, then i is a dictator for $\rho_\nu = \rho_\mu^c$ and conversely.*

Example 3.5.14 *Let $N = \{1,2\}$ and $X = \{x,y\}$. Let $(\langle \rho_{1_\mu}, \rho_{1_\nu} \rangle, \langle \rho_{2_\mu}, \rho_{2_\nu} \rangle) \in T^2$ be such that*

$$\rho_{1_\mu}(x,y) = \frac{1}{2}, \rho_{1_\mu}(y,x) = 0, \rho_{2_\mu}(x,y) = 1, \rho_{2_\mu}(y,x) = \frac{1}{2},$$

$$\rho_{1_\nu}(x,y) = \frac{1}{2}, \rho_{1_\nu}(y,x) = \frac{1}{4}, \rho_{2_\nu}(x,y) = 0, \rho_{2_\mu}(y,x) = \frac{1}{2}.$$

Suppose player 1 is a dictator for ρ_μ. Then $\rho_{1_\mu}(x,y) > \rho_{1_\mu}(y,x)$ and so $\pi_{1_\mu}(x,y) > 0$. Hence $\pi_\mu(x,y) > 0$ since 1 is a dictator for ρ_μ. Now $\pi_{1_\nu}(x,y) = 1$ and $\pi_{1_\nu}(y,x) < 1$. Thus if player 1 were a dictator for ρ_ν, then $\pi_\nu(y,x) < 1$. But we should have $\pi_\nu(x,y) < 1$ if player 1 is to be a dictator for $(\langle \rho_{1_\mu}, \rho_{1_\nu} \rangle, \langle \rho_{2_\mu}, \rho_{2_\nu} \rangle)$. (Note that it is not the case that $\rho_{1_\mu}(x,y) > \rho_{1_\mu}(y,x) \Leftrightarrow \rho_{1_\nu}(x,y) < \rho_{1_\nu}(y,x)$.)

Let c be an involutive fuzzy complement. We have $\forall x, y \in X$,

$$\rho(x,y) > 0 \Leftrightarrow \rho^c(x,y) < 1, \quad \rho(x,y) = 0 \Leftrightarrow \rho^c(x,y) = 1, \quad \rho(x,y) = 1$$
$$\Leftrightarrow \rho^c(x,y) = 0, \rho(x,y) = \rho(y,x) \Leftrightarrow \rho^c(x,y) = \rho^c(y,x),$$

and $Supp(\rho) = Cosupp(\rho^c)$. Then ρ is asymmetric $\Leftrightarrow \rho^c$ is c-asymmetric.

Proposition 3.5.15 [15, *Proposition 2.2, p.18*] *Let* $\rho, \pi, \iota \in \mathcal{FR}(X)$. *Suppose that* (i) $Supp(\rho) = Supp(\pi) \cup Supp(\iota)$, (ii) π *is asymmetric,* (iii) *is symmetric. Then*

(1) $\forall x, y \in X$, $\rho(x, y) > 0$ *and* $\rho(y, x) = 0$ *if and only if* $\iota(x, y) = \iota(y, x) = 0$.

(2) $\forall x, y \in X$, $\rho(x, y) > 0$ *and* $\rho(y, x) > 0$ *if and only if* $\iota(x, y) = \iota(y, x) > 0$.

Then the following proposition follows from Proposition 3.5.15 and the discussion immediately preceding it.

Proposition 3.5.16 *Let* $\rho, \pi, \iota \in \mathcal{FR}(X)$. *Suppose that*
(i) $Cosupp(\rho^c) = Cosupp(\pi^c) \cup Cosupp(\iota^c)$, (ii) π^c *is c-asymmetric,* (iii) ι^c *is symmetric. Then*
(1) $\forall x, y \in X$, $\rho^c(x, y) < 1$ *and* $\rho^c(y, x) = 1$ *if and only if* $\iota^c(x, y) = \iota^c(y, x) = 1$.
(2) $\forall x, y \in X$, $\rho^c(x, y) < 1$ *and* $\rho^c(y, x) < 1$ *if and only if* $\iota^c(x, y) = \iota^c(y, x) < 1$.

Since ρ is an arbitrary element of $\mathcal{FR}(X)$ and f is a one-to-one function from $\mathcal{FR}(X)$ onto $\mathcal{FR}(X)$, we have that ρ^c is also an arbitrary element from $\mathcal{FR}(X)$. Hence the following result is true.

Proposition 3.5.17 *Let* $\rho, \pi, \iota \in \mathcal{FR}(X)$. *Suppose that*
(i) $Cosupp(\rho) = Cosupp(\pi) \cup Cosupp(\iota)$, (ii) π *is c-asymmetric,* (iii) ι *is symmetric. Then*
(1) $\forall x, y \in X$, $\rho(x, y) < 1$ *and* $\rho(y, x) = 1$ *if and only if* $\iota(x, y) = \iota(y, x) = 1$.
(2) $\forall x, y \in X$, $\rho(x, y) < 1$ *and* $\rho(y, x) < 1$ *if and only if* $\iota(x, y) = \iota(y, x) < 1$.

Consider an intuitionistic fuzzy subset $\langle \rho_\mu, \rho_\nu \rangle$ of X. If Proposition 3.5.15 holds for ρ_μ, then Proposition 3.5.17 holds for ρ_ν (and conversely).

Example 3.5.18 *Let* $x = \{x, y\}$. *Let* ρ_μ *and* ρ_ν *be fuzzy binary relations on* X *defined as follows:* $\forall w \in X$, $\rho_\mu(w, w) = 1$ *and* $\rho_\nu(w, w) = 0$,

$$\rho_\mu(x, y) = \frac{1}{2}, \rho_\mu(y, x) = \frac{3}{8}, \rho_\nu(x, y) = \frac{3}{8}, \rho_\nu(y, x) = \frac{1}{4}.$$

Then $\rho_\mu(u, v) + \rho_\nu(u, v) < 1$ $\forall u, v \in X$. *Let* $\oplus = \vee$ *and* $\otimes = \wedge$. *Then* $\rho_\mu = \pi_\mu \vee \iota_\mu$ *and* $\rho_\nu = \pi_\nu \wedge \iota_\nu$, *where* $\iota_\mu(u, v) = \rho_\mu(u, v) \wedge \rho_\mu(v, u)$ *and* $\iota_\nu(u, v) = \rho_\nu(u, v) \vee \rho_\nu(v, u)$ *for all* $u, v \in X$ *and*

$$\pi_\mu(x, y) = \frac{1}{2}, \pi_\mu(y, x) = 0, \pi_\nu(x, y) = 1, \pi_\nu(y, x) = \frac{1}{4}.$$

We have that $\pi_\mu(x, y) + \pi_\nu(x, y) > 1$.

Norms

Standard intersection: $a \otimes b = a \wedge b$

Bounded difference: $a \otimes_2 b = 0 \vee (a + b - 1)$

Drastic intersection: $a \otimes_d b = \begin{cases} a \text{ if } b = 1 \\ b \text{ if } a = 1 \\ 0 \text{ otherwise} \end{cases}$

Algebraic product: $a \otimes_A b = ab$

Yager: $a \otimes_w b = \begin{cases} 1 - ((1-a)^w + (1-b)^w)^{1/w} \\ \qquad \text{if } (1-a)^w + (1-b)^w \in [0,1] \\ 0 \text{ otherwise} \end{cases}$

Nilpotent: $a \otimes_n b = \begin{cases} a \wedge b \text{ if } a + b > 1 \\ 0 \text{ otherwise} \end{cases}$

Einstein: $a \otimes_{H_2} b = ab/(2 - a - b + ab)$

Schweizer and Sklar: $a \otimes_p b = \begin{cases} 0 \vee (a^p + b^p - 1)^{1/p} \text{ if } 2 - a^p - b^p \in [0,1] \\ 0 \text{ otherwise} \end{cases}$

Frank: $a \otimes_s b = \log_s \left[1 + \frac{(s^a - 1)(s^b - 1)}{s - 1} \right], s > 0, s \neq 1$

Dombi: $a \otimes b = \left[1 + \left[(\frac{1}{a} - 1)^\lambda + (\frac{1}{b} - 1)^\lambda \right]^{\frac{1}{\lambda}} \right]^{-1}, \lambda > 0$

Hamacher: $a \otimes b = ab/[r + (1-r)(a + b - ab], r > 0$

Schweizer and Sklar 2: $a \otimes b = 1 - [(1-a)^p + (1-b)^p - (1-a)^p(1-b)^p]^{1/p}$, $p > 0$

Schweizer and Sklar 3: $a \otimes b = e^{-(|\ln a|^p + |\ln b|^p)^{1/p}}$, $p > 0$

Schweizer and Sklar 4: $a \otimes b = ab/[a^p + b^p - a^p b^p]^{1/p}$, $p > 0$

Dubois and Prade: $a \otimes b = ab/a \vee b \vee \alpha$, $\alpha \in [0,1]$

Weber: $a \otimes b = 0 \vee (a + b + \lambda ab - 1)/(1 + \lambda)$, $\lambda > -1$

Yu: $a \otimes b = 0 \vee [(1+\lambda)(a + b - 1) - \lambda ab]$, $\lambda > -1$

We now derive factorizations for t-norms. We give direct proofs for several results. Later we use factorization results for t-conorms to derive factorization results for t-norms.

Proposition 3.5.19 *Let $\otimes = \wedge$. Let $\rho \in \mathcal{FR}(X)$. Suppose $\rho = \pi \otimes \iota$. Then $\forall x, y \in X$,*

$$\pi(x,y) = \begin{cases} \rho(x,y) & \text{if } \rho(x,y) < \rho(y,x) \\ 1 & \text{otherwise.} \end{cases}$$

Theorem 3.5.20 *Let* $\otimes = \otimes_A$. *Let* $\rho \in \mathcal{FR}(X)$. *Suppose* $\rho = \pi \otimes \iota$. *Then* $\forall x, y \in X$,

$$\pi(x, y) = \begin{cases} 1 & \text{if } \rho(x, y) \geq \rho(y, x) \\ \frac{\rho(x,y)}{\rho(y,x)} & \text{if } \rho(x, y) < \rho(y, x). \end{cases}$$

Proof. Let $x, y \in X$. Then $\rho(x, y) = \pi(x, y) \otimes \iota(x, y)$. Since $\iota(x, y) = \rho(x, y) \vee \rho(y, x)$, we have $\pi(x, y) = 1$ if $\rho(x, y) \geq \rho(y, x)$. Suppose $\rho(x, y) < \rho(y, x)$. Then $\rho(x, y) = \pi(x, y) \otimes \rho(y, x) = \pi(x, y)\rho(y, x)$. Thus $\pi(x, y) = \rho(x, y)/\rho(y, x)$.

Let c be the standard fuzzy complement. Then $\oplus_A$ and $\otimes_A$ are dual with respect to c. Suppose $\oplus = \oplus_A$ and $\rho(x, y) = \pi(x, y) \oplus \iota(x, y)$ as in Theorem 3.5.4. Then $\rho^c(x, y) = \pi^c(x, y) \otimes \iota^c(x, y)$, where $\otimes = \otimes_A$. Suppose $\rho(x, y) > \rho(y, x)$. Then $\rho^c(x, y) < \rho^c(y, x)$. By Theorem 3.5.4, $\pi^c(x, y) = (\rho^c(x, y) - \rho^c(y, x))/(1 - \rho^c(y, x)) = (1 - \rho(x, y) - 1 + \rho(y, x))/\rho(y, x) = (\rho(y, x) - \rho(x, y))/\rho(y, x)$. Thus $\pi(x, y) = 1 - (\rho(y, x) - \rho(x, y))/\rho(y, x) = \rho(x, y)/\rho(y, x)$ which agrees with Theorem 3.5.20. ∎

Proposition 3.5.21 *Let* $\otimes_2$ *be the t-norm defined by* $\forall a, b \in [0, 1]$, $a \otimes_2 b = 0 \vee (a + b - 1)$. *Let* $\rho \in \mathcal{FR}(X)$. *Suppose* $\rho = \pi \otimes_2 \iota$. *Then* $\forall x, y \in X$,

$$\pi^\varnothing(x, y) = \begin{cases} 1 & \text{if } \rho(x, y) \geq \rho(y, x) \\ 1 + \rho(x, y) - \rho(y, x) & \text{if } \rho(x, y) < \rho(y, x). \end{cases}$$

Proof. If $\rho(x, y) \geq \rho(y, x)$, then $\iota(x, y) = \rho(x, y)$ and so $\pi(x, y) = 1$. Suppose $\rho(x, y) < \rho(y, x)$. Then $\iota(x, y) = \rho(y, x)$. Thus $\rho(x, y) = 0 \vee (\pi(x, y) + \rho(y, x) - 1)$. If $\pi(x, y) + \rho(y, x) - 1 \geq 0$, then $\pi(x, y) = 1 + \rho(x, y) - \rho(y, x)$. If $\pi(x, y) + \rho(y, x) - 1 \leq 0$, then $\rho(x, y) = 0$ and so $\pi(x, y) \leq 1 - \rho(y, x) + \rho(x, y)$. In this case, $\pi(x, y) = 1 - \rho(y, x)$ is the maximal solution. ∎

Proposition 3.5.22 *Let* $\otimes = \otimes_p$. *Let* $\rho \in \mathcal{FR}(X)$. *Suppose* $\rho = \pi \otimes \iota$. *Then* $\forall x, y \in X$,

$$\pi(x, y) = \begin{cases} 1 & \text{if } \rho(x, y) \geq \rho(y, x) \\ (1 + \rho(x, y)^p - \rho(y, x)^p)^{1/p} & \text{if } \rho(x, y) < \rho(y, x) \end{cases}$$

Proof. Suppose $\rho(x, y) \geq \rho(y, x)$. Then $\rho(x, y) = 0 \vee ((t^p + \rho(x, y)^p - 1)^{1/p}$ since $\iota(x, y) = \rho(x, y)$. Thus $t = 1$. Suppose $\rho(x, y) < \rho(y, x)$. Then $\rho(x, y) = t \otimes \rho(y, x) = 0 \vee (t^p + \rho(y, x)^p - 1)^{1/p} = \rho(x, y)$ for $t = (1 + \rho(x, y)^p - \rho(y, x)^p)^{1/p}$.

∎

(Note $2 - t^p - \rho(y, x)^p = 2 - (1 + \rho(x, y)^p - \rho(y, x)^p) - \rho(y, x)^p \in [0, 1]$.)

Proposition 3.5.23 *Let* $\otimes = \otimes_w$. *Let* $\rho \in \mathcal{FR}(X)$. *Suppose* $\rho = \pi \otimes \iota$. *Then* $\forall x, y \in X$,

$$\pi^{\oslash}(x,y) = \begin{cases} 1 & \text{if } \rho(x,y) \ge \rho(y,x) \\ 1 - [(1 - \rho(x,y))^w - (1 - \rho(y,x))^w]^{1/w} & \text{if } 0 < \rho(x,y) < \rho(y,x). \end{cases}$$

Proof. Let $x,y \in X$. Suppose $0 < \rho(x,y) < \rho(y,x)$. Then $\rho(x,y) = t \otimes \rho(y,x) = 1 - ((1-t)^w + (1-\rho(y,x))^w)^{1/w}$ and so

$$(1-t)^w = (1-\rho(x,y))^w - (1 - (\rho(y,x))^w,$$
$$t = 1 - [(1-\rho(x,y))^w - (1 - \rho(y,x))^w]^{1/w}.$$

■

(Note that if $\rho(x,y) = 0, \rho(y,x) = \frac{7}{8}$ and $w = 1$, then any $t \in [0, \frac{1}{8}]$ is a solution.)

Remark 3.5.24 *Let $\otimes = \otimes_s$. Let $\rho \in \mathcal{FR}(X)$. Suppose $\rho = \pi \otimes \iota$. Then $\forall x, y \in X$,*

$$\pi(x,y) = \log_s[1 + (s-1)(s^{\rho(x,y)} - 1)/(s^{\rho(y,x)} - 1)] \quad \text{if } \rho(x,y) < \rho(y,x).$$

Proof. Suppose $\rho(x,y) < \rho(y,x)$. Then $\rho(x,y) = t \otimes \rho(y,x) = \log_s[1 + (s^t-1)(s^{\rho(y,x)}-1)/(s-1)]$. Thus $s^{\rho(x,y)} - 1 = (s^t-1)(s^{\rho(y,x)}-1)/(s-1)$. Hence $s^t - 1 = (s-1)(s^{\rho(x,y)} - 1)/(s^{\rho(y,x)} - 1)$. Thus $t = \log_s[1 + (s-1)(s^{\rho(x,y)} - 1)/(s^{\rho(y,x)} - 1)]$. ■

Example 3.5.25 *Let $\otimes = \otimes_n$. Then there exists $\rho \in \mathcal{FR}(X)$ such that ρ does not factor: Let $X = \{x, y\}$ and let ρ be such that $\rho(x,y) = \frac{1}{4}$ and $\rho(y,x) = \frac{1}{2}$. Then $\frac{1}{4} = t \otimes \frac{1}{2} \ne t \wedge \frac{1}{2}$ if $t + \frac{1}{2} > 1$ and $\frac{1}{4} = t \otimes \frac{1}{2} \ne 0$ otherwise.*

Proposition 3.5.26 *Let $\otimes = \otimes_n$. Let $\rho \in \mathcal{FR}(X)$. Suppose $\rho(x,y) < \rho(y,x)$ and $\rho(x,y) = \pi(x,y) \otimes \rho(y,x)$. Then $\pi(x,y) = \rho(x,y)$.*

Proof. Since $\rho(x,y) = \pi(x,y) \otimes \rho(y,x)$, there exists $t \in [0,1]$ such that $t + \rho(y,x) > 1$ or $\rho(x,y) = 0$ and $\rho(x,y) = t \otimes \rho(y,x)$ for $t \in [0, 1 - \rho(y,x)]$. In either case, $\pi(x,y) = \rho(x,y)$. ■

The dual to Proposition 3.3.6 is the following result.

Proposition 3.5.27 *Let $\otimes = \otimes_n$. Let $\rho \in \mathcal{FR}(X)$. Then ρ has a factorization $\rho = \pi \otimes \iota$ if and only if $\forall x, y \in X$, either (1) $\rho(x,y) = 0$ or $\rho(y,x) = 0$ or (2) $\rho(x,y) = \rho(y,x)$ or (3) $\rho(x,y) + \rho(y,x) > 1$ and $\rho(x,y) \ne \rho(y,x)$.*

Proof. $\rho(x,y) + \rho(y,x) < 1$ if and only if $1 - \rho(x,y) + 1 - \rho(y,x) > 1$. ■

Proposition 3.5.28 *Let* $\otimes = \otimes_{H_2}$. *Let* $\rho \in \mathcal{FR}(X)$. *Suppose* $\rho = \pi \otimes \iota$. *Then* $\forall x, y \in X$,

$$\pi(x, y) = \begin{cases} (2\rho(x, y) - \rho(x, y)\rho(y, x))/(\rho(x, y) + \rho(y, x) - \rho(x, y)\rho(y, x)) \\ \quad if\ \rho(x, y) \geq \rho(y, x) \\ 1\ \ if\ \rho(x, y) < \rho(y, x). \end{cases}$$

Proof. Let t be a solution to Eq. (3.2). Then $\rho(x, y) = t\iota(x, y)/(2 - t - \iota(x, y) + t\iota(x, y))$ and so $2\rho(x, y) - t\rho(x, y) - \rho(x, y)\iota(x, y) + t\rho(x, y)\iota(x, y) = t\iota(x, y)$. Thus $t\iota(x, y) + t\rho(x, y) - t\rho(x, y)\iota(x, y) = 2\rho(x, y) - \rho(x, y)\iota(x, y)$. Hence

$$t = (2\rho(x, y) - \iota(x, y)\rho(x, y))/(\iota(x, y) + \rho(x, y) - \rho(x, y)\iota(x, y)).$$

The desired result now follows easily. ∎

Proposition 3.5.29 *Let* $\otimes = \otimes_d$. *Let* $\rho \in \mathcal{FR}(X)$. *Suppose* $\rho(x, y) < \rho(y, x)$ *and* $\rho(x, y) = \pi(x, y) \otimes \rho(y, x)$. *Then*

$$\pi(x, y) = \begin{cases} \rho(x, y)\ \ if\ \ \rho(x, y) = 1, \\ 0\ \ if\ \rho(x, y) < 1. \end{cases}$$

Proof. If $\rho(x, y) < 1$, then $\rho(x, y) < 1$ and since $\rho(x, y) = \pi(x, y) \otimes \rho(y, x)$, we have $\rho(x, y) = 0$ and $\pi(x, y) = 0$. If $\rho(y, x) = 1$, then the desired result is immediate. ∎

In the remainder of the chapter, factorization results for $\otimes$ are obtained from known factorization results for $\oplus$ and the notion of duality involving an involutive fuzzy complement.

Proposition 3.5.30 *Let* $\otimes = \otimes_p$. *Let* $\rho \in \mathcal{FR}(X)$. *Suppose* $\rho(x, y) = \pi(x, y) \otimes \rho(y, x)$. *Then*

$$\pi(x, y) = \begin{cases} [\rho(x, y)^p - \rho(y, x)^p + 1]^{1/p}\ \ if\ \ \rho(x, y) < \rho(y, x), \\ 1\ \ otherwise. \end{cases}$$

Proof. For $\oplus = \oplus_p$, we have for $\rho(x, y) = \pi(x, y) \otimes \rho(y, x)$,

$$\pi(x, y) = \begin{cases} 1 - [(1 - \rho(x, y))^p - (1 - \rho(y, x))^p + 1]^{1/p}\ if\ \rho(x, y) > \rho(y, x) \\ 0\ \text{otherwise} \end{cases}$$

Thus

$$\pi^c(x, y) = \begin{cases} 1 - [1 - [(1 - \rho(x, y))^p - (1 - \rho(y, x))^p + 1]^{1/p}]\ if\ \rho(x, y) > \rho(y, x) \\ 1\ \text{otherwise.} \end{cases}$$

Hence

$$\pi^c(x,y) = \begin{cases} [(1-\rho(x,y))^p - (1-\rho(y,x))^p + 1]^{1/p} & \text{if } \rho(x,y) > \rho(y,x) \\ 1 & \text{otherwise.} \end{cases}$$

Thus

$$\pi^c(x,y) = \begin{cases} [\rho^c(x,y)^p - \rho^c(y,x)^p + 1]^{1/p} & \text{if } \rho^c(x,y) < \rho^c(y,x) \\ 1 & \text{otherwise.} \end{cases}$$

∎

Proposition 3.5.31 *Let* $\otimes$ *denote the Dubois and Prade norm. Let* $\rho \in \mathcal{FR}(X)$. *Suppose* $\rho(x,y) > \rho(y,x)$ *and* $\rho(x,y) = \pi(x,y) \otimes \rho(y,x)$. *Then* $\forall x,y \in X$,

$$\pi(x,y) = \begin{cases} \rho(x,y) & \text{if } \alpha \le \rho(y,x) \\ \alpha(\rho(x,y)/\rho(y,x)) & \text{if } \alpha \ge \rho(y,x). \end{cases}$$

Proof. For the Dubois and Prade conorm $\oplus$ with $\rho(x,y) < \rho(y,x)$ and $\rho(x,y) = \pi(x,y) \oplus \rho(y,x)$, we have

$$\pi(x,y) = \begin{cases} \rho(x,y) & \text{if } \alpha \le 1 - \rho(y,x) \\ 1 - [\alpha(1-\rho(x,y))]/[1-\rho(y,x)] & \text{if } \alpha \ge 1 - \rho(y,x). \end{cases}$$

Thus

$$\pi^c(x,y) = \begin{cases} 1 - \rho(x,y) & \text{if } \alpha \le 1 - \rho(y,x) \\ 1 - [1 - [\alpha(1-\rho(x,y))]/[1-\rho(y,x)] & \text{if } \alpha \ge 1 - \rho(y,x). \end{cases}$$

Hence

$$\pi^c(x,y) = \begin{cases} \rho^c(x,y) & \text{if } \alpha \le \rho^c(y,x) \\ \alpha(\rho^c(x,y))/\rho^c(y,x) & \text{if } \alpha \ge \rho^c(y,x). \end{cases}$$

∎

Proposition 3.5.32 *Let* $\otimes$ *denote the Schweizer and Sklar 2 norm. Let* $\rho \in \mathcal{FR}(X)$. *Suppose* $\rho(x,y) < \rho(y,x)$ *and* $\rho(x,y) = \pi(x,y) \oplus \rho(y,x)$. *Then* $\forall x,y \in X$,

$$\pi(x,y) = 1 - \left[\frac{(1-\rho(x,y))^p - (1-\rho(y,x))^p}{1-(1-\rho(y,x))^p} \right]^{1/p}.$$

Proof. For the Schweizer and Sklar 2 conorm $\oplus$ with $\rho \in \mathcal{FR}(X)$, $\rho(x,y) > \rho(y,x)$, and $\rho(x,y) = \pi(x,y) \oplus \rho(y,x)$ we have $\forall x,y \in X$,

$$\pi(x,y) = \left[\frac{\rho(x,y)^p - \rho(y,x)^p}{1-\rho(y,x)^p} \right]^{1/p}.$$

Thus

$$\pi^c(x,y) = 1 - \left[\frac{\rho(x,y)^p - \rho(y,x)^p}{1-\rho(y,x)^p} \right]^{1/p}.$$

Hence for $\rho^c(x,y) > \rho^c(y,x)$

$$\pi^c(x,y) = 1 - \left[\frac{(1-\rho^c(x,y))^p - (1-\rho^c(y,x))^p}{1-(1-\rho^c(y,x))^p}\right]^{1/p}.$$

■

Proposition 3.5.33 *Let* $\oplus$ *denote the Schweizer and Sklar 4 conorm. Let* $\rho \in \mathcal{FR}(X)$. *Suppose* $\rho(x,y) > \rho(y,x)$ *and* $\rho(x,y) = \pi(x,y) \oplus \rho(y,x)$. *Then* $\forall x, y \in X$,

$$\pi(x,y) = \left[\frac{\rho(x,y)^p\rho(y,x)^p}{\rho(y,x)^p + \rho(x,y)^p\rho(y,x)^p - \rho(x,y)^p}\right]^{1/p}.$$

Proof. For the Schweizer and Sklar 4 conorm $\oplus$ with $\rho \in \mathcal{FR}(X)$, $\rho(x,y) > \rho(y,x)$, and $\rho(x,y) = \pi(x,y) \oplus \rho(y,x)$ we have $\forall x, y \in X$,

$$\pi(x,y) = 1 - \left[\frac{(1-\rho(x,y))^p(1-\rho(y,x))^p}{(1-\rho(y,x))^p + (1-\rho(x,y))^p(1-\rho(y,x))^p - (1-\rho(x,y))^p}\right]^{1/p}.$$

Thus

$$\pi^c(x,y) = \left[\frac{\rho^c(x,y)^p\rho^c(y,x)^p}{\rho^c(y,x)^p + \rho^c(x,y)^p\rho^c(y,x)^p - \rho^c(x,y)^p}\right]^{1/p}$$

for $\rho^c(x,y) < \rho^c(y,x)$. ■

Proposition 3.5.34 *Let* $\oplus$ *denote the Schweizer and Sklar 3 conorm. Let* $\rho \in \mathcal{FR}(X)$. *Suppose* $\rho(x,y) > \rho(y,x)$ *and* $\rho(x,y) = \pi(x,y) \oplus \rho(y,x)$. *Then* $\forall x, y \in X$,

$$\pi(x,y) = \begin{cases} 0 \text{ if } \rho(x,y) = 0 \\ 1 - e^{-\left[(\ln(\frac{1}{\rho(x,y)}))^p - |\ln(\rho(y,x)|^p\right]^{1/p}} \text{ if } \rho(x,y) > 0. \end{cases}$$

Proof. For the Schweizer and Sklar 3 conorm $\oplus$ with $\rho \in \mathcal{FR}(X)$, $\rho(x,y) > \rho(y,x)$, and $\rho(x,y) = \pi(x,y) \oplus \rho(y,x)$ we have $\forall x, y \in X$,

$$\pi(x,y) = \begin{cases} 1 \text{ if } \rho(x,y) = 1 \\ 1 - e^{-\left[(\ln(\frac{1}{1-\rho(x,y)}))^p - |\ln(1-\rho(y,x)|^p\right]^{1/p}} \text{ if } \rho(x,y) < 1. \end{cases}$$

Thus for $\rho^c(x,y) < \rho^c(y,x)$,

$$\pi^c(x,y) = \begin{cases} 0 \text{ if } \rho^c(x,y) = 0 \\ e^{-\left[(\ln(\frac{1}{\rho^c(x,y)}))^p - |\ln(\rho^c(y,x)|^p\right]^{1/p}} \text{ if } \rho^c(x,y) > 0. \end{cases}$$

■

Proposition 3.5.35 *Let $\otimes$ denote the Yu norm. Let $\rho \in \mathcal{FR}(X)$. Suppose $\rho(x, y) < \rho(y, x)$ and $\rho(x, y) = \pi(x, y) \oplus \rho(y, x)$. Then $\forall x, y \in X$,*

$$\pi(x, y) = 1 - \frac{\rho(y, x) - \rho(x, y)}{1 + \lambda(1 - \rho(y, x))} \quad \text{if } \rho(x, y) > 0$$

and $\pi^{\oslash}(x, y) = 0$ if $\rho(x, y) = 0$.

Proof. For the Yu conorm $\oplus$ with $\rho \in \mathcal{FR}(X)$, $\rho(x, y) > \rho(y, x)$, and $\rho(x, y) = \pi(x, y) \oplus \rho(y, x)$ we have $\forall x, y \in X$,

$$\pi(x, y) = \frac{\rho(x, y) - \rho(y, x)}{1 + \lambda\rho(y, x)} \quad \text{if } \rho(x, y) < 1$$

and $\pi^{\ominus}(x, y) = 1$ if $\rho(x, y) = 1$. Thus for $\rho^c(x, y) < \rho^c(y, x)$,

$$\pi^c(x, y) = 1 - \frac{(1 - \rho^c(x, y)) - (1 - \rho^c(y, x))}{1 + \lambda(1 - \rho^c(y, x))} \quad \text{if } \rho^c(x, y) > 0$$

and $\pi^{c\oslash}(x, y) = 0$ if $\rho^c(x, y) = 0$. ∎

The next result follows from Proposition 3.3.14 and a dual argument.

Proposition 3.5.36 *Let $\otimes$ be the Dombi norm. Then it is not the case that $\forall x, y \in X, \forall \rho \in \mathcal{FR}(X)$ such that $\rho(x, y) < \rho(y, x), \rho(x, y) = t \otimes \rho(y, x)$ has a solution for t.*

Theorem 3.5.37 *Let $\otimes$ be the Hamacher norm. Then $\forall x, y \in X, \forall \rho \in \mathcal{FR}(X)$ such that $\rho(x, y) < \rho(y, x), \rho(x, y) = t \otimes \rho(y, x)$ has a solution for t if and only if $r = 1$.*

Proof. For the Hamacher conorm $\oplus$, we have $\forall x, y \in X, \forall \rho \in \mathcal{FR}(X)$ such that $\rho(x, y) > \rho(y, x), \rho(x, y) = t \oplus \rho(y, x)$ has a solution for t if and only if $r = 1$. Now $\rho(x, y) > \rho(y, x)$ and $\rho(x, y) = t \oplus \rho(y, x)$ if and only if $\rho^c(x, y) < \rho^c(y, x)$ and $\rho^c(x, y) = 1 - (t \otimes (1 - \rho^c(y, x)))$. ∎

Lemma 3.5.38 *Let $\otimes$ denote the Weber norm. Let $\rho \in \mathcal{FR}(X)$. Let $x, y \in X$. Suppose $\rho(x, y) < \rho(y, x)$ and $\rho(x, y) = \pi(x, y) \otimes \rho(y, x)$. Then*

$$\pi(x, y) = 1 - \frac{\rho(y, x) - \rho(x, y)}{1 - \frac{\lambda}{1-\lambda}(1 - \rho(y, x))} \quad \text{if } -1 < \lambda \le 0$$

$$\text{or } (0 < \lambda < 1 \text{ and } \lambda \le \frac{1 + \rho(x, y) - \rho(y, x)}{1 + 2(1 - \rho(y, x)) - (1 - \rho(x, y))}) \text{ or } (\lambda > 1).$$

Proof. For the Weber conorm $\oplus$, we have with $\rho \in \mathcal{FR}(X)$, $\rho(x, y) > \rho(y, x)$, and $\rho(x, y) = \pi(x, y) \oplus \rho(y, x)$ and $\forall x, y \in X$,

$$\pi(x, y) = \frac{\rho(x, y) - \rho(y, x)}{1 - \frac{\lambda}{1-\lambda}\rho(y, x)} \quad \text{if } -1 < \lambda \le 0$$

$$\text{or } (0 < \lambda < 1 \text{ and } \lambda \le \frac{1 - \rho(x, y) + \rho(y, x)}{1 + 2\rho(y, x) - \rho(x, y)}) \text{ or } (\lambda > 1).$$

Hence

$$\pi^c(x,y) = 1 - \frac{\rho(x,y) - \rho(y,x)}{1 - \frac{\lambda}{1-\lambda}\rho(y,x)} \text{ if } -1 < \lambda \le 0$$

$$\text{or } (0 < \lambda < 1 \text{ and } \lambda \le \frac{1 - \rho(x,y) + \rho(y,x)}{1 + 2\rho(y,x) - \rho(x,y)}) \text{ or } (\lambda > 1).$$

Thus

$$\pi^c(x,y) = 1 - \frac{(1 - \rho^c(x,y)) - (1 - \rho^c(y,x))}{1 - \frac{\lambda}{1-\lambda}(1 - \rho^c(y,x))} \text{ if } -1 < \lambda \le 0$$

$$\text{or } (0 < \lambda < 1 \text{ and } \lambda \le \frac{1 - (1 - \rho^c(x,y)) + (1 - \rho^c(y,x))}{1 + 2(1 - \rho^c(y,x)) - (1 - \rho^c(x,y))}) \text{ or } (\lambda > 1).$$

∎

Proposition 3.5.39 *Let* $\otimes$ *denote the Weber norm. Let* $\rho \in \mathcal{FR}(X)$. *Let* $x, y \in X$. *Suppose* $\rho(x,y) < \rho(y,x)$ *and* $\rho(x,y) = \pi(x,y) \otimes \rho(y,x)$. *Then*

$$\pi(x,y) = 1 - \frac{\rho(y,x) - \rho(x,y)}{1 - \frac{\lambda}{1-\lambda}(1 - \rho(y,x))} \text{ if } -1 < \lambda \le 0 \text{ or } \lambda > 1.$$

Proof. The proof is immediate by Lemma 3.5.38. ∎

Example 3.5.40 *Let* $\langle \rho_\mu, \rho_\nu \rangle$ *be an intuitionistic fuzzy set. Let* $\oplus$ *denote the algebraic sum and* $\otimes$ *the algebraic product. Then* $\forall x, y \in X$,

$$\pi_\mu(x,y) = \begin{cases} \frac{\rho_\mu(x,y) - \rho_\mu(y,x)}{1 - \rho_\mu(y,x)} & \text{if } \rho_\mu(x,y) > \rho_\mu(y,x) \\ 0 & \text{if } \rho_\mu(x,y) \le \rho_\mu(y,x) \end{cases}$$

and

$$\pi_\nu(x,y) = \begin{cases} 1 & \text{if } \rho_\nu(x,y) \ge \rho_\nu(y,x) \\ \frac{\rho_\nu(x,y)}{\rho_\nu(y,x)} & \text{if } \rho_\nu(x,y) < \rho_\nu(y,x). \end{cases}$$

Let $x, y \in X$ *be such that* $\rho_\mu(x,y) = 3/4, \rho_\mu(y,x) = 1/4, \rho_\nu(x,y) = 1/4$, *and* $\rho_\nu(y,x) = 1/2$. *Then* $\rho_\mu(x,y) + \rho_\nu(x,y) \le 1$ *and* $\rho_\mu(y,x) + \rho_\nu(y,x) \le 1$ *and*

$$\begin{aligned}
\pi_\mu(x,y) + \pi_\nu(x,y) &= \frac{\rho_\mu(x,y) - \rho_\mu(y,x)}{1 - \rho_\mu(y,x)} + \frac{\rho_\nu(x,y)}{\rho_\nu(y,x)} \\
&= \frac{[\rho_\mu(x,y) - \rho_\mu(y,x)]\rho_\nu(y,x) + [1 - \rho_\mu(y,x)]\rho_\nu(x,y)}{[1 - \rho_\mu(y,x)]\rho_\nu(y,x)} \\
&= \frac{[3/4 - 1/4]1/2 + [1 - 1/4]1/4}{[1 - 1/4]1/2} = \frac{7/16}{3/8} = \frac{7}{6} > 1.
\end{aligned}$$

Now let $\rho_\mu(x, y) = 1/4$, $\rho_\mu(y, x) = 1/8$, $\rho_\nu(x, y) = 1/4$, and $\rho_\nu(y, x) = 3/4$. Then $\rho_\mu(x, y) + \rho_\nu(x, y) \leq 1$ and $\rho_\mu(y, x) + \rho_\nu(y, x) \leq 1$ and

$$
\begin{aligned}
\pi_\mu(x, y) + \pi_\nu(x, y) &= \frac{\rho_\mu(x, y) - \rho_\mu(y, x)}{1 - \rho_\mu(y, x)} + \frac{\rho_\nu(x, y)}{\rho_\nu(y, x)} \\
&= \frac{1/4 - 1/8}{1 - 1/8} + \frac{1/4}{3/4} = \frac{1/8}{7/8} + \frac{1}{3} \\
&= \frac{1}{7} + \frac{1}{3} < 1.
\end{aligned}
$$

In this section, we provided factorization results for intuitionistic binary relations $\langle \rho_\mu, \rho_\nu \rangle$. We showed that the results for ρ_μ and ρ_ν are dual. A further research project would be to apply the results to Arrowian-like conditions.

3.6 Exercises

1. [15] Let $\rho \in \mathcal{FR}(X)$. Let $h_\rho : X^2 \to [0, 1]^2$ be defined by $\forall x, y \in X$, $h_\rho(x, y) = (\rho(x, y), \rho(y, x))$ and let $\iota_\rho(x, y) = \rho(x, y) \wedge \rho(y, x)$. Let $U = \{\oplus \mid \oplus \text{ is a continuous strict } t\text{-conorm}\}$. Let $P = \{\varphi \mid \varphi : [0, 1]^2 \to [0, 1]\}$. Prove that there exists an injective function $f : P \to U$ defined by $f(\oplus) = \varphi$ such that $\rho(x, y) = \varphi(h_\rho(x, y) \oplus \iota_\rho(x, y))$ $\forall x, y \in X$. Conclude that $\rho = \varphi \circ h_\rho \oplus \iota_\rho$ is a factorization of ρ.

2. [17] Let ρ, π, and ι be fuzzy binary relations on X. Suppose ρ is reflexive and strongly complete. Then (π, ι) is called an axiomatic factorization of ρ if (1) π is irreflexive and asymmetric, (2) ι is reflexive and symmetric, (3) $\pi \cap \iota = \theta$, and (4) $\rho = \pi \cup \iota$. Prove that if (π, ι) is an axiomatic factorization of ρ and $*$ is a continuous t-norm without zero divisors, then ρ is exact.

   Let T be a t-norm and S the dual t-conorm of T, i.e. $S(x, y) = 1 - T(1 - x), 1 - y)$ for all $x \in [0, 1]$. A function $\phi : [0, 1] \to [0, 1]$ is called an order isomorphism if it is bijective and increasing. An order isomorphism is called reciprocal if $\phi(1 - x) = 1 - \phi(x)$. The ϕ-transform of t is a t-norm T_ϕ defined by $T_\phi(x, y) = \phi^{-1}(T(\phi(x), \phi(y)))$. For example $W_\phi(x, y) = \phi^{-1}(0 \vee (\phi(x) + \phi(y) - 1))$ and its dual t-conorm is $W'(x, y) = 1 - \phi^{-1}(0 \vee (\phi(1 - x) + \phi(1 - y) - 1))$, where W is the Lukasiewicz t-norm.

3. [17] Let T be a continuous non-Archimedian t-norm with zero divisors and S its dual conorm. Prove that there exists $s \in (0, 1)$ and an order isomorphism ϕ such that $\forall x, y \in [0, 1]$,

$$
S(x, y) = 1 \Leftrightarrow \begin{cases} x = 1 \text{ or} \\ y = 1 \text{ or} \\ (1 - x, 1 - y) \in (0, s)^2 \text{ and } \phi(\frac{1-x}{s}) + \phi(\frac{1-y}{s}) \leq 1. \end{cases}
$$

4. [17] Let ρ be a fuzzy weak preference relation and (π, ι) an axiomatic factorization of ρ. If T is a continuous non-Archimedian t-norm with zero divisors, then prove that there exists $s \in (0,1)$ such that $\rho(x,y) \in \{0,1\}$ or $\rho(x,y) > 1 - s$ for all $x, y \in X$.

5. [17] Prove that a t-norm T is a continuous Archimedean t-norm with zero divisors if and only if there exists an order isomorphism ϕ such that $T = W_\phi$.

6. [17] Let ρ be a fuzzy weak preference relation and (π, ι) an axiomatic factorization of ρ. If ϕ is a reciprocal order automorphism and $T = W_\phi$, then prove that for all $x, y \in X$,

$$\pi(x,y) = \phi^{-1}(\phi(\rho(x,y)) - \phi(\iota(x,y)))$$

and

$$\phi^{-1}\left(\frac{\phi(\rho(x,y)) + \phi(\rho(y,x)) - 1}{2}\right) \leq \iota(x,y) \leq \rho(x,y) \wedge \rho(y,x).$$

7. [17] Let ρ be a fuzzy weak preference relation and (π, ι) an axiomatic factorization of ρ. If ϕ is a reciprocal order automorphism, $T = W_\phi$, and $\pi = \rho \cap (\rho^{-1})^c$, then prove that for all $x, y \in X$,

$$\pi(x,y) = \phi^{-1}(\phi(\rho(x,y)) - \phi(\rho(y,x))) \vee 0,$$
$$\iota(x,y) = \rho(x,y) \wedge \rho(y,x),$$

where $\rho^c(x,y) = 1 - \rho(x,y)$.

8. [17] Let ρ be a fuzzy weak preference relation and (π, ι) an axiomatic factorization of ρ. If ϕ is a reciprocal order automorphism, $T = W_\phi$, then prove that $\iota = \rho \cap \rho^{-1}$ is equivalent to $\pi = (\rho^{-1})^c$ and when one of these relationships holds,

$$\pi(x,y) = 1 - \rho(y,x)$$
$$\iota(x,y) = \phi^{-1}(\phi(\rho(x,y)) + \phi(\rho(y,x)) - 1)$$

for all $x, y \in X$.

3.7 References

1. K. T. Atanassov, Intuitionistic fuzzy sets, *Fuzzy Sets and Systems*, 20 (1986) 87–96.

2. K. T. Atanassov, *Intuitionistic Fuzzy Sets*, Studies in Fuzziness and Soft Computing, Physica-Verlag, Heidelberg, 1999.

3. A. Banerjee, Fuzzy preferences and Arrow-type problems in social choice, *Soc. Choice Welfare,* 11 (1994) 121–130.

4. C. R. Barrett, P. K. Pattanaik, and M. Salles, On the structure of fuzzy social welfare functions, *Fuzzy Sets and Systems,* 19 (1986) 1–10.

5. A. Bufardi, On the construction of fuzzy preference relations, *J. Multicriteria Decision Anal.,* 7 (1998) 169–175.

6. C. Cornelis, G. Deschrijver, and E. E. Kerre, Implication in intuitionistic fuzzy intervalvalued fuzzy set theory: Construction, classification, application, *International Journal of Approximate Reasoning,* 35 (2004) 55–95.

7. M. Dasgupta and R. Deb, Factoring fuzzy transitivity, *Fuzzy Sets and Systems,* 118 (2001) 489–502.

8. B. De Baets and E. E. Kerre, Fuzzy preference relations and their characterization, *J. Fuzzy Math.,* 3 (1995) 373–381.

9. B. De Baets, B. Van de Walle, and E. E. Kerre, Fuzzy preference structures without incomparability, *Fuzzy Sets and Systems,* 76 (1995) 333–348.

10. D. Dimitrov, Intuitionistic fuzzy preferences: A factorization, *Advanced Studies in Contemporary Mathematics,* 5 (2002) 93–104.

11. B. Dutta, Fuzzy preference relations, *Math. Social Sci.,* 13 (1987) 215–229.

12. L. A. Fono and N. G. Andjiga, Fuzzy strict preference and social choice, *Fuzzy Sets and Systems,* 155 (2005) 372–389.

13. L. A. Fono, G. N. Nana, M. Salles, and H. Gwet, A binary intuitionistic fuzzy relation: Some new results, a general factorization, and two properties of strict components, *International Journal of Mathematics and Mathematical Sciences,* Hindawi Publishing Corporation Volume 2009, Article ID 580918, 38 pages dos:10.1155/2009/580918.

14. P. Hajek, *Metamathematics of Fuzzy Logic,* Kluwer, Dordrecht, 1998.

15. J. E. Herr and J. N. Mordeson, Factorization of fuzzy preference relations, *Advances in Fuzzy Sets and Systems,* 9 (2011) 17–35.

16. G. J. Klir and Bo Yuan, *Fuzzy Sets and Fuzzy Logic: Theory and Application,* Prentice Hall, Upper Saddle River, New Jersey, 1995.

17. B. Llamazares, Characterization of fuzzy preference structures through Lukasiewicz triplets, *Fuzzy Sets and Systems,* 136 (2003) 217–234.

18. B. Llamazares, Factorization of fuzzy preferences, *Social Choice Welfare,* 24 (2005) 475–496.

19. J. N. Mordeson and T. D. Clark, Fuzzy Arrow's theorem, *New Mathematics and Natural Computation,* 5 (2009) 371–383.

20. J. N. Mordeson, T. D. Clark, and K. Albert, Factorization of intuitionistic fuzzy preference relations, *New Mathematics and Natural Computation,* 10 (2014) 1–25.

21. G. N. Nana and L. A. Fono, Arrow-type results under intuitionistic fuzzy preferences, *New Mathematics and Natural Computation,* 9 (2013) 97–123.

22. S. Orlovsky, Decision making with a fuzzy preference relation, *Fuzzy Sets and Systems,* 1 (1978) 155–167.

23. S. Ovchinnikov, Structure of fuzzy binary relations, *Fuzzy Sets and Systems,* 6 (1981) 169–195.

24. G. Richardson, The structure of fuzzy preference: social choice implications, *Soc. Choice Welf.,* 15 (1998) 359–369.

25. B. Schweizer and A. Sklar, *Probabilistic Metric Spaces*, North-Holland, New York, 1983.

26. F. F. Tang, Fuzzy preferences and social choice, *Bull. Res.,* 46 (1994) 263–269.

Chapter 4

Fuzzy Non-Arrow Results

In this chapter, we focus on one of Arrow's conditions, independence of irrelevant alternatives to obtain nondictatorial results. We give special consideration to auxiliary functions considered by Garcia-Lapresta and Llamazares [21]. It was shown in [21] that neutral aggregation rules based on quadratic arithmetic means generalize simple majority when individuals have ordinary preferences, and collective preferences are reciprocal and that a weighted average is a nondictatorial fuzzy aggregation rule. We also show that it is not the automorphic image of the ordinary average.

4.1 Nondictatorial Fuzzy Aggregation Rules

The results in this section are based on Dutta [18], Richardson [39], Banerjee [3], and Mordeson, Giblisco, Clark [28].

Let T be a subset of $\mathcal{FR}(X)$, where X is a finite set.

Definition 4.1.1 *A function $\tilde{f} : T^n \to T$ is called a **fuzzy aggregation rule**.*

In the crisp case, an independent of irrelevant alternatives aggregation rule requires the social preference between alternatives x and y to depend only on the individual preferences between x and y. It is known that the Borda rule fails to satisfy the independence of irrelevant alternatives property.

Definition 4.1.2 *Let $\tilde{f} : T^n \to T$ be a fuzzy aggregation rule. Then $\tilde{f}$ satisfies*
*(1) **Independence of irrelevant alternatives** (**IIA1**) if $\forall(\rho_1, \ldots, \rho_n)$, $(\rho'_1, \ldots, \rho'_n) \in T^n$, $\forall x, y \in X$, $[\forall i \in N, \rho_i(x,y) = \rho'_i(x,y) \Rightarrow \rho(x,y) = \rho'(x,y)]$.*
*(2) **Pareto condition** (**PC**) with respect to π if $\forall(\rho_1, \ldots, \rho_n) \in T^n$, $\forall x, y \in X, \pi(x,y) \geq \wedge\{\pi_i(x,y) \mid i \in N\}$.*

(3) **Positive responsiveness (PR)** *with respect to* π *if* $\forall(\rho_1, \ldots, \rho_n)$, $(\rho_1', \ldots, \rho_n') \in T^n$, $\forall x, y \in X$, $\rho_i = \rho_i'$ $\forall i \neq j$, $\rho(x,y) = \rho(y,x)$, *and* $(\pi_j(x,y) = 0$ *and* $\pi_j'(x,y) > 0$ *or* $\pi_j(y,x) > 0$ *and* $\pi_j'(y,x) = 0)$ *implies* $\pi'(x,y) > 0$.

An independent of irrelevant alternatives (IIA1) fuzzy aggregation rule requires an alternative x to be socially preferred to an alternative y to the same degree whenever all individuals prefer x to y to the same degree. The Pareto condition requires x to be socially preferred to y to a degree at least as large as the smallest degree of the strict preference of the individuals whenever all individuals strictly prefer x to y. If a fuzzy aggregation rule is positive responsive, then by changing at least one individual's, say j, preference from $\pi_j(x,y) = 0$ to $\pi_j'(x,y) > 0$ or $\pi_j(y,x) > 0$ to $\pi_j'(y,x) = 0$, then x becomes strictly preferred to y with the change. Thus with positive responsive fuzzy aggregation rules, individuals can break ties $(\rho(x,y) = \rho(y,x))$.

Let $\rho = (\rho_1, \ldots, \rho_n) \in T^n$. Let $\widetilde{f}$ be a fuzzy aggregation rule. In the following, when a strict preference of a particular type for $\rho = \widetilde{f}(\rho)$ is assumed, then it is assumed that strict preference for ρ_i is of the same type, $i = 1, \ldots, n$.

Proposition 4.1.3 *Let* $\rho \in T$. *Let* π, τ *be two different types of strict preference with respect to* ρ. *Suppose* $\forall x, y \in X, \pi(x,y) > 0$ *if and only if* $\tau(x,y) > 0$. *Let* $\widetilde{f} : T^n \to T$ *be a fuzzy aggregation rule. Let* $\rho \in T^n$. *Then* $\widetilde{f}$ *satisfies PR with respect to* π *if and only if* $\widetilde{f}$ *satisfies PR with respect to* τ.

Proof. Suppose $\widetilde{f}$ satisfies PR with respect to π. Suppose $\forall(\rho_1, \ldots, \rho_n)$, $(\rho_1', \ldots, \rho_n') \in T^n$, $\forall x, y \in X$, $\rho_i = \rho_i'$ $\forall i \neq j$, $\rho(x,y) = \rho(y,x)$, and $(\tau_j(x,y) = 0$ and $\tau'(x,y) > 0$ or $\tau_j(y,x) > 0$ and $\tau_j'(y,x) = 0)$. Then $\forall(\rho_1, \ldots, \rho_n)$, $(\rho_1', \ldots, \rho_n') \in T^n$, $\forall x, y \in X, \rho_i = \rho_i'$ $\forall i \neq j$, $\rho(x,y) = \rho(y,x)$, and $(\pi_j(x,y) = 0$ and $\pi_j'(x,y) > 0$ or $\pi_j(y,x) > 0$ and $\pi_j'(y,x) = 0)$ since $\tau_j(x,y) > 0$ if and only if $\pi_j(x,y) > 0$ and $\tau_j'(x,y) > 0$ if and only if $\pi_j'(x,y) > 0$ for all $x, y \in X$. Thus $\pi(x,y) > 0$ since $\widetilde{f}$ satisfies PR with respect to π. Thus $\tau(x,y) > 0$ by hypothesis and so $\widetilde{f}$ satisfies PR with respect to τ. ∎

Corollary 4.1.4 *Let* $\widetilde{f} : T^n \to T$ *be a fuzzy aggregation rule. Then* $\widetilde{f}$ *satisfies PR with respect to* $\pi_{(3)}$ *if and only if* $\widetilde{f}$ *satisfies PR with respect to* $\pi_{(1)}$.

Proof. It suffices to show that $\forall x, y \in X$, $\pi_{(3)}(x,y) > 0$ if and only if $\pi_{(1)}(x,y) > 0$. However, this is immediate from the definitions since $\rho(x,y) = \rho(y,x) > 0$ if and only if $\rho(x,y) - \rho(y,x) > 0$. ∎

In the following definition, we define a fuzzy aggregation rule which we show later is nondictatorial, but yet satisfies certain reasonable properties.

Definition 4.1.5 *Define the fuzzy aggregation rule* $\widetilde{f} : T^n \to T$ *as follows:* $\forall \rho = (\rho_1, \ldots, \rho) \in T^n$, $\forall x, y \in X$,

$$\widetilde{f}(\rho)(x, y) = \sum_{i=1}^{n} w_i \rho_i(x, y),$$

where $\sum_{i=1}^{n} w_i = 1$ *and* $w_i > 0$, $i = 1, \ldots, n$.

Proposition 4.1.6 *Let* $\widetilde{f}$ *be defined as in Definition 4.1.5. Then* $\widetilde{f}$ *satisfies PR with respect to* $\pi_{(3)}$ *and* $\pi_{(1)}$.

Proof. Let $\rho, \rho' \in T^n$. Suppose $\rho_i = \rho_i'$, $i = 1, \ldots, n$, $i \neq j$. Let $x, y \in X$. Suppose $\rho(x, y) = \rho'(x, y)$. Suppose also that either (1) $(\pi_j(x, y) = 0$ and $\pi_j'(x, y) > 0)$ or (2) $(\pi_j(y, x) > 0$ and $\pi_j'(y, x) = 0)$, where strict preference is of type $\pi_{(3)}$. Then $\pi'(x, y) = 0 \vee (\rho'(x, y) - \rho'(y, x))$ and

$$
\begin{aligned}
& \rho'(x, y) - \rho'(y, x) \\
= & \sum_{i=1}^{n}(w_i \rho_i'(x, y) - w_i \rho_i'(y, x)) \\
= & \sum_{i=1, i \neq j}^{n}(w_i \rho_i(x, y) - w_i \rho_i(y, x)) + w_j \rho_j'(x, y) - w_j \rho_j'(y, x) \\
= & \sum_{i=1}^{n}(w_i \rho_i(x, y) - w_i \rho_i(y, x)) - w_j(\rho_j(x, y) - \rho_j(y, x)) \\
& + w_j(\rho_j'(x, y) - \rho_j'(y, x)) \\
= & -w_j(\rho_j(x, y) - \rho_j(y, x)) + w_j(\rho_j'(x, y) - \rho_j'(y, x)) \\
> & \ 0,
\end{aligned}
$$

where the inequality holds if either (1) or (2) hold. Hence $\widetilde{f}$ satisfies PR with respect to $\pi_{(3)}$. The desired result for $\pi_{(1)}$ follows from Corollary 4.1.4. ∎

Proposition 4.1.7 *Let* $\widetilde{f}$ *be the fuzzy aggregation rule defined in Definition 4.1.5. Then* $\forall x, y \in X, \pi_{(2)}(x, y) = \sum_{i=1}^{n} w_i \pi_i(x, y)$.

Proof. Let $x, y \in X$. Then $\pi_{(2)}(x, y) = 1 - \rho(x, y) = 1 - \sum_{i=1}^{n} w_i \rho_i(x, y) = 1 - \sum_{i=1}^{n} w_i(1 - \pi_i(x, y)) = 1 - \sum_{i=1}^{n} w_i + \sum_{i=1}^{n} w_i \pi_i(x, y)$. ∎

Example 4.1.8 *Let* $\widetilde{f}$ *be the fuzzy aggregation rule defined in Definition 4.1.5. Then* $\widetilde{f}$ *does not satisfy PR with respect to* $\pi_{(2)}$. *Let* $X = \{x, y\}$ *and* $N = \{1, 2\}$. *Define* ρ_1, ρ_2 *and* ρ_1', ρ_2' *on* X *as follows:* $\rho_i(x, x) = \rho_i(y, y) = 1$, $i = 1, 2$, $\rho_1(x, y) = 1/2$, $\rho_1(y, x) = 1$, $\rho_2(x, y) = 3/4$, $\rho_2(y, x) = 1/2$ *and* $\rho_i'(x, x) = \rho_i'(y, y) = 1$, $i = 1, 2$, $\rho_1'(x, y) = 1/2$, $\rho_1'(y, x) = 1$, $\rho_2'(x, y) = 1$, $\rho_2'(y, x) = 1$. *Let* $w_1 = 1/3$ *and* $w_2 = 2/3$. *Now* $\rho_1 = \rho_1'$. *Also,*

$$\rho(x, y) = 1/3\rho_1(x, y) + 2/3\rho_2(x, y) = 1/3 \cdot 1/2 + 2/3 \cdot 3/4 = 2/3$$

and

$$\rho(y, x) = 1/3\rho_1(y, x) + 2/3\rho_2(y, x) = 1/3 \cdot 1 + 2/3 \cdot 1/2 = 2/3.$$

Thus $\rho(x, y) = \rho(y, x)$. However,

$$\pi_2(y, x) = 1 - \rho_2(x, y) = 1 - 3/4 = 1/4 > 0$$

and

$$\pi'_2(y, x) = 1 - \rho'_2(x, y) = 1 - 1 = 0.$$

Thus $[\pi_2(y, x) > 0$ and $\pi'_2(y, x) = 0] \not\Rightarrow \pi'_{(2)}(x, y) > 0$.

Proposition 4.1.9 *Let $\widetilde{f}$ be the fuzzy aggregation rule defined in Definition 4.1.5. Then $\widetilde{f}$ satisfies PC with respect to $\pi_{(3)}$.*

Proof. Let $x, y \in X$ and $\rho \in T^n$. Let $m = m_{x,y} = \wedge\{\pi_i(x, y) \mid i = 1, \ldots, n\}$. There is no loss in generality in assuming $m = \pi_1(x, y)$. Suppose $m > 0$. Then

$$1 \leq w_1 + w_2 + \ldots + w_n + w_2\left(\frac{\pi_2(x, y)}{m} - 1\right) + \ldots + w_n\left(\frac{\pi_n(x, y)}{m} - 1\right)$$

$$= w_1\frac{\pi_1(x, y)}{m} + \ldots + w_n\frac{\pi_n(x, y)}{m}.$$

Hence

$$\pi_1(x, y) = m \leq w_1\pi_1(x, y) + \ldots + w_n\pi_n(x).$$

Since $m > 0, \pi_i(x, y) > 0 \; \forall i \in N$. Thus $\rho_i(x, y) > \rho_i(y, x) \; \forall i \in N$. Hence $\sum_{i=1}^n w_i\rho_i(x, y) > \sum_{i=1}^n w_i\rho_i(y, x)$. Thus $\sum_{i=1}^n w_i[\rho_i(x, y) - \rho_i(y, x)] > 0$. However, $\rho_i(x, y) - \rho_i(y, x) = \pi_i(x, y), i = 1, \ldots, n$. Hence

$$0 < \rho_1(x, y) - \rho_1(y, x) \leq \sum_{i=1}^n w_i[\rho_i(x, y) - \rho_i(y, x)]$$

$$= \sum_{i=1}^n w_i\rho_i(x, y) - \sum_{i=1}^n w_i\rho_i(y, x),$$

or $\wedge\{\pi_i(x, y) \mid i \in N\} \leq \rho(x, y) - \rho(y, x) = \pi(x, y)$. Hence $\widetilde{f}$ satisfies PC with respect to $\pi_{(3)}$. If $m = 0$, then clearly $\widetilde{f}$ satisfies PC with respect to $\pi_{(3)}$. ∎

Definition 4.1.10 *Let ρ be a fuzzy binary relation on X. Then ρ is said to satisfy type T_2-**transitivity** if $\forall x, y, z \in X, \rho(x, z) \geq \rho(x, y) + \rho(y, z) - 1$.*

Definition 4.1.11 *Let $\widetilde{f}$ be a fuzzy aggregation rule. Then*
    *(1) $\widetilde{f}$ is said to be **max-min transitive** if $\forall \rho \in T^n, \widetilde{f}(\rho)$ is max-min transitive;*
    *(2) $\widetilde{f}$ is said to be T_2-**transitive** if $\forall \rho \in T^n, \widetilde{f}(\rho)$ is T_2-transitive.*

Proposition 4.1.12 *Let $\widetilde{f}$ be the fuzzy aggregation rule defined in Definition 4.1.5. Let $\rho = (\rho_1, \ldots, \rho_n) \in \mathcal{FR}^n(X)$. If ρ_i is T_2-transitive $\forall i \in N$, then $\widetilde{f}(\rho)$ is T_2-transitive.*

Proof. Let $x, y, z \in X$. Then $\rho_i(x, z) \geq \rho_i(x, y) + \rho_i(y, z) - 1 \; \forall i \in N$. Thus

$$\rho(x, z) = \sum_{i=1}^{n} w_i \rho_i(x, z) \geq \sum_{i=1}^{n} w_i \rho_i(x, y) + \sum_{i=1}^{n} w_i \rho_i(y, z) - \sum_{i=1}^{n} w_i$$
$$= \rho(x, y) + \rho(y, z) - 1.$$

■

Let $H_2 = \{\rho \in \mathcal{FR}(X) \mid \rho \text{ is reflexive, strongly complete, and } T_2\text{-transitive}\}$.

Let C be a nonempty subset of N. Then C is called a **coalition**.
Let $\rho \in \mathcal{FR}^n$ and $x, y \in X$. Then

$$P(x, y; \rho) = \{i \in N \mid \pi_i(x, y) > 0\}.$$

Definition 4.1.13 *Let $\tilde{f}$ be a fuzzy aggregation rule. Then*
*(1) $\tilde{f}$ is called **nondictatorial** if it is not the case that there exists $i \in N$ such that for all $\rho \in T^n$, $\forall x, y \in X$, $\pi_i(x, y) > 0$ implies $\pi(x, y) > 0$;*
*(2) $\tilde{f}$ is said to be **weakly Paretian** if $\forall \rho \in T^n$, $\forall x, y \in X$ [$\forall i \in N$, $\pi_i(x, y) > 0$ implies $\pi(x, y) > 0$];*
*(3) $\tilde{f}$ is called **neutral** if $\forall x, y, z, w \in X$, $P(x, y; \rho) = P(z, w; \rho')$ and $P(y, x; \rho) = P(w, z; \rho')$ imply $\tilde{f}(\rho)(x, y) > 0$ if and only if $\tilde{f}(\rho')(z, w) > 0$.*

A fuzzy aggregation rule is called **dictatorial** if it is not nondictatorial.

The weakly Paretian condition requires an alternative x to be strictly preferred to an alternative y whenever all individuals strictly prefer x to y. In the crisp case, the Pareto condition and the weakly Paretian condition are equivalent. A neutral fuzzy aggregation rule treats all pairwise comparisons the same in that the labeling of alternatives is immaterial. All that matters are the individuals' preferences. We see that a similar situation holds in the fuzzy case.

Definition 4.1.14 *Let $\tilde{f}$ be a fuzzy aggregation rule.*
*(1) A coalition C is called an **oligarchy** if for all distinct $x, y \in X$ and all $\rho = (\rho_1, \ldots, \rho_n) \in T^n$, (i) $\pi_i(x, y) > 0 \; \forall i \in C \Rightarrow \pi(x, y) > 0$ and (ii) $\forall j \in C, [\pi_j(x, y) > 0 \Rightarrow \pi(y, x) = 0]$.*
*(2) An individual $j \in N$ is called a **dictator** if for all distinct $x, y \in X$ and all $\rho = (\rho_1, \ldots, \rho_n) \in T^n$, $\pi_j(x, y) > 0 \Rightarrow \pi(x, y) > 0$.*
*(3) An individual $j \in N$ is called a **vetoer** if for all distinct $x, y \in X$ and all $\rho = (\rho_1, \ldots, \rho_n) \in T^n$, $\pi_j(x, y) > 0 \Rightarrow \pi(y, x) = 0$.*

A fuzzy aggregation rule is called **oligarchic** if there exists $L \subseteq N$ such that every member of L has a veto and L is an oligarchy. It is known in the crisp case that there exist aggregation rules that acyclic, but not oligarchic. Such an example is the rule employed by the United Nations Security Council before 1965.

Theorem 4.1.15 *Let strict preference be defined by $\pi_{(3)}$. Then there exists a nondictatorial $\tilde{f} : H_2^n \to H_2$ satisfying IIA1, PR, and PC.*

Proof. Let $\tilde{f}$ be the fuzzy aggregation rule defined in Definition 4.1.5. Clearly, $\tilde{f}$ is not dictatorial. By Propositions 4.1.6, 4.1.9, and 4.1.12, it only remains to show that $\tilde{f}$ satisfies IIA1. However, this is immediate. ∎

Definition 4.1.16 *Let $\rho, \rho' \in T$. Let $Im(\rho) = \{s_1, \ldots, s_k\}$ and $Im(\rho') = \{t_1, \ldots, t_m\}$ be such that $s_1 < \ldots < s_k$ and $t_1 < \ldots < t_m$. Then ρ and ρ' are said to be **equivalent**, written $\rho \sim \rho'$, if $s_1 = 0 \Leftrightarrow t_1 = 0$, $k = m$, and $\rho_{s_i} = \rho'_{t_i}$ for $i = 1, \ldots, k$.*

If the fuzzy binary relations ρ and ρ' in Definition 4.1.16 are reflexive, then $s_k = 1 = t_m$.

Proposition 4.1.17 *Let $\rho, \rho' \in T$. Let $x, y \in X$. Suppose $\rho \sim \rho'$. Then $\rho(x, y) > \rho(y, x)$ if and only if $\rho'(x, y) > \rho'(y, x)$.*

Proof. Suppose $\rho(x, y) > \rho(y, x)$. Let $\rho(x, y) = s_i$. Then $s_i > \rho(y, x)$. Hence $(y, x) \notin \rho_{s_i}$. Thus $(y, x) \notin \rho'_{t_i}$. Since $(x, y) \in \rho_{s_i}$, we have $(x, y) \in \rho'_{t_i}$. Hence $\rho'(x, y) \geq t_i > \rho'(y, x)$. ∎

For $x, y \in X$, let $\Delta_{x,y} = \{(x, x), (y, y), (x, y), (y, x)\}$.

Definition 4.1.18 *Let $\tilde{f}$ be a fuzzy aggregation rule. Then $\tilde{f}$ is said to be **independent of irrelevant alternatives** (**IIA3**) if $\forall x, y \in X$, $[\forall \rho, \rho' \in T^n$, $\rho_i|_{\Delta_{x,y}} \sim \rho'_i|_{\Delta_{x,y}} \; \forall i \in N \Rightarrow \tilde{f}(\rho)|_{\Delta_{x,y}} \sim \tilde{f}(\rho')|_{\Delta_{x,y}}]$.*

In the following, we let t denote a function from T^n into $(0, 1]$. Some examples of t are $t(\rho) = \vee\{\sum_{i=1}^{n} w_i \rho_i(x, y) \mid x, y \in X, x \neq y\}$ and $t(\rho) = \wedge\{\sum_{i=1}^{n} w_i \rho_i(x, y) > 0 \mid x, y \in X\}$, where $\sum_{i=1}^{n} w_i = 1$ and $w_i > 0$, $i = 1, \ldots, n$.

Definition 4.1.19 *Define the fuzzy aggregation rule $\tilde{f} : T^n \to T$ as follows: $\forall \rho = (\rho_1, \ldots, \rho) \in T^n, \forall x, y \in X$,*

$$\tilde{f}(\rho)(x, y) = \begin{cases} 1 & \text{if } x = y, \\ 1 & \text{if } \pi_i(x, y) > 0 \; \forall i \in N, \\ t(\rho) & \text{otherwise.} \end{cases}$$

where $\sum_{i=1}^{n} w_i = 1$ and $w_i > 0$, $i = 1, \ldots, n$.

Proposition 4.1.20 *Let strict preference be defined by $\pi_{(1)}$ or $\pi_{(3)}$. Let $\tilde{f}$ be the fuzzy aggregation rule defined in Definition 4.1.19. Suppose $Im(t) \subseteq (0, 1)$. Then $\tilde{f}$ is independent of irrelevant alternatives IIA3.*

Proof. Let $\rho, \rho' \in T^n$ and $x, y \in X$. Suppose $\rho_i|_{\Delta_{x,y}} \sim \rho'_i|_{\Delta_{x,y}} \forall i \in N$. Let $i \in N$. Then $\rho_i|_{\Delta_{x,y}}(x,y) > \rho_i|_{\Delta_{x,y}}(y,x) \Leftrightarrow \rho'_i|_{\Delta_{x,y}}(x,y) > \rho'_i|_{\Delta_{x,y}}(y,x)$ by Proposition 4.1.17. Thus $\rho_i(x,y) > \rho_i(y,x) \Leftrightarrow \rho'_i(x,y) > \rho'_i(y,x)$. Hence by definition of $\widetilde{f}$, we have $\widetilde{f}(\rho)(x,y) = 1 \Leftrightarrow \widetilde{f}(\rho')(x,y) = 1$ and so $\widetilde{f}(\rho)(x,y) = t(\rho) \Leftrightarrow \widetilde{f}(\rho')(x,y) = t(\rho')$. Thus $\widetilde{f}(\rho)|_{\Delta_{x,y}} \sim \widetilde{f}(\rho')|_{\Delta_{x,y}}$. ∎

If t is a constant function in Proposition 4.1.20, then we can conclude $\widetilde{f}(\rho)|_{\Delta_{x,y}} = \widetilde{f}(\rho')|_{\Delta_{x,y}}$ in the proof. If $t(\rho) \geq 1/2$, then we can conclude the $\widetilde{f}(\rho)$ is strongly complete.

Proposition 4.1.21 *Let strict preference be defined by $\pi_{(1)}$ or $\pi_{(3)}$. Suppose $Im(t) \subseteq (0,1)$. Let $\widetilde{f}$ be the fuzzy aggregation rule defined in Definition 4.1.19. Then $\widetilde{f}$ is weakly Paretian.*

Proof. Let $x, y \in X$. If $\wedge_{i \in N} \pi_i(x,y) = 0$, then clearly $\pi(x,y) \geq 0$. Suppose $\wedge_{i \in N} \pi_i(x,y) > 0$. Then $\pi_i(x,y) > 0 \ \forall i \in N$. Hence $\rho_i(x,y) > \rho_i(y,x) \ \forall i \in N$. Thus $\widetilde{f}(\rho)(x,y) = 1$ and $\widetilde{f}(\rho)(y,x) = t(\rho) < 1$. Hence $\pi(x,y) \geq \wedge_{i \in N} \pi_i(x,y)$ for $\pi = \pi_1$ and $\pi(x,y) > 0$ for $\pi = \pi_{(3)}$. ∎

Let $\rho \in T$. Recall that ρ is called max-min quasi-transitive or just quasi-transitive if $\forall x, y, z \in X, \pi(x,z) \geq \pi(x,y) \wedge \pi(y,z)$, Definition 2.2.16.

Proposition 4.1.22 *Let π be defined by $\pi_{(1)}$. Let $\rho \in T$. If ρ is max-min transitive, then ρ is max-min quasi-transitive.*

Proof. Suppose there exists $x, y, z \in X$ such that

$$\rho(x,z) \geq \rho(x,y) \wedge \rho(y,z) \tag{4.1}$$

and

$$\pi(x,z) < \pi(x,y) \wedge \pi(y,z). \tag{4.2}$$

Then

$$0 < \pi(x,y) = \rho(x,y) > \rho(y,x) \tag{4.3}$$

and

$$0 < \pi(y,z) = \rho(y,z) > \rho(z,y) \tag{4.4}$$

By (4.3) and (4.4), $\rho(x,y) > 0$ and $\rho(y,z) > 0$ and so by (4.1) $\rho(x,z) > 0$. Suppose $0 < \pi(x,z)$. Then $\rho(x,z) > \rho(z,x)$. Thus $\pi(x,z) = \rho(x,z)$ and so $\pi(x,z) = \rho(x,z) \geq \rho(x,y) \wedge \rho(y,z) = \pi(x,y) \wedge \pi(y,z)$, where the latter equality holds by (4.3) and (4.4). Since this contradicts (4.2), $\pi(x,z) = 0$. Hence

$$\rho(z,x) \geq \rho(x,z) \tag{4.5}$$

Case 1: Suppose

$$\rho(x,y) \wedge \rho(y,z) = \rho(x,y). \tag{4.6}$$

Then
$$\rho(x, z) \geq \rho(x, y). \tag{4.7}$$

By max-min transitivity, $\rho(y, x) \geq \rho(y, z) \wedge \rho(z, x) \geq \rho(x, y) \wedge \rho(x, z) = \rho(x, y)$. However, this contradicts (4.3).

Case 2: Suppose
$$\rho(x, y) \wedge \rho(y, z) = \rho(y, z). \tag{4.8}$$

Then by (4.1),
$$\rho(x, z) \geq \rho(y, z). \tag{4.9}$$

By max-min transitivity, either $\rho(z, y) \geq \rho(z, x)$ or $\rho(z, y) \geq \rho(x, y)$, If $\rho(z, y) \geq \rho(z, x)$, then $\rho(z, y) \geq \rho(z, x) \geq \rho(x, z) \geq \rho(y, z)$ by (4.5) and (4.9). However, this contradicts (4.4). If $\rho(z, y) \geq \rho(x, y)$, then $\rho(z, y) \geq \rho(x, y) \geq \rho(y, z)$ by (4.8). However, this also contradicts (4.4). Thus ρ is max-min quasi-transitive. ∎

Theorem 4.1.23 *Let π be defined by $\pi_{(1)}$. Then there exists a nondictatorial fuzzy aggregation rule that is max-min transitive, independent of irrelevant alternatives IIA3, and weakly Paretian.*

Proof. Let $\widetilde{f}$ be defined as in Definition 4.1.19. Clearly, $\widetilde{f}$ is reflexive and complete. By Propositions 4.1.20 and 4.1.21, $\widetilde{f}$ is independent of irrelevant alternatives IIA3 and weakly Paretian. By definition, $\widetilde{f}(\rho)$ is reflexive $\forall \rho \in T^n$. It remains to be shown that $\widetilde{f}$ is max-min transitive. Let $\rho \in T^n$. Let $x, y, z \in X$. Suppose $\widetilde{f}(\rho)(x, y) \wedge \widetilde{f}(\rho)(y, z) = 1$. Assume $x \neq y \neq z$. Then $\forall i \in N, \pi_i(x, y) > 0$ and $\pi_i(y, z) > 0$. By Proposition 4.1.22, we have $\forall i \in N$ that $\pi_i(x, z) \geq \pi_i(x, y) \wedge \pi_i(y, z)$ since ρ_i is max-min transitive. Thus $\pi_i(x, z) > 0 \; \forall i \in N$. Hence $\rho_i(x, z) > \rho_i(z, x) \; \forall i \in N$. Thus $\widetilde{f}(\rho)(x, z) = 1$. If $\widetilde{f}(\rho)(x, y) \wedge \widetilde{f}(\rho)(y, z) = t(\rho)$, then clearly $\widetilde{f}(\rho) \geq \widetilde{f}(\rho)(x, y) \wedge \widetilde{f}(\rho)(y, z)$. Assume $x = y \neq z$. Then $\forall i \in N, \pi_i(y, z) > 0$. Now $\forall i \in N, \pi_i(x, z) = \pi_i(y, z) > 0$. Thus $\widetilde{f}(\rho)(x, z) = 1$. A similar argument holds for $x \neq y = z$. If $x = y = z$, then the result is immediate. The result is also immediate if $\widetilde{f}(\rho)(x, y) \wedge \widetilde{f}(\rho)(y, z) = t(\rho)$. ∎

Example 4.1.24 *Let $X = \{x, y, z\}$. Define the fuzzy binary relation ρ on X as follows:*

$$\rho(x, x) = \rho(y, y) = \rho(z, z) = 1,$$
$$\rho(x, y) = \rho(y, x) = \rho(y, z) = \rho(z, y) = 1/2, \rho(x, z) = 0, \rho(z, x) = 1.$$

Then

$$
\begin{aligned}
\rho(x,z) &= 0 \geq 1/2 + 1/2 - 1 = \rho(x,y) + \rho(y,z) - 1, \\
\rho(z,x) &= 1 \geq 1/2 + 1/2 - 1 = \rho(z,y) + \rho(y,x) - 1, \\
\rho(x,y) &= 1/2 \geq 0 + 1/2 - 1 = \rho(x,z) + \rho(z,y) - 1, \\
\rho(y,x) &= 1/2 \geq 1/2 + 1 - 1 = \rho(y,z) + \rho(z,x) - 1, \\
\rho(y,z) &= 1/2 \geq 1/2 + 0 - 1 = \rho(y,x) + \rho(x,z) - 1, \\
\rho(z,y) &= 1/2 \geq 1 + 1/2 - 1 = \rho(z,x) + \rho(x,y) - 1.
\end{aligned}
$$

Also, $\forall a, b \in X$,

$$
\begin{aligned}
\rho(a,b) &= \rho(a,a) + \rho(a,b) - 1, \\
\rho(a,b) &= \rho(a,b) + \rho(b,b) - 1, \\
\rho(a,a) &\geq \rho(a,b) + \rho(b,a) - 1.
\end{aligned}
$$

Thus ρ is T_2-transitive. However, $\pi_{(2)}(x,y) = 1 - \rho(y,x) = 1 - 1/2 > 0, \pi_{(2)}(y,z) = 1 - \rho(z,y) = 1 - 1/2$, but $\pi_{(2)}(x,z) = 1 - \rho(z,x) = 1 - 1 = 0$. Hence ρ is not max-min quasi-transitive with respect to $\pi_{(2)}$.

Note also that ρ is partially transitive since $\rho(u,v) > 0 \; \forall u, v \in X$. However, $\pi_{(2)}(x,y) > 0, \pi_{(2)}(y,z) > 0$, but $\pi_{(2)}(x,z) = 0$ so ρ is not partially quasi-transitive (Definition 2.2.16) with respect to $\pi_{(2)}$.

Proposition 4.1.25 *There exists a weakly Paretian (with respect to $\pi_{(2)}$) fuzzy aggregation rule $\tilde{f} : H_2^n \to H_2$ that satisfies IIA3 and is nondictatorial.*

Proof. Define $\tilde{f} : H_2^n \to H_2$ by $\forall \rho \in H_2^n$, $\forall x, y \in X$,

$$
\tilde{f}(\rho)(x,y) = \begin{cases} 1 & \text{if } x = y, \\ t(\rho) & \text{if } x \neq y. \end{cases}
$$

Then clearly $\tilde{f}(\rho)$ is reflexive and complete. Let $x, y, z \in X$. Then for $x \neq z$,

$$
\tilde{f}(\rho)(x,z) = t(\rho) \text{ and } \tilde{f}(\rho)(x,y) + \tilde{f}(\rho)(y,z) - 1 \leq t(\rho)
$$

since both $\tilde{f}(\rho)(x,y) = 1$ and $\tilde{f}(\rho)(y,z) = 1$ is impossible. For $x = z$, $\tilde{f}(\rho)(x,z) = 1 \geq \tilde{f}(\rho)(x,y) + \tilde{f}(\rho)(y,z) - 1$. Thus $\tilde{f}(\rho)$ is T_2-transitive. Let $\pi = \pi_{(2)}$. Suppose $\pi_i(x,y) > 0 \forall i \in N$. Then $x \neq y$. Hence $\pi(x,y) = 1 - \tilde{f}(\rho)(y,x) = 1 - t(\rho) > 0$. Thus $\tilde{f}$ is weakly Paretian with respect $\pi_{(2)}$. Clearly, $\tilde{f}$ satisfies IIA3 by a similar argument as in Proposition 4.1.20 . ∎

4.2 Auxiliary Functions

In this section, we show how neutrality can be used to classify certain fuzzy aggregation rules.

Theorem 4.2.1 *Let $\widetilde{f}$ be a fuzzy aggregation rule. Then the following conditions are equivalent:*

(1) $\widetilde{f}$ *is neutral;*

(2) *there exists a unique function $f_n : [0,1]^n \to [0,1]$ such that $\forall x, y \in X$,*
$\forall \rho = (\rho_1, \ldots, \rho_n) \in \mathcal{FR}^n$, $f_n((\rho_1(x,y), \ldots, \rho_n(x,y)) = \widetilde{f}(\rho)(x,y)$.

Proof. Suppose condition (1) holds. Let $x, y \in X$. Define $f_n : [0,1]^n \to [0,1]$ as follows: Let $(a_1, \ldots, a_n) \in [0,1]^n$. Then there exists $\rho = (\rho_1, \ldots, \rho_n) \in \mathcal{FR}^n$ such that $\rho_i(x,y) = a_i, i = 1, \ldots, n$. Define f_n by $f_n((a_1, \ldots, a_n)) = \widetilde{f}(\rho)(x,y)$. It remains to be shown that f_n is single-valued. Let $w, z \in X$. Then there exists $\sigma = (\sigma_1, \ldots, \sigma_n) \in \mathcal{FR}^n$ such that $\sigma_i(w,z) = a_i, i = 1, \ldots, n$. Thus $\rho_i(x,y) = \sigma_i(w,z), i = 1, \ldots, n$. Since $\widetilde{f}$ is neutral, $f_n(\rho)(x,y) = f_n(\sigma)(w,z)$. Hence f_n is single-valued. The uniqueness of f_n follows by its construction.

Suppose condition (2) holds. Let $\rho, \sigma \in \mathcal{FR}^n$ and $x, y, w, z \in X$. Suppose $\rho_i(x,y) = \sigma_i(w,z), i = 1, \ldots, n$. Then $\widetilde{f}(\rho)(x,y) = f_n((\rho_1(x,y), \ldots, \rho_n(x,y)))$ $= f_n((\sigma_1(w,z)), \ldots, \sigma_n(w,z))) = f_n(\sigma)(w,z)$. Thus $\widetilde{f}$ is neutral.   ∎

We call f_n the **auxiliary function** associated with $\widetilde{f}$.

Definition 4.2.2 *Let $\widetilde{f}$ be a fuzzy aggregation rule and let f_n be the auxiliary function associated with $\widetilde{f}$. Then $\widetilde{f}$ is said to be **linearly decomposable** if $\forall (a_1, \ldots, a_n) \in [0,1]^n$,*

$$f_n((a_1, \ldots, a_n)) = a_1 f_n((1, 0, \ldots, 0) + \ldots + a_n f_n((0, \ldots, 0, 1)).$$

Note:

$$f_n((a_1, \ldots, a_n)) = a_1 f_n((1, 0, \ldots, 0)) + \ldots + a_n f_n((0, \ldots, 0, 1))$$
$$\Leftrightarrow \quad f_n((a_1, \ldots, a_n)) = f_n((a_1, 0, \ldots, 0) + \ldots + f_n((0, \ldots, 0, a_n))$$

and

$$f_i((0, \ldots, 0, a_i, 0, \ldots, 0)) = a_i f_i(0, \ldots, 0, 1, 0, \ldots, 0) \; for \; i = 1, \ldots, n.$$

Proposition 4.2.3 *Let $\widetilde{f}$ be a fuzzy aggregation rule and let f_n be the auxiliary function associated with $\widetilde{f}$. If $\widetilde{f}$ is linearly decomposable, then $\widetilde{f}$ is additive.*

Proof. Let $(a_1, \ldots, a_n), (b_1, \ldots, b_n) \in [0,1]^n$ be such that $a_i + b_i \in [0,1]$, $i = 1, \ldots, n$. Then

$$
\begin{aligned}
&f_n((a_1, \ldots, a_n) + (b_1, \ldots, b_n)) \\
=\; &f_n((a_1 + b_1, \ldots, a_n + b_n)) \\
=\; &(a_1 + b_1) f_n((1, 0, \ldots, 0)) + \ldots + (a_n + b_n) f_n((0, \ldots, 0, 1)) \\
=\; &a_1 f_n((1, 0, \ldots, 0)) + \ldots + a_n f_n((0, \ldots, 0, 1)) + \\
&b_1 f_n((1, 0, \ldots, 0)) + \ldots + b_n f_n((0, \ldots, 0, 1)) \\
=\; &f_n((a_1, \ldots, a_n)) + f_n((b_1, \ldots, b_n)).
\end{aligned}
$$

∎

The next two results show that a neutral and linearly decomposable aggregation rule must be a weighted mean.

Theorem 4.2.4 *Let $\widetilde{f}$ be a fuzzy aggregation rule and let f_n be the auxiliary function associated with $\widetilde{f}$. If $\widetilde{f}$ is linearly decomposable, then there exists a unique linear transformation of $\mathbb{R}^n$ into $\mathbb{R}$ such that $\widehat{f_n}|_{[0,1]^n} = f_n$.*

Proof. For $i = 1, \ldots, n$, let $\overline{1}_i = (u_1, \ldots, u_n)$, where $u_j = 0$, $j = 1, \ldots, n$; $j \neq i$, and $u_i = 1$. Then there exists a unique linear transformation $\widehat{f_n}$ of $\mathbb{R}^n$ onto $\mathbb{R}$ such that $\widehat{f_n}(\overline{1}_i) = w_i$, where $w_i = f_n(\overline{1}_i)$. Since f_n is additive, $\sum_{i=1}^n w_i = \sum_{i=1}^n f_n(\overline{1}_i) = f_n((1, \ldots, 1)) \leq 1$. Now $\widehat{f_n}(\sum_{i=1}^n c_i \overline{1}_i) = \sum_{i=1}^n c_i \widehat{f_n}(\overline{1}_i)$, where $c_i \in \mathbb{R}$, $i = 1, \ldots, n$. Thus if $c_i \in [0,1]$ for $i = 1, \ldots, n$, then $f_n(\sum_{i=1}^n c_i \overline{1}_i) = \sum_{i=1}^n c_i f_n(\overline{1}_i) \in [0,1]$ since $\sum_{i=1}^n w_i \leq 1$. Let $(a_1, \ldots, a_n) \in [0,1]^n$. Then $\widehat{f_n}|_{[0,1]^n}((a_1, \ldots, a_n)) = \widehat{f_n}((a_1, \ldots, a_n)) = \sum_{i=1}^n a_i \widehat{f_n}(\overline{1}_i) = \sum_{i=1}^n a_i f_n(\overline{1}_i) = f_n((a_1, \ldots, a_n))$ since $\widetilde{f}$ is linearly decomposable. ∎

Corollary 4.2.5 *Let $\widetilde{f}$ be a fuzzy aggregation rule and let f_n be the auxiliary function associated with $\widetilde{f}$. If $\widetilde{f}$ is linearly decomposable, then $\forall \rho = (\rho_1, \ldots, \rho_n) \in \mathcal{FR}^n(X)$ and $\forall x, y \in X$, $\widetilde{f}(\rho)(x,y) = \sum_{i=1}^n w_i \rho_i(x,y)$, where $w_i = f_n(\overline{1}_i)$ for $i = 1, \ldots, n$.*

The previous result applies to committees or voting bodies where individuals do not contribute equally to social preference choices. This may happen, for example, when seniority or rank of individuals is involved.

Definition 4.2.6 *A fuzzy aggregation rule $\widetilde{f}$ is called **anonymous** if for all $(\rho_1, \ldots, \rho_n) \in \mathcal{FR}^n(X)$ and for all permutations σ of N,*

$$\widetilde{f}(\rho_{\sigma(1)}, \ldots, \rho_{\sigma(n)}) = \widetilde{f}(\rho_1, \ldots, \rho_n).$$

Corollary 4.2.7 *Let $\widetilde{f}$ be a fuzzy aggregation rule and let f_n be the auxiliary function associated with $\widetilde{f}$. If $\widetilde{f}$ is linearly decomposable, then $w_i = \frac{1}{n}$ for $i = 1, \ldots, n$ if and only if $\widetilde{f}$ is anonymous.*

Example 4.2.8 *Let $N = \{1, 2\}$ and $X = \{x, y\}$. Let ρ_1 and ρ_2 be WFPR on X defined by $\rho_1(x,x) = \rho_1(y,y) = 1 = \rho_2(x,x) = \rho_2(y,y)$ and $\rho_1(x,y) = 2/3, \rho_1(y,x) = 1/3, \rho_2(x,y) = 1/3, \rho_2(y,x) = 2/3$. Let $w_1 = 1/4$ and $w_2 = 3/4$. Suppose $\widetilde{f}$ is a fuzzy aggregation rule such that $\forall u, v \in X$,*

$$\widetilde{f}(\rho)(u,v) = w_1 \rho_1(u,v) + w_2 \rho_2(u,v),$$

where $\rho = (\rho_1, \rho_2)$. Let f_2 be the auxiliary function associated with $\widetilde{f}$. Then $f_2(\rho_1(u,v), \rho_2(u,v)) = \widetilde{f}(\rho)(u,v)$. Let ϕ be an automorphism of $[0,1]$. Suppose that $f_2((a_1, a_2)) = \phi^{-1}\left(\frac{\phi(a_1)+\phi(a_2)}{2}\right)$ for all $a_1, a_2 \in [0,1]$. Then $\phi(f_2((a_1, a_2))) = \frac{\phi(a_1)+\phi(a_2)}{2}$ for all $a_1, a_2 \in [0,1]$. Thus

$$
\begin{aligned}
&\phi(f_2((\rho_1(x,y), \rho_2(x,y)))) \\
&= \frac{\phi(\rho_1(x,y)) + \phi(\rho_2(x,y))}{2} = \frac{\phi(2/3) + \phi(1/3)}{2} \\
&= \frac{\phi(1/3) + \phi(2/3)}{2} = \frac{\phi(\rho_1(y,x)) + \phi(\rho_2(y,x))}{2} \\
&= \phi(f_2((\rho_1(y,x), \rho_2(y,x)))).
\end{aligned}
$$

Also,

$$
\begin{aligned}
&\phi(f_2((\rho_1(x,y), \rho_2(x,y)))) \\
&= \phi(\widetilde{f}(\rho)(x,y)) = \phi\left(\frac{1}{4}\cdot\frac{2}{3} + \frac{3}{4}\cdot\frac{1}{3}\right) = \phi\left(\frac{5}{12}\right)
\end{aligned}
$$

and

$$
\begin{aligned}
&\phi(f_2((\rho_1(y,x), \rho_2(y,x)))) \\
&= \phi(\widetilde{f}(\rho)(y,x)) = \phi\left(\frac{1}{4}\cdot\frac{1}{3} + \frac{3}{4}\cdot\frac{2}{3}\right) = \phi\left(\frac{7}{12}\right).
\end{aligned}
$$

Hence $\phi\left(\frac{5}{12}\right) = \phi\left(\frac{7}{12}\right)$. However, this contradicts the fact that ϕ is one-to-one. Thus no such automorphism exists. That is, we cannot obtain $\sum_{i=1}^{n} w_i \rho_i(x,y)$ from $\sum_{i=1}^{n} \frac{1}{n}\rho_i(x,y)$ by an automorphism of $[0,1]$.

4.3   Arrow's Theorem and Max-Star Transitivity

In this section, we consider the important work of Duddy, Perote-Pena, and Piggins [17]. Max-* transitivity plays a central roll in fuzzy preference theory. Attention is restricted to fuzzy aggregation rules that satisfy counterparts of unanimity and independence of irrelevant alternatives. In this section, the set of triangular norms that permit preference aggregation to be non-dictatorial is characterized. This set contains exactly those norms that contain a zero divisor.

Many triangular norms do not possess zero divisors, e.g., max-min and max-product. The Lukasiewicz triangular norm does possess zero divisors, e.g., $T_L(x,y) = 0$ when $x = \frac{1}{2}$, $y = \frac{1}{2}$.

A survey of the literature on social choice theory can be found in Salles [40]. Various results can be found in [5, 12, 17, 31, 3, 9, 39, 18, 20, 34, 16,

25, 26, 27]. Differences between impossibility and possibility theorems can be explained by the fact that each model uses different assumptions, particularly with regard to how transitivity is modelled. This was seen in the previous two sections. As seen in Chapter 3, the appropriate way of factoring a fuzzy weak preference relation into a fuzzy strict preference relation and a fuzzy indifference relation can also play a role.

The main result in this section states that an aggregation rule satisfying certain criteria is dictatorial if and only if the triangular norm used in the formulation of the transitivity condition has no zero divisor. Consequently, max-min transitivity leads to dictatorship, whereas Lukasiewicz transitivity does not.

Fuzzy aggregation rules that satisfy counterparts of unanimity and independence of irrelevant alternatives are assumed in the characterization of the set of triangular norms that permit preference aggregation to be non-dictatorial.

Recall that X is a set of alternatives with $|X| \geq 3$ and $N = \{1, 2, \ldots, n\}$ with $n \geq 2$ is a finite set of individuals. For $\rho \in \mathcal{FR}(X)$, we say that ρ is **connected** if for all $x, y \in X$, $\rho(x, y) = 0$ implies $\rho(y, x) = 1$. Let $\mathcal{S}$ be the set of all $\rho \in \mathcal{FR}(X)$ such that ρ is reflexive, connected, and max-$*$ transitive.

A **fuzzy aggregation rule (FAR)** is a function $\widetilde{f} : \mathcal{S}^n \to \mathcal{S}$. We often write $\rho = \widetilde{f}(\rho_1, \ldots, \rho_n)$, where $\widetilde{f}$ is an FAR.

Definition 4.3.1 *Let $\widetilde{f}$ be an FAR. Then $\widetilde{f}$ is said to satisfy*

(1) **Independence** *(I) if for all $(\rho_1, \ldots, \rho_n)$, $(\rho_1', \ldots, \rho_n') \in \mathcal{S}^n$ and for all $x, y \in X$, $\rho_j(x, y) = \rho_j'(x, y)$ for all $j \in N$ implies $\rho(x, y) = \rho'(x, y)$;*

(2) **Unanimity** *(U) if for all $(\rho_1, \ldots, \rho_n) \in \mathcal{S}^n$, for all $x, y \in X$, and for all $v \in [0, 1]$, $\rho_j(x, y) = v$ for all $j \in N$ implies $\rho(x, y) = v$;*

(3) **Neutrality** *if for all $(\rho_1, \ldots, \rho_n)$, $(\rho_1', \ldots, \rho_n') \in \mathcal{S}^n$ and for all $x, y, z, w \in X$, $\rho_j(x, y) = \rho_j'(z, w)$ for all $j \in N$ implies $\rho(x, y) = \rho'(z, w)$.*

An FAR $\widetilde{f}$ is said to be **dictatorial** if there exists an individual $i \in N$ such that for all $x, y \in X$, and for every $(\rho_1, \ldots, \rho_n) \in \mathcal{S}^n$, $\rho_i(x, y) = \rho(x, y)$.

The properties I, U and the dictatorship condition are stronger than the conditions commonly used in the literature. The reader will easily observe this in Chapter 5. There are two reasons for using them according to Duddy, Perot-Pena, and Piggins [17]. In other work, I and U follow from the requirement that a non-constant FAR cannot be manipulated, [34, 35, 36, 16]. Thus if one accepts the formulation of strategy-proofness presented in these articles, then I and U follow as logical consequences. Second, the weaker conditions would require one to take a position on the "factorization" issue that was mentioned earlier which is something that Duddy, Pena, and Piggins chose to avoid. Statements of the standard conditions can be found in [40] and [4].

The purpose of this section is to prove the next theorem.

Theorem 4.3.2 (*Duddy, Perote-Pena, and Piggins* [17]) *If $*$ has no zero divisors, then any FAR satisfying I and U is dictatorial. Moreover, if $*$ has a zero divisor, then a non-dictatorial FAR exists that satisfies I and U.*

We first prove sufficiency. The following lemma holds for any triangular norm. The lemma generalizes Lemma 1 in [34]. See also [14, 15].

Lemma 4.3.3 ([17]) *Let $\widetilde{f}$ be an FAR. If $\widetilde{f}$ satisfies I and U, then $\widetilde{f}$ is neutral under any triangular norm.*

Proof. We consider five cases.

(1) If $(a, b) = (c, d)$, then the result follows immediately from the fact that $\widetilde{f}$ satisfies Independence.

(2) Let $(a, b), (a, c) \in X \times X$. Let $(\rho_1, \ldots, \rho_n) \in \mathcal{S}^n$ be such that $\rho_j(b, c) = 1$ for all $j \in N$. Since $\widetilde{f}$ satisfies Unanimity, $\rho(b, c) = 1$, where $\rho = \widetilde{f}(\rho)$. Since ρ is max-$*$ transitive, we have $\rho(a, c) \geq \rho(a, b)$. Since $\rho_j(b, c) = 1$ for all $j \in N$ and the ρ_j are max-$*$ transitive, it follows that $\rho_j(a, c) \geq \rho_j(a, b)$ for all $j \in N$. Let $(\overline{\rho}_1, \ldots, \overline{\rho}_n) \in \mathcal{S}^n$ be such that $\overline{\rho}_j(b, c) = 1$ and $\overline{\rho}_j(c, b) = 1$ for all $j \in N$. From the above argument, we have that $\overline{\rho}(a, c) \geq \overline{\rho}(a, b)$ and $\overline{\rho}_j(a, c) \geq \overline{\rho}_j(a, b)$ for all $j \in N$. However, an identical argument shows that $\overline{\rho}(a, b) \geq \overline{\rho}(a, c)$ and $\overline{\rho}_j(a, b) \geq \overline{\rho}_j(a, c)$ for all $j \in N$. Thus it must be the case that $\overline{\rho}(a, b) = \overline{\rho}(a, c)$ and $\overline{\rho}_j(a, b) = \overline{\rho}_j(a, c)$ for all $j \in N$. Since $(\overline{\rho}_1, \ldots, \overline{\rho}_n) \in \mathcal{S}^n$ is arbitrary, this condition holds for all profiles $(\rho_1, \ldots, \rho_n) \in \mathcal{S}^n$ such that $\rho_j(b, c) = 1$ and $\rho_j(c, b) = 1$ for all $j \in N$. Let $\mathcal{F}^n$ denote the set of all such profiles. Let $(\widehat{\rho}_1, \ldots, \widehat{\rho}_n) \in \mathcal{S}^n$ be such that $\widehat{\rho}_j(a, b) = \widehat{\rho}_j(a, c)$ for all $j \in N$. Then there exists a profile $(\rho'_1, \ldots, \rho'_n) \in \mathcal{F}^n$ such that $\widehat{\rho}_j(a, b) = \widehat{\rho}_j(a, c) = \rho'_j(a, b) = \rho'_j(a, c)$ for all $j \in N$. Since $\widetilde{f}$ satisfies I, we have $\widehat{\rho}(a, b) = \widehat{\rho}(a, c) = \rho'(a, b) = \rho'(a, c)$. Consider distinct profiles $(\rho''_1, \ldots, \rho''_n), (\rho^*_1, \ldots, \rho^*_n) \in \mathcal{S}^n$ such that $\rho''_j(a, b) = \rho^*_j(a, c)$ for all $j \in N$. Then there exists a profile $(\rho^{**}_1, \ldots, \rho^{**}_n) \in \mathcal{F}^n$ such that $\rho''_j(a, b) = \rho^*_j(a, c) = \rho^{**}_j(a, b) = \rho^{**}_j(a, c)$ for all $j \in N$. Since $\widetilde{f}$ satisfies I, we have $\rho''(a, b) = \rho^*(a, c) = \rho^{**}(a, b) = \rho^{**}(a, c)$.

(3) $(a, b), (c, b) \in X \times X$. Let $(\rho_1, \ldots, \rho_n) \in \mathcal{S}^n$ be such that $\rho_j(a, c) = 1$ for all $j \in N$. Since $\widetilde{f}$ satisfies U, we have $\rho(a, c) = 1$. Since ρ is max-$*$ transitive, we have $\rho(a, b) \geq \rho(c, b)$. Since $\rho_j(a, c) = 1$ for all $j \in N$ and individual preferences are max-$*$ transitive, it follows that $\rho_j(a, b) \geq \rho_j(c, b)$ for all $j \in N$. Let $(\overline{\rho}_1, \ldots, \overline{\rho}_n) \in \mathcal{S}^n$ be such that $\overline{\rho}_j(a, c) = 1$ and $\overline{\rho}_j(c, a) = 1$ for all $j \in N$. From the argument above, we have that $\overline{\rho}(a, b) \geq \overline{\rho}(c, b)$ and $\overline{\rho}_j(a, b) \geq \overline{\rho}_j(c, b)$ for all $j \in N$. However, an identical argument shows that $\overline{\rho}(c, b) \geq \overline{\rho}(a, b)$ and $\overline{\rho}_j(c, b) \geq \overline{\rho}_j(a, b)$ for all $j \in N$. Thus $\overline{\rho}(a, b) = \overline{\rho}(c, b)$ and $\overline{\rho}_j(a, b) = \overline{\rho}_j(c, b)$ for all $j \in N$. Since $(\overline{\rho}_1, \ldots, \overline{\rho}_n) \in \mathcal{S}^n$ is arbitrary, this condition holds for all profiles $(\rho_1, \ldots, \rho_n) \in \mathcal{S}^n$ such that $\rho_j(a, c) = 1$ and $\rho_j(c, a) = 1$ for all $j \in N$. Let $\mathcal{G}^n$ denote the set of such profiles. Let $(\widehat{\rho}_1, \ldots, \widehat{\rho}_n) \in \mathcal{S}^n$ be such that $\widehat{\rho}_j(a, b) = \widehat{\rho}_j(c, b)$ for all $j \in N$. Then there exists $(\rho'_1, \ldots, \rho'_n) \in \mathcal{G}^n$ such that $\widehat{\rho}_j(a, b) = \widehat{\rho}_j(c, b) = \rho'_j(a, b) = \rho'_j(c, b)$ for

all $j \in N$. Since $\widetilde{f}$ satisfies I, we have that $\widehat{\rho}(a,b) = \widehat{\rho}(c,b) = \rho'(a,b) = \rho'(c,b)$. Consider distinct profiles $(\rho_1'', \ldots, \rho_n''), (\rho_1^*, \ldots, \rho_n^*) \in \mathcal{S}^n$ such that $\rho_j''(a,b) = \rho_j^*(c,b)$ for all $j \in N$. Then there exists $(\rho_1^{**}, \ldots, \rho_n^{**}) \in \mathcal{G}^n$ such that $\rho_j''(a,b) = \rho_j^*(c,b) = \rho_j^{**}(a,b) = \rho_j^{**}(c,b)$ for all $j \in N$. Since $\widetilde{f}$ satisfies I, we have that $\rho''(a,b) = \rho^*(c,b) = \rho^{**}(a,b) = \rho^{**}(c,b)$.

(4) $(a,b), (c,d) \in X \times X$ with a,b,c,d distinct. Let $(\rho_1, \ldots, \rho_n) \in \mathcal{S}^n$ be such that $\rho_j(b,d) = \rho_j(d,b) = \rho_j(a,c) = \rho_j(c,a) = 1$ for all $j \in N$. By U, we have that $\rho(d,b) = 1$. Since ρ is max-$*$ transitive, we have $\rho(a,b) \geq \rho(a,d)$. An identical argument shows that $\rho(a,d) \geq \rho(a,b)$ and so $\rho(a,b) = \rho(a,d)$. Since $\rho_j(d,b) = \rho_j(b,d) = 1$ for all $j \in N$ and individual preferences are max-$*$ transitive, it follows that $\rho_j(a,b) = \rho_j(a,d)$ for all $j \in N$. This argument can be repeated to show that $\rho(a,b) = \rho(c,d)$ and $\rho_j(a,b) = \rho_j(c,d)$ for all $j \in N$. Since $(\rho_1, \ldots, \rho_n) \in \mathcal{S}^n$ is arbitrary, this condition holds for all profiles $(\rho_1, \ldots, \rho_n) \in \mathcal{S}^n$ such that $\rho_j(b,d) = \rho_j(d,b) = \rho_j(a,c) = \rho_j(c,a) = 1$ for all $j \in N$. Let $\mathcal{G}^n$ be the set of all such profiles. Let $(\widehat{\rho}_1, \ldots, \widehat{\rho}_n) \in \mathcal{S}^n$ be such that $\widehat{\rho}_j(a,b) = \widehat{\rho}_j(c,d)$ for all $j \in N$. Then there exists $(\rho_1', \ldots, \rho_n') \in \mathcal{G}^n$ such that $\widehat{\rho}_j(a,b) = \widehat{\rho}_j(c,d) = \rho_j'(a,b) = \rho_j'(c,d)$ for all $j \in N$. By I, we have that $\widehat{\rho}(a,b) = \widehat{\rho}(c,d) = \rho'(a,b) = \rho'(c,d)$. Consider distinct profiles $(\rho_1'', \ldots, \rho_n''), (\rho_1^*, \ldots, \rho_n^*) \in \mathcal{S}^n$ such that $\rho_j''(a,b) = \rho_j^*(c,d)$ for all $j \in N$. Then there exists $(\rho_1^{**}, \ldots, \rho_n^{**}) \in \mathcal{G}^n$ such that $\rho_j''(a,b) = \rho_j^*(c,d) = \rho_j^{**}(a,b) = \rho_j^{**}(c,d)$ for all $j \in N$. By I, we have that $\rho''(a,b) = \rho^*(c,d) = \rho^{**}(a,b) = \rho^{**}(c,d)$.

(5) $(a,b), (c,d) \in X \times X$ with $a = d$ or $b = c$. Since $|X| \geq 3$, there exists $e \in X$ such that $a \neq e \neq c$. Let $(\rho_1, \ldots, \rho_n) \in \mathcal{S}^n$ be such that $\rho_j(a,b) = \rho_j(a,e) = \rho_j(c,e) = \rho_j(c,d) = 1$ for all $j \in N$. By cases (2) and (3), we have that $\rho(a,b) = \rho(a,e) = \rho(c,e) = \rho(c,d)$. Let $\mathcal{W}^n$ denote the set of such profiles. Let $(\overline{\rho}_1, \ldots, \overline{\rho}_n) \in \mathcal{S}^n$ be such that $\overline{\rho}_j(a,b) = \overline{\rho}_j(c,d) = 1$ for all $j \in N$. Then there exists $(\rho_1', \ldots, \rho_n') \in \mathcal{W}^n$ such that $\overline{\rho}_j(a,b) = \overline{\rho}_j(c,d) = \rho_j'(a,b) = \rho_j'(c,d)$ for all $j \in N$. By I, we have that $\overline{\rho}(a,b) = \overline{\rho}(c,d) = \rho'(a,b) = \rho'(c,d)$. Consider distinct profiles $(\rho_1'', \ldots, \rho_n''), (\rho_1^*, \ldots, \rho_n^*) \in \mathcal{S}^n$ such that $\rho_j''(a,b) = \rho_j^*(c,d)$ for all $j \in N$. Then there exists $(\rho_1^{**}, \ldots, \rho_n^{**}) \in \mathcal{W}^n$ such that $\rho_j''(a,b) = \rho_j^*(c,d) = \rho_j^{**}(a,b) = \rho_j^{**}(c,d)$ for all $j \in N$. Then I implies that $\rho''(a,b) = \rho^*(c,d) = \rho^{**}(a,b) = \rho^{**}(c,d)$. ∎

Lemma 4.3.4 ([17]) *If $*$ has no zero divisor, then every FAR $\widetilde{f}$ satisfying I and U is dictatorial.*

Proof. By Lemma 4.3.3, $\widetilde{f}$ is neutral. Let $(\rho_1, \ldots, \rho_n) \in \mathcal{S}^n$ be such that $\rho_i(a,b) = 0$ for all $i \in N$. Since $\widetilde{f}$ satisfies U, we have that $\rho(a,b) = 0$. Let $(\rho_1', \ldots, \rho_n') \in \mathcal{S}^n$ be such that $\rho_i'(a,b) = 1$ for all $i \in N$. By U, we have that

$\rho'(a, b) = 1$. Consider the following sequence of profiles:

$$\rho^{(0)} = (\rho_1, \ldots, \rho_n),$$
$$\rho^{(1)} = (\rho_1', \rho_2, \ldots, \rho_n),$$
$$\rho^{(2)} = (\rho_1', \rho_2', \rho_3, \ldots, \rho_n),$$
$$\vdots$$
$$\rho^{(n)} = (\rho_1', \ldots, \rho_n').$$

At some point in this sequence, the social value of (a, b) rises from 0 to some positive number. Assume that this happens at $\rho^{(2)}$ when individual 2 changes his or her preferences from $\rho(a, b)$ to $\rho'(a, b)$. We prove that this individual is a dictator.

Consider a profile $(\rho_1', \rho_2, \rho_3', \ldots, \rho_n') \in \mathcal{S}^n$. We now show that at this profile the social value of (a, b) is zero. Let $(\rho_1^*, \ldots, \rho_n^*) \in \mathcal{S}^n$ be as follows: At this profile, every individual's (a, c) preference is the same as their (a, b) preference at $\rho^{(1)}$. Everyone's (a, b) preference is the same as their (a, b) preference at $(\rho_1', \rho_2, \rho_3', \ldots, \rho_n')$. Finally everyone's (b, c) preference is the same as their (a, b) preference at $\rho^{(2)}$. Max-$*$ transitivity implies that $\rho^*(a, c) \geq \rho^*(a, b) * \rho^*(b, c)$. Since $\widetilde{f}$ is neutral, $0 \geq *(\rho^*(a, b), \alpha)$, where $\alpha > 0$. If $\alpha = 1$, then $\rho^*(a, b) = 0$. If $\alpha < 1$, then since $*$ contains no zero divisor, $\rho^*(a, b) = 0$. By I, we have that at $(\rho_1', \rho_2, \rho_3', \ldots, \rho_n') \in \mathcal{S}^n$ the social value of (a, b) is zero.

At this profile, connectedness implies that $\rho_2(b, a) = 1$ and also that the social value of (b, a) must be equal to 1. This is true irrespective of everyone else's (b, a) values. Neutrality therefore implies that for all $(\rho_1, \ldots, \rho_n) \in \mathcal{S}^n$ and for all $(a, b) \in X \times X$, $\rho_2(a, b) = 1$ implies $\rho(a, b) = 1$.

$[\widetilde{f}(\rho^{(1)})(a, b) = 0, f(\rho^{(2)})(a, b) > 0, \rho_1^*(a, c) = \rho_1'(a, b), \rho_i^*(a, c) = \rho_i(a, b), i = 2, \ldots, n,$

$\rho_2^*(a, c) = \rho_2(a, b) = 0,$

$\rho_1^*(a, b) = \rho_1'(a, b), \rho_2^*(a, b) = \rho_2(a, b) = 0, \rho_i^*(a, b) = \rho_i'(a, b), i = 3, \ldots, n,$

$\rho_1^*(b, c) = \rho_1'(a, b), \rho_2^*(b, c) = \rho_2'(a, b), \rho_i^*(b, c) = \rho_i(a, b), i = 3, \ldots, n.$
(Thus $\rho^*(b, c) = \widetilde{f}(\rho^{(2)})(a, b) > 0$ by neutrality.)
Neutrality yields $\rho^*(a, c) = \widetilde{f}(\rho^{(1)}(a, b)) = 0$. Hence
$0 = \rho^*(a, c) \geq \rho^*(a, b) * \rho^*(b, c), \rho^*(b, c) = f(\rho^{(2)})(a, b) > 0$ by neutrality.
Hence
$\rho^*(a, b) = 0$ since $*$ has no zero divisors. $\rho^*(a, b) = 0$ so $\rho^*(b, a) = 1.]$

Let $(\rho_1, \ldots, \rho_n) \in \mathcal{S}^n$ be such that $\rho_2(c, b) = \rho_2(b, c) = 1$ and $\rho_i(a, c) = \rho_2(a, b)$ for all $i \in N$. From the above argument, it follows that $\rho(c, b) = \rho(b, c) = 1$ and that U implies $\rho(a, c) = \rho_2(a, b)$. Since ρ is max-$*$ transitive, we have $\rho(a, c) * \rho(c, b) \leq \rho(a, b)$ and $\rho(a, b) * \rho(b, c) \leq \rho(a, c)$. In other words, $\rho_2(a, b) * 1 \leq \rho(a, b)$ and $\rho(a, b) * 1 \leq \rho_2(a, b)$. Thus $\rho(a, b) = \rho_2(a, b)$. By neutrality, we have that for all $(\widehat{\rho}_1, \ldots, \widehat{\rho}_n) \in \mathcal{S}^n$ and for all $(a, b) \in X \times X$, $\widehat{\rho}(a, b) = \widehat{\rho}_2(a, b)$. ∎

We note that if $*$ is a triangular norm with a zero divisor, then there exists a zero divisor x such that $x * x = 0$: There exists $x, y \in (0, 1)$ such that $x * y = 0$. Suppose that $x \leq y$. Then $x * x = 0$ from the definition of t-norm.

Let $M(a, b) = \{x \in [0, 1] \mid$ at $(\rho_1, \ldots, \rho_n) \in \mathcal{S}^n$ there exists an $i \in N$ such that $\rho_i(a, b) = x$ and $\rho_i(a, b) \geq \rho_j(a, b)$ for all $j \in N\}$.

Let $m(a, b) = \{x \in [0, 1] \mid$ at $(\rho_1, \ldots, \rho_n) \in \mathcal{S}^n$ there exists an $i \in N$ such that $\rho_i(a, b) = x$ and $\rho_j(a, b) \geq \rho_i(a, b)$ for all $j \in N\}$.

Lemma 4.3.5 ([17]) *If $*$ has a zero divisor, then there exists a non-dictatorial FAR satisfying I and U.*

Proof. Define the function $\tilde{f} : \mathcal{S}^n \to \mathcal{S}$ as follows. For all $a, b \in X$ and all $(\rho_1, \ldots, \rho_n) \in \mathcal{S}^n$, let $\rho(a, b)$ be equal to the median value of the three numbers $M(a, b), x,$ and $m(a, b)$, where x is a zero divisor with the property $x * x = 0$. Then $\tilde{f}$ satisfies I and U and is non-dictatorial. It only remains to prove is that $\tilde{f}$ takes values in $\mathcal{S}$. Clearly, $\tilde{f}$ is reflexive and connected. It remains to prove that $\tilde{f}$ satisfies max-$*$ transitivity.

Assume, by way of contradiction that $\tilde{f}$ does not satisfy max-$*$ transitivity. Then there exists $(\rho_1, \ldots, \rho_n) \in \mathcal{S}^n$ and $a, b, c \in X$ such that $\rho(a, b) * \rho(b, c) > \rho(a, c)$. We first show that $\rho(a, b) \leq x$ and $\rho(b, c) \leq x$ is impossible. Since $x * x = 0$, we have that if $\rho(a, b) \leq x$ and $\rho(b, c) \leq x$, then $\rho(a, b) * \rho(b, c) = 0$. However, this contradicts the assumption that $\rho(a, b) * \rho(b, c) > \rho(a, c)$.

Second, we show that $\rho(a, b) > x$ and $\rho(b, c) > x$ is impossible. Suppose it is the case that $\rho(a, b) > x$ and $\rho(b, c) > x$. By the definition of $\tilde{f}$, $\rho(a, b) > x$ and $\rho(b, c) > x$ imply that $\rho(a, b) = m(a, b)$ and $\rho(b, c) = m(b, c)$. Let $j \in N$ be such that $\rho_j(a, c) = m(a, c)$. Since individual j's preferences are max-$*$ transitive, we have that $\rho_j(a, b) * \rho_j(b, c) \leq \rho_j(a, c)$. From the definition of m, it follows that $\rho_j(a, b) \geq m(a, b)$ and $\rho_j(b, c) \geq m(b, c)$, and so $\rho_j(a, b) \geq \rho(a, b)$ and $\rho_j(b, c) \geq \rho(b, c)$. Consequently, $\rho(a, b) * \rho(b, c)$ is less than or equal to $\rho_j(a, b) * \rho_j(b, c)$. Hence $\rho(a, b) * \rho(b, c) \leq \rho_j(a, c)$. We have that $\rho_j(a, b) = m(a, c)$ and that $m(a, c) \leq \rho(a, c) \leq M(a, c)$. Thus $\rho(a, b) * \rho(b, c) \leq \rho(a, c)$. This contradicts the assumption that $\rho(a, b) * \rho(b, c) > \rho(a, c)$.

It remains to consider two other possibilities. Either (i) $\rho(a, b) > x$ and $\rho(b, c) \leq x$, or (ii) $\rho(a, b) \leq x$ and $\rho(b, c) > x$. Assume that (i) is true. Then $\rho(a, b) * x$ is greater than or equal to $\rho(a, b) * \rho(b, c)$. Therefore, given our earlier assumption that $\rho(a, b) * \rho(b, c) > \rho(a, c)$, it must be the case that $\rho(a, b) * x > \rho(a, c)$. Now $1 * x = x$. Since $\rho(a, b) \leq 1$, we have $\rho(a, b) * x \leq x$. Since $\rho(a, b) * x > \rho(a, c)$ and $\rho(a, b) = m(a, b)$, it follows that $x > \rho(a, c)$. By the definition of $\tilde{f}$, $x > \rho(a, c)$ implies $\rho(a, c) = M(a, c)$. It also holds that $\rho(a, b) > x$ implies $\rho(a, b) = m(a, b)$. Thus for all $i \in N$, $\rho_i(a, b) \geq \rho(a, b)$ and $\rho_i(a, c) \leq \rho(a, c)$. Hence there exists $k \in N$ such that $\rho_k(a, b) \geq \rho(a, b), \rho_k(a, c) \leq \rho(a, c)$ and $\rho_k(b, c) = M(b, c)$. Since ρ_k is max-$*$ transitive, we have that $\rho_k(a, b) * \rho_k(b, c) \leq \rho_k(b, c)$. We also have that $m(b, c) \leq \rho(b, c) \leq M(b, c)$. Thus $\rho(b, c) \leq M(b, c) = \rho_k(b, c)$. Since

$\rho(a, b) \leq \rho_k(a, b)$ and $\rho(b, c) \leq \rho_k(b, c)$, it follows that $\rho(a, b) * \rho(b, c)$ is less than or equal to $\rho_k(a, b) * \rho_k(b, c)$. Hence $\rho(a, b) * \rho(b, c) \leq \rho_k(a, c)$. Thus since $\rho_k(a, c) \leq \rho(a, c)$, we have $\rho(a, b) * \rho(b, c) \leq \rho(a, c)$. However, this contradicts the assumption that $\rho(a, b) * \rho(b, c) > \rho(a, c)$. ∎

We consider the intuitive meaning that can be given to the requirement that a triangular norm contains, or does not contain, a zero divisor. The max-* transitivity for a fuzzy binary relation states that for all $x, y, x \in X$, $\rho(x, z) \geq \rho(x, y) * \rho(y, z)$. If this triangular norm contains a zero divisor, then (under one interpretation of fuzzy preferences) an individual or society can believe that the proposition "x is at least as good as y" is not false i.e., $\rho(x, y) > 0$ and that the proposition "y is at least as good as z" is not false, i.e., $\rho(y, z) > 0$ and yet still believe that the proposition "x is at least as good as z" is false, i.e., $\rho(x, z) = 0$. These beliefs are not inconsistent. However, if the triangular norm does not contain a zero divisor, then the proposition "x is at least as good as z" cannot be false, i.e., $\rho(x, z) > 0$. This is a difference between the two conditions.

4.4   Exercises

Let ρ be a fuzzy preference relation. Then ρ is called **minimally transitive** (T_m) if $\forall x, y, z \in X, \rho(x, y) = 1$ and $\rho(y, z) = 1$ imply $\rho(x, z) = 1$. Let π be the strict fuzzy preference relation associated with ρ. π is called **partially quasitransitive** if $\pi(x, y) > 0$ and $\pi(y, z) > 0$ imply $\pi(x, z) > 0$. π is called **negatively transitive** if $\forall x, y, z \in X, \pi(x, y) > 0$ implies $\pi(x, z) > 0$ or $\pi(z, y) > 0$.

1. Let ρ be a fuzzy preference relation. Show that if ρ is not strongly complete, then $\pi_{(2)}$ is not asymmetric.

2. If ρ is T_M and strongly connected and π is regular, prove that π is partially quasitransitive.

3. If ρ is T_M and strongly connected and π is regular, prove that π is negatively transitive.

4.5   References

1. K. J. Arrow, *Social Choice and Individual Values*, Wiley, New York 1951.

2. A. Banerjee, Rational choice under fuzzy preferences: The Orlovsky choice function, *Fuzzy Sets and Systems*, 53 (1993) 295–299.

3. A. Banerjee, A Fuzzy preferences and arrow-type problems in social choice, *Soc. Choice Welf.*, 11 (1994) 121–130.

4. C. R. Barrett and M. Salles, Social choice with fuzzy preferences, Working paper. Centre for Research in Economics and Management, UMR CNRS 6211, University of Cane, 2006.

5. C. R. Barrett, P. K. Pattanaik, and M. Salles, On the structure of fuzzy social welfare functions, *Fuzzy Sets and Systems,* 19 (1986) 1–10.

6. C. R. Barrett, P. K. Pattanaik, and M. Salles, Rationality and aggregation of preferences in an ordinally fuzzy framework, *Fuzzy Sets and Systems,* 49 (1992) 9–13.

7. K. Basu, Fuzzy revealed preference, *J. Econ. Theory,* 32 (1984) 212–227.

8. K. Basu, R. Deb, and P. K. Pattanaik, Soft sets: An ordinal reformulation of vagueness with some applications to the theory of choice, *Fuzzy Sets and Systems,* 45 (1992) 45–58.

9. A. Billot, *Economic Theory of Fuzzy Equilibria,* Springer, Berlin 1995.

10. M. Dasgupta and R. Deb, Fuzzy choice functions, *Soc. Choice Welf.,* 8 (1991) 171–182.

11. M. Dasgupta and R. Deb, Transitivity and fuzzy preferences, *Soc. Choice Welf.,* 13 (1996) 305–318.

12. M. Dasgupta and R. Deb, An impossibility theorem with fuzzy preferences. In: de Swart H (ed) Logic, game theory and social choice: Proceedings of the International Conference, LGS '99, May 13–16, 1999, Tilburg University Press 1999.

13. M. Dasgupta and R. Deb, Factoring fuzzy transitivity, *Fuzzy Sets and Systems,* 118 (2001) 489–502.

14. F. Dietrich and C. List, The aggregation of propositional attitudes: towards a general theory, forthcoming in *Oxford Studies in Epistemology* 2009.

15. C. Duddy and A. Piggins, Many-valued judgement aggregation: Characterizing the possibility/impossibility boundary for an important class of agendas, Working paper, Department of Economics, National University of Ireland, Galway 2009.

16. C. Duddy, J. Perote-Peña, and A. Piggins, Manipulating an aggregation rule under ordinally fuzzy preferences, *Soc. Choice Welf.,* 34 (2010) 411–428.

17. C. Duddy, J. Perote-Peña, and A. Piggins, Arrow's theorem and max-∗ transitivity, *Social Choice and Welfare,* 36 (2011) 25–34.

18. B. Dutta, Fuzzy preferences and social choice. *Math Soc. Sci.,* 13 (1987) 215–229.

19. B. Dutta, S. C. Panda, and P. K. Pattanaik, Exact choice and fuzzy preferences, *Math Soc. Sci.,* 11 (1986) 53–68.

20. L. A. Fono and N. G. Andjiga, Fuzzy strict preferences and social choice, *Fuzzy Sets and Systems,* 155 (2005) 372–389.

21. J. L. Garcia-Lapresta and B. Llamazares, Aggregation of fuzzy preferences: Some results of the mean, *Social Choice and Welfare,* 17 (2009) 673–690.

22. J. A. Goguen, L-fuzzy sets, *J. Math. Anal. Appl.,* 18 (1967) 145–174.

23. N. Jain, Transitivity of fuzzy relations and rational choice, *Ann. Oper. Res.,* 23 (1990) 265–278.

24. E. P. Klement, R. Mesiar, and E. Pap, *Triangular norms,* Kluwer Academic Publishers, Dordrecht, 2000.

25. B. Leclerc, Efficient and binary consensus functions on transitively valued relations, *Math. Soc. Sci.,* 8 (1984) 45–61.

26. B. Leclerc, Aggregation of fuzzy preferences: A theoretic Arrow-like approach, *Fuzzy Sets and Systems,* 43 (1991) 291–309.

27. B. Leclerc and B. Monjardeet, Lattical theory of consensus. In: Barnett W., Moulin H., Salles M., and Schofeld N. (eds.) *Social Choice, Welfare and Ethics,* Cambridge University Press, Cambridge, 1995.

28. J. N. Mordeson, M. B. Gibilisco, and T. D. Clark, Independence of irrelevant alternatives and fuzzy Arrow's theorem, *New Mathematics and Natural Computation,* 8 (2012) 219–237.

29. S. A. Orlovsky, Decision-making with a fuzzy preference relation, *Fuzzy Sets and Systems,* 1 (1978) 155–167.

30. S. V. Ovchinnikov, Structure of fuzzy binary relations, *Fuzzy Sets and Systems,* 6 (1981) 169–195.

31. S. V. Ovchinnikov, Social choice and Lukasiewicz logic, *Fuzzy Sets and Systems,* 43 (1991) 275–289.

32. S. V. Ovchinnikov and M. Roubens, On strict preference relations, *Fuzzy Sets and Systems,* 43 (1991) 319–326.

33. S. V. Ovchinnikov and M. Roubens, On fuzzy strict preference, indifference and incomparability relations, *Fuzzy Sets and Systems,* 49 (1992) 15–20.

34. J. Perote-Peña and A. Piggins, Strategy-proof fuzzy aggregation rules, *J. Math. Econ.*, 43 (2007) 564–580.

35. J. Perote-Peña and A. Piggins, Non-manipulative social welfare functions when preferences are fuzzy, *J. Log. Comput.*, 19 (2009) 503–515.

36. J. Perote-Peña and A. Piggins, Social choice, fuzzy preferences and manipulation. In. Boylan T. Gekker R. (eds), *Economics, rational choice, and normative philosophy*. Routledge, London, 2009.

37. A. Piggins and M. Salles, Instances of indeterminacy, *Analyse und Kritik*, 29 (2007) 311–328.

38. C. Ponsard, Some dissenting views on the transitivity of individual preferences, *Ann. Oper. Res.*, 23 (1990) 279–288.

39. G. Richardson, The structure of fuzzy preferences: Social choice implications, *Soc. Choice Welf.*, 15 (1990) 359–369.

40. M. Salles, Fuzzy utility. In: Barbara S., Hammond P.J., and Seidl C. (eds), *Handbook of utility theory, Vol. 1: Principles*, Kluwer, Dordrecht, 1998.

41. A. K. Sen, Interpersonal aggregation and partial comparability, *Econometrica*, 38 (1970) 393–409.

.

Chapter 5

Fuzzy Arrow's Theorem

5.1 Dictatorial Fuzzy Preference Aggregation Rules

Arrow's Theorem [1] is one of the most important discoveries in social science theory. Arrow asks if there are any methods for aggregating the preferences of individuals that meet several reasonable conditions: universal admissibility, transitivity, unanimity (weak Paretianism), and independence from irrelevant alternatives. Through a series of formal proofs he concludes that the only methods for achieving all four of these conditions are dictatorial, that is, the social choice is perfectly aligned with the preferences of one individual.

A voting system is a function that maps the voting preferences of the voters for the candidates to a ranking of the candidates. We list certain reasonable assumptions one might think a perfect voting system should satisfy:

(i) The voting system preserves rationality, that is, the output of the voting system is a total ordering.

(ii) The output of the voting system is determined only by the ranking preferences of the voters; no other factors are allowed.

(iii) If all voters prefer one candidate to another, then the output of the voting system is the favored candidate.

(iv) All candidates are treated equally.

(v) The ranking of two candidates in the output of the voting system is independent of the voters' preferences for a third candidate.

(vi) There is no dictator.

Arrow's theorem says that these six axioms are inconsistent.

In the situation where fuzzy preference aggregation rules are allowed, there are many types of transitivity, independence of irrelevant alternatives, strict fuzzy preference relations, and so on. We examine how various combinations of these concepts yield impossibility theorems.

We begin by considering a certain set of definitions for Arrow's conditions. This set of definitions does not lead to a different conclusion than that reached by Arrow under crisp logic. We next consider applications of the fuzzy version of Arrow's Theorem involving representation rules, oligarchies, and veto players. We also consider an approach that uses the notion of filters to further strengthen the fuzzy version of Arrow's Theorem. In the last three sections, we consider approaches by Dutta, Banerjee, and Richardson.

Definition 5.1.1 *Let ρ be a fuzzy binary relation on X. Then ρ is called a* ***fuzzy weak order*** *on X if it is reflexive, complete, and partially transitive.*

Definition 5.1.2 *Let ρ be a fuzzy binary relation on X. Define the fuzzy subset ι of $X \times X$ by $\forall x, y \in X$, $\iota(x,y) = \rho(x,y) \wedge \rho(y,x)$.*

Assume that each individual i has a fuzzy weak order ρ_i on X. Let $\mathcal{FWR}$ denote the set of all fuzzy weak orders on X. When there is no indifference, i.e., $\iota_i(x,y) = 0$ for all $x, y \in X$, $x \neq y$, individual i's preferences π_i are said to be strict. Let $\mathcal{FWP}$ denote the set of all strict orders on X. A fuzzy preference profile on X is a n-tuple of fuzzy weak orders $\rho = (\rho_1, \ldots, \rho_n)$ describing the fuzzy preferences of all individuals. Let $\mathcal{FWR}^n$ denote the set of all fuzzy preference profiles. Recall $\mathcal{P}(X)$ denotes the power set of X. For any $\rho \in \mathcal{FWR}^n$ and for all $S \in \mathcal{P}(X)$, let $\rho\rceil_S = (\rho_1\rceil_S, \ldots, \rho_n\rceil_S)$, where $\rho_i\rceil_S = \rho_i\rceil_{S \times S}, i = 1, \ldots, n$. For all $\rho \in \mathcal{FWR}^n$ and $\forall x, y \in X$, let

$$P(x, y; \rho) = \{i \in N \mid \pi_i(x, y) > 0\} \text{ and } R(x, y; \rho) = \{i \in N \mid \rho_i(x, y) > 0\}.$$

Let $\mathcal{FB}$ denote the set of all reflexive and complete fuzzy binary relations on X.

Definition 5.1.3 *A function $\widetilde{f} : \mathcal{FWR}^n \to \mathcal{FB}$ is called a* ***fuzzy preference aggregation rule***.

We sometimes suppress the underlying fuzzy preference aggregation rule and write $\rho(x, y)$ for $\widetilde{f}(\rho)(x, y)$ and $\pi(x, y) > 0$ for $\widetilde{f}(\rho)(x, y) > 0$ and $\widetilde{f}(\rho)(y, x) = 0$.

Definition 5.1.4 *Let $\widetilde{f}$ be a fuzzy preference aggregation rule.*

    (1) *$\widetilde{f}$ is said to be a* ***simple majority rule*** *if $\forall \rho \in \mathcal{FWR}^n$, $\forall x, y \in X$, $\widetilde{f}(\rho)(x, y) > 0$ if and only if $|P(x, y; \rho)| > n/2$.*

    (2) *Let strict preferences be regular. $\widetilde{f}$ is said to be a* ***Pareto extension rule*** *if $\forall \rho \in \mathcal{FWR}^n$, $\forall x, y \in X$, $\pi(x, y) > 0$ if and only if $\widehat{R}(x, y; \rho) = N$ and $P(x, y; \rho) \neq \emptyset$, where $\widehat{R}(x, y; \rho) = \{i \in N \mid \rho_i(x, y) \geq \rho_i(y, x)\}$.*

For the Pareto extension rule, there are two underlying rules: (1) if everyone in the society is indifferent between two alternatives x and y, then the society should be indifferent also and (2) if at least one individual strictly prefers x to y and every individual regards x to be at least as good as y, then the society should prefer x to y. Consider rule (1). Suppose everyone in the society is indifferent between two alternatives x and y. Then $R(x, y; \rho) = N$ and $P(x, y; \rho) = \emptyset$. Hence by the contrapositive of the Pareto extension rule, we conclude that it is not the case that $\widetilde{f}(\rho)(x, y) > 0$, $\widetilde{f}(\rho)(y, x) = 0$ for some x and y. For $\widetilde{f}(\rho)$ complete, we thus have $\widetilde{f}(\rho)(x, y) > 0$, $\widetilde{f}(\rho)(y, x) > 0$. In the crisp case, this yields that the society is indifferent between x and y while in the fuzzy case, there is a positive degree with which the society is indifferent.

We sometimes repeat a definition involving fuzzy preference aggregation rules when the context is changed, i.e., its domain is changed.

Definition 5.1.5 *Let $\widetilde{f}$ be a fuzzy preference aggregation rule. Then*
   (1) *$\widetilde{f}$ is said to be **nondictatorial** if it is not the case that $\exists i \in N$ such that $\forall \rho \in \mathcal{FWR}^n$, $\forall x, y \in X$, $\pi_i(x, y) > 0$ implies $\pi(x, y) > 0$;*
   (2) *$\widetilde{f}$ is said to be **weakly Paretian** if $\forall \rho \in \mathcal{FWR}^n$, $\forall x, y \in X$ [$\forall i \in N$, $\pi_i(x, y) > 0$ implies $\pi(x, y) > 0$];*
   (3) *$\widetilde{f}$ is said to be **independent of irrelevant alternatives IIA2** if $\forall \rho, \rho' \in \mathcal{FR}^n$, $\forall x, y \in X$, $Supp(\rho_i\rceil_{\{x,y\}}) = Supp(\rho'_i\rceil_{\{x,y\}})$ $\forall i \in N$ implies $Supp(\widetilde{f}(\rho)\rceil_{\{x,y\}}) = Supp(\widetilde{f}(\rho')\rceil_{\{x,y\}})$.*

In the 1992 U. S. election, Clinton won the election with approximately 43 percent of the vote while Bush had approximately 38 percent and Perot 19 percent. Had all Perot voters voted for Bush if Perot had not run, then Bush would have won the election. This is a violation of the condition of independence of irrelevant alternatives in the crisp case.

Definition 5.1.6 *Let $\widetilde{f}$ be a fuzzy preference aggregation rule. Let $(x, y) \in X \times X$. Let λ be a fuzzy subset of N. Then*
   (1) *λ is called **semidecisive** for x against y, written $x\widetilde{D}_\lambda y$, if $\forall \rho \in \mathcal{FWR}^n$, [$\pi_i(x, y) > 0$ $\forall i \in Supp(\lambda)$ and $\pi_j(y, x) > 0$ $\forall j \notin Supp(\lambda)$] implies $\pi(x, y) > 0$;*
   (2) *λ is called **decisive** for x against y, written $xD_\lambda y$, if $\forall \rho \in \mathcal{FWR}^n$, [$\pi_i(x, y) > 0$ $\forall i \in Supp(\lambda)$] implies $\pi(x, y) > 0$.*

Definition 5.1.7 *Let $\widetilde{f}$ be a fuzzy preference aggregation rule. Let λ be a fuzzy subset of N. Then λ is called **semidecisive (decisive)** for $\widetilde{f}$ if $\forall (x, y) \in X \times X$, λ is semidecisive (decisive) for x against y.*

Proposition 5.1.8 *Let $\widetilde{f}$ be a fuzzy preference aggregation rule. Suppose $\widetilde{f}$ is dictatorial with dictator i. Let λ be a fuzzy subset of N. Suppose λ is a coalition, i.e., $|Supp(\lambda)| \geq 2$. Then λ is decisive if and only if $i \in Supp(\lambda)$.*

Proof. Suppose λ is decisive. Suppose $i \notin Supp(\lambda)$. Let $x, y \in X$ be such that $x \neq y$. Then $\exists \rho \in \mathcal{FWR}^n$ such that $\pi_j(x, y) > 0 \; \forall j \in Supp(\lambda)$ and $\pi_i(y, x) > 0$. Thus $\pi(x, y) > 0$ and $\pi(y, x) > 0$, a contradiction. Hence $i \in Supp(\lambda)$. The converse is immediate. ∎

In the next definition we restate Definition 2.2.16 since we are interested in $* = \wedge$ in this chapter.

Definition 5.1.9 *Let $\rho \in \mathcal{FR}^*(X)$. Then ρ is said to be **partially quasi-transitive** if $\forall x, y, z \in X$, $\pi(x, y) \wedge \rho(y, z) > 0$ implies $\pi(x, z) > 0$, where ρ is a fuzzy strict preference relation associated with ρ.*

Definition 5.1.10 *Let $\widetilde{f}$ be a fuzzy preference aggregation rule. Then*
(1) *$\widetilde{f}$ is said to be **(partially) transitive** if $\forall \rho \in \mathcal{FR}^n$, $\widetilde{f}(\rho)$ is (partially) transitive;*
(2) *$\widetilde{f}$ is said to be **(partially) quasi-transitive** if $\forall \rho \in \mathcal{FR}^n$, $\widetilde{f}(\rho)$ is (partially) quasi-transitive;*
(3) *$\widetilde{f}$ is said to be **(partially) acyclic** if $\forall \rho \in \mathcal{FR}^n$, $\widetilde{f}(\rho)$ is (partially) acyclic.*

In the remainder of the section, we assume that $|X| \geq 3$ unless otherwise specified.

Proposition 5.1.11 *Let ρ be a fuzzy binary relation on X. If ρ is partially transitive, then ρ is partially quasi-transitive with respect to $\pi = \pi_{(0)}$.*

Proof. Let $x, y, z \in X$. Suppose $\pi(x, y) > 0$ and $\pi(y, z) > 0$. Then $\rho(x, y) > 0, \rho(y, x) = 0$ and $\rho(y, z) > 0, \rho(z, y) = 0$. Hence $\rho(x, z) > 0$. Suppose $\pi(x, z) = 0$. Then $\rho(z, x) > 0$. Thus since $\rho(x, y) > 0, \rho(z, y) > 0$ by partial transitivity, a contradiction. Thus $\pi(x, z) > 0$. ∎

Example 5.1.12 *(Partial transitivity $\nRightarrow$ partial quasi-transitivity for $\pi = \pi_{(1)}$ or $\pi = \pi_{(3)}$) Let $X = \{x, y, z\}$. Define the fuzzy relation ρ on X as follows:*

$$\rho(x, x) = \rho(y, y) = \rho(z, z) = 1, \; \rho(x, y) = \rho(y, z) = 2/3,$$
$$\rho(y, x) = \rho(z, y) = 1/3, \; \rho(x, z) = \rho(z, x) = 1/2.$$

Then ρ is partially transitive since $\rho(u, v) > 0 \; \forall u, v \in X$. However, $\pi(x, y) > 0, \pi(y, z) > 0$, but $\pi(x, z) = 0$.

Definition 5.1.13 *Let $\widetilde{f}$ be a fuzzy aggregation rule. Then $\widetilde{f}$ is said to be **independent of irrelevant alternatives** IIA4 if $\forall x, y \in X$, $[\forall \rho, \rho' \in \mathcal{FR}^n(X), Cosupp(\rho_i|_{\{x,y\}}) = Cosupp(\rho'_i|_{\{x,y\}}) \; \forall i \in N \Rightarrow Cosupp(f(\rho)|_{\{x,y\}}) = Cosupp(f(\rho')|_{\{x,y\}})]$.*

Proposition 5.1.14 *Let $\rho, \rho' \in \mathcal{FR}(X)$. Let $x, y \in X$. Then $Cosupp(\rho\rvert_{\{x,y\}})$ $= Cosupp(\rho'\rvert_{\{x,y\}})$ if and only if $(\forall u, v \in \{x, y\}, \rho(u, v) < 1 \Leftrightarrow \rho'(u, v) < 1)$.*

Proof. Suppose $Cosupp(\rho\rvert_{\{x,y\}}) = Cosupp(\rho'\rvert_{\{x,y\}})$. Let $u, v \in \{x, y\}$. Then $\rho(u, v) < 1 \Leftrightarrow (u, v) \in Cosupp(\rho\rvert_{\{x,y\}}) \Leftrightarrow (u, v) \in Cosupp(\rho'\rvert_{\{x,y\}}) \Leftrightarrow \rho'(u, v) < 1$. Conversely, suppose $\forall u, v \in \{x, y\}$, $\rho(u, v) < 1 \Leftrightarrow \rho'(u, v) < 1$. Then $(u, v) \in Cosupp(\rho\rvert_{\{x,y\}}) \Leftrightarrow \rho(u, v) < 1 \Leftrightarrow \rho'(u, v) < 1 \Leftrightarrow Cosupp(\rho'\rvert_{\{x,y\}})$. ∎

For $\pi = \pi_{(2)}$ and $x, y \in X$, $\pi(x, y) > 0$ if and only if $\rho(y, x) < 1$. For $\pi = \pi_{(1)}$, or $\pi_{(3)}$, or $\pi_{(4)}$, or $\pi_{(5)}$, π has the property that $\forall x, y \in X$, $\pi(x, y) > 0$ if and only if $\rho(x, y) > \rho(y, x)$. This property plays a key role in the proof of the next result.

Lemma 5.1.15 *Let λ be a fuzzy subset of N. Let $\tilde{f}$ be a partially quasi-transitive fuzzy preference aggregation rule that is weakly Paretian and either independent of irrelevant alternatives IIA2 with strict preferences of type $\pi_{(0)}$ or independent of irrelevant alternatives IIA3 with regular strict preferences or is independent of irrelevant alternatives IIA4 with strict preferences of type $\pi_{(2)}$. If λ is semidecisive for x against y w.r.t π, then $\forall (v, w) \in X \times X$, λ is decisive for v against w.*

Proof. Suppose λ is semidecisive for x against y. Let ρ be any fuzzy preference profile such that $\pi_i(x, z) > 0$ $\forall i \in Supp(\lambda)$, where $z \notin \{x, y\}$. For IIA3 and $\pi = \pi_{(1)}$, or $\pi_{(3)}$, or $\pi_{(4)}$, or $\pi_{(5)}$, let ρ' be a fuzzy preference profile such that

$$\rho_i'(x, z) = \rho_i(x, z) \text{ and } \rho_i'(z, x) = \rho_i(z, x) \; \forall i \in N,$$
$$\pi_i'(x, y) > 0 \; \forall i \in Supp(\lambda) \text{ and } \pi_j'(y, x) > 0 \; \forall j \in N \backslash Supp(\lambda),$$
$$\pi_i'(y, z) > 0 \; \forall i \in N.$$

Since $\pi_i(x, z) > 0$ $\forall i \in Supp(\lambda)$, $\pi_i'(x, z) > 0$ $\forall i \in Supp(\lambda)$ by the definition of ρ'. Since $x\tilde{D}_\lambda y$, $\pi'(x, y) > 0$. Since $\tilde{f}$ is weakly Paretian, $\pi'(y, z) > 0$. Since $\tilde{f}$ is partially quasi-transitive, $\pi'(x, z) > 0$. Hence $\rho'(x, z) > \rho'(z, x)$. Since $\rho_i\rvert_{\{x,z\}} = \rho_i'\rvert_{\{x,z\}}$ $\forall i \in N$ and $\tilde{f}$ is IIA3, $\rho\rvert_{\{x,z\}} \sim \rho'\rvert_{\{x,z\}}$. Thus $\rho(x, z) > \rho(z, x)$. (For strict preferences of type $\pi_{(0)}$ and the crisp case, we write: Hence $\rho'(x, z) > 0$ and $\rho'(z, x) = 0$. Since $Supp(\rho_i\rvert_{\{x,z\}}) = Supp(\rho_i'\rvert_{\{x,z\}})$ $\forall i \in N$ and $\tilde{f}$ is IIA2, $Supp(\rho\rvert_{\{x,z\}}) = Supp(\rho'\rvert_{\{x,z\}})$. Thus $\rho(x, z) > 0$ and $\rho(z, x) = 0$.) (For strict preferences of type $\pi_{(2)}$, we write: Hence $\rho'(z, x) < 1$. Since $Cosupp(\rho_i\rvert_{\{x,z\}}) = Cosupp(\rho_i'\rvert_{\{x,z\}})$ $\forall i \in N$ and $\tilde{f}$ is IIA4, $Cosupp(\rho\rvert_{\{x,z\}}) = Cosupp(\rho'\rvert_{\{x,z\}})$. Thus $\rho(z, x) < 1$.) Hence $\pi(x, z) > 0$. Since ρ was arbitrary, $x D_\lambda z$. Thus since z was arbitrary in $X \backslash \{x, y\}$,

$$\forall z \in X \backslash \{x, y\}, \; x\tilde{D}_\lambda y \Rightarrow x D_\lambda z. \tag{5.1}$$

Since λ is decisive for x against z implies λ is semidecisive for x against z, interchanging y and z in the above argument implies λ is decisive for x against y.

For IIA3 and regular strict preferences, let ρ'' be any fuzzy preference profile with $\pi_i''(y,z) > 0 \ \forall i \in \mathrm{Supp}(\lambda)$ and let ρ^+ be a fuzzy preference profile such that

$$
\begin{aligned}
\rho_i^+(y,z) &= \rho_i''(y,z) \text{ and } \rho_i^+(z,y) = \rho_i''(z,y) \ \forall i \in N \\
\pi_i^+(y,x) &> 0 \ \forall i \in N, \\
\pi_i^+(x,z) &> 0 \ \forall i \in \mathrm{Supp}(\lambda) \text{ and } \pi_j^+(z.x) > 0 \ \forall j \in N\backslash\mathrm{Supp}(\lambda).
\end{aligned}
$$

Then $\pi_i^+(y,z) > 0 \ \forall i \in \mathrm{Supp}(\lambda)$. Since $xD_\lambda z$, $\pi^+(x,z) > 0$. Since $\widetilde{f}$ is weakly Paretian, $\pi^+(y,x) > 0$. Since $\widetilde{f}$ is partially quasi-transitive, $\pi^+(y,z) > 0$. Since $\rho_i^+\rceil_{\{y,z\}} = \rho_i''\rceil_{\{y,z\}} \ \forall i \in N$ and $\widetilde{f}$ is IIA3, $\rho^+\rceil_{\{y,z\}} \sim \rho''\rceil_{\{y,z\}}$. Since $\pi^+(y,z) > 0$, $\rho^+(y,z) > \rho^+(z,y)$. Hence $\rho''(y,z) > \rho''(z,y)$. (For $\pi = \pi_{(0)}$ and the crisp case, we write: Since $\rho_i^+\rceil_{\{y,z\}} = \rho_i''\rceil_{\{y,z\}} \ \forall i \in N$ and $\widetilde{f}$ is IIA2, $\mathrm{Supp}(\rho^+\rceil_{\{y,z\}}) = \mathrm{Supp}(\rho''\rceil_{\{y,z\}})$. Since $\pi^+(y,z) > 0$, $\rho^+(y,z) > 0$ and $\rho^+(z,y) = 0$. Hence $\rho''(y,z) > 0$ and $\rho''(z,y) = 0$.) (For $\pi = \pi_{(2)}$, we write: Since $\rho_i^+\rceil_{\{y,z\}} = \rho_i''\rceil_{\{y,z\}} \ \forall i \in N$ and $\widetilde{f}$ is IIA4, $\mathrm{Cosupp}(\rho^+\rceil_{\{y,z\}}) = \mathrm{Cosupp}(\rho''\rceil_{\{y,z\}})$. Since $\pi^+(y,z) > 0$, $\rho^+(z,y) < 1$. Hence $\rho''(z,y) < 1$.) Thus $\pi''(y,z) > 0$ and so $yD_\lambda z$. Hence since z is arbitrary in $X\backslash\{x,y\}$, the preceding two steps yield

$$
\forall z \notin \{x,y\}, \ x\widetilde{D}_\lambda y \Rightarrow yD_\lambda z. \tag{5.2}
$$

Now λ decisive for y against z implies λ is semidecisive for y against z. Thus by (5.1), λ is decisive for y against x. We have $\forall(v,w) \in X \times X$,

$$
x\widetilde{D}_\lambda y \Rightarrow xD_\lambda v \text{ (by (5.1))} \Rightarrow x\widetilde{D}_\lambda v \Rightarrow vD_\lambda w \text{ (by (5.2)) with } v \text{ replacing } y.
$$

∎

We next present our first impossibility theorem.

Theorem 5.1.16 (*Fuzzy Arrow's Theorem*) *Let $\widetilde{f}$ be a fuzzy aggregation rule. Suppose strict preferences are of type $\pi_{(0)}$. Let $\widetilde{f}$ be weakly Paretian, partially transitive, and independent of irrelevant alternatives IIA2. Then $\widetilde{f}$ is dictatorial.*

Proof. Since $\widetilde{f}$ is partially transitive, $\widetilde{f}$ is partially quasi-transitive since $\pi = \pi_{(0)}$. By Lemma 5.1.15, it suffices to show that $\exists i \in N$, $\exists x,y \in X$ such that $\{i\}$ is semidecisive for x against y. (That is, $\exists \lambda$ with $\mathrm{Supp}(\lambda) = \{i\}$ such that λ is semidecisive for x against y.) This follows by Lemma 5.1.15 because then λ is decisive for x against y for all $x,y \in X$, where $\mathrm{Supp}(\lambda) = \{i\}$. Hence $\forall \rho \in \mathcal{FR}^n(X)$, $\forall x,y \in X$, $\pi_i(x,y) > 0$ implies $\pi(x,y) > 0$.

Since $\widetilde{f}$ is weakly Paretian, $\exists$ a decisive λ for any pair of alternatives, namely $\lambda = 1_N$. For all $(u, v) \in X \times X$, let $m(u, v)$ denote the size of the smallest $|\text{Supp}(\lambda)|$ for λ semidecisive for u against v. Let $m = \wedge\{m(u, v) \mid (u, v) \in X \times X\}$. Without loss of generality suppose λ is semidecisive for x against y with $|\text{Supp}(\lambda)| = m$. If $m = 1$, the proof is complete. Suppose $m > 1$. Let $i \in \text{Supp}(\lambda)$. Consider any fuzzy profile $\rho \in \mathcal{FR}^n$ such that

$$\pi_i(x, y) > 0, \ \pi_i(y, z) > 0, \ \pi_i(x, z) > 0,$$
$$\forall j \in \text{Supp}(\lambda)\backslash\{i\}, \ \pi_j(z, x) > 0, \pi_j(x, y) > 0, \ \pi_j(z, y) > 0,$$
$$\forall k \notin \text{Supp}(\lambda), \ \pi_k(y, z) > 0, \pi_k(z, x) > 0, \ \pi_k(y, x) > 0.$$

Since λ is semidecisive for x against y and $\pi_j(x, y) > 0 \ \forall j \in \text{Supp}(\lambda)$, $\pi(x, y) > 0$. Since $|\text{Supp}(\lambda)| = m$, it is not the case that $\pi(z, y) > 0$ otherwise λ' is semidecisive for z against y, where $\lambda'(j) = \lambda(j) \ \forall j \in \text{Supp}(\lambda)\backslash\{i\}$ and $\lambda'(i) = 0$. (Suppose $\pi(z, y) > 0$. $\text{Supp}(\rho_i]_{\{z,y\}}) = \text{Supp}(\rho'_i]_{\{z,y\}}) \ \forall i \in N, \forall \rho' \in \mathcal{FR}^n(X) \Rightarrow \text{Supp}(\widetilde{f}(\rho)]_{\{z,y\}}) = \text{Supp}(\widetilde{f}(\rho')]_{\{z,y\}}) \ \forall \rho' \in \mathcal{FR}^n(X)$ by independence of irrelevant alternatives $\Rightarrow (z, y) \in \text{Supp}(\widetilde{f}(\rho')), (y, z) \notin \text{Supp}(\widetilde{f}(\rho')) \ \forall \rho' \in \mathcal{FR}^n(X)$ since this is the case for $\widetilde{f}(\rho)$ and $\pi_j(z, y) > 0$ $\forall j \in \text{Supp}(\lambda')$ and $\pi_j(y, z) > 0$ otherwise.) However, this contradicts the minimality of m since $|\text{Supp}(\lambda')| = m - 1$. Thus $\widetilde{f}(\rho)(y, z) > 0$ since $\widetilde{f}(\rho)$ is complete. Since $\widetilde{f}$ is partially transitive, $\widetilde{f}(\rho)(x, z) > 0$. Suppose $\pi(x, z) = 0$. Then $\widetilde{f}(\rho)(z, x) > 0$. Since $\widetilde{f}(\rho)(y, z) > 0$, we have $\widetilde{f}(\rho)(y, x) > 0$ by partial transitivity. However, this contradicts the fact that $\pi(x, y) > 0$. Hence $\pi(x, z) > 0$. By independence of irrelevant alternatives, λ^* is semidecisive for x against z, where $\lambda^*(i) = \lambda(i)$ and $\lambda^*(j) = 0$ for $j \in N\backslash\{i\}$. However, this contradicts the fact that $m > 1$. ∎

Definition 5.1.17 *Let ρ be a fuzzy relation on X. Then ρ is said to be **weakly transitive** if $\forall x, y, z \in X$, $\rho(x, y) \geq \rho(y, x)$ and $\rho(y, z) \geq \rho(z, y)$ implies $\rho(x, z) \geq \rho(z, x)$.*

Example 5.1.18 *Let $X = \{x, y, z\}$.*
(a) Define the fuzzy relation on X as follows:

$$\rho(x, x) = \rho(y, y) = \rho(z, z) = 1,$$
$$\rho(x, y) = \rho(y, z) = \rho(x, z) = 1/2,$$
$$\rho(y, x) = \rho(z, y) = 1/2, \ \rho(z, x) = 1.$$

Then ρ is max-min transitive, but ρ is not weakly transitive since $\rho(x, z) \not\geq \rho(z, x)$.
(b) Define the fuzzy relation ρ on X as follows:

$$\rho(x, x) = \rho(y, y) = \rho(z, z) = 1,$$
$$\rho(x, y) = \rho(y, z) = \rho(x, z) = 1/2,$$
$$\rho(y, x) = \rho(z, y) = 1/4, \ \rho(z, x) = 1/8.$$

Then ρ is weakly transitive, but not max-min transitive since $\rho(z,x) \not\geq \rho(z,y) \wedge \rho(y,x)$.

(c) Define the fuzzy relation ρ on X as follows:

$$\rho(x,x) = \rho(y,y) = \rho(z,z) = 1,$$
$$\rho(x,y) = \rho(y,z) = 1, \quad \rho(x,z) = 0,$$
$$\rho(y,x) = \rho(z,y) = \rho(z,x) = 0.$$

Then ρ is weakly transitive, but not transitive in the crisp sense. Note that ρ is not complete.

Proposition 5.1.19 *Suppose ρ is exact. If ρ is weakly transitive and complete, then ρ is transitive in the crisp sense.*

Proof. Suppose $\rho(x,y) = 1$ and $\rho(y,x) = 1$. Since ρ is weakly transitive, $\rho(x,z) \geq \rho(z,x)$. Since ρ is complete, $\rho(x,z) = 1$. ∎

Proposition 5.1.20 *Let ρ be a fuzzy relation on X. Then the following properties are equivalent:*

(1) ρ is weakly transitive.

(2) For all $x,y,z \in X, \rho(x,y) \geq \rho(y,x)$ and $\rho(y,z) \geq \rho(z,y)$ with strict inequality holding at least once, then $\rho(x,z) > \rho(z,x)$.

Proof. Suppose (1) holds. Assume that $\rho(x,y) \geq \rho(y,x)$ and $\rho(y,z) > \rho(z,y)$. Then $\rho(x,z) \geq \rho(z,x)$. Suppose $\rho(z,x) \geq \rho(x,z)$. Then $\rho(z,y) \geq \rho(y,z)$ by (1), a contradiction. Thus $\rho(x,z) > \rho(z,x)$. A similar argument shows that $\rho(x,y) > \rho(y,x)$ and $\rho(y,z) \geq \rho(z,y)$ implies $\rho(x,z) > \rho(z,x)$.

Suppose (2) holds. Let $x,y,z \in X$. Suppose $\rho(x,y) \geq \rho(y,x)$ and $\rho(y,z) \geq \rho(z,y)$. Suppose $\rho(z,x) > \rho(x,z)$. Then by (2), $\rho(z,x) > \rho(x,z)$ and $\rho(x,y) \geq \rho(y,x)$, we have $\rho(z,y) > \rho(y,z)$, a contradiction. Hence $\rho(x,z) \geq \rho(z,x)$. ∎

Corollary 5.1.21 *Let ρ be a fuzzy relation on X. If ρ is weakly transitive, then ρ is partially quasi-transitive.*

We next consider an impossibility result when strict preferences are regular and with different types of transitivity and independence of irrelevant alternative conditions.

Theorem 5.1.22 *(Fuzzy Arrow's Theorem) Let $\widetilde{f}$ be a fuzzy aggregation rule. Suppose strict preferences are regular. Let $\widetilde{f}$ be weakly Paretian, weakly transitive, and independent of irrelevant alternatives IIA3. Then $\widetilde{f}$ is dictatorial.*

Proof. Since $\widetilde{f}$ is weakly transitive, $\widetilde{f}$ is partially quasi-transitive by the Corollary 5.1.21. By Lemma 5.1.15, it suffices to show that $\exists i \in N, \exists x,y \in X$ such that $\{i\}$ is semidecisive for x against y. (That is, $\exists \lambda$ with $\text{Supp}(\lambda) = \{i\}$

such that λ is semidecisive for x against y.) This follows by Lemma 5.1.15 because then λ is decisive for x against y for all $x, y \in X$, where $\text{Supp}(\lambda) = \{i\}$. Hence $\forall \rho \in \mathcal{FR}^n(X)$, $\forall x, y \in X$, $\pi_i(x, y) > 0$ implies $\pi(x, y) > 0$.

Since $\widetilde{f}$ is weakly Paretian, $\exists$ a decisive λ for any pair of alternatives, namely $\lambda = 1_N$. For all $(u, v) \in X \times X$, let $m(u, v)$ denote the size of the smallest $|\text{Supp}(\lambda)|$ for λ semidecisive for u against v. Let $m = \wedge\{m(u, v) \mid (u, v) \in X \times X\}$. Without loss of generality suppose λ is semidecisive for x against y with $|\text{Supp}(\lambda)| = m$. If $m = 1$, the proof is complete. Suppose $m > 1$. Let $i \in \text{Supp}(\lambda)$. Consider any fuzzy profile $\rho \in \mathcal{FR}^n$ such that

$$\pi_i(x, y) > 0, \ \pi_i(y, z) > 0, \ \pi_i(x, z) > 0,$$
$$\forall j \in \text{Supp}(\lambda)\backslash\{i\}, \ \pi_j(z, x) > 0, \pi_j(x, y) > 0, \ \pi_j(z, y) > 0,$$
$$\forall k \notin \text{Supp}(\lambda), \ \pi_k(y, z) > 0, \ \pi_k(z, x) > 0, \ \pi_k(y, x) > 0.$$

Since λ is semidecisive for x against $y, \pi_i(x, y) > 0 \ \forall i \in \text{Supp}(\lambda)$, and $\pi_j(y, x) > 0 \ \forall j \notin \text{Supp}(\lambda)$, we have that $\pi(x, y) > 0$. Since $|\text{Supp}(\lambda)| = m$, it is not the case that $\pi(z, y) > 0$ otherwise λ' is semidecisive for z against y, where $\lambda'(j) = \lambda(j) \ \forall j \in \text{Supp}(\lambda)\backslash\{i\}$ and $\lambda'(i) = 0$. (Suppose $\pi(z, y) > 0$. Then $\rho_i]_{\{z,y\}} \sim \rho_i']_{\{z,y\}}) \ \forall i \in N, \forall \rho' \in \mathcal{FR}^n(X) \Rightarrow \widetilde{f}(\rho)]_{\{z,y\}} \sim \widetilde{f}(\rho')]_{\{z,y\}} \ \forall \rho' \in \mathcal{FR}^n(X)$ by independence of irrelevant alternatives IIA3 $\Rightarrow \widetilde{f}(\rho')(z, y) > \widetilde{f}(\rho')(y, z)$ since $\widetilde{f}(\rho)(z, y) > \widetilde{f}(\rho)(y, z)$. Thus $\pi'(z, y) > 0$.) However, this contradicts the minimality of m since $|\text{Supp}(\lambda')| = m - 1$. Since $\pi(z, y) = 0$, $\widetilde{f}(\rho)(y, z) \geq \widetilde{f}(\rho)(z, y)$. Since $\widetilde{f}(\rho)(x, y) > \widetilde{f}(\rho)(y, x)$, $\widetilde{f}(\rho)(x, z) > \widetilde{f}(\rho)(z, x)$ by weak transitivity. Hence $\pi(x, z) > 0$. By independence of irrelevant alternatives IIA3, λ^* is semidecisive for x against z, where $\lambda^*(i) = \lambda(i)$ and $\lambda^*(j) = 0$ for $j \in N\backslash\{i\}$. However, this contradicts the fact that $m > 1$. ∎

5.2 Representation Rules, Veto Players, Oligarchies, and Collegiums

Suppose the set of players must select a member to be its representative with respect to certain decisions. For the fuzzy situation, we consider a representative rule to be a function from the set of fuzzy preference profiles into the set of individual fuzzy singletons.

Definition 5.2.1 *Let $i \in N$ and $t \in (0, 1]$. Define the fuzzy subset i_t of N by $\forall j \in N, i_t(j) = t$ if $j = i$ and $i_t(j) = 0$ otherwise. (Then i_t is called a* ***fuzzy singleton.****) If $g : \mathcal{FR}^n \to \{i_t \mid i \in N, t \in (0, 1]\}$, then g is called a* ***representation rule.***

Let g be a representation rule. Then $\text{Supp}(g) = \{i_0\}$ for some $i_0 \in N$. By the notation, $\rho_{\text{Supp}(g(\rho))}$ we mean ρ_{i_0}.

Definition 5.2.2 *Let g be a representation rule. Then*

(1) *g is called **dictatorial** if and only if $\forall \rho, \rho' \in \mathcal{FR}^n$, $Supp(g(\rho)) = Supp(g(\rho'))$;*

(2) *g is called **independent of irrelevant alternatives IIA2** if $\forall \rho, \rho' \in \mathcal{FR}^n$, $\forall x, y \in X$, $Supp(\rho_i\rceil_{\{x,y\}}) = Supp(\rho_i'\rceil_{\{x,y\}}) \forall i \in N$ implies*

$$Supp(\rho_{Supp(g(\rho))}\rceil_{\{x,y\}}) = Supp(\rho_{Supp(g(\rho'))}\rceil_{\{x,y\}}).$$

(3) *g is called **independent of irrelevant alternatives IIA3** if $\forall \rho, \rho' \in \mathcal{FBR}^n$, $\forall x, y \in X$, $\rho_i\rceil_{\{x,y\}} \sim \rho_i'\rceil_{\{x,y\}} \forall i \in N$ implies*

$$\rho_{Supp(g(\rho))}\rceil_{\{x,y\}} \sim \rho_{Supp(g(\rho'))}\rceil_{\{x,y\}}.$$

The following theorem is a companion to Theorem 5.1.16.

Theorem 5.2.3 *Let strict preferences be of type $\pi_{(0)}$. Let g be a representation rule. Then g is independent of irrelevant alternatives IIA2 if and only if g is dictatorial.*

Proof. Let $\widetilde{f}$ be a fuzzy aggregation rule determined by g. Let $\rho \in \mathcal{FR}^n$. Then $\widetilde{f}(\rho) = \rho_{Supp(g(\rho))}$. Since ρ_i is partially transitive $\forall i \in N$, $\rho_{Supp(g(\rho))}$ is partially transitive. Thus $\widetilde{f}(\rho)$ is partially transitive. Hence $\widetilde{f}$ is partially transitive. Let $x, y \in X$. Suppose $\pi_i(x, y) > 0 \ \forall i \in N$. Then $\pi_{Supp(g(\rho))}(x, y) > 0$, but $\pi_{Supp(g(\rho))}(x, y) = \widetilde{f}(\rho)(x, y)$. Thus $\widetilde{f}$ is weakly Paretian. Since $\widetilde{f}(\rho) = \rho_{Supp(g(\rho))}$ and $\widetilde{f}(\rho') = \rho'_{Supp(g(\rho'))}$, $\widetilde{f}$ is independent of irrelevant alternatives if and only if g is. Thus if g is independent of irrelevant alternatives, then $\widetilde{f}$ is dictatorial by fuzzy Arrow's Theorem. Thus g is dictatorial. Conversely, if g is dictatorial, then it follows easily that g is independent of irrelevant alternatives. ∎

Theorem 5.2.4 *Let strict preferences be regular. Let g be a representation rule. Then g is independent of irrelevant alternatives IIA2 if and only if g is dictatorial.*

Proof. Let $\widetilde{f}$ be a fuzzy aggregation rule determined by g, i.e., $\widetilde{f}(\rho)(x, y) = \rho_{Supp(g(\rho))}$. Let $\rho \in \mathcal{FWR}^n$. Then $\widetilde{f}(\rho) = \rho_{Supp(g(\rho))}$. Since ρ_i is weakly transitive $\forall i \in N$, $\rho_{Supp(g(\rho))} = \rho_{i_0}$ is partially transitive. Thus $\widetilde{f}(\rho)$ is weakly transitive. Hence $\widetilde{f}$ is weakly transitive. Let $x, y \in X$. Suppose $\pi_i(x, y) > 0 \ \forall i \in N$. Then $\pi_{Supp(g(\rho))}(x, y) > 0$. Thus $\rho_{i_0}(x, y) > \rho_{i_0}(y, x)$, i.e., $\widetilde{f}(\rho)(x, y) > \widetilde{f}(\rho)(y, x)$. Hence $\pi(x, y) > 0$. Thus $\widetilde{f}$ is weakly Paretian. Since $\widetilde{f}(\rho) = \rho_{Supp(g(\rho))}$ and $\widetilde{f}(\rho') = \rho'_{Supp(g(\rho'))}$, $\widetilde{f}$ is independent of irrelevant alternatives IIA2 if and only if g is. Thus if g is independent of irrelevant alternatives, then $\widetilde{f}$ is dictatorial by fuzzy Arrow's Theorem. Thus g is dictatorial. Conversely, if g is dictatorial, then it follows easily that g is independent of irrelevant alternatives. ∎

The Pareto extension rule is nondictatorial, weakly Paretian, independent of irrelevant alternatives, and quasi-transitive. Such rules evade Arrow's theorem. However, if there are as many individuals as alternatives and each alternative is preferred by some individual, then the Pareto extension rule says that every alternative is as good as any other.

Definition 5.2.5 *Let* $x, y \in X$. *An individual* $i \in N$ *is said to have a* **veto** *for* x *against* y, *written* $xV_i y$, *if for every* $\rho \in \mathcal{FR}^n$, $\pi_i(x, y) > 0$ *implies not* $\pi(y, x) > 0$. *An element* $i \in N$ *is said to have a* **veto** *if for all* $x, y \in X$, i *has a veto for* x *against* y.

Definition 5.2.6 *A fuzzy aggregation rule is called* **oligarchic** *if there exits* $\lambda \in \mathcal{FN}$ *(called an* **oligarchy***) such that*
(1) *every member of* $\text{Supp}(\lambda)$ *has a veto;*
(2) λ *is decisive.*

Definition 5.2.7 *Let* $x, y \in X$. *Let strict preferences be of type* $\pi_{(0)}$. *An individual* $i \in N$ *is said to have a* **semiveto** *for* x *against* y, *written* $x\widetilde{V}_i y$, *if* $\forall \rho \in \mathcal{FR}^n$, $\pi_i(x, y) > 0$ *and* $\pi_j(y, x) > 0 \; \forall j \neq i$ *implies not* $\pi(y, x) > 0$.

Lemma 5.2.8 *Let strict preferences be of type* $\pi_{(0)}$. *Suppose* ρ *is partially quasi-transitive. Then*
(1) $\pi(x, y) > 0$ *and* $\rho(y, z) > 0$ *implies* $\rho(x, z) > 0$;
(2) $\rho(x, y) > 0$ *and* $\pi(y, z) > 0$ *implies* $\rho(x, z) > 0$.

Proof. (1) Suppose not $\rho(x, z) > 0$. Since ρ is complete, $\pi(z, x) > 0$. Since also $\pi(x, y) > 0$, we have by partial quasi-transitivity that $\pi(z, y) > 0$. However, this is impossible since $\rho(y, z) > 0$.
(2) The proof here is similar to that in (1). ∎

Lemma 5.2.9 *Let strict preferences be of type* $\pi_{(0)}$. *Let* $\widetilde{f}$ *be a fuzzy aggregation rule which is partially quasi-transitive, weakly Paretian, and independent of irrelevant alternatives IIA2. If* $i \in N$ *has a semiveto for* x *against* y *for some* $x, y \in X$, *then* i *has a veto over all ordered pairs* $(v, w) \in X \times X$.

Proof. Let $\rho \in \mathcal{FR}^n$ be such that

$$\pi_i(x, y) > 0, \; \pi_i(x, z) > 0, \; \pi_i(y, z) > 0,$$
$$\forall j \neq i, \; \pi_j(y, x) > 0, \; \pi_j(y, z) > 0.$$

Since i has a semiveto for x against y, not $\pi(y, x) > 0$. Since $\widetilde{f}$ is complete $\rho(x, y) > 0$. Since $\widetilde{f}$ is weakly Paretian, $\pi(y, z) > 0$. Since $\widetilde{f}$ is partially quasi-transitive, $\rho(x, z) > 0$ since $\pi = \pi_{(0)}$. Hence not $\pi(z, x) > 0$. By independence of irrelevant alternatives IIA2, i has a veto for x against z. (We have not

$\pi(z, x) > 0$ for this ρ, but we need independence to get it for all ρ. Let ρ' be such that $\pi'_i(x, z) > 0$. Then not $\pi'(z, x) > 0$ by independence.) Thus we have

$$\forall z \notin \{x, y\}, \ x\widetilde{V}_i y \Rightarrow xV_i z. \tag{5.3}$$

Now since i has a veto for x against z, i has a semiveto for x against z. Thus switching y and z in the above argument implies i has a veto for x against y. Let ρ'' be any fuzzy preference profile such that $\pi''_i(y, z) > 0$. Let ρ^+ be a fuzzy preference profile such that

$$\pi_i^+(y, x) \ > \ 0, \ \pi_i^+(y, z) > 0, \ \pi_i^+(x, z) > 0$$
$$\forall j \ \neq \ i, \ \pi_j^+(z, x) > 0, \ \pi_j^+(y, x) > 0.$$

Since i has a semiveto for x against z, not $\pi^+(z, x) > 0$. Thus $\rho^+(x, z) > 0$. Since $\widetilde{f}$ is weakly Paretian, $\pi^+(y, x) > 0$. Since $\widetilde{f}$ is partially quasi-transitive, $\rho^+(y, z) > 0$. Hence not $\pi^+(z, y) > 0$ since $\pi = \pi_{(0)}$. Since only the preferences for i are specified and $\text{Supp}(\rho_i^+|_{\{y,z\}}) = \text{Supp}(\rho_i''|_{\{y,z\}})$, independence of irrelevant alternatives IIA2 implies not $\pi''(z, y) > 0$ and so i has a veto for y against z. Thus

$$\forall z \notin \{x, y\}, \ x\widetilde{V}_i y \Rightarrow yV_i z. \tag{5.4}$$

Now i has a veto for y against z, i has a semiveto for y against z. Thus by (5.3), i has veto for y against x. We have $\forall (v, w) \in X \times X$,

$$x\widetilde{V}_i y \Rightarrow xV_i v \ (\text{by (5.3)}) \Rightarrow x\widetilde{V}_i v \Rightarrow vV_i w \ (\text{by (5.4)}) \text{ with } v \text{ replacing } y.$$

∎

The following theorem is a companion to Theorems 5.1.16 and 5.2.3.

Theorem 5.2.10 *Let strict preferences be of type $\pi_{(0)}$. If a fuzzy aggregation rule is partially quasi-transitive, weakly Paretian, and independent of irrelevant alternatives IIA2, then it is oligarchic.*

Proof. Since $\widetilde{f}$ is weakly Paretian, there exists $\lambda \in \mathcal{FP}(N)$ such that λ is semidecisive, namely $\lambda = 1_N$. Let λ be such that $\text{Supp}(\lambda)$ is smallest for which λ is semidecisive. Let $m = |\text{Supp}(\lambda)|$. By Lemma 5.1.15, λ is decisive for all ordered pairs $(u, v) \in X^2$. If $m = 1$, then the proof is complete. Suppose $m > 1$. Let $x, y, z \in X$ be distinct. Let ρ be a fuzzy preference profile such that for some $i \in \text{Supp}(\lambda)$,

$$\pi_i(x, y) > 0, \ \pi_i(x, z) > 0, \ \pi_i(y, z) > 0,$$

$$\forall j \in \text{Supp}(\lambda)\backslash\{i\}, \ \pi_j(z, x) > 0, \ \pi_j(z, y) > 0, \ \pi_j(x, y) > 0, \tag{5.5}$$

$$\forall k \notin \text{Supp}(\lambda), \ \pi_k(y, z) > 0, \ \pi_k(y, x) > 0, \ \pi_k(z, x) > 0.$$

Since $i \in \text{Supp}(\lambda)$ and λ is decisive, $\pi(x, y) > 0$. Since $|\text{Supp}(\lambda)| = m$, there is such a ρ such that not $\pi(z, y) > 0$ else $\text{Supp}(\lambda)\backslash\{i\}$ would be the support of some fuzzy subset of N semidecisive for z over y and thus decisive, contradicting the minimality of m. (Now for $\rho, \rho' \in \mathcal{FR}^n$ satisfying (5.5), $\text{Supp}(\rho_i|_{\{y,z\}}) = \text{Supp}(\rho'_i|_{\{y,z\}})$, $i = 1, \ldots, n$.) Hence for any such ρ, $\widetilde{f}(\rho)(y, z) > 0$ by completeness. Since $\widetilde{f}$ is partially quasi-transitive, $\widetilde{f}(\rho)(x, z) > 0$ and hence not $\pi(z, x) > 0$. Hence i has a semiveto for x against z. Thus i has a veto over all ordered pairs $(u, v) \in X^2$ by Lemma 5.2.9. Hence $\widetilde{f}$ is oligarchic since i was arbitrary.   ∎

In Theorem 5.2.10, the condition of partial transitivity for a fuzzy aggregation rule of Theorem 5.1.16 is relaxed to that of partial quasi-transitivity. The result of this change is a dispersion of the decision making from a dictator to an oligarchy.

Lemma 5.2.11 *Let strict preferences be regular. Let $\widetilde{f}$ be a fuzzy aggregation rule which is partially transitive, weakly Paretian, and independent of irrelevant alternatives IIA3. If $i \in N$ has a semiveto for x against y for some $x, y \in X$, then i has a veto over all ordered pairs $(v, w) \in X \times X$.*

Proof. Let $\rho \in \mathcal{FR}^n$ be such that

$$\pi_i(x, y) > 0, \ \pi_i(x, z) > 0, \ \pi_i(y, z) > 0,$$
$$\forall j \neq i, \ \pi_j(y, x) > 0, \ \pi_j(y, z) > 0.$$

Since i has a semiveto for x against y, not $\pi(y, x) > 0$. Thus $\rho(x, y) \geq \rho(y, x)$. Since $\widetilde{f}$ is complete $\rho(x, y) > 0$. Since $\widetilde{f}$ is weakly Paretian, $\pi(y, z) > 0$. Since $\widetilde{f}$ is partially quasi-transitive by Lemma 5.2.8, $\rho(x, z) > 0$. In fact, since $\rho(y, z) > \rho(z, y)$ and $\rho(x, y) \geq \rho(y, x)$, we have $\rho(x, z) > \rho(z, x)$ by Proposition 5.1.20. Hence not $\pi(z, x) > 0$. By independence of irrelevant alternatives, i has a veto for x against z. (We have not $\pi(z, x) > 0$ for this ρ, but we need independence to get it for all ρ. Let ρ' be such that $\pi'_i(x, z) > 0$. Then not $\pi'(z, x) > 0$ by independence.) Thus we have

$$\forall z \notin \{x, y\}, \ x\widetilde{V}_i y \Rightarrow xV_i z. \tag{5.6}$$

Now since i has a veto for x against z, i has a semiveto for x against z. Thus switching y and z in the above argument implies i has a veto for x against y. Let ρ'' be any fuzzy preference profile such that $\pi''_i(y, z) > 0$. Let ρ^+ be a fuzzy preference profile such that

$$\pi_i^+(y, x) > 0, \ \pi_i^+(y, z) > 0, \ \pi_i^+(x, z) > 0$$
$$\forall j \neq i, \ \pi_j^+(z, x) > 0, \ \pi_j^+(y, x) > 0.$$

Since i has a semiveto for x against z, not $\pi^+(z, x) > 0$. Thus $\rho^+(x, z) > 0$. Since $\widetilde{f}$ is weakly Paretian, $\pi^+(y, x) > 0$. Since $\widetilde{f}$ is weakly transitive, $\rho^+(y, z) > \rho^+(z, y)$. Thus $\pi^+(y, z) > 0$. Hence not $\pi^+(z, y) > 0$. Since only the preferences for i are specified and $\rho_i^+|_{\{y,z\}} \sim \rho_i''|_{\{y,z\}}$, independence of irrelevant alternatives IIA3 implies not $\pi''(z,y) > 0$ and so i has a veto for y against z. Thus

$$\forall z \notin \{x,y\}, \ x\widetilde{V}_i y \Rightarrow yV_i z. \tag{5.7}$$

Now i has a veto for y against z, i has a semiveto for y against z. Thus by (5.6), i has veto for y against x. We have $\forall(v,w) \in X \times X$,

$$x\widetilde{V}_i y \Rightarrow xV_i v \text{ (by (5.6))} \Rightarrow x\widetilde{V}_i v \Rightarrow vV_i w \text{ (by (5.7)) with } v \text{ replacing } y.$$

∎

Let $\widetilde{f}$ be a fuzzy aggregation rule and let $\mathcal{L}(\widetilde{f})$ denote the set of decisive coalitions associated with $\widetilde{f}$.

Theorem 5.2.12 *Suppose strict preferences are regular. If a fuzzy aggregation rule is partially quasi-transitive, weakly Paretian, and independent of irrelevant alternatives IIA2, then it is oligarchic.*

Proof. Since $\widetilde{f}$ is weakly Paretian, there exists $\lambda \in \mathcal{F}\mathcal{P}(N)$ such that λ is semidecisive, namely $\lambda = 1_N$. Let λ be such that $\text{Supp}(\lambda)$ is smallest for which λ is semidecisive. Let $m = |\text{Supp}(\lambda)|$. By Lemma 5.1.15, λ is decisive for all ordered pairs $(u,v) \in X^2$. If $m = 1$, then the proof is complete. Suppose $m > 1$. Let $x,y,z \in X$ be distinct. Let ρ be a fuzzy preference profile such that for some $i \in \text{Supp}(\lambda)$,

$$\pi_i(x,y) > 0, \pi_i(x,z) > 0, \pi_i(y,z) > 0,$$

$$\forall j \in \text{Supp}(\lambda)\backslash\{i\}, \ \pi_j(z,x) > 0, \ \pi_j(z,y) > 0, \ \pi_j(x,y) > 0, \tag{5.8}$$

$$\forall k \notin \text{Supp}(\lambda), \ \pi_k(y,z) > 0, \ \pi_k(y,x) > 0, \ \pi_k(z,x) > 0.$$

Since $i \in \text{Supp}(\lambda)$ and λ is decisive, $\pi(x,y) > 0$. Since $|\text{Supp}(\lambda)| = m$, there is such a ρ such that not $\pi(z,y) > 0$ else $\text{Supp}(\lambda)\backslash\{i\}$ would be the support of some fuzzy subset of N semidecisive for z over y and thus decisive, contradicting the minimality of m. (Now for $\rho, \rho' \in \mathcal{F}R^n$ satisfying (5.8), $\text{Supp}(\rho_i]_{\{y,z\}}) = \text{Supp}(\rho'_i]_{\{y,z\}})$, $i = 1,\ldots,n$.) Hence for any such ρ, $\widetilde{f}(\rho)(y,z) > \widetilde{f}(\rho)(z,y)$. Thus $\widetilde{f}(\rho)(y,z) > 0$ by completeness. Since $\widetilde{f}$ is partially quasi-transitive, $\widetilde{f}(\rho)(x,z) > 0$ and hence not $\pi(z,x) > 0$. Hence i has a semiveto for x against z. Thus i has a veto over all ordered pairs $(u,v) \in X^2$ by Lemma 5.2.11. Hence $\widetilde{f}$ is oligarchic since i was arbitrary. ∎

Let $\mathcal{L}(\widetilde{f})$ denote the set of all decisive coalitions associated with the aggregation rule $\widetilde{f}$.

Definition 5.2.13 *Let $\mathcal{L} \subseteq \mathcal{F}\mathcal{P}(N)$ be a family of coalitions.*
   (1) $\mathcal{L}$ *is called* **monotonic** *if $\forall\lambda \in \mathcal{L}$, $\forall\lambda' \in \mathcal{F}\mathcal{P}(N)$, $1_N \supseteq \lambda' \supset \lambda$ implies $\lambda' \in \mathcal{L}$;*
   (2) $\mathcal{L}$ *is called* **proper** *if $\forall\lambda \in \mathcal{L}$, $\forall\lambda' \in \mathcal{F}\mathcal{P}(N)\backslash\mathcal{L}$ if $\text{Supp}(\lambda) \cap \text{Supp}(\lambda') = \emptyset$.*

Lemma 5.2.14 *Let $\widetilde{f}$ be a fuzzy aggregation rule. Then $\mathcal{L}(\widetilde{f})$ is monotonic and proper.*

Proof. Suppose $\lambda \in \mathcal{L}$ and $\lambda' \in \mathcal{FP}(N)$ are such that $\lambda \subset \lambda'$. Suppose λ' is not decisive for $\widetilde{f}$. Then there exists $\rho \in \mathcal{FR}^n$ and $x, y \in X$ such that $\pi_i(x, y) > 0$ for all $i \in \text{Supp}(\lambda')$, but not $\pi(x, y) > 0$. Thus $\pi_i(x, y) > 0$ for all $i \in \text{Supp}(\lambda)$, but not $\pi(x, y) > 0$. Hence λ is not decisive for $\widetilde{f}$.

Let $\lambda \in \mathcal{L}(\widetilde{f})$. Let $x, y \in X$ and $\rho \in \mathcal{FR}^n$ be such that $\pi_i(x, y) > 0$ for all $i \in \text{Supp}(\lambda)$ and $\pi_i(y, x) > 0 \ \forall i \in N\backslash\text{Supp}(\lambda)$. Then $\pi(x, y) > 0$ and so $\pi(y, x) > 0$ is impossible. Hence λ' is not decisive for $\widetilde{f}$ for all $\lambda' \in \mathcal{FP}(N)$ such that $\text{Supp}(\lambda) \cap \text{Supp}(\lambda') = \emptyset$. ∎

Definition 5.2.15 *Let $\widetilde{f}$ be a fuzzy aggregation rule. If $\forall \mathcal{L}^*(\widetilde{f}) \subseteq \mathcal{L}(\widetilde{f})$ such that $0 < |\mathcal{L}^*(\widetilde{f})| < \infty$, $\cap_{\lambda \in \mathcal{L}^*(\widetilde{f})} \lambda \neq \theta$, then $\widetilde{f}$ is called **collegial**.*

Theorem 5.2.16 *Suppose $|X| \geq n$, where $|N| = n$. Let $\widetilde{f}$ be a fuzzy aggregation rule. If $\widetilde{f}$ is partially acyclic and weakly Paretian, then $\widetilde{f}$ is collegial.*

Proof. Since $\widetilde{f}$ is weakly Paretian, $1_N \in \mathcal{L}(\widetilde{f})$ so $\mathcal{L}(\widetilde{f}) \neq \emptyset$. Suppose $\widetilde{f}$ is not collegial. Then there exists $\mathcal{L}^*(\widetilde{f}) \subseteq L(\widetilde{f})$ such that $0 < |\mathcal{L}^*(\widetilde{f})| < \infty$ and $(\cap_{\lambda \in \mathcal{L}^*(\widetilde{f})} \lambda)(i) = \theta(i) = 0 \ \forall i \in N$ and so $\wedge_{\lambda \in \mathcal{L}^*(\widetilde{f})} \lambda(i) = 0$. Thus $\forall i \in N$, there exists $\lambda \in \mathcal{L}(\widetilde{f})$ such that $i \notin \text{Supp}(\lambda)$. By Lemma 5.2.14, $\mathcal{L}(\widetilde{f})$ is monotonic and so in particular $\forall i \in N$, $1_{N\backslash\{i\}} \in \mathcal{L}(\widetilde{f})$ since $1_{N\backslash\{i\}} \supseteq \lambda$. Let the fuzzy preference profile ρ be such that

$$\pi_1(x_i, x_{i+1}) > 0 \text{ for } i = 1, \ldots, n-1;$$
$$\pi_2(x_i, x_{i+1}) > 0 \text{ for } i = 2, \ldots, n-1; \ \pi_2(x_n, x_1) > 0;$$
$$\pi_3(x_i, x_{i+1}) > 0 \text{ for } i = 3, \ldots, n-1; \ \pi_3(x_n, x_1) > 0, \ \pi_3(x_1, x_2) > 0;$$

$$\cdots$$

$$\pi_n(x_n, x_1) > 0, \ \pi_n(x_i, x_{i+1}) > 0 \text{ for } i = 1, \ldots, n-2.$$

Thus for $j = 2, \ldots, n$, $P(x_{j-1}, x_j; \rho) = N\backslash\{j\}$ implying since $1_{N\backslash\{j\}} \in \mathcal{L}(\widetilde{f})$ and $\text{Supp}(1_{N\backslash\{j\}}) = N\backslash\{j\}$ that $\pi_j(x_{j-1}, x_j) > 0$. Also $P(x_n, x_1; \rho) = N\backslash\{1\}$ implying since $1_{N\backslash\{1\}} \in \mathcal{L}(\widetilde{f})$ and $\text{Supp}(1_{N\backslash\{1\}}) = N\backslash\{1\}$ that $\pi(x_n, x_1) > 0$. Hence we have a cycle $\pi(x_n, x_1) > 0, \pi(x_1, x_2) > 0, \ldots, \pi(x_{n-1}, x_n) > 0$. Thus $\widetilde{f}$ is not partially acyclic. (Suppose $\pi = \pi_{(0)}$. Then $\pi(x_1, x_2) > 0, \ldots, \pi(x_{n-1}, x_n) > 0 \nRightarrow \rho(x_1, x_n) > 0$ since $\pi(x_n, x_1) > 0$. For $\pi = \pi_{(1)}$, $\pi_{(3)}, \pi_{(4)}$, or $\pi_{(5)}$, $\pi(x_1, x_2) > 0, \ldots, \pi(x_{n-1}, x_n) > 0 \nRightarrow \rho(x_1, x_n) \leq \rho(x_n, x_1)$ since $\pi(x_n, x_1) > 0$.) ∎

It is known that the converse of the previous result is not true. In Austen-Smith and Banks [2], it is stated that there exist weakly Paretian and collegial aggregation rules that are not acyclic, e.g., the post-1965 United Nations Security Council.

Lemma 5.2.17 *Suppose $|X| > n$. Let $\tilde{f}$ be a fuzzy aggregation rule. If $\tilde{f}$ is partially acyclic, weakly Paretian, and independent of irrelevant alternatives IIA2 for strict preferences of type $\pi_{(0)}$ and IIA3 for regular strict preferences, then there exists $i \in N$ and $x, y \in X$ such that i has a veto for x against y.*

Proof. Let $X = \{x_1, \ldots, x_m\}$, where $m > n = |N|$. Suppose it is not the case that there exist $i \in N$ and $x, y \in X$ such that i has a veto for x against y. Then $\forall i \in N$, i has no veto for x_{i+1} against x_i. Since $\tilde{f}$ is independent of irrelevant alternatives, there exists $\rho^{(i)}$ that depends exclusively on $\rho^{(i)}|_{\{x_i, x_{i+1}\}}$ such that $\pi_i^{(i)}(x_{i+1}, x_i) > 0$ and $\pi^{(i)}(x_i, x_{i+1}) > 0$. Now let $\rho^{(0)}$ be such that $\forall i \in N$, $\pi_i^{(0)}(x_{n+1}, x_1) > 0$. Pick any $\rho \in \mathcal{FR}^n$ such that $\rho|_{\{x_1, \ldots, x_{n+1}\}}$ is given as follows: for $\pi = \pi_{(0)}$,

$$\mathrm{Supp}(\rho|_{\{x_i, x_{i+1}\}}) = \mathrm{Supp}(\rho^{(i)}|_{\{x_i, x_{i+1}\}}) \; i = 1, \ldots, n \text{ and}$$
$$\mathrm{Supp}(\rho|_{\{x_1, x_{n+1}\}}) = \mathrm{Supp}(\rho^{(0)}|_{\{x_1, x_{n+1}\}}),$$

and for π regular,

$$\rho|_{\{x_i, x_{i+1}\}} \sim \rho^{(i)}|_{\{x_i, x_{i+1}\}}, \; i = 1, \ldots, n \text{ and } \rho|_{\{x_1, x_{n+1}\}} \sim \rho^{(0)}|_{\{x_1, x_{n+1}\}}.$$

Since $\tilde{f}$ is independent of irrelevant alternatives IIA3 and by the construction of ρ, we have that

$$\pi(x_1, x_2) > 0, \ldots, \pi(x_{n-1}, x_n) > 0, \pi(x_n, x_{n+1}) > 0.$$

Since $\tilde{f}$ is weakly Paretian and $\pi_i^{(0)}(x_{n+1}, x_1) > 0 \; \forall i \in N$, we have that $\pi(x_{n+1}, x_1) > 0$. However, for $\pi = \pi_{(0)}$, this contradicts the fact that $\pi(x_1, x_{n+1}) > 0$ since $\tilde{f}$ is partially acyclic. For regular strict preferences, this contradicts the fact that $\rho(x_{n+1}, x_1) \leq \rho(x_1, x_{n+1})$ since $\tilde{f}$ is partially acyclic. ∎

Definition 5.2.18 *Let strict preferences be of type $\pi_{(0)}$. Let $\tilde{f}$ be a fuzzy aggregation rule. Then $\tilde{f}$ is called **neutral** if $\forall \rho, \rho' \in \mathcal{FR}^n$, $\forall x, y, u, v \in X$, $P(x, y; \rho) = P(u, v; \rho')$ and $P(y, x; \rho) = P(v, u; \rho')$ imply $\tilde{f}(\rho)(x, y) > 0$ if and only if $\tilde{f}(\rho')(u, v) > 0$.*

Neutral fuzzy aggregation rules ignore the names of alternatives, whereas anonymous fuzzy aggregation rules ignore the names of individuals.

Definition 5.2.19 *Suppose strict preferences are regular. Let $\tilde{f}$ be a fuzzy aggregation rule. Then $\tilde{f}$ is called **regularly neutral** if $\forall \rho, \rho' \in \mathcal{FR}^n$, $\forall x, y, u, v \in X$, $P(x, y; \rho) = P(u, v; \rho')$ and $P(y, x; \rho) = P(v, u; \rho')$ imply $\tilde{f}(\rho)(x, y) > \tilde{f}(\rho)(y, x)$ if and only if $\tilde{f}(\rho')(u, v) > \tilde{f}(\rho')(v, u)$.*

Theorem 5.2.20 *Suppose* $|X| > n$. *Let* $\widetilde{f}$ *be a fuzzy aggregation rule.*

(1) *Let strict preferences be of type* $\pi_{(0)}$. *If* $\widetilde{f}$ *is partially acyclic, weakly Paretian, and neutral, then there exists* $i \in N$ *such that* i *has a veto.*

(2) *Suppose strict preferences are regular. If* $\widetilde{f}$ *is weakly transitive, weakly Paretian, and regularly neutral, then there exists* $i \in N$ *such that* i *has a veto.*

Proof. By setting $u = x$ and $v = y$ in Definition 5.2.18, we see that $\widetilde{f}$ neutral implies $\widetilde{f}$ is independent of irrelevant alternatives. Thus it follows by Lemma 5.2.17 that some i has a veto for some pair of alternatives x, y. Hence by Lemma 5.2.9 for (1) and Lemma 5.2.11 for (2), i has a veto for every pair of alternatives u, v. ∎

5.3 Decisive Sets, Filters, and Fuzzy Arrow's Theorem

We use the concepts developed in the previous section to further demonstrate variants of Fuzzy Arrow's Theorem. We begin by considering decisive sets and filters. Let S be a set and $\mathcal{F}(S)$ be collection of fuzzy subsets of S. Recall θ denotes the fuzzy subset of S defined by $\theta(i) = 0$ for all $i \in S$. We consider the following properties:

(P1) $1_S \in \mathcal{F}(S), \theta \notin \mathcal{F}(S)$;

(P2) $\tau_1 \in \mathcal{F}(S)$ and $\tau_1 \subseteq \tau_2$ implies $\tau_2 \in \mathcal{F}(S)$, where τ_2 is a fuzzy subset of S;

(P3) $\tau_1, \tau_2 \in \mathcal{F}(S)$ implies $\tau_1 \cap \tau_2 \in \mathcal{F}(S)$;

(P4) $\tau_1, \tau_2, \dots, \tau_r \in \mathcal{F}(T)$ implies $\cap_{i=1}^{r} \tau_i \neq \theta$;

(P5) $\tau \notin \mathcal{F}(S)$ implies $1_{S \setminus \mathrm{Supp}(\tau)} \in \mathcal{F}(S)$, where τ is a fuzzy subset of S.

Clearly, (P1) and (P3) imply (P4).

Note that if we restrict (P1) - (P5) to characteristic functions, then (P1) - (P5) become the crisp case.

Definition 5.3.1 *Let* S *be a set and* $\mathcal{F}(S)$ *be a collection of fuzzy subsets of* S. *Then*

(1) $\mathcal{F}(S)$ *is called a **filter** if* $\mathcal{F}(S)$ *satisfies* (P1), (P2), *and* (P3);

(2) $\mathcal{F}(S)$ *is called a **prefilter** if* $\mathcal{F}(S)$ *satisfies* (P1), (P2), *and* (P4);

(3) $\mathcal{F}(S)$ *is called an ultrafilter if* $\mathcal{F}(S)$ *satisfies* (P1), (P2), (P3), *and* (P5).

Let $\widetilde{f}$ be a fuzzy aggregation rule. If $|X| \geq n$ and $\widetilde{f}$ is weakly Paretian, then it follows that $\mathcal{L}(\widetilde{f})$ is a prefilter.

Let $\mathcal{L}(\widetilde{f}) = \{\lambda \in \mathcal{FP}(N) \mid \lambda \text{ is decisive for } \widetilde{f}\}$.

Let $\mathcal{FQ} = \{\rho \in \mathcal{FB} \mid \rho \text{ is partially quasi-transitive}\}, \mathcal{FR} = \{\rho \in \mathcal{FB} \mid \rho \text{ is partially transitive}\}$, and $\mathcal{FR}' = \{\rho \in \mathcal{FB} \mid \rho \text{ is weakly transitive}\}$.

Theorem 5.3.2 *Suppose $\widetilde{f}$ is a fuzzy preference aggregation rule. Suppose that $\widetilde{f}$ is weakly Paretian and independent of irrelevant alternatives IIA2. Then the following properties hold:*

(1) *If $\widetilde{f}(\mathcal{FR}^n) \subseteq \mathcal{FQ}$, then $\mathcal{L}(\widetilde{f})$ is a filter.*

(2) *If strict preferences are of type $\pi_{(0)}$ and $\widetilde{f}(\mathcal{FR}^n) \subseteq \mathcal{FR}$, then $\mathcal{L}(\widetilde{f})$ is an ultrafilter. If strict preferences are regular and $\widetilde{f}(\mathcal{FR}^n) \subseteq \mathcal{FR}'$, then $\mathcal{L}(\widetilde{f})$ is an ultrafilter.*

Proof. Since $\widetilde{f}$ is weakly Paretian, $1_N \in \mathcal{L}(\widetilde{f})$. Now $\forall \rho \in \mathcal{FR}^n$ such that $\forall i \in N, \pi_i(x, y) > 0$, we have that not $\pi(y, x) > 0$ since $\widetilde{f}$ is weakly Paretian. Thus $\theta \notin \mathcal{L}(\widetilde{f})$. Hence (P1) holds. Now (P2) holds since $\mathcal{L}(\widetilde{f})$ is decisive.

(1) Suppose $\widetilde{f}(\mathcal{FR}^n) \subseteq \mathcal{FQ}$. Suppose $\lambda, \mu \in \mathcal{L}(\widetilde{f})$. Let $x, y, z \in X$. Let ρ be such that

$$\forall i \in \operatorname{Supp}(\lambda) \cap \operatorname{Supp}(\mu), \ \pi_i(x, z) > 0, \ \pi_i(z, y) > 0, \ \pi_i(x, y) > 0,$$
$$\forall j \in \operatorname{Supp}(\lambda) \backslash \operatorname{Supp}(\mu), \ \pi_j(x, z) > 0, \ \pi_j(y, z) > 0,$$
$$\forall k \in N \backslash \operatorname{Supp}(\lambda), \ \pi_k(z, x) > 0, \ \pi_k(z, y) > 0.$$

Since $\lambda \in \mathcal{L}(\widetilde{f})$ and $\forall i \in \operatorname{Supp}(\lambda), \pi_i(x, z) > 0$, we have $\pi(x, z) > 0$. Since $\mu \in \mathcal{L}(\widetilde{f})$ and $\forall i \in \operatorname{Supp}(\mu), \pi_i(z, y) > 0$, we have $\pi(z, y) > 0$. Since $\widetilde{f}(\mathcal{FR}^n) \subseteq \mathcal{FQ}, \pi(x, y) > 0$ by partial quasi-transitivity. Since $\widetilde{f}$ is independent of irrelevant alternatives IIA2, $\lambda \cap \mu$ is decisive for x against y. Since x and y are arbitrary and $\operatorname{Supp}(\lambda \cap \mu) = \operatorname{Supp}(\lambda) \cap \operatorname{Supp}(\mu), \lambda \cap \mu \in \mathcal{L}(\widetilde{f})$. Hence (P3) holds.

(2) Since $\mathcal{FR} \subseteq \mathcal{FQ}$, (P3) holds by (1). Thus it remains only to show that (P5) holds. Let $\lambda \in \mathcal{FP}(N)$. Let $x, y, z \in X$ and let ρ be such that $\forall i \in \operatorname{Supp}(\lambda), \pi_i(x, z) > 0$. Suppose $1_{N \backslash L}$ is not decisive for y against x, where $L = \operatorname{Supp}(\lambda)$, i.e., not $y \mathcal{D}_{1_{N \backslash L}} x$. Since not $y \mathcal{D}_{1_{N \backslash L}} x$, there exists a fuzzy preference profile ρ^1 such that $\forall j \in N \backslash L, \pi_j^1(y, x) > 0$ and $\widetilde{f}(\rho^1)(x, y) > 0$. Let ρ^2 be a fuzzy preference profile such that $\rho^2|_{\{x,y\}} = \rho^1|_{\{x,y\}}$ and $\rho^2|_{\{x,z\}} = \rho|_{\{x,z\}}$ and

$$\forall i \in L, \ \pi_i^2(x, z) > 0, \ \pi_i^2(y, z) > 0,$$
$$\forall j \in N \backslash L, \ \pi_j^2(y, x) > 0, \ \pi_j^2(y, z) > 0.$$

Since $\widetilde{f}$ is independent of irrelevant alternatives IIA2, $\widetilde{f}(\rho^2)(x, y) > 0$. Since $\widetilde{f}$ is weakly Paretian, $\pi^2(y, z) > 0$. Since $\widetilde{f}(\mathcal{FR}^n) \subseteq \mathcal{FR}$ for $\pi = \pi_{(0)}, \pi^2(x, z) > 0$. (For regular strict preferences, $\widetilde{f}(\mathcal{FR}^n) \subseteq \mathcal{FR}'$ implies $\pi^2(x, z) > 0$ by Proposition 5.1.20.) Since $\widetilde{f}$ is independent of irrelevant alternatives IIA2, λ is decisive for x against z, $x \mathcal{D}_\lambda z$. Let ρ' be such that $\forall i \in \operatorname{Supp}(\lambda), \pi_i'(z, y) > 0$. Consider the fuzzy preference profile ρ^3 such that $\rho^3|_{\{x,y\}} = \rho^1|_{\{x,y\}}$ and $\rho^3|_{\{y,z\}} = \rho'|_{\{y,z\}}$ and also

$$\forall i \in \operatorname{Supp}(\lambda), \ \pi_i^3(z, y) > 0, \ \pi_i^3(z, x) > 0,$$
$$\forall j \in N \backslash \operatorname{Supp}(\lambda), \ \pi_j^3(y, x) > 0, \ \pi_j^3(z, x) > 0.$$

Since $\widetilde{f}$ is independent of irrelevant alternatives IIA2 and $\rho^3|_{\{x,y\}} = \rho^1|_{\{x,y\}}$, $\widetilde{f}(\rho^3)(x,y) > 0$. Since $\widetilde{f}$ is weakly Paretian, $\pi^3(z,x) > 0$. Since $\widetilde{f}(\mathcal{FR}^n) \subseteq \mathcal{FR}$ for $\pi = \pi_{(0)}$, $\pi^3(z,y) > 0$. (For regular strict preferences, $\widetilde{f}(\mathcal{FR}^n) \subseteq \mathcal{FR}'$ implies $\pi^3(z,y) > 0$ by Proposition 5.1.20.) Since $\widetilde{f}$ is independent of irrelevant alternatives IIA2, λ is decisive for z against y, $zD_\lambda y$. Since z is arbitrary up to $z \notin \{x,y\}$, we have by a similar argument as above that the following holds, where $L = \text{Supp}(\lambda)$:

$$\text{not } yD_{1_{N\backslash L}}x \;\Rightarrow\; \forall z \notin \{x,y\},\; xD_\lambda z \text{ and } zD_\lambda y \Rightarrow \text{not } zD_{1_{N\backslash L}}x$$
$$\Rightarrow\; \forall w \notin \{x,z\},\; xD_\lambda w \text{ and } wD_\lambda z \Rightarrow \text{not } zD_{1_{N\backslash L}}w$$
$$\Rightarrow\; \forall v \notin \{w,z\},\; wD_\lambda v \text{ and } vD_\lambda z \Rightarrow \text{not } vD_{1_{N\backslash L}}w$$
$$\Rightarrow\; \forall u \notin \{v,w\},\; wD_\lambda u \text{ and } uD_\lambda v.$$

That is, if $1_{N\backslash L}$ is not decisive for some ordered pair of alternatives $(y,x) \in X \times X$, then λ must be decisive for all ordered pairs of alternatives. Therefore, since $\widetilde{f}$ is weakly Paretian and independent of irrelevant alternatives IIA2, $\widetilde{f}(\mathcal{FR}^n) \subseteq \mathcal{FR}$ implies that (P5) holds. ∎

In the crisp case, the previous theorem shows the trade-off between the concentration of power in a society and the extent to which social preferences are rational. A fuller discussion can be found in Austen-Smith and Banks [2, p. 48].

Theorem 5.3.3 *(Fuzzy Arrow's Theorem) Let $\widetilde{f}$ be a fuzzy aggregation rule that is weakly Paretian. If either strict preferences are of type $\pi_{(0)}$ and $\widetilde{f}$ is IIA2 and partially transitive or strict preferences are regular and $\widetilde{f}$ is IIA3 and weakly transitive, then $\widetilde{f}$ is dictatorial.*

Proof. By (2) of Theorem 5.3.2, $\mathcal{L}(\widetilde{f})$ is an ultrafilter. Since N is finite and $\text{Supp}(\lambda)$ is finite for all $\lambda \in \mathcal{FP}(N)$, $\{\text{Supp}(\lambda) \mid \lambda \in \mathcal{L}(\widetilde{f})\}$ is finite. Thus there exists a finite number $\lambda_1, \dots, \lambda_k \in \mathcal{L}(\widetilde{f})$ such that $\cap_{i=1}^k \text{Supp}(\lambda_i) = \cap_{\lambda \in \mathcal{L}(\widetilde{f})} \text{Supp}(\lambda)$. Now $\cap_{i=1}^k \lambda_i \neq \theta$ by (P3). Thus

$$\cap_{\lambda \in \mathcal{L}(\widetilde{f})} \text{Supp}(\lambda) = \cap_{i=1}^k \text{Supp}(\lambda_i) \neq \emptyset.$$

Let $i \in \cap_{\lambda \in \mathcal{L}(\widetilde{f})} \text{Supp}(\lambda)$. Suppose that $1_{\{i\}} \notin \mathcal{L}(\widetilde{f})$. Then by (P5), $1_{N\backslash\{i\}} \in \mathcal{L}(\widetilde{f})$. Hence $i \in \cap_{\lambda \in \mathcal{L}(\widetilde{f})} \text{Supp}(\lambda) \subseteq N\backslash\{i\}$, a contradiction. Thus $1_{\{i\}} \in \mathcal{L}(\widetilde{f})$. Therefore $\widetilde{f}$ is dictatorial. ∎

We have shown that a fuzzy Arrow's Theorem holds under an assumption of partial transitivity. The only condition under which any method for aggregating preferences is both weakly Paretian and independent of irrelevant alternatives is if it is dictatorial. This is true even when the definitions of weakly Paretian, independence, and dictatorship incorporate fuzzy logic.

These definitions are not terribly restrictive. For an aggregation function to be dictatorial, the degree to which the group prefers a choice need not be at the same intensity level as that of the individual whose choice is always preferred by the group. Hence the group might conceivably prefer a lower level or even higher level of intensity. In analogous fashion, weakly Paretian merely requires that every member of a group prefers a certain alternative over another at various levels of intensity, the group need not prefer the alternative at any fixed intensity level relative to the individual preferences as long as the preference is positive. Similarly, independence from irrelevant alternatives is released from any requirement that the degree of intensity with which one alternative is preferred to another across two profiles be associated with the degree of intensity of the social choice outcomes.

5.4 Fuzzy Preferences and Social Choice

We next turn our attention to some of the work of Dutta. In [11], Dutta demonstrate that a fuzzy aggregation rule is oligarchic rather than dictatorial when the definition of transitivity is relatively strong. However, if positive responsiveness is assumed, then the dictatorship result holds. Moreover, under a weaker transitivity condition, these impossibility results, including oligarchics, do not hold.

Let ρ be an *FWPR* on X. Recall that ρ is said to be T_2-transitive if for all $x, y, z \in X, \rho(x, z) \geq \rho(x, y) + \rho(y, z) - 1$. Recall also that

$$H_1 = \{\rho \in \mathcal{FR}(X) \mid \rho \text{ is reflexive, strongly complete and}$$
$$\text{max-min transitive}\},$$
$$H_2 = \{\rho \in \mathcal{FR}(X) \mid \rho \text{ is reflexive, strongly complete and } T_2\text{-transitive}\}.$$

A fuzzy weak preference relation ρ on X is called **exact (EWPR)** if $\text{Im}(\rho) \subseteq \{0, 1\}$. Let R be an exact fuzzy preference relation on X. Let $(x, y) \in R$. Then $(x, y) \in P$ or $(x, y) \in I$ and so $R = P \cup I$, where P and I are asymmetric and symmetric relations, respectively. Also, $P \cap I = \emptyset$.

Proposition 5.4.1 *Let ρ be a strongly complete FWPR on X. Suppose ρ satisfies the following conditions.*
  (1) $\rho = \pi \cup \iota$, *where* $\cup$ *denotes maximum.*
  (2) ι *is symmetric;*
  (3) π *is asymmetric;*
  (4) $\pi \cap \iota = \theta.$
  Then either ρ is an exact fuzzy preference relation on X or for all $x, y \in X$,
$\rho(x, y) = \rho(y, x) = \iota(x, y) = \iota(y, x).$

  Proof. From (1) we have for all $x, y \in X$ that
$$\rho(x, y) = \pi(x, y) \vee \iota(x, y)$$
$$\rho(y, x) = \pi(y, x) \vee \iota(y, x).$$

Suppose $\pi(x, y) > \iota(x, y)$. Then $\rho(x, y) = \pi(x, y)$. Thus $\iota(x, y) = 0$ by (4). Since ι is symmetric, $\iota(y, x) = 0$. Also, $\pi(x, y) = 0$ from the asymmetry of π. Hence $\rho(y, x) = 0$. But by the strong completeness of ρ, $\rho(x, y) + \rho(y, x) \geq 1$. Hence $\rho(x, y) = 1$, $\rho(y, x) = 0$. Thus ρ is an EWPR.

Suppose $\iota(x, y) > \pi(x, y)$. By (4), $\pi(x, y) = 0$. Also, $\iota(y, x) = \iota(x, y) > 0$. Hence by (4), $\pi(y, x) = 0$. Thus $\rho(x, y) = \rho(y, x) = \iota(x, y) = \iota(y, x)$. Finally, if $\iota(x, y) = \pi(x, y)$, then $\rho(x, y) = \iota(x, y) = \pi(x, y) = 0$ by (1) and (4). From the strong completeness of ρ, we have $\rho(y, x) = \pi(y, x) = 1$. Thus ρ is exact.

∎

Due to the previous proposition we will not require the condition that $\pi \cap \iota = \theta$.

Recall Proposition 3.1.9. The derivation of π and ι there was suggested by Ovchinnikov [24].

There have been other suggestions in the literature concerning the derivation of the π and ι relations from an FWPR. One possibility is the following for all $x, y \in X$,

$$\pi(x, y) = 0 \vee (\rho(x, y) - \rho(y, x)) \text{ and } \iota(x, y) = \rho(x, y) \wedge \rho(y, x). \qquad (5.9)$$

Another possibility is for all $x, y \in X$,

$$\pi(x, y) = 1 - \rho(y, x) \text{ and } \iota(x, y) = \rho(x, y) \wedge \rho(y, x). \qquad (5.10)$$

Both (5.9) and (5.10) fail to satisfy the requirements of $\rho = \pi \cup \iota$ and $\pi \cap \iota = \theta$. In view of Propositions 5.4.1 and 3.1.9, Dutta chose to use the derivation of π and ι outlined by (1) and (2) in Proposition 3.1.9. (See Propositions 3.1.8 and 3.1.11.) Thus we make the same choice in this section.

We next present some of these transitivity properties of fuzzy weak preference relations.

Definition 5.4.2 *Let ρ be an FWPR, with π and ι its asymmetric and symmetric components. Let $*$ be a t-norm. Then ρ satisfies*
(1) **PP-transitivity** *if for all $x, y, z \in X$, $\pi(x, z) \geq \pi(x, y) * \pi(y, z)$;*
(2) **IP-transitivity** *if for all $x, y, z \in X$, $\pi(x, z) \geq \iota(x, y) * \pi(y, z)$;*
(3) **PI-transitivity** *if for all $x, y, z \in X$, $\pi(x, z) \geq \pi(x, y) * \iota(y, z)$;*
(4) **II-transitivity** *if for all $x, y, z \in X$, $\iota(x, z) \geq \iota(x, y) * \iota(y, z)$.*

Example 5.4.3 *Let ρ be a FWPR, with π and ι its asymmetric and symmetric components, respectively.*
(1) *max-$*$ transitivity of ρ does not imply PI or IP transitivity.*
(2) *T_2-transitivity of ρ does not imply PP-transitivity.*
For (1), Ovchinnikow [24] shows that PI as well as IP implies $\pi \cap \iota = \theta$. Since $\pi \cap \iota \neq \theta$ in our construction, the proposition follows.
For (2), Let $\rho(x, x) = \rho(y, y) = \rho(z, z) = 1$; $\rho(x, y) = \rho(y, z) = \rho(z, x) = \frac{2}{3}$; $\rho(x, z) = \rho(y, x) = \rho(z, y) = \frac{1}{3}$. Then, the corresponding π values are:

$\pi(x, x) = \pi(y, y) = \pi(z, z) = 0;\ \pi(x, y) = \pi(y, z) = \pi(z, x) = \frac{2}{3};\ \pi(x, z) = \pi(y, x) = \pi(z, y) = 0.$

Proposition 5.4.4 (*Dutta* [11]) *Let ρ be a FWPR, with π and ι its asymmetric and symmetric components. Then max–min transitivity of ρ implies PP-transitivity.*

Proof. Suppose ρ satisfies max-min transitivity, but not PP-transitivity. Then for some $x, y, z \in X$, we must have:

$$\rho(x, z) \geq \rho(x, y) \wedge \rho(y, z) \tag{5.11}$$

and

$$\pi(x, z) < \pi(x, y) \wedge \pi(y, z). \tag{5.12}$$

Thus

$$0 < \pi(x, y) = \rho(x, y) > \rho(y, x), \tag{5.13}$$

$$0 < \pi(y, z) = \rho(y, z) > \rho(z, y). \tag{5.14}$$

By (5.11) and (5.12), we have

$$0 = \pi(x, z) \Rightarrow \rho(x, z) \leq \rho(z, x). \tag{5.15}$$

We consider two cases.

(1) Suppose

$$\rho(x, y) \wedge \rho(y, z) = \rho(x, y). \tag{5.16}$$

By (5.11),

$$\rho(x, z) \geq \rho(x, y) \tag{5.17}$$

By the max–min transitivity of ρ,

$$\rho(y, x) \geq \rho(y, z) \wedge \rho(z, x). \tag{5.18}$$

However, (5.16) and $\rho(y, x) \geq \rho(y, z) \Rightarrow \rho(y, x) \geq \rho(x, y)$ contradicting (5.13). Similarly, (5.15), (5.19) and $\rho(y, x) \geq \rho(z, x) \Rightarrow \rho(y, x) \geq \rho(x, y)$. Hence $\rho(y, z) \wedge \rho(z, x) > \rho(y, x)$, contradicting (5.18). Thus case (1) is not possible.

(2) Suppose

$$\rho(x, y) \wedge \rho(y, z) = \rho(y, z). \tag{5.19}$$

Then from (5.11),

$$\rho(x, z) \geq \rho(y, z). \tag{5.20}$$

By the max-min transitivity of ρ;

$$\rho(z, y) \geq \rho(z, x) \tag{5.21}$$

or

$$\rho(z, y) \geq \rho(x, y). \tag{5.22}$$

Now (5.21), (5.16) and (5.20) imply $\rho(z,y) \geq \rho(y,z)$. Similarly, (5.22) and (5.19) imply $\rho(z,y) \geq \rho(y,z)$. In either case, (5.14) is violated. ∎

Example 5.4.3 and Proposition 5.4.4 play an important role in the possibility theorems that follow.

Let H_E denote the set of reflexive, strongly complete, transitive exact fuzzy preference relations over X. Nonempty subsets of N are called coalitions at times.

We next present the fuzzy counterpart of Gibbard's oligarchy result. Its proof is analogous to the proof of the corresponding result in [6, Theorem 3.5] and also similar to the proof of Gibbard's theorem in the exact framework. For a sketch of the proof of the latter, see for instance Sen, [28].

Proposition 5.4.5 *Let* $\widetilde{f} : H_1^n \to H_1$ *be an FAR satisfying IIA1 and PC. Then there exists a unique oligarchy C.*

Proposition 5.4.5 assumes that individual and social preferences are max–min transitive. As shown in [6], the proposition is true under any transitivity condition on FWPRs such that the corresponding strict preference relations satisfy for all $x, y, z \in X$,

$$\pi(x,y) > 0 \text{ and } \pi(y,z) > 0 \Rightarrow \pi(x,z) > 0. \tag{5.23}$$

By Proposition 5.4.4(2), it follows that max–min transitivity implies that (5.23) holds.

From Proposition 5.4.1, it follows that the above proposition is true when the range of $\widetilde{f}$ is expanded to include FWPRs which are max-star transitive under any triangular t-norm. However, a corresponding expansion of the domain of $\widetilde{f}$ may not be permissible because the proof requires PP-transitivity of ρ_i, and this is not necessarily valid under every notion of max-star transitivity as can be seen by Example 5.4.3.

If preferences are assumed to be exact, then the conditions imposed here lead to Arrow's [1] result. The fuzzy framework allows for the construction of nondictatorial FARs as we saw in Chapter 4. Of course, $\widetilde{f}$ is oligarchic, N being the unique oligarchy.

One reason the Arrow Impossibility Theorem can be avoided in the fuzzy framework is due to transitivity. Arrow's result uses the fact that transitivity of R implies PI and IP transitivity. However, as Example 5.4.3 shows, PI and IP do not follow as corollaries of transitivity of the FWPR.

The following discussion is from Dutta [11]. In an intuitive sense, strict preference under $\widetilde{f}$ coincides with the Pareto dominance relation. This is clear if it is assumed that individual preferences are exact, while allowing social preferences to be fuzzy. The problem with the Pareto dominance relation is that it is not strongly complete. Suppose that for some $x, y \in X$, and for some

n-tuple of individual preferences, neither x nor y Pareto dominates the other. Then $\widetilde{f}(\rho)(x,y) = \widetilde{f}(\rho)(y,x) = \beta \geq \frac{1}{2}$ and ρ is strongly complete. Thus $\widetilde{f}$ leads to a possibility result (see Theorem 4.1.23). The counterpart of $\widetilde{f}$ in the crisp case would not be strongly complete since $\rho(x,y) = \rho(y,x) = 0$ in this case. However, there is no nondictatorial FAR satisfying positive responsiveness in addition to the conditions like those imposed in Theorem 4.1.23.

Proposition 5.4.6 *Let $n \geq 3$, and $\widetilde{f} : H_1^n \to H_1$ be a FAR satisfying IIA1, PC and PR. Then $\widetilde{f}$ is dictatorial.*

Proof. By Proposition 5.4.5, there is a unique oligarchy C. If C consists of a single individual, then that individual is a dictator. Thus let $i, j \in C$.

Let x, y be distinct alternatives in X. Consider $(\rho_1, \ldots, \rho_n) \in H_i^n$ such that $\rho_i(x,y) > \rho_i(y,x)$ and $\rho_j(y,x) > \rho_j(x,y)$. Since i and j are vetoers, $\pi(x,y) = \pi(y,x) = 0$. ∎

Proposition 5.4.6 is the fuzzy counterpart of a result in Mas-Collel and Sonnenschein [18].

The results proved so far in this section have assumed that individual and social preferences are max-min transitive. T_2-transitivity allows one to avoid impossibility results. The Gibbard oligarchy result is avoided in this case because T_2-transitivity does not ensure that (5.23) is satisfied.

Theorem 4.1.23 and the previous results show that if we allow judgements about social welfare to be fuzzy, Arrow-type impossibility theorems can be avoided. Some scholars feel that the final social choice must be exact even if social preferences are permitted to be fuzzy. Thus we next examine the implications of these results here and in Chapter 7 for exact social choice. We next consider the relationship between choice and the fuzzy social preferences. Dutta et al. [12] discuss several alternative notions of rationalisability of exact choice by fuzzy preferences in the context of revealed preference theory. We present below a notion discussed by Dutta [11].

Definition 5.4.7 *A function $C : \mathcal{P}^*(X) \to \mathcal{P}^*(X)$ is called an **exact choice function (ECF)** if for all $A \in \mathcal{P}^*(X)$, $\emptyset \neq C(A) \subset A$.*

Definition 5.4.8 *Let ρ be any FWPR and $A \in P^*(X)$. The ρ-greatest set in A is defined to be the function $G(A, \rho) : P^*(X) \to [0,1]$ such that for all $x \in X \backslash A$, $G(A, \rho)(x) = 0$ and for all $x \in A$, $G(A, \rho)(x) = \wedge \{\rho(x,y) \mid y \in A\}$. Let $B(A, \rho) = \{x \in A \mid G(A, \rho)(x) \geq G(A, \rho)(y), \text{ for all } y \in A\}$.*

Definition 5.4.9 *An ECF C is H-**rationalizable** in terms of an FWPR ρ if for all $A \in X$, $C(A) = B(A, \rho)$.*

Under H-rationalizability, the player chooses those elements in A which score the highest with respect to the function $G(A, \rho)$.

Definition 5.4.10 *Let* $\tilde{f} : T^n \to T$ *be an FAR and C an ECF. Then* $\tilde{f}$ *H-generates C if and only if for all* $(\rho_1, \ldots, \rho_n) \in T^n$, *C is H-rationalizable in terms of* $\tilde{f}(\rho_1, \ldots, \rho_n)$.

The following condition, first proposed by Bordes [8], is a well-know rationality condition in social choice.

Definition 5.4.11 *An ECF C satisfies* **Property** $(\beta+)$ *if for all* $x, y \in X$, *and all* $A, B \subseteq X$, $(x, y \in A \subseteq B$ *and* $y \in C(A)$ *and* $x \in C(B)) \to (y \in C(B))$.

The following lemma, proved in Dutta et al. [12], shows the importance of Property $(\beta+)$ in the present context.

Lemma 5.4.12 *If C is an ECF which is H-rationalizable in terms of some* $\rho \in H_1$, *then C satisfies Property* $(\beta+)$.

Proposition 5.4.13 *(Dutta [11]) Let* $\tilde{f} : H_1^n \to H_1$ *be a nondictatorial FAR satisfying IIA and PC and C an ECF which is H-generated by* $\tilde{f}$. *Then there exists* $A \subseteq X$ *and* $(\rho_1, \ldots, \rho_n) \in H_1^n$ *such that* (i) $x, y \in A$ *and* $\pi_i(x, y) = 1$ *for all* $i \in N$ *and* (ii) $y \in C(A)$.

Proof. Let $A = \{x, y, z\}$, $\tilde{f}$ and C satisfy the hypotheses of the proposition. By Proposition 5.4.5, there is a unique oligarchy G. Since f is nondictatorial, $|G| \geq 1$.

Without loss of generality, assume that $\{1, 2\} \subseteq G$. Construct $(\rho_1, \rho_2, \ldots, \rho_n)$ as follows:

        (i) for all $i \in N$, $\rho_i(x, y) = 1$ and $\rho_i(y, x) < 1$,
        (ii) $\rho_1(y, z) > \rho_1(z, y)$ and $\rho_1(x, z) > \rho_1(z, x)$,
        (iii) $\rho_2(z, y) > \rho_2(y, z)$ and $\rho_2(z, x) > \rho_2(x, z)$,
        (iv) for all $i \in N \backslash \{1, 2\}$, $\rho_i(y, z) = \rho_i(z, y) = \rho_i(z, x) = \rho_i(x, z)$.

It follows that there exists $(\rho_1, \ldots, \rho_n) \in H^n$ satisfying restrictions $(i) - (iv)$.

Since f satisfies PC, $\pi(x, y) = 1$. Hence $G(\{x, y\}, \rho)(x) = 1$. Thus $G(\{x, y\}, \rho)(y) = 0$. Since $1 \in C$, (ii) implies that $\pi(z, y) = \pi(z, x) = 0$. Similarly, (iii) implies that $\pi(y, z) = \pi(x, z) = 0$. Also, $\pi(z, y) = \pi(y, z) = 0$ implies $\rho(y, z) = \rho(z, y)$, and $\pi(z, x) = \pi(x, z) = 0$ implies that $\rho(z, x) = \rho(x, y)$. Hence $G(\{y, z\}, \rho)(y) = G(\{y, z\})(z)$ and $G(\{x, z\}, \rho)(x) = G(\{x, z\}, \rho)(z)$.

Since C is H-generated by ρ, $G(\{x, y\}, \rho)(x) > G(\{x, y\}, \rho)(y) \Rightarrow C(\{x, y\}) = \{x\}$. Also, $C(\{y, z\}, \rho)(y) = G(\{y, z\}, \rho)(z) \Rightarrow C(\{y, z\}) = \{y, z\}$ and $G(\{x, z\}, \rho)(x) = G(\{x, z\}, \rho)(z) \Rightarrow C(\{x, z\}) = \{x, z\}$.

Suppose $y \notin C(A)$. Since C satisfies Property $(\beta+)$, and $y \in C(\{y, z\})$, we have that $z \notin C(A)$. However, $z \in C(\{x, z\})$. Hence, if $z \notin C(A)$, then $x \notin C(A)$.

Thus if $y \notin C(A)$, then $C(A) = \emptyset$. Hence $y \in C(A)$ and the desired result holds. ∎

In contrast to Proposition 5.4.13, Propositions 5.4.5 and Theorem 4.1.23 show that the class of FARs (with domain H_1^n and range H_1) satisfying IIA and PC must be oligarchic, but not necessarily dictatorial. However, if the nondictatorial FARs in this class are further constrained to H-generated exact choice functions, then in some situations, an alternative which is Pareto-dominated will also be chosen. Thus the necessity to generate exact social choice may under certain circumstances lead to strong impossibility results even in a fuzzy framework as commented by Dutta.

As noted by Dutta, the concept of H-rationability in terms of max-min transitive FWPRs requires the exact choice function to satisfy Property $(\beta+)$. Suppose the domain of the aggregation rule is restricted to H_E^n, i.e., individual preferences are assumed to be exact. If IIA and the Pareto condition are rephrased in choice-functional terms, and if the aggregation rule is defined to be a mapping from H_E^n to the set of choice functions, then it is known that property $(\beta+)$ leads to the existence of a dictator. Thus this result shows that a dictator on the domain H_E^n remains a dictator on the expanded domain H_1^n.

A corresponding result for T_2-transitivity is not necessarily true. The restrictions imposed on exact choice functions by the concept of H-rationalizability in terms of T_2-transitive FWPRs are not known. These restrictions could, however, be weaker than known rationality conditions such as Property $(\beta+)$. Hence an impossibility result, even when social choice is assumed to be exact, is not inevitable in a fuzzy framework.

In contrast to [6], Dutta has permitted individual and social preferences to be fuzzy weak preference relations. Dutta has argued that differences in the structure of fuzzy binary relations from that of exact binary relations become apparent only if the binary relations are allowed to be weak. Also, qualitative differences in the nature of the possibility theorems in the two frameworks do follow form these differences in the structure of binary relations.

Dutta goes on to state that with a relatively strong version of the transitivity condition on FWPRs, the counterparts of the conditions assumed by Arrow [1] lead to an oligarchy in the fuzzy framework. The dictatorship result is restored if Positive Responsiveness is imposed in addition to the other conditions. However, with a weaker form of transitivity, all these conditions can be satisfied without leading to oligarchic rules.

5.5   Fuzzy Preferences and Arrow-Type Problems

Here we consider some interesting results of Banerjee [4]. In the special case, where the individual preference relations are exact but the social preference relation is permitted to be fuzzy, it is possible to distinguish between different degrees of power of the dictator, [4]. This power increases with the strength of the transitivity requirement.

In the previous section, it is shown that if a relatively strong version of transitivity is used, fuzzy aggregation rules which satisfy fuzzy analogs of Arrow's independence and Pareto conditions and whose range consists of fuzzy binary relations would be oligarchic, but not necessarily dictatorial. If weaker versions of transitivity are used, oligarchies can be avoided. However, if positive responsiveness is assumed, then dictatorship is obtained.

To get around this problem, Banerjee looked for an alternative way of defining the union operator.

There are still other definitions of the union operator in the literature. The choice between the alternative definitions should be decided with reference to the context. In the present context it seems to Banerjee that $\cup_2$ would be pertinent in that it fulfills certain intuitive requirements.

However, if we take the $\cup_2$ union operator, then Propositions 3.1.1 and 3.1.5 are unable to define π and ι uniquely on the basis of a given ρ. To get a unique definition, we add two further conditions.

In this section, we assume that for all $x, y \in X, \pi(x, y) = 1 - \rho(y, x)$ and $\iota(x, y) = \rho(x, y) \wedge \rho(y, x)$. Note that π here is $\pi_{(2)}$ defined in Definition 1.1.6.

Most of the theorems on social choice depend strongly on the transitivity properties of the individual and the social preference relations. Applications of fuzzy set theory to choice theory have made use of various definitions of transitivity. Basu [7] introduced the following definition:

Definition 5.5.1 *A fuzzy weak preference relation ρ on X^2 is called T-transitive if for all $x, y, z \in X, \rho(x, z) \geq \frac{1}{2}\rho(x, y) + \frac{1}{2}\rho(y, z)$ for all $z \in X \backslash \{x, y\}$ such that $\rho(x, y) \neq 0$ and $\rho(y, z) \neq 0$.*

Proposition 5.5.2 *If a weak preference relation R is T_2-transitive, then for all $x, y, z \in X, [\pi(x, y) > 0$ and $\pi(y, z) > 0] \Rightarrow \pi(x, z) > 0$, where we recall fuzzy preferences are of type $\pi_{(2)}$.*

Let N be a finite set of individuals with cardinality not less than 2. X is a set of alternatives. Let H, H_1, and H_2 be the set of reflexive and strongly complete fuzzy weak preference relations satisfying T, max-min and T_2-transitivity, respectively.

Definition 5.5.3 *Let $\widetilde{f}$ be a fuzzy aggregation rule. A subset $M \subseteq N$ is almost decisive over $x, y \in X$ if for all $(\rho_1, \rho_2, \ldots, \rho_n) \in T^n, [\pi_j(x, y) > 0$ for all $j \in M$ and $\pi_j(y, x) > 0$ for all $j \notin M] \Rightarrow \pi(x, y) > 0$.*

In the propositions below, we consider H, H_1, or H_2.

Proposition 5.5.4 (*Banerjee* [4]) *Let* $\tilde{f} : H_2^n \to H_2$ *be a fuzzy aggregation rule satisfying IIA1 and PC. Then there exists a dictator.*

Proof. If a group of individuals G is almost decisive between a pair $(x, y) \in X$, then it is decisive. The proof which uses Proposition 5.5.2 is similar to the proof of the Field Expansion Lemma in the exact case (see Sen [28]).

We show that if G is a decisive group, a proper subset of G is a decisive group. Even though the proof of this part is similar to that of the Group Contraction Lemma in the exact case (see Sen 1985), we present it here for later reference. Let G be decisive. Let $G = G_1 \cup G_2$, where G_1 and G_2 are non-empty. Suppose that for all $i \in G_1$, $\pi_i(x, y) > 0$ and $\pi_i(y, z) > 0$ and for all $i \in G_2$, $\pi_i(y, z) > 0$ and $\pi_i(z, x) > 0$. Suppose also for all $i \in N \backslash G$ that $\pi_i(z, x) > 0$ and $\pi_i(x, y) > 0$, where $x, y, z \in X$. If $\pi(x, z) > 0$, G_1 is almost decisive over (x, z). Hence, it is decisive. If $\pi(z, x) > 0$, $\pi(x, z) = 0$. Thus $\rho(z, x) = 1 - \pi(x, z) = 1$.

It follows that $\pi(y, x) > 0$ since if $\pi(y, x) = 0$, $\rho(x, y) = 1$ and T_2-transitivity would then imply that $\rho(z, y) = 1$ and so $\pi(y, z) = 0$. However, this contradicts the fact that G is decisive. Thus in this case G_2 is decisive. (The roles of IIA and PC are similar to their roles in the exact case.)

As in the exact case, the proof of the proposition is completed by repeated application of this "group contraction" result. ∎

Corollary 5.5.5 *Proposition 5.5.4 remains valid if H_2 is replaced by either H or H_1.*

Proof. $H \subseteq H_1 \subseteq H_2$. ∎

It is very important to notice that the proof of Proposition 5.5.4 depends on the fact that for all $x, y \in X$, $\pi(x, y) = 1 - \rho(y, x)$. This proof does not apply if, given ρ, we define π by the relation: for all $x, y \in X$, $\pi(x, y) = \rho(x, y)$ if $\rho(x, y) > \rho(y, x)$ and $\pi(x, y) = 0$, otherwise. In fact, the counterexample constructed in Dutta [11] for the case H_1, provides a counterexample to the proposition since $H_1 \subseteq H_2$.

Consider the special case where the individual preference relations are exact, but the social preference relation is fuzzy. Banerjee notes that it can be considered to be the natural first step in extending the traditional exact framework. For this case, the earlier definitions have to be restated. We assume that individual exact preference relations are transitive. Denote the set of transitive exact preference relations by U.

Definition 5.5.6 *A **fuzzy aggregation rule** is a function* $\tilde{f} : U^n \to \mathcal{FR}(X)$.

Definition 5.5.7 *Let $\tilde{f}$ be a fuzzy aggregation rule. Then $\tilde{f}$ is said to satisfy the **exact Pareto condition** (EPC) if for all $(\rho_1, \rho_2, \ldots, \rho_n) \in U^n$ and all distinct $x, y \in X$, $\pi(x, y) = 1$ if $\pi_i(x, y) = 1$ for all $i \in N$.*

In modifying the definition of a dictator, a complication arises. An individual i can be defined to be a dictator if for all $x, y \in X, \pi(x, y) > 0$ whenever $\pi_i(x, y) = 1$. It is easily seen that this definition of a dictator would reduce to the traditional definition for U since in this case for all $x, y \in X, 1$ is the only allowable positive value of $\pi(x, y)$. However, Banerjee raises an interesting question concerning the *power* of a dictator. How powerful is individual i as dictator if for all $x, y \in X, \pi(x, y)$ is positive but small, and $\iota(x, y)$ close to 1, whenever $\pi_i(x, y) = 1$?

This leads to the following definition which distinguishes between different degrees of dictatorships. Definition 5.5.8(1) is a modification of Definition 4.1.13(1) and that following Definition 4.3.1.

Definition 5.5.8 *Let $\widetilde{f}$ be a fuzzy aggregation rule.*

*(1) An individual $j \in N$ is a **dictator** if for all distinct $x, y \in X$ and for all $(\rho_1, \rho_2, \ldots, \rho_n) \in U^n$, $\pi_j(x, y) = 1 \Rightarrow \pi(x, y) > 0$.*

(2) An individual $j \in N$ is an almost perfect dictator if for all distinct $x, y \in X$ and for all $(\rho_1, \rho_2, \ldots, \rho_n) \in U^n$ and for any positive number $\alpha < 1$, $\pi_j(x, y) = 1 \Rightarrow \pi(x, y) \geq \alpha$.

*(3) An individual $j \in N$ is a **perfect dictator** if for all distinct $x, y \in X$ and for all $(\rho_1, \rho_2, \ldots, \rho_n) \in U^n$, $\pi_j(x, y) = 1 \Rightarrow \pi(x, y) = 1$.*

Clearly, a dictator is not necessarily either an almost perfect dictator or a perfect dictator. An almost perfect dictator is a dictator, but not necessarily a perfect dictator.

The following result follows from Proposition 5.5.4.

Proposition 5.5.9 *If $\widetilde{f} : U^n \to H_2$ is a fuzzy aggregation rule satisfying IIA1 and EPC, there exists a dictator in the sense of Definition 5.5.8.*

However, there does not necessarily exist a perfect dictator as the following result shows.

Proposition 5.5.10 (*Banerjee* [4]) *There exists a fuzzy aggregation rule $\widetilde{f} : U^n \to H_2$ which satisfies IIA1 and EPC and which is not perfectly dictatorial.*

Proof. Define the fuzzy aggregation rule $\widetilde{f}$ as follows:

For all $x \in X$ and for all $(\rho_1, \ldots, \rho_n) \in U^n$, $\rho(x, x) = 1$;

For all distinct $x, y \in X$ and for all $(\rho_1, \ldots, \rho_n) \in U^n$,

(i) $\rho(x, y) = 1$ and $\rho(y, x) = 0$ if $\rho_i(x, y) = 1$ and $\rho_i(y, x) = 0$ for all $i = 1, \ldots, n$;

(ii) $\rho(x, y) = \beta$ and $\rho(y, x) = 1 - \beta$, where $\beta \in (0, 1)$, if $\rho_1(x, y) = 1$ and $\rho_1(y, x) = 0$ and there exists $j \neq 1$ for which $\rho_j(y, x) = 1$;

(iii) $\rho(x, y) = 0$ and $\rho(y, x) = 1$ otherwise.

It follows that ρ is strongly complete for all $(\rho_1, \ldots, \rho_n) \in U^n$. It is easily seen that $\widetilde{f}$ satisfies $IIA1$. Condition EPC follows from Case (i). Case (ii), where $\pi(x, y) = 1 - \rho(y, x) = \beta < 1$ shows that individual 1 is not a perfect dictator.

It remains to be shown that for all $(\rho_1, \ldots, \rho_n) \in U^n$, $\rho \in H_2$. However, for later reference, we prove the stronger preposition that $\rho \in H_1$, i.e., for all $x, y, z \in X$, $\rho(x, z) \geq \rho(x, y) \wedge \rho(y, z)$. Note that $\rho(x, y)$ can take any of the three values $1, \beta$ or 0. Similarly for $\rho(y, z)$. The inequality can be established for all the combinations of values of $\rho(x, y)$ and $\rho(y, z)$. For instance, let $\rho(x, y) = 1$ and $\rho(y, z) = \beta$. Here $\rho(x, y) \wedge \rho(y, z) = \beta$. Also $\rho_i(x, y) = 1$ for all $i \in N$ and $\rho_1(y, z) = 1$ so that $\rho_1(x, z) = 1$. Thus $\rho(x, z)$ equals either 1 or β. Hence the desired inequality follows. The other cases are similar. ∎

Definition 5.5.11 *Let $\widetilde{f}$ be a fuzzy aggregation rule.*

(1) *A group of individuals $M \subseteq N$ is called **almost decisive of degree** α over (x, y), where $x, y \in X$ if for all $(\rho_1, \ldots, \rho_n) \in U^n$, $[\pi_i(x, y) = 1$ for all $i \in M$ and $\pi_j(y, x) = 1$ for all $j \notin M] \Rightarrow \pi(x, y) \geq \alpha$.*

(2) *A group of individuals $M \subseteq N$ is called **decisive of degree** α if for all distinct $x, y \in X$ and for all $(\rho_1, \ldots, \rho_n) \in U^n$, $[\pi_i(x, y) = 1$ for all $i \in M] \Rightarrow \pi(x, y) \geq \alpha$.*

The proof of the following result is similar to the crisp case.

Proposition 5.5.12 *(Banerjee [4]) Let $\widetilde{f} : U^n \to H_1$ be a fuzzy aggregation rule satisfying IIA1 and EPC. Then there exists a unique individual who is an almost perfect dictator.*

The example constructed in the proof of Proposition 5.5.10 can be used to establish the following result.

Proposition 5.5.13 *There exists a fuzzy aggregation rule $\widetilde{f} : U^n \to H_1$ which satisfies IIA1 and EPC and which is not perfectly dictatorial.*

The existence of a perfect dictator follows for H, however.

Definition 5.5.14 *Let $\widetilde{f}$ be a fuzzy aggregation rule.*

(1) *A group $M \subseteq N$ is **almost decisive** over (x, y), if for all $x, y \in X$ for all $(\rho_1, \ldots, \rho_n) \in U^n$, $(\pi_i(x, y) = 1$ for all $i \in M$ and $\pi_i(y, x) = 1$ for all $i \in N \backslash M) \Rightarrow \pi(x, y) = 1$.*

(2) *M is **perfectly decisive** if for all $x, y \in X$ and for all $(\rho_1, \ldots, \rho_n) \in U^n$, $(\pi_i(x, y) = 1$ for all $i \in M) \Rightarrow \pi(x, y) = 1$.*

Proposition 5.5.15 *(Banerjee [4]) Let $\widetilde{f} : U^n \to H$ be a fuzzy aggregation rule satisfying IIA1 and EPC. Then there exists a perfect dictator.*

Proof. The proof is once again similar to the crisp case. We consider the proof of the Group contraction part. Let M be a decisive group. Let $M = M_1 \cup M_2$, where M_1 and M_2 are not empty. Suppose that there exists $x, y, z \in X$ such that

$$\begin{aligned}
\pi_i(x, y) &= 1 \text{ and } \pi_i(y, z) = 1 \text{ for all } i \in M_1; \\
\pi_i(y, z) &= 1 \text{ and } \pi_i(z, x) = 1 \text{ for all } i \in M_2; \\
\pi_i(z, x) &= 1 \text{ and } \pi_i(x, y) = 1 \text{ for all } i \in N \backslash M.
\end{aligned}$$

Suppose $\pi(x, z) = 1$. Then M_1 is almost perfectly decisive over (x, z). Hence it is a decisive group. Suppose $\pi(x, z) < 1$. Then $\rho(z, x) > 0$. Hence we have $\rho(z, y) \geq \frac{1}{2}\rho(z, x) + \frac{1}{2}\rho(x, y) > 0$ contradicting the fact that M is decisive so that $\pi(y, z) = 1$ and $\rho(z, y) = 1 - \pi(y, z) = 0$. Thus $\rho(x, y) = 0$. (Note that by the definition of T-transitivity, the weak inequality does not apply in this case.) Hence $\pi(y, x) = 1$. Thus M_2 is almost decisive over (y, x). Hence it is decisive. ∎

It has been shown here that when individual preference relations are exact, the power of the dictator increases if the transitivity property of the social preference relation is strengthened. That is, T_2-transitivity, max-min transitivity, and T-transitivity restrictions on the social preference relation imply respectively, the existence of a dictator, an almost perfect dictator and a perfect dictator.

Banerjee notes the following. Similar results can be obtained when the domain of $\tilde{f}$ is H_1^n and H^n instead of U^n. If the definitions of a dictator, an almost perfect dictator and a perfect dictator in Definition 5.5.8 are kept unchanged, these similar results are immediately established. However, in these cases the definition that is accepted in the literature declares individual j as a dictator if $\pi_j(x, y) > 0$ implies $\pi(x, y) > 0$ for all $x, y \in X$ and for all profiles of individual preference relations. In this case, if we wish to define, say, a perfect dictator, a natural definition will assign that status to individual j if $\pi_j(x, y) = \alpha$ implies $\pi(x, y) = \alpha$ for all values of α lying between 0 and 1, for all $x, y \in X$, and for all profiles of individual preference relations. However, this strong a form of the result is not true, as counterexamples are easily constructed.

5.6   The Structure of Fuzzy Preferences: Social Choice Implications

In this section we consider some results of Richardson [26] involving strong connectedness, Definition 5.6.1. Fuzzy preference relations are assumed to be reflexive.

Banerjee [4] objects to this version of strict preference, $\pi_{(1)}$, on the intuitive grounds that if $\iota(x, y) > 0$, $\pi(x, y)$ should be less than $\rho(x, y)$. Banerjee states that it seems unreasonable that the two situations $(\rho(x, y) = 1, \rho(y, x) = .99)$ and $(\rho(x, y) = 1, \rho(y, x) = 0)$ should yield exactly the same value for $\pi(x, y)$. Richardson adds that the discontinuity of $\pi_{(1)}$ is also an intuitive difficulty; while in the previous pair of cases that values of π are too similar, it seems rather unreasonable for the values of π to be so different in the two situations $(\rho(x, y) = 1, \ \rho(y, x) = .99)$ and $(\rho(x, y) = 1, \rho(y, x) = 1)$.

Richardson also states the following. In the absence of strong connectedness, the relation $\pi_{(2)}(x, y) = 1 - \rho(y, x)$ is not asymmetric. Its insensitivity to $\rho(x, y)$ also becomes an intuitive problem if ρ is not strongly connected; $\pi_{(2)}(x, y)$ takes exactly the same value in the two situations $(\rho(x, y) = 1.0, \rho(y, x) = .99)$ and $(\rho(x, y) = .01, \rho(y, x) = .99)$. This is not to say that strong connectedness is an undesirable restriction in all situations. It has proven to be useful in [9] and [22].

Recall Proposition 3.1.8. Let ρ, ι, and π satisfy the following conditions:

(i) $\rho = \iota \cup_2 \pi$;

(ii) $\forall x, y \in X, \ \iota(x, y) + \pi(x, y) \leq 1$;

(iii) ι is symmetric;

(iv) π is asymmetric.

Then $\forall x, y \in X$,

$$\iota(x, y) = \rho(x, y) \wedge \rho(y, x) \tag{5.24}$$

and

$$\pi(x, y) = \pi_{(3)}(x, y) = 0 \vee (\rho(x, y) - \rho(y, x)). \tag{5.25}$$

Clearly, $\pi_{(3)}$ and $\pi_{(2)}$ coincide when ρ is strongly connected. This version of strict preference, $\pi_{(3)}(x, y) = 0 \vee (\rho(x, y) - \rho(y, x))$, satisfies Banerjee's objection to $\pi_{(1)}$ and is sensitive to the values of both $\rho(x, y)$ and $\rho(y, x)$. A further advantage of $\pi_{(3)}$ over $\pi_{(1)}$ is its continuity ($\pi_{(1)}(x, y) = \rho(x, y)$ if $\rho(x, y) > \rho(y, x)$ and 0 otherwise), as a function of $\rho(x, y)$ and $\rho(y, x)$.

The use of $\pi_{(3)}(x, y)$ as the specification of strict preference implies

$$\forall x, y \in X, \pi(x, y) = 1 \Rightarrow \rho(y, x) = 0, \tag{5.26}$$

which is similar in appearance to the contrapositive of Banerjee's condition,

$$\forall x, y \in X, \pi(x, y) = 0 \Rightarrow \rho(y, x) = 1. \tag{5.27}$$

For a given FWPR ρ, $\pi_{(1)}$ and $\pi_{(3)}$ are clearly regular and, when ρ is strongly connected, so is $\pi_{(2)}$.

The general requirement for extending any concept from exact sets to fuzzy sets is that the fuzzy extension must still agree with the original concept when fuzzy sets are restricted to be exact. Therefore any fuzzy extension of the concept of transitivity must be at least as strong as the following.

Definition 5.6.1 ρ is **minimally (TM) transitive** if $\forall x, y, z \in X, \rho(x, y) = 1$ and $\rho(y, z) = 1 \Rightarrow \rho(x, z) = 1$.

Definition 5.6.2 π is **negatively transitive** if $\forall x, y, z \in X, \pi(x, y) > 0$ or $\pi(z, y) > 0$.

Note that π is negatively transitive if and only if

$$\forall x, y, z \in X, \pi(x, z) = 0 \text{ and } \pi(z, y) = 0 \Rightarrow \pi(x, y) = 0. \tag{5.28}$$

Proposition 5.6.3 *If ρ is TM and strongly connected, then any regular π is partially quasitransitive.*

Proof. If ρ is strongly connected, regularity becomes

$$\forall x, y \in X, \rho(y, x) = 1 \iff \pi(x, y) = 0. \tag{5.29}$$

Now suppose $\pi(x, y) > 0$ and $\pi(y, z) > 0$, but that $\pi(x, z) = 0$. Then by (5.29), $\rho(z, x) = 1$. Also, $\pi(x, y) > 0 \Rightarrow \pi(y, x) = 0$. Thus $\rho(x, y) = 1$. However, since ρ is also TM, $\rho(z, x) = 1$ and $\rho(x, y) = 1 \Rightarrow \rho(z, y) = 1$. Hence by (5.29), $\pi(y, z) = 0$, a contradiction. ∎

By requiring strong connectedness and replacing T_2-transitivity with the even weaker condition TM, Proposition 5.6.3 clarifies and extends Proposition 5.5.2 by which T_2-transitivity implies the quasitransitivity of $\pi_{(2)}$.

Proposition 5.6.4 *If ρ is TM and strongly connected, then any regular π is negatively transitive.*

Proof. By (5.28) and (5.29), we have that $\forall x, y, z \in X, \pi(x, z) = 0$ and $\pi(z, y) = 0 \Rightarrow \rho(z, x) = 1$ and $\rho(y, z) = 1 \Rightarrow \rho(y, x) = 1 \Rightarrow \pi(x, y) = 0$. ∎

We next consider social choice implications. Given two sets of fuzzy preference relations W and V, a **fuzzy aggregation rule** (FAR) is a function $f : W^n \to V$. The Pareto Condition (PC) and decisiveness involve the use of strict preference. Hence they are particularly relevant in this section, and need to be restated.

Definition 5.6.5 *A fuzzy aggregation rule is said to satisfy the **Pareto condition (PC)** if $\forall \rho \in W^n$ and $\forall x, y \in X, \pi(x, y) \geq \wedge\{\pi_i(x, y) \mid i \in N\}$.*

Definition 5.6.6 *For a given FAR, a coalition of voters $C \subseteq N$ is **decisive** if $\forall \rho \in U^n$ and $\forall x, y \in X, \pi_C(x, y) > 0 \Rightarrow \pi(x, y) > 0$, where $\pi_C(x, y) > 0$ means $\pi_i(x, y) > 0$ for all $i \in C$.*

Definition 5.6.7 *Let $x, y \in X$. For a given FAR, a coalition $C \subseteq N$ is **almost decisive** over (x, y) if $\exists \rho \in W^n$ such that $\pi_C(x, y) > 0, \pi_{N\setminus C}(y, x) > 0$, and $\pi(x, y) > 0$.*

Let

H_1 = the set of strongly complete FWPRs satisfying max-min transitivity.

H_2 = the set of strongly complete FWPRs satisfying T_2-transitivity.

S_2 = the set of strongly connected FWPRs satisfying T_2-transitivity.

S_M = the set of strongly connected FWPRs satisfying TM-transitivity.

E = the set of exact, strongly connected, and transitive weak preference relations.

In Proposition 5.4.4, it was shown with strict preferences of type $\pi_{(1)}$ that only FAR $\widetilde{f} : H_1^n \to H_1$ satisfying IIA1, PC and PR is a dictatorship. There exists a nondictatorial FAR $\widetilde{f} : H_2^n \to H_2$ that satisfies IIA1, PC and PR (see the proof of Theorem 4.1.15). Weakening the transitivity requirement seems to create some possibility for bypassing the dictatorship. However, in Proposition 5.4.4 it is proved that if $\pi_{(2)}$ were substituted for $\pi_{(1)}$, the only FAR that satisfies IIA1 and PC is a dictatorship, even when the version of transitivity used is only T_2. As we have seen, however, the use of $\pi_{(2)}$ implies strong connectedness (and Banerjee's proof of this proposition uses facts about $\pi_{(2)}$ that are only true under strong connectedness). Hence the range and domain of the FAR in Banerjee's theorem are S_2 and S_2^n, respectively.

The following proposition demonstrates that it does not seem to be the choice of specification of strict preference that generates the dictatorship result, but rather it is the restriction to strong connectedness. Furthermore, Richardson noticed that when strong connectedness is assumed, the dictatorship result stands even if the transitivity requirement is further weakened beyond T_2-transitivity to the weakest possible version of fuzzy transitivity, TM.

Theorem 5.6.8 *(Richardson [26]) Let strict preference be defined by any regular π. If $\widetilde{f} : S_M^n \to S_M$ satisfies IIA1 and PC, there exists a dictator.*

Proof. As in [26], we sketch the proof since the proof follows known arguments. It is customary to prove an Arrowian impossibility theorem in two stages ([15], [27]). First, a "field expansion lemma" shows that a group of voters that are almost decisive over one pair of alternatives will be decisive over any pair of alternatives. It is known that this lemma is implied by quasitransitivity. In fuzzy preference terms, Proposition 5.6.3 demonstrates that TM-transitivity is all that is required to obtain partial quasitransitivity of any regular strict fuzzy preference relation, including $\pi_{(1)}, \pi_{(2)}$ and $\pi_{(3)}$, if strong connectedness is assumed.

Secondly, a group contraction lemma is used to show that if a group of voters containing more than one member is decisive, it must have a proper subset that is also decisive. The dictatorship result then follows from the finiteness of N and the fact that N is decisive. ∎

Proof of the *"Group Contraction Lemma."* Let D be a decisive group of at least two members, and let D_1 and D_2 be non-empty and disjoint subsets of D such that $D_1 \cup D_2 = D$. Define the preference profile $\rho = (\rho_1, \ldots, \rho_n)$ as follows:

$$i \in D_2 \Rightarrow \pi_i(x, y) > 0, \; \pi_i(z, y) > 0, \; \pi_i(z, x) > 0,$$
$$i \in D_1 \Rightarrow \pi_i(x, y) > 0, \; \pi_i(y, z) > 0, \; \pi_i(x, z) > 0,$$
$$i \notin D \;\; \Rightarrow \pi_i(y, x) > 0, \; \pi_i(y, z) > 0, \; \pi_i(z, x) > 0.$$

Since D is decisive, $\pi(x, y) > 0$. By Proposition 5.6.4, π is negatively transitive. Thus we must have either $\pi(x, z) > 0$ in which case D_1 is decisive, or $\pi(z, y) > 0$ in which case D_2 is decisive.

Corollary 5.6.9 *Let strict preference be defined by any regular π. If $\widetilde{f} : E^n \to S_M$ satisfies IIA1 and PC, there exists a dictator.*

Proof. The proof follows the pattern of that in Theorem 5.6.8. ∎

Since no additional condition such as PR is required, and since the set of exact transitive, and connected weak preference relations is a proper subset of S_M, Corollary 5.6.9 is a generalization of Arrow's Theorem.

Without strong connectedness, the possibilities for nondictatorial FARs when the transitivity requirement is T_2 (or weaker) reappear, even when $\pi_{(1)}$ is replaced by $\pi_{(3)}$.

5.7 Exercises

1. Prove Lemma 5.4.12.

2. Prove that T-transitivity $\Rightarrow$ max-min transitivity $\Rightarrow$ T_2-transitivity.

3. Prove Proposition 5.5.2.

4. Prove Proposition 5.5.9.

5. Prove Proposition 5.5.13.

5.8 References

1. K. J. Arrow, *Social Choice and Individual Values*, New York, Wiley 1951.

2. D. Austen-Smith and J. S. Banks, *Positive Political Theory I: Collective Preference*, Michigan Studies in Political Science, The University Michigan Press 2000.

3. D. Austen-Smith and J. S. Banks, *Positive Political Theory II: Strategy and Structure*, Michigan Studies in Political Science, The University Michigan Press 2005.

4. A. Banerjee, Fuzzy preferences and Arrow-type problems in social choice, *Soc Choice Welfare,* 11 (1994) 121–130.

5. S. Barbera and H. Sonnenschein, Preference aggregation with randomized social orderings, *Journal of Economic Theory,* 18 (1978) 244–254.

6. C. R. Barrett, P. K. Pattanaik and M. Salles, On the structure of fuzzy social welfare functions, *Fuzzy Sets and Systems,* 19 (1986) 1–10.

7. K. Basu, Fuzzy revealed preference theory, *Journal of Economic Theory,* 32 (1984) 212–227.

8. G. Bordes, Consistency, rationality and collective choice, *Review of Economic Studies,* 43 (1976) 447–457

9. M. Dasgupta and R. Deb, Fuzzy choice functions, *Soc Choice Welfare,* 8 (1991)171–182.

10. M. Dasgupta and R. Deb, Transitivity and fuzzy preferences, *Soc Choice Welfare,* 13 (1996) 305–318.

11. B. Dutta, Fuzzy preferences and social choice, *Math. Soc. Sci.,* 13 (1987) 215–229.

12. B. Dutta, S.C. Panda and P. K. Pattanaik, Exact choice and fuzzy preferences, to appear in *Mathematical Social Science* (1986).

13. A. Gibbard, Intransitive Social Indifference and the Arrow Dilemma, mimeographed (1969).

14. R. A. Heiner and P. K. Pattanaik, The structure of general probabilistic group decision rules, in: P. K Pattanaik and M. Salles, eds., *Social Choice and Welfare*, North-Holland, Amsterdam, 1983.

15. J. S. Kelly, *Arrow Impossibility Theorems*, Academic Press, New York, 1978.

16. G. Klir and Bo Yuan, *Fuzzy Sets and Fuzzy Logic: Theory and Applications,* Prentice Hall, Englewood Cliffs, 1985.

17. A. Kaufman, *Introduction to the Theory of Fuzzy Sets*, Academic Press, New York, 1975.

18. A. Mas-Collel and H. Sonnenschein, General possibility theorems for group decisions, *Review of Economic Theory,* 29 (1972) 185–192.

19. A. McLennan, Randomized preference aggregation: Additivity of power and strategy proofness, *Journal of Economic Theory,* 22 (1972) 185–192.

20. J. N. Mordeson and T. D. Clark, Fuzzy Arrow's theorem, *New Mathematics and Natural Computation,* 5 (2009) 371–383.

21. J. N. Mordeson, M. B. Gibilisco, and T. D. Clark, Independence of irrelevant alternatives and fuzzy Arrow's theorem, *New Mathematics and Natural Computation,* 8 (2012) 219–237.

22. Nguyen Hung T and E. A. Walker, *A First Course in Fuzzy Logic,* CRC Press, Boca Raton, FL, 1997.

23. S. A. Orlovsky, Decision making with a fuzzy binary relation, *Fuzzy Sets and Systems,* 1 (1978) 155–167.

24. S. V. Ovchinnikov, Structure of fuzzy binary relations, *Fuzzy Sets and Systems,* 6 (1981) 169–195.

25. S. V. Ovchinnikov, Representations of transitive fuzzy relations, in: H.J. Skala, S. Termini and E. Trillas, eds., *Aspects of Vagueness,* D. Reidel, Dordrecht, 1984.

26. G. Richardson, The structure of fuzzy preferences: Social choice implications, *Soc. Choice Welf.,* 15 (1990) 359–369.

27. A. K. Sen, *Collective Choice and Social Welfare,* Holden-Day, San Francisco, 1970.

28. A. K. Sen, Social Choice Theory, to appear in: K.J. Arrow and M.J. Intriligator, eds., *Handbook of Mathematical Economics, Vol.* 3, North-Holland, Amsterdam, 1985.

29. H. J. Skala, Arrow's impossibility theorem: Some new aspects, *Decision Theory and Social Ethics: Issues in Social Choice,* Eds.: H. W. Gottinger and W. Leinfellner, Vol. 17, 1978, 215–225.

Chapter 6

Single Peaked Fuzzy Preferences: Black's Median Voter Theorem

The median voter theorem states that a majority rule voting system will select the outcome most preferred by the median voter. The theorem assumes that voters can place election alternatives along a one-dimensional spectrum. This is often not the case since each party will have its own policy on each of many issues. Similarly, in the case of a referendum, the alternatives may cover more than one issue. The theorem also assumes that voters' preferences are single-peaked, that is voters choose the alternative closest to their own view. The theorem also assumes that voters always vote for their true preferences which actually is not always the case. The results in this chapter were influenced by [23, 24].

A troubling issue in spatial models is the degree to which they are susceptible to majority cycling. Under majority rules, spatial models can not assure the existence of a maximal set, that is, a set of alternatives each of which is not strictly dominated by some other alternative. Every alternative in the maximal set is no less preferred by a majority than any other alternative in X. In the absence of a maximal set, spatial models tend to be badly behaved. Any alternative in X is as likely as any other and no prediction can be made [2].

Single dimensional models with preference profiles that are single peaked comprise an important exception to this general rule. If there exists some alignment of the alternatives in X along a single continuum such that the Euclidean preferences of each political actor making a majority decision over those alternatives descend monotonically from their respective ideal points, then a maximal set is guaranteed to lie at the median ideal point. Known as Black's Median Voter Theorem, [4], the theorem has been widely used in models predicting the outcomes of numerous political "games," to include the outcome in presidential vetoes and the likelihood that a legislative committee will report a bill to the floor of the parent assembly. (See for example [19, 10].)

We show that the Median Voter Theorem holds for fuzzy preferences. Our approach considers the degree to which players prefer options in binary relations. We demonstrate that Black's Median Voter Theorem holds for fuzzy single-dimensional spatial models even though there are significant differences between conventional models and their fuzzy counterparts. In contrast to crisp spatial models, where preferences in space are most often represented by a single ideal point, in the fuzzy case, a political actor's ideal policy position encompasses a range of alternatives. Moreover, contrary to the conventional approach, preferences are not assumed to monotonically decrease from the actor's ideal point.

6.1 Fuzzy Aggregation Preference Rules

In this section, we give some basic properties of fuzzy aggregation rules. We first recall some notation. A **weak order** on X is a fuzzy binary relation ρ on X that is reflexive, complete, and (max-min) transitive. Assume that each individual i has a weak order ρ_i on X. Let $\mathcal{FTR}$ denote the set of all weak orders on X. A **fuzzy preference profile** on X is a n-tuple of weak orders $\rho = (\rho_1, \ldots, \rho_n)$ describing the preferences of all individuals. Let $\mathcal{FTR}^n$ denote the set of all fuzzy preference profiles. For all $\rho \in \mathcal{FTR}^n$ and $\forall x, y \in X$, let $P(x, y; \rho) = \{i \in N \mid \pi_i(x, y) > 0\}$ and $R(x, y; \rho) = \{i \in N \mid \rho_i(x, y) > 0\}$. Let $\mathcal{FB}$ denote the set of all reflexive and complete fuzzy binary relations on X. In this chapter, a **fuzzy preference aggregation rule** is a function $\widetilde{f} : \mathcal{FTR}^n \to \mathcal{FB}$.

We assume throughout the chapter that fuzzy strict preferences are of type $\pi_{(0)}$.

Definition 6.1.1 Let λ be a fuzzy subset of N. Then λ is called a **coalition** if $|Supp(\lambda)| \geq 2$.

Definition 6.1.2 Let $\widetilde{f}$ be a fuzzy preference aggregation rule. Let $\mathcal{L}(\widetilde{f}) = \{\lambda \in \mathcal{FP}(N) \mid \lambda \text{ is decisive for } \widetilde{f}\}$.

If $\lambda \in \mathcal{L}(\widetilde{f})$, then λ provides the degree to which a set of players in N constitute a winning coalition.

Proposition 6.1.3 Let $\widetilde{f}$ be a fuzzy preference aggregation rule. Suppose $\widetilde{f}$ is dictatorial with dictator i. Let λ be a fuzzy subset of N. Suppose λ is a coalition. Then λ is decisive if and only if $i \in Supp(\lambda)$.

Proof. Suppose λ is decisive. Suppose $i \notin \mathrm{Supp}(\lambda)$. Let $x, y \in X$ be such that $x \neq y$. Then $\exists \rho \in \mathcal{F}\mathcal{R}^n$ such that $\pi_j(x, y) > 0 \ \forall j \in \mathrm{Supp}(\lambda)$ and $\pi_i(y, x) > 0$. Thus $\pi(x, y) > 0$ and $\pi(y, x) > 0$, a contradiction. Hence $i \in \mathrm{Supp}(\lambda)$. The converse is immediate. ∎

Proposition 6.1.3 says that if making collective decisions over a set of alternatives results is possible with an individual i being a dictator, then coalition λ is decisive if and only if that individual is at least partially a member of the coalition.

Definition 6.1.4 *Let $\widetilde{f}$ be a fuzzy preference aggregation rule on $\mathcal{F}\mathcal{T}\mathcal{R}^n$. Define $f : \mathcal{T}\mathcal{R}^n \to \mathcal{B}$ by $\forall \rho = (R_1, \ldots, R_n) \in \mathcal{T}\mathcal{R}^n$, where $\mathcal{T}\mathcal{R}$ is the set of all crisp reflexive, complete, and transitive relations on X and $\mathcal{B}$ is the set of all reflexive and complete crisp relations on X.*

$$f(\rho) = \{(x, y) \in X \times X \mid \widetilde{f}(1_{R_1}, \ldots, 1_{R_n})(x, y) > 0\}.$$

*Then f is called the **preference aggregation rule associated with** $\widetilde{f}$.*

We have that $\forall (x, y) \in X \times X$,

$$(x, y) \in f(R_1, \ldots, R_n) \Leftrightarrow (x, y) \in \mathrm{Supp}(\widetilde{f}(1_{R_1}, \ldots, 1_{R_n})).$$

Since $\widetilde{f}(1_{R_1}, \ldots, 1_{R_n}) \in \mathcal{F}\mathcal{B}$, it follows that $(x, x) \in \mathrm{Supp}(\widetilde{f}(1_{R_1}, \ldots, 1_{R_n}))$ and so $(x, x) \in f(\rho)$ for all $x \in X$. Also, $\forall x, y \in X$, either $(x, y) \in f(\rho)$ or $(y, x) \in f(\rho)$ since either $(x, y) \in \mathrm{Supp}(\widetilde{f}(1_{R_1}, \ldots, 1_{R_n}))$ or $(y, x) \in \mathrm{Supp}(\widetilde{f}(1_{R_1}, \ldots, 1_{R_n}))$ because $\widetilde{f}(1_{R_1}, \ldots, 1_{R_n}) \in \mathcal{F}\mathcal{B}$. That is, $f(\rho) \in \mathcal{B}$.

Let $\rho \in \mathcal{F}\mathcal{T}\mathcal{R}^n$. Let $\widetilde{f}(\rho)$ be denoted by ρ. Let $\rho \in \mathcal{T}\mathcal{R}^n$ and $f(\rho)$ be denoted by R. Let $\widetilde{\rho} = (1_{R_1}, \ldots, 1_{R_n})$, where $\rho = (R_1, \ldots, R_n)$. Then $P(x, y; \widetilde{\rho}) = P(x, y, \rho)$ since $1_{R_i}(x, y) > 0$ if and only if $(x, y) \in R_i, i = 1, \ldots, n$.

In the following result, we make a strong connection between the crisp and fuzzy case.

Proposition 6.1.5 *Let $\widetilde{f}$ be a fuzzy preference aggregation rule and f be the preference aggregation rule associated with $\widetilde{f}$. Let λ be a fuzzy subset of N. If λ is decisive with respect to $\widetilde{f}$, then $\mathrm{Supp}(\lambda)$ is decisive with respect to f.*

Proof. Let $\rho = (R_1, \ldots, R_n) \in \mathcal{T}\mathcal{R}^n$ and $(x, y) \in X \times X$. Suppose $(x, y) \in P_i \ \forall i \in \mathrm{Supp}(\lambda)$. Then $1_{P_i}(x, y) > 0 \ \forall i \in \mathrm{Supp}(\lambda)$. Since λ is decisive with respect $\widetilde{f}, \pi(x, y) > 0$. Thus $\widetilde{f}(1_{R_1}, \ldots, 1_{R_n})(x, y) > 0$ and $\widetilde{f}(1_{R_1}, \ldots, 1_{R_n})(y, x) = 0$. Hence $(x, y) \in \mathrm{Supp}(\widetilde{f}(1_{R_1}, \ldots, 1_{R_n})$ and $(y, x) \notin \mathrm{Supp}(\widetilde{f}(1_{R_1}, \ldots, 1_{R_n})$. Thus $(x, y) \in f(R_1, \ldots, R_n)$ and $(y, x) \notin f(R_1, \ldots, R_n)$. Hence xPy. ∎

Definition 6.1.6 *Let $\widetilde{f}$ be a fuzzy preference aggregation rule. Then*

(1) *$\widetilde{f}$ is called a **simple majority rule** if $\forall \rho \in \mathcal{FR}^n$, $\forall x, y \in X$, $\pi(x, y) > 0$ if and only if $|P(x, y; \rho| > n/2$;*

(2) *$\widetilde{f}$ is called a **Pareto extension rule** if $\forall \rho \in \mathcal{FR}^n$, $\forall x, y \in X$, $\pi(x, y) > 0$ if and only if $R(x, y; \rho) = N$ and $P(x, y, \rho) \neq \emptyset$.*

A slightly different version of the Pareto extension rule for fuzzy aggregation rules was presented in Definition 5.1.4. It was discussed there that this rule was based on two rules. For the first, no one cares which of the two alternatives is chosen. Also it is in no one's interest to select y rather than x and it is of interest for someone to strictly prefer x to y. Further discussion of the Pareto extension rule can be found in Sen [28].

Proposition 6.1.7 *Suppose $\widetilde{f}$ is a simple majority rule. Let λ be a fuzzy subset of N. If λ is decisive with respect to $\widetilde{f}$, then $|Supp(\lambda)| > n/2$.*

Proof. Since $\widetilde{f}$ is a simple majority rule, it follows by comments preceding Proposition 6.1.5 that f is a simple majority rule. Since λ is decisive with respect to $\widetilde{f}$, we have that $\text{Supp}(\lambda)$ is decisive with respect to f by Proposition 6.1.5, where f is the aggregation rule associated with $\widetilde{f}$. Hence $|\text{Supp}(\lambda)| > n/2$. ∎

Proposition 6.1.8 *Suppose $\widetilde{f}$ is a Pareto extension rule. Let λ be a fuzzy subset of N. If λ is decisive with respect to $\widetilde{f}$, then $Supp(\lambda) = N$.*

Proof. Let f be the aggregation rule associated with $\widetilde{f}$. Since $\widetilde{f}$ is a Pareto extension rule, it follows that f is a Pareto extension rule. Since λ is decisive with respect to $\widetilde{f}$, $\text{Supp}(\lambda)$ is decisive with respect to f by Proposition 6.1.5. The desired result now follows. ∎

Definition 6.1.9 *Let $\mathcal{L} \subseteq \mathcal{FP}(N)$ be such that $\forall \lambda \in \mathcal{L}, \lambda$ is a coalition. Then*

(1) *$\mathcal{L}$ is said to be **monotonic** if $\forall \lambda \in \mathcal{L}$, $\forall \lambda' \in \mathcal{FP}(N), N \supseteq Supp(\lambda') \supseteq Supp(\lambda)$ implies $\lambda' \in \mathcal{L}$;*

(2) *$\mathcal{L}$ is said to be **proper** if $\forall \lambda \in \mathcal{L}$, $\forall \lambda' \in \mathcal{FP}(N)$, $Supp(\lambda') \cap Supp(\lambda) = \emptyset$ implies $\lambda' \notin \mathcal{L}$.*

Suppose $\mathcal{L} = \mathcal{L}(\widetilde{f})$ for some fuzzy aggregation rule. By monotonicity, it follows that adding individuals to a decisive fuzzy subset yields another decisive fuzzy subset. If $\mathcal{L}$ is proper, then two decisive fuzzy subsets must have an individual that belongs to their support.

Proposition 6.1.10 *Let $\widetilde{f}$ be a fuzzy preference aggregation rule. Then*

(1) *$\mathcal{L}(\widetilde{f})$ is monotonic;*

(2) *$\mathcal{L}(\widetilde{f})$ is proper.*

Proof. (1) Suppose $\lambda, \lambda' \in \mathcal{FP}(N)$ are such that $\text{Supp}(\lambda) \subset \text{Supp}(\lambda')$. Suppose λ' is not decisive for $\tilde{f}$. Then there exists $\rho \in \mathcal{FTR}^n$ and $x, y \in X$ such that $\pi_i(x, y) > 0$ for all $i \in \text{Supp}(\lambda')$, but not $\pi(x, y) > 0$. Thus $\pi_i(x, y) > 0$ for all $i \in \text{Supp}(\lambda)$, but not $\pi(x, y) > 0$. Hence λ is not decisive for $\tilde{f}$.

(2) Let $\lambda \in \mathcal{L}(\tilde{f})$. Let $x, y \in X$ and $\rho \in \mathcal{FTR}^n$ be such that $\pi_i(x, y) > 0$ for all $i \in \text{Supp}(\lambda)$ and $\pi_i(y, x) > 0$ for all $i \in N \backslash \text{Supp}(\lambda)$. Then $\pi(x, y) > 0$. Let $\lambda' \in \mathcal{FP}(N)$. Suppose that $\text{Supp}(\lambda) \cap \text{Supp}(\lambda') = \emptyset$. If $\lambda' \in \mathcal{L}(\tilde{f})$, then $\pi(y, x) > 0$, a contradiction. Hence λ' is not decisive for $\tilde{f}$ for all $\lambda' \in \mathcal{FP}(N)$ such that $\text{Supp}(\lambda') \cap \text{Supp}(\lambda) = \emptyset$. Thus $\mathcal{L}(\tilde{f})$ is proper. $\blacksquare$

Definition 6.1.11 Let $\tilde{f}$ be a fuzzy aggregation rule. Then $\tilde{f}$ is called **collegial** if $\forall \mathcal{L}^*(\tilde{f}) \subseteq \mathcal{L}(\tilde{f})$ such that $0 < |\mathcal{L}^*(\tilde{f})| < \infty, \cap_{\lambda \in \mathcal{L}^*(\tilde{f})} \lambda \neq \theta$.

Since the members of a fuzzy collegium are at least partially members of every fuzzy decisive set for a given fuzzy preference aggregation rule, their agreement is needed for a decision to be made. Thus we may concentrate on the set of alternatives which the fuzzy collegium favors in order to predict a likely outcome. We will show that the fuzzy collegium for a fuzzy aggregation rule in single peaked, single-dimensional models is the individual whose most preferred alternatives lie at the median.

Definition 6.1.12 Let $\tilde{f}$ be a fuzzy preference aggregation rule. Then
(1) $\tilde{f}$ is called **acyclic** if $\tilde{f}(\rho)$ is acyclic $\forall \rho \in \mathcal{FTR}^n$;
(2) $\tilde{f}$ is called **partially acyclic** if $\tilde{f}(\rho)$ is partially acyclic $\forall \rho \in \mathcal{FTR}^n$.

The following result allows to obtain the result that a fuzzy preference aggregation rule which is partially acyclic weakly Paretian is also collegial, Proposition 6.1.14.

Proposition 6.1.13 Let $\tilde{f}$ be a fuzzy aggregation rule and f the aggregation rule associated with $\tilde{f}$.
(1) If $\tilde{f}$ is partially acyclic, then f is acyclic.
(2) If $\tilde{f}$ is weakly Paretian, then f is weakly Paretian.

Proof. (1) Since $\tilde{f}$ is partially acyclic, it follows that $\forall x_1, \ldots, x_k \in X$, $1_P(x_1, x_2) > 0, \ldots, 1_P(x_{k-1}, x_k) > 0$ implies $1_R(x_1, x_k) > 0$, where P and R correspond to f. Thus $x_1 P x_2, \ldots, x_{k-1} P x_k$ implies $x_1 R x_k$ since $1_P(x_{i-1}, x_i) > 0 \Leftrightarrow (x_{i-1}, x_i) \in \text{Supp}(1_P) = P, i = 2, \ldots, n$.
(2) Since $\tilde{f}$ is weakly Paretian, $\forall \rho \in \mathcal{TR}^n$, we have that $\forall x, y \in X$, $(\forall i \in N, 1_{P_i}(x, y) > 0$ implies $1_P(x, y) > 0)$. Thus $\forall i \in N$, $x P_i y$ implies $x P y$. $\blacksquare$

The following result states that any fuzzy preference aggregation rule that is partially acyclic and weakly Paretian results in a fuzzy collegium.

Proposition 6.1.14 *If $\widetilde{f}$ is partially acyclic and weakly Paretian, then $\widetilde{f}$ is collegial.*

Proof. By Proposition 6.1.13, f is acyclic and weakly Paretian. Thus $\cap_{\lambda \in \mathcal{L}(\widetilde{f})} \mathrm{Supp}(\lambda) \neq \emptyset$ by Proposition 6.1.5 and [2, Theorem 2.4, p. 43]. Hence $\cap_{\lambda \in \mathcal{L}^*(\widetilde{f})} \lambda \neq \theta \; \forall \mathcal{L}^*(\widetilde{f}) \subseteq \mathcal{L}(\widetilde{f})$ such that $0 < |\mathcal{L}^*(\widetilde{f})| < \infty$. ∎

6.2   Fuzzy Voting Rules

Recall that we let π be of type $\pi_{(0)}$ throughout the chapter. We consider the case for π regular in the exercises at the end of the chapter.

Simple rules form a subset of the set of voting rules. Voting rules include majority rules, super majority rules, and the unanimity rule. We define the characteristics of each class. We show that fuzzy simple rules produce a maximal set located at the median, which is generalizable to the core in n-dimensional space, $n > 1$, Theorem 6.3.9. Consequently, fuzzy single-dimensional models predict an outcome at the median, in a manner analogous to conventional models, when fuzzy simple rules are used to aggregate the preferences of individuals.

Let $R \subseteq X \times X$. We let $\mathrm{Symm}(R)$ denote the symmetric closure of R, i.e., the smallest subset of $X \times X$ that contains R and is symmetric. One reason for the form the following definition takes is for it to correspond with the fact that strict preferences are of type by $\pi_{(0)}$. In the exercises at the end of this chapter, the form is different in order to accommodate strict preferences which are regular.

Definition 6.2.1 *Let $\widetilde{f}$ be a fuzzy aggregation rule. Let $\mathcal{L} \subseteq \mathcal{FP}(N)$. Let $\rho \in \mathcal{FTR}^n$ and set $\widetilde{\mathcal{P}}(\rho) = \{(x, y) \in X \times X \mid \exists \lambda \in \mathcal{L}, \forall i \in \mathrm{Supp}(\lambda), \pi_i(x, y) > 0\}$. Define $\widetilde{f}_\mathcal{L} : \mathcal{FTR}^n \to \mathcal{FB}$ by $\forall \rho \in \mathcal{FR}^n, \forall x, y \in X$,*

$$\widetilde{f}_\mathcal{L}(\rho)(x, y) = \begin{cases} 1 & \text{if } (x, y) \notin \mathrm{Symm}(\widetilde{\mathcal{P}}(\rho)), \\ \vee\{\pi_i(x, y) \mid \exists \lambda \in \mathcal{L}, \forall i \in \mathrm{Supp}(\lambda), \pi_i(x, y) > 0\} \\ \qquad \text{if } (x, y) \in \widetilde{\mathcal{P}}(\rho), \\ 0 & \text{if } (x, y) \in \mathrm{Symm}(\widetilde{\mathcal{P}}(\rho)) \backslash \widetilde{\mathcal{P}}(\rho). \end{cases}$$

In the following example, we illustrate the definition of $\widetilde{f}_\mathcal{L}$. It becomes clear how the definition corresponds to the crisp case in Austen-Smith and Banks [2, p. 59]. For example, those elements not in $\mathrm{Symm}(\widetilde{\mathcal{P}}(\rho))$ must include all elements of the form (x, x) and we wish $\widetilde{f}_\mathcal{L}$ to be reflexive.

Example 6.2.2 *Let* $X = \{x, y, z\}$, $N = \{1, 2, 3\}$, *and* $\rho = (\rho_1, \rho_2, \rho_3)$, *where* $\rho_i(x, x) = \rho_i(y, y) = \rho_i(z, z) = 1$, $i = 1, 2, 3$ *and*

$$\rho_1(x, y) = 0.7, \ \rho_1(x, z) = 0.2, \ \rho_1(y, z) = 0.5,$$
$$\rho_2(x, y) = 0.5, \ \rho_2(x, z) = 0.3, \ \rho_2(z, y) = 0.6,$$
$$\rho_3(y, x) = 0.7, \ \rho_3(z, x) = 0.5, \ \rho_3(y, z) = 0.2.$$

Each ρ_i *is defined to be 0 otherwise,* $i = 1, 2, 3$. *Let* $\mathcal{L} = \{\lambda, \lambda'\}$, *where* $\lambda(1) = 0.5$, $\lambda(2) = 0.2$, $\lambda(3) = 0$ *and* $\lambda'(1) = 0$, $\lambda'(2) = 0.7$, $\lambda'(3) = 0.5$. *Now* $Supp(\lambda) = \{1, 2\}$ *and* $Supp(\lambda') = \{2, 3\}$. *Hence* $\widetilde{\mathcal{P}}(\rho) = \{(x, y), (x, z)\}$ *and* $Symm(\widetilde{\mathcal{P}}(\rho)) = \{(x, y), (y, x), (x, z), (z, x)\}$. *Thus it follows that*

$$\widetilde{f}_{\mathcal{L}}(\rho)(x, y) = 0.7 \vee 0.5 = 0.7, \ \widetilde{f}_{\mathcal{L}}(\rho)(x, z) = 0.2 \vee 0.3 = 0.3.$$

We also have that $\widetilde{f}_{\mathcal{L}}(\rho)(y, z) = \widetilde{f}_{\mathcal{L}}(\rho)(z, y) = 1$, $\widetilde{f}_{\mathcal{L}}(\rho)(y, x) = \widetilde{f}_{\mathcal{L}}(\rho)(z, x) = 0$, *and* $\widetilde{f}_{\mathcal{L}}(\rho)(x, x) = \widetilde{f}_{\mathcal{L}}(\rho)(y, y) = \widetilde{f}_{\mathcal{L}}(\rho)(z, z) = 1$.

We recall from the crisp situation that the knowledge of decisive sets associated with the aggregation rule is equivalent to knowledge of the rule itself. This leads us to the following definition.

Definition 6.2.3 *Let* $\widetilde{f}$ *be a fuzzy aggregation rule. Then* $\widetilde{f}$ *is called a **partial fuzzy simple rule** if* $\forall \rho \in \mathcal{FR}^n$, $\forall x, y \in X$, $\widetilde{f}(\rho)(x, y) > 0$ *if and only if* $\widetilde{f}_{\mathcal{L}(\widetilde{f})}(\rho)(x, y) > 0$.

The following result supports the fact preceding the definition and also helps explain our definition of a partial fuzzy simple rule.

Proposition 6.2.4 *Let* $\widetilde{f}$ *be a fuzzy aggregation rule and* f *the aggregation rule associated with* $\widetilde{f}$. *If* $\widetilde{f}$ *is a partial fuzzy simple rule, then* f *is a simple rule.*

Proof. Let $\rho \in \mathcal{FTR}^n$ and let $x, y \in X$. Recall that $\widetilde{\mathcal{P}}(\rho) = \{(x, y) \in X \times X \mid \exists \lambda \in \mathcal{L}(\widetilde{f}), \forall i \in Supp(\lambda), \pi_i(x, y) > 0\}$ and

$$\mathcal{L}(\widetilde{f})$$
$$= \{\lambda \in \mathcal{FP}(N) \mid \lambda \text{ is decisive for} \widetilde{f}\}$$
$$= \{\lambda \in \mathcal{FP}(N) \mid \forall \rho \in \mathcal{FR}^n, \ \pi_i(x, y) > 0 \ \forall i \in Supp(\lambda) \Rightarrow \pi(x, y) > 0\}.$$

Let $\rho = (1_{R_1}, \ldots, 1_{R_n})$, where $R_i \in \mathcal{R}, i = 1, \ldots, n$. Then

$$\widetilde{f}_{\mathcal{L}(\widetilde{f})}(\rho)(x, y) = \begin{cases} 1 \text{ if } (x, y) \notin Symm(\widetilde{\mathcal{P}}(\rho)), \\ \vee \{1_{P_i}(x, y) \mid \exists \lambda \in \mathcal{L}(\widetilde{f}), \ \forall i \in Supp(\lambda), 1_{P_i}(x, y) = 1\} \\ \quad \text{if } (x, y) \in \widetilde{\mathcal{P}}(\rho), \\ 0 \text{ if } (x, y) \in Symm(\widetilde{\mathcal{P}}(\rho)) \backslash \widetilde{\mathcal{P}}(\rho). \end{cases}$$

Thus for $\rho = (1_{R_1}, \ldots, 1_{R_n})$,

$$\widetilde{f}_{\mathcal{L}(\widetilde{f})}(\rho)(x, y) = \begin{cases} 1 \text{ if } (x, y) \in \widetilde{\mathcal{P}}(\rho) \cup (X \times X \backslash \text{Symm}(\widetilde{\mathcal{P}}(\rho))), \\ 0 \text{ if } (x, y) \in \text{Symm}(\widetilde{\mathcal{P}}(\rho)) \backslash \widetilde{\mathcal{P}}(\rho). \end{cases}$$

Now $f(R_1, \ldots, R_n) = \{(x, y) \mid \widetilde{f}(1_{R_1}, \ldots, 1_{R_n})(x, y) > 0\} = \text{Supp}(\widetilde{f}(1_{R_1}, \ldots, 1_{R_n})) = \text{Supp}(\widetilde{f}_{\mathcal{L}(\widetilde{f})}(1_{R_1}, \ldots, 1_{R_n})) = \widetilde{\mathcal{P}}(\rho) \cup (X \times X \backslash \text{Symm}(\widetilde{\mathcal{P}}(\rho)))$. By definition,

$$f_{\mathcal{L}(f)}(R_1, \ldots, R_n) = \mathcal{P}(\rho) \cup (X \times X \backslash \text{Symm}(\mathcal{P}(\rho))),$$

where $\mathcal{P}(\rho) = \{(x, y) \mid \exists L \in \mathcal{L}(f), \forall i \in L, x P_i y\}$ and $\rho = (R_1, \ldots, R_n)$. For $\rho = (1_{R_1}, \ldots, 1_{R_n})$, $\exists \lambda \in \mathcal{L}(\widetilde{f}), \forall i \in \text{Supp}(\lambda), x P_i y$ (P_i corresponds to 1_{R_i}) $\Leftrightarrow \exists L \in \mathcal{L}(f), \forall i \in L, x P_i y$ (P_i corresponds to R_i), i.e., given λ, let $L = \text{Supp}(\lambda)$ and given L, let $\lambda = 1_L$. Hence $\widetilde{\mathcal{P}}((1_{R_1}, \ldots, 1_{R_n})) = \mathcal{P}((R_1, \ldots, R_n))$. Thus $\widetilde{\mathcal{P}}(1_{R_1}, \ldots, 1_{R_n}) = \mathcal{P}(R_1, \ldots, R_n)$. Therefore, $f = f_{\mathcal{L}(f)}$. ∎

Definition 6.2.5 *Let $\widetilde{f}$ be a fuzzy aggregation rule. Then*
(1) $\widetilde{f}$ *is called **decisive** if $\forall \rho, \rho' \in \mathcal{FR}^n$ and $\forall x, y \in X, [P(x, y; \rho) = P(x, y; \rho')$ and $\pi(x, y) > 0]$ imply $\pi'(x, y) > 0$;*
(2) $\widetilde{f}$ *is called **monotonic** if $\forall \rho, \rho' \in \mathcal{FR}^n$ and $\forall x, y \in X, [P(x, y; \rho) \subseteq P(x, y; \rho'), R(x, y; \rho) \subseteq R(x, y; \rho')$ and $\pi(x, y) > 0]$ imply $\pi'(x, y) > 0$.*

A fuzzy aggregation rule is decisive if the set of individuals strictly preferring x to y remains the same under a different preference profile, then the social choice remains the same in that x is strictly preferred to y.

An aggregation rule is monotonic if x is preferred to y, then increasing support for x over y or decreasing the opposition does not change the social preference.

We next characterize partial fuzzy simple rules.

Theorem 6.2.6 *Let $\widetilde{f}$ be a fuzzy aggregation rule. Then $\widetilde{f}$ is a partial fuzzy simple rule if and only if $\widetilde{f}$ is decisive, neutral, and monotonic.*

Proof. Let $x, y \in X$. Suppose $\pi_{\mathcal{L}(\widetilde{f})}(x, y) > 0$. Then $\exists \lambda \in \mathcal{L}(\widetilde{f})$ such that $\forall i \in \text{Supp}(\lambda), \pi_i(x, y) > 0$. Thus $\pi(x, y) > 0$. Suppose $\widetilde{f}$ is decisive, neutral, and monotonic. Let $\rho \in \mathcal{FTR}^n$. Suppose $\pi(x, y) > 0$. In order to show $\widetilde{f}$ is a partial fuzzy simple rule, it suffices to show $P(x, y; \rho) \in \mathcal{L}(\widetilde{f})$ for then $\pi_{\mathcal{L}(\widetilde{f})}(x, y) > 0$. Let $u, v \in X$. Let ρ^* be any fuzzy preference profile such that $\forall i \in P(x, y; \rho), \pi_i^*(u, v) > 0$. Let $L^+ = P(u, v; \rho^*) \backslash P(x, y; \rho)$. Let ρ' be a fuzzy preference profile such that $P(u, v; \rho') = P(x, y; \rho)$ and $P(v, u; \rho') = P(y, x; \rho)$. Since $\widetilde{f}$ is neutral, we have that $\pi'(u, v) > 0$. Let $\rho'' \in \mathcal{FTR}^n$ be defined by $\rho_i''|_{\{u, v\}} = \rho_i^*|_{\{u, v\}}$ if and only if $i \in L^+ \cup P(x, y; \rho)$ and $\rho_j''|_{\{u, v\}} = \rho_j'|_{\{u, v\}}$

otherwise. Then individuals that ρ'' and ρ' differ from 0 on u, v must come from L^+. Thus $\pi''(u, v) > 0$ since $\widetilde{f}$ is monotonic. Since $P(u, v; \rho^*) = P(u, v; \rho'')$ and $\widetilde{f}$ is decisive, $\pi^*(u, v) > 0$. Hence since ρ and u, v are arbitrary except for $\pi_i^*(u, v) > 0$ if $i \in P(x, y; \rho)$ and $\pi^*(u, v) > 0$, it follows that $P(x, y; \rho) \in \mathcal{L}(\widetilde{f})$.

Conversely, suppose that $\widetilde{f}$ is a partial fuzzy simple rule. Then $\widetilde{f}$ is neutral since $\mathcal{L}(\widetilde{f})$ is defined without regard to alternatives. Monotonicity follows from Proposition 6.1.10 and the definition of $\widetilde{f}_{\mathcal{L}(\widetilde{f})}$. That $\widetilde{f}$ is decisive follows directly from the definitions of $\mathcal{L}(\widetilde{f})$ and $\widetilde{f}_{\mathcal{L}(\widetilde{f})}$. ∎

Definition 6.2.7 *Let $\widetilde{f}$ be a fuzzy aggregation rule. The **decisive structure** of $\widetilde{f}$, denoted $\mathcal{D}(\widetilde{f})$, is defined to be the set $\mathcal{D}(\widetilde{f}) = \{(\sigma, \omega) \in \mathcal{FP}(N) \times \mathcal{FP}(N) \mid Supp(\sigma) \subseteq Supp(\omega)$ and $\forall x, y \in X, \forall \rho \in \mathcal{FTR}^n, \pi_i(x, y) > 0 \ \forall i \in Supp(\sigma)$ and $\rho_j(x, y) > 0 \ \forall j \in Supp(\omega)$ implies $\pi(x, y) > 0\}$.*

It is worth recalling the crisp situation. An aggregation rule is a plurality rule if $\forall x, y \in X$, alternative x is chosen over alternative y when the number of individuals strictly preferring x to y exceeds the number strictly preferring y to x. The plurality rule is a voting rule, but not a simple rule. Majority rule is both a simple rule and a voting rule.

Definition 6.2.8 *Let $\mathcal{D} \subseteq \mathcal{FP}(N) \times \mathcal{FP}(N)$. Then $\mathcal{D}$ is called **monotonic** if $(\sigma, \omega) \in \mathcal{D}, Supp(\sigma) \subseteq Supp(\sigma') \subseteq Supp(\omega')$ and $Supp(\sigma) \subseteq Supp(\omega) \subseteq Supp(\omega')$ imply $(\sigma', \omega') \in \mathcal{D}$ for all $(\sigma', \omega') \in \mathcal{FP}(N) \times \mathcal{FP}(N)$.*

The intuition for the definition of $\widetilde{f}_{\mathcal{D}}$ in the next definition is similar to that given for the definition of $\widetilde{f}_{\mathcal{L}}$ in Definition 6.2.1.

Definition 6.2.9 *Let $\widetilde{f}$ be a fuzzy aggregation rule. Let $\mathcal{D} \subseteq \mathcal{FP}(N) \times \mathcal{FP}(N)$ be such that $(\sigma, \omega) \in \mathcal{D}$ implies $Supp(\sigma) \subseteq Supp(\omega)$. Let $\rho \in \mathcal{FTR}^n$ and set $\mathcal{FP}_{\mathcal{D}}(\rho) = \{(x, y) \in X \times X \mid \exists (\sigma, \omega) \in \mathcal{D}, \forall i \in Supp(\sigma), \pi_i(x, y) > 0$ and $\forall j \in Supp(\omega), \rho_j(x, y) > 0\}$. Define $\widetilde{f}_{\mathcal{D}} : \mathcal{FTR}^n \to \mathcal{FB}$ by $\forall \rho \in \mathcal{FTR}^n, \forall x, y \in X$,*

$$\widetilde{f}_{\mathcal{D}}(\rho)(x, y) = \begin{cases} 1 \text{ if } (x, y) \notin Symm(\mathcal{FP}_{\mathcal{D}}(\rho)) \\ \vee\{\pi_i(x, y) \wedge \rho_j(x, y) \mid \exists (\sigma, \omega) \in \mathcal{D}, \ \pi_i(x, y) > 0, \forall i \in Supp(\sigma), \\ \qquad \rho_j(x, y) > 0 \ \forall j \in Supp(\rho_j)\} \quad \text{if } (x, y) \in \mathcal{FP}_{\mathcal{D}}(\rho), \\ 0 \ \text{ if } (x, y) \in Symm(\mathcal{FP}_{\mathcal{D}}(\rho)) \backslash \mathcal{FP}_{\mathcal{D}}(\rho). \end{cases}$$

Example 6.2.10 *Let $X = \{x, y, z\}$, $N = \{1, 2, 3\}$, and $\rho = (\rho_1, \rho_2, \rho_3)$, where $\rho_i(x, x) = \rho_i(y, y) = \rho_i(z, z) = 1, \ i = 1, 2, 3$ and*

$$\rho_1(x, y) = 0.7, \quad \rho_1(x, z) = 0.2, \quad \rho_1(y, z) = 0.5,$$
$$\rho_2(x, y) = 0.5, \quad \rho_2(x, z) = 0.3, \quad \rho_2(z, y) = 0.6,$$
$$\rho_3(x, y) = 0.1, \quad \rho_3(y, x) = 0.7, \quad \rho_3(z, x) = 0.5, \quad \rho_3(y, z) = 0.2.$$

Each ρ_i is defined to be 0 otherwise, $i = 1, 2, 3$. Let $\mathcal{D} = \{(\sigma, \omega), (\sigma', \omega')\}$, where $\sigma(1) = 0.5$, $\sigma(2) = 0.2$, $\sigma(3) = 0$, $\omega(1) = 0.5$, $\omega(2) = 0.2$, $\omega(3) = 0.3$, $\sigma'(1) = 0$, $\sigma'(2) = 0.7$, $\sigma'(3) = 0.5$, and $\omega' = \sigma'$. Now $\text{Supp}(\sigma) = \{1, 2\}$, $\text{Supp}(\omega) = N$, and $\text{Supp}(\sigma') = \{2, 3\} = \text{Supp}(\omega')$. Now $\pi_i(x, y) > 0 \ \forall i \in \text{Supp}(\sigma)$ and $\rho_i(x, y) > 0 \ \forall i \in \text{Supp}(\omega)$. Thus $(x, y) \in \mathcal{FP_D}(\rho)$. Now $\pi_i(x, z) > 0 \ \forall i \in \text{Supp}(\sigma)$, but $\rho_3(x, z) = 0$ and $3 \in \text{Supp}(\omega)$. Hence $(x, z) \notin \mathcal{FP_D}(\rho)$. Note also that $(y, z) \notin \mathcal{FP_D}(\rho)$ since it is not the case that $\pi_i(y, z) > 0$ $\forall i \in \text{Supp}(\sigma)$ or $\pi'_i(y, z) > 0 \ \forall i \in \text{Supp}(\sigma')$. Thus it follows that $\mathcal{FP_D}(\rho) = \{(x, y)\}$, $\text{Symm}(\mathcal{FP_D}(\rho)) = \{(x, y), (y, x)\}$ and

$$
\begin{aligned}
\widetilde{f}(\rho)(x, y) &= \vee\{\pi_i(x, y) \wedge \rho_j(x, y) \mid i = 1, 2; j = 1, 2, 3\} \\
&= \vee\{0.7 \wedge 0.7, 0.7 \wedge 0.5, 0.7 \wedge 0.1, 0.5 \wedge 0.7, 0.5 \wedge 0.5, 0.5 \wedge 0.1\} \\
&= 0.7.
\end{aligned}
$$

We also have that $\widetilde{f}_\mathcal{D}(\rho)(u, v) = 1$ if $(y, x) \neq (u, v) \neq (x, y)$ and that $\widetilde{f}(y, x) = 0$.

Definition 6.2.11 Let $\widetilde{f}$ be a fuzzy aggregation rule. Then $\widetilde{f}$ is called a **partial fuzzy voting rule** if $\forall \rho \in \mathcal{FTR}^n$ and $\forall x, y \in X$, $\widetilde{f}(\rho)(x, y) > 0$ if and only if $\widetilde{f}_{\mathcal{D}(\widetilde{f})}(\rho)(x, y) > 0$.

In the crisp case, voting rules are completely characterized by their decisive structure. Partial fuzzy voting rules are characterized in a similar manner as can be seen by the next two propositions.

Proposition 6.2.12 Let $\widetilde{f}$ be a fuzzy aggregation rule and f the aggregation rule associated with $\widetilde{f}$. If $\widetilde{f}$ is a partial fuzzy voting rule, then f is a voting rule.

Proof. Let $\rho = (R_1, \dots, R_n) \in R^n$ and let $\rho = (1_{R_1}, \dots, 1_{R_n})$. Then

$$
\begin{aligned}
&(x, y) \in \mathcal{FP}_{\mathcal{D}(\widetilde{f})}(\rho) \\
\Leftrightarrow \ & \exists(\sigma, \omega) \in \mathcal{D}(\widetilde{f}), \ \forall i \in \text{Supp}(\sigma), \ 1_{P_i}(x, y) = 1, \\
& \forall j \in \text{Supp}(\omega), \ 1_{R_j}(x, y) = 1 \\
\Leftrightarrow \ & \exists(\sigma, \omega) \in \mathcal{D}(\widetilde{f}), \ \forall i \in \text{Supp}(1_{\text{Supp}(\sigma)}), \ 1_{P_i}(x, y) = 1, \\
& \forall j \in \text{Supp}(1_{\text{Supp}(\omega)}), 1_{R_j}(x, y) = 1 \\
\Leftrightarrow \ & \exists S, W \in \mathcal{D}(f), \ \forall i \in S, \ (x, y) \in P_i, \ \forall j \in W, \ (x, y) \in R_j,
\end{aligned}
$$

where $S = \text{Supp}(1_{\text{Supp}(\sigma)})$, $W = \text{Supp}(1_{\text{Supp}(\omega)})$ for the implication $\Rightarrow$ and $\sigma = 1_S$ and $\omega = 1_W$ for the implication $\Leftarrow$. Thus

$$
\begin{aligned}
&(x, y) \in f(R_1, \dots, R_n) \\
\Leftrightarrow \ & (x, y) \in \text{Supp}(\widetilde{f}(1_{R_1}, \dots, 1_{R_n})) \\
\Leftrightarrow \ & (x, y) \in (X \times X \backslash \text{Symm}(\mathcal{FP}_{\mathcal{D}(\widetilde{f})}(\rho))) \cup (\mathcal{FP}_{\mathcal{D}(\widetilde{f})}(\rho)) \\
\Leftrightarrow \ & (x, y) \in (X \times X \backslash \text{Symm}(\mathcal{P}_{\mathcal{D}(f)}(R_1, \dots, R_n))) \cup (\mathcal{P}_{\mathcal{D}(f)}(R_1, \dots, R_n)) \\
\Leftrightarrow \ & (x, y) \in f_{\mathcal{D}(f)}(R_1, \dots, R_n).
\end{aligned}
$$

Therefore, $f = f_{\mathcal{D}(f)}$. ∎

We next characterize partial fuzzy voting rules.

Theorem 6.2.13 *Let $\widetilde{f}$ be a fuzzy aggregation rule. Then $\widetilde{f}$ is a partial fuzzy voting rule if and only if it is neutral and monotonic.*

Proof. Suppose $\widetilde{f}$ is a partial fuzzy voting rule. We show that $\widetilde{f}$ is monotonic. We first note that $\mathcal{D}(\widetilde{f})$ is monotonic. Let $(\sigma, \omega) \in \mathcal{D}(\widetilde{f})$ and $\sigma', \omega' \in \mathcal{FP}(N)$ be such that $\text{Supp}(\sigma) \subseteq \text{Supp}(\sigma') \subseteq \text{Supp}(\omega')$ and $\text{Supp}(\sigma) \subseteq \text{Supp}(\omega) \subseteq \text{Supp}(\omega')$. Since $\text{Supp}(\sigma) \subseteq \text{Supp}(\sigma')$ and $\text{Supp}(\omega) \subseteq \text{Supp}(\omega')$, clearly $(\sigma', \omega') \in \mathcal{D}(\widetilde{f})$. Let $\rho, \rho' \in \mathcal{FTR}^n$ and $x, y \in X$ be such that $P(x, y; \rho) \subseteq P(x, y; \rho'), R(x, y; \rho) \subseteq R(x, y; \rho')$, and $\pi(x, y) > 0$. By Definitions 6.2.9 6.2.9 and 6.2.11 6.2.11, we have that since $\pi(x, y) > 0$, $\exists (\sigma, \omega) \in \mathcal{D}(\widetilde{f})$ such that $\text{Supp}(\sigma) \subseteq \text{Supp}(\omega)$ and $\forall x, y \in X, \forall \rho \in \mathcal{FTR}^n, \pi_i(x, y) > 0$ $\forall i \in \text{Supp}(\sigma)$ and $\rho_j(x, y) > 0$ $\forall j \in \text{Supp}(\omega)$. By hypothesis,

$$\{i \in N \mid \pi_i(x, y) > 0\} \subseteq \{i \in N \mid \pi_i'(x, y) > 0\},$$
$$\{j \in N \mid \rho_j(x, y) > 0\} \subseteq \{j \in N \mid \rho_j'(x, y) > 0\}.$$

Since $(\sigma', \omega') \in \mathcal{D}(\widetilde{f})$, $\pi'(x, y) > 0$. Thus $\widetilde{f}$ is monotonic. We now show $\widetilde{f}$ is neutral. Let $\rho, \rho' \in \mathcal{FR}^n$ and $x, y, z, w \in X$ be such that

$$\{i \in N \mid \pi_i(x, y) > 0\} = \{i \in N \mid \pi_i'(z, w) > 0\},$$
$$\{i \in N \mid \pi_i(y, x) > 0\} = \{i \in N \mid \pi_i'(w, z) > 0\}.$$

Then it follows easily that $\pi_i(x, y) > 0 \Leftrightarrow \pi_i'(z, w) > 0$ $\forall i \in N$. Suppose $\widetilde{f}(\rho)(x, y) > 0$. Then $(x, y) \in (X \times X \backslash \text{Symm}(\mathcal{FP}_{\mathcal{D}(\widetilde{f})}(\rho)) \cup \mathcal{FP}_{\mathcal{D}(\widetilde{f})}(\rho)$, say $(x, y) \in \mathcal{FP}_{\mathcal{D}(\widetilde{f})}(\rho)$. Then $\exists (\sigma, \omega) \in \mathcal{D}(\widetilde{f})$ such that $\forall i \in \text{Supp}(\sigma), \pi_i(x, y) > 0, \forall j \in \text{Supp}(\omega), \rho_i(x, y) > 0\}$. By hypothesis, it follows that $(z, w) \in \mathcal{FP}_{\mathcal{D}(\widetilde{f})}(\rho')$. Hence $\widetilde{f}(\rho')(z, w) > 0$ and in fact $\pi'(z, w) > 0$. Suppose $(x, y) \in X \times X \backslash \text{Symm}(\mathcal{FP}_{\mathcal{D}(\widetilde{f})}(\rho))$. By the argument just given, $(z, w) \notin \mathcal{FP}_{\mathcal{D}(\widetilde{f})}(\rho')$ else $(x, y) \in \mathcal{FP}_{\mathcal{D}(\widetilde{f})}(\rho)$. Thus $(z, w) \in X \times X \backslash \mathcal{FP}_{\mathcal{D}(\widetilde{f})}(\rho')$. Hence $\widetilde{f}(\rho')(z, w) > 0$. Thus $\widetilde{f}$ is neutral.

For the converse, we first show that $\pi_{\mathcal{D}(\widetilde{f})}(x, y) > 0$ implies $\pi(x, y) > 0$. Suppose $\widetilde{f}_{\mathcal{D}(\widetilde{f})}(\rho)(x, y) > 0$ for $\rho \in \mathcal{FR}^n$ and $x, y \in X$. Then $(x, y) \in (X \times X \backslash \text{Symm}(\mathcal{FP}_{\mathcal{D}(\widetilde{f})}(\rho)) \cup \mathcal{FP}_{\mathcal{D}(\widetilde{f})}(\rho)$, say $(x, y) \in \mathcal{FP}_{\mathcal{D}(\widetilde{f})}(\rho)$. Thus $\exists (\sigma, \omega) \in \mathcal{D}(\widetilde{f})$ such that $\forall i \in \text{Supp}(\sigma), \pi_i(x, y) > 0, \forall j \in \text{Supp}(\omega), \rho_i(x, y) > 0\}$. Since $(\sigma, \omega) \in \mathcal{D}(\widetilde{f}), \widetilde{f}(\rho)(x, y) > 0$. Suppose $(x, y) \in (X \times X) \backslash \text{Symm}(\mathcal{FP}_{\mathcal{D}(\widetilde{f})}(\rho))$. Then $(y, x) \in (X \times X) \backslash \text{Symm}(\mathcal{FP}_{\mathcal{D}(\widetilde{f})}(\rho))$ and so $\widetilde{f}_{\mathcal{D}(\widetilde{f})}(\rho)(y, x) = \widetilde{f}_{\mathcal{D}(\widetilde{f})}(\rho)(x, y) = 1$. However, this contradicts the assumption that $\pi_{\mathcal{D}(\widetilde{f})}(x, y) > 0$. Thus $(y, x) \in \text{Symm}(\mathcal{FP}_{\mathcal{D}(\widetilde{f})}(\rho)) \backslash \mathcal{FP}_{(\mathcal{D}(\widetilde{f})}(\rho))$. Hence it follows that $\widetilde{f}(\rho)(y, x) = 0$. Now suppose that $\widetilde{f}$ is neutral and monotonic. Let $x, y \in X$ and $\rho \in \mathcal{FR}^n$.

Suppose that $\pi(x, y) > 0$. Let $(\sigma, \omega) \in \mathcal{FP}(N) \times \mathcal{FP}(N)$, $\mathrm{Supp}(\sigma) \subseteq \mathrm{Supp}(\omega)$, be such that $\mathrm{Supp}(\sigma) = S$ and $\mathrm{Supp}(\omega) = W$, where $S = P(x, y; \rho)$ and $W = R(x, y; \rho)$. We wish to show $(\sigma, \omega) \in \mathcal{D}(\tilde{f})$. Let $z, w \in X$. Let $\rho' \in \mathcal{FTR}^n$ be such that $\pi_i'(z, w) > 0 \Leftrightarrow i \in S$ and $\rho_i'(z, w) > 0 \Leftrightarrow i \in W$, i.e., $P(z, w; \rho') = S$ and $R(z, w; \rho') = W$. Since $\tilde{f}$ is neutral, $\pi'(z, w) > 0$. Now let $\rho'' \in \mathcal{FTR}^n$ be such that $P(z, w; \rho') \subseteq P(z, w; \rho'')$ and $R(z, w; \rho') \subseteq R(z, w; \rho'')$. Since $\tilde{f}$ is monotonic, $\pi''(z, w) > 0$. Thus we have that $\pi_i(z, w) > 0 \; \forall i \in S$ and $\rho_i(z, w) > 0 \; \forall i \in W$ implies $\pi(z, w) > 0$. Hence $(\sigma, \omega) \in \mathcal{D}(\tilde{f})$ and so $\pi_{D(\tilde{f})}(x, y) > 0$. ∎

6.3   Single-Peaked Fuzzy Profiles

We first consider single-peakedness. We then develop the argument relating fuzzy $\tilde{f}$-medians to single-peakedness in order to introduce the notion of a core, a maximal set (set of undominated alternatives) under some fuzzy preference aggregation procedure. Our main result, Theorem 6.3.9, says that the core is located at the median if preferences of all political actors are single-peaked in single-dimensional policy space.

A **strict ordering** on a set X is a relation on X that is irreflexive, transitive, and asymmetric.

Definition 6.3.1 *Suppose $|X| = n \geq 2$. Suppose Q is a strict ordering of X. Label X so that $\forall a_t \in X$, $a_{t+1} Q a_t, t = 1, 2, \ldots, n-1$. Let $\pi \in \mathcal{FTR}$. Then π is called **single-peaked** on X with respect to Q if*

$$\pi(a_t, a_{t+1}) > 0, \pi(a_{t+1}, a_{t+2}) > 0, \ldots, \pi(a_{n-1}, a_n) > 0 \quad \text{and}$$
$$\pi(a_t, a_{t-1}) > 0, \pi(a_{t-1}, a_{t-2}) > 0, \ldots, \pi(a_2, a_1) > 0.$$

Recall that π in Definition 6.3.1 is max-min transitive.

An approach to the problem of cyclical majorities is that of single-peakedness. If the set of individual preferences satisfy a certain unimodal pattern, called single-peaked preference, then majority decisions must be transitive regardless of the number individuals holding any of the possible orderings, as long as the total number of persons is odd.

Definition 6.3.2 *Let $\rho \in \mathcal{FTR}^n$. Then ρ is said to be **single-peaked** on X if there exists a strict ordering Q of X such that for all $i \in N$, ρ_i is single-peaked on X with respect to Q.*

Let $\mathcal{FS}$ denote the set of all single-peaked fuzzy profiles.

Definition 6.3.3 *Let $\rho \in \mathcal{FTR}^n$. Suppose ρ is single-peaked with respect to a strict ordering Q. Define x_i to be that element of X such that $\pi_i(x_i, y) > 0$ $\forall y \in X$ $(x_i \neq y)$, $i = 1, \ldots, n$. Define $\tilde{L}^-, \tilde{L}^+ : X \to \mathcal{FP}(N)$ by $\forall z \in X$, $\forall i \in N$,*

$$\tilde{L}^-(z)(i) = \begin{cases} \pi_i(x_i, z) & \text{if } zQx_i, \\ 0 & \text{otherwise,} \end{cases} \qquad \tilde{L}^+(z)(i) = \begin{cases} \pi_i(x_i, z) & \text{if } x_iQz, \\ 0 & \text{otherwise.} \end{cases}$$

For any $\rho \in \mathcal{FTR}^n$, let x_i denote individual i's most preferred alternative from X and for all $z \in X, \tilde{L}^-(z)(i)$ is the degree to which x_i is strictly preferred to z when zQx_i. Similar comments hold for $\tilde{L}^+(z)(i)$. Then $\tilde{L}^-(z)$ is the fuzzy subset of X which gives the degree of the individuals whose ideal points lie to the left of z and $\tilde{L}^+(z)$ denotes the fuzzy subset of X which gives the degree of the individuals whose ideal points lie to the right of z.

Definition 6.3.4 *Let $\tilde{f}$ be a fuzzy preference aggregation rule. Let $\rho \in \mathcal{FTR}^n$. Suppose ρ is single-peaked with respect to a strict ordering Q. Then an element z of X is called an $\tilde{f}$-median if $Supp(\tilde{L}^-(z)) \notin \{Supp(\lambda) \mid \lambda \in \mathcal{L}(\tilde{f})\}$ and $Supp(\tilde{L}^+(z)) \notin \{Supp(\lambda) \mid \lambda \in \mathcal{L}(\tilde{f})\}$.*

Definition 6.3.5 *Let $F : [0,1]^2 \to [0,1]$. $\forall \rho \in \mathcal{FR}^*(X), \forall \mu \in \mathcal{FP}^*(X)$, $\forall x \in X$, define the fuzzy subset $M_F(\rho, \mu)$ of X by $\forall x \in X$,*

$$M_F(\rho, \mu)(x) = \vee\{t \in [0,1] \mid F(\rho(x,y), \mu(x)) \geq t \; \forall y \in Supp(\mu)\}.$$

*The fuzzy subset $M_F(\rho, \mu)$ is called the **maximal fuzzy subset** of ρ and μ with respect to F.*

Definition 6.3.6 *Let $\tilde{f}$ be a fuzzy preference aggregation rule. Let $F : [0,1]^2 \to [0,1]$. Define $\forall \rho \in \mathcal{FS}$ the fuzzy subset $\mu_{F,\tilde{f}}(\rho; Q)$ of X by $\forall z \in X$,*

$$\mu_{F,\tilde{f}}(\rho, Q)(z) = \begin{cases} \wedge\{F(\tilde{f}(\rho)(z,y), 1) \mid y \in X\} & \text{if } z \text{ is an } \tilde{f}\text{-median} \\ 0 & \text{otherwise.} \end{cases}$$

In Definition 6.3.6, we write $\mu_{F,\tilde{f}}(\rho)$ for $\mu_{F,\tilde{f}}(\rho, Q)$ when the strict order Q is understood. $\mu_{F,\tilde{f}}(\rho)$ is the fuzzy subset of $\tilde{f}$-medians given ρ with respect to F.

Definition 6.3.7 *Let $F : [0,1]^2 \to [0,1]$. For all $\rho \in \mathcal{FTR}^n$, define $C_{F,\tilde{f}}(\rho)(x) = M_F(\tilde{f}(\rho), 1_X)(x)$ for all $x \in X$. Then $C_{F,\tilde{f}}(\rho)$ is called the **fuzzy core** of $\tilde{f}$ at ρ with respect to F.*

Before presenting our main result, we provide some intuition for the median voter theorem. Suppose $N = \{1, 2, 3\}$ and $X = \{x, y, z\}$, where the alternatives of X are ordered, say $x < y < z$. Suppose player 1 favors alternative x, player 2 alternative y, and player 3 alternative z. Each player prefers the alternative closest to its ideal choice. The following table presents the majority decision of the players.

Options	Player 1 Votes	Player 2 Votes	Player 3 Votes	Results
x vs y	x	y	y	y
x vs z	x	x or z	z	x or z
y vs z	y	y	z	y

We see that alternative y is chosen if the voting system is majoritarian. Thus the median voter y votes in favor of the outcome that wins the election.

If $F = \wedge$, we use the notation $C_{\tilde{f}}(\rho)$ for $C_{F,\tilde{f}}(\rho)$, $\mu_{\tilde{f}}(\rho)(x)$ for $\mu_{F,\tilde{f}}(\rho)(x)$, and M for M_F.

Example 6.3.8 *Let* $F = \wedge$. *Let* $X = \{x, y, z\}$ *and* $N = \{1, 2, 3\}$. *Let* $\rho = (\rho_1, \rho_2, \rho_3) \in \mathcal{FR}^3$ *be such that*

$$\pi_1(x, y) = 0.1, \quad \pi_1(y, z) = 0.7, \quad \pi_1(x, z) = 0.9,$$
$$\pi_2(y, x) = 0.2, \quad \pi_2(x, z) = 0.6, \quad \pi_2(y, z) = 0.9,$$
$$\pi_3(z, y) = 0.5, \quad \pi_3(y, x) = 0.7, \quad \pi_3(z, x) = 1.$$

Let Q *be such that* $a_3 Q a_2 Q a_1$, *where* $a_3 = z$, $a_2 = y$, *and* $a_1 = x$. *It follows that* ρ *is single-peaked with respect to* Q, *where for* π_1, $t = 1$, *for* π_2, $t = 2$, *and for* $\pi_3, t = 3$. *Now* $x_1 = x$, $x_2 = y$, *and* $x_3 = z$ *since* $\pi_1(x, w) > 0 \; \forall w \in X$, $w \neq x$, $\pi_2(y, w) > 0 \; \forall w \in X$, $w \neq y$, $\pi_3(z, w) > 0 \; \forall w \in X$, $w \neq z$. *Hence*

$\tilde{L}^-(x)(1) = 0$ *since not* xQx, $\tilde{L}^-(x)(2) = 0$ *since not* xQy,
$\tilde{L}^-(x)(3) = 0$ *since not* xQz, $\tilde{L}^-(y)(1) = \pi_1(x, y)$ *since* yQx,
$\tilde{L}^-(y)(2) = 0$ *since not* yQy, $\tilde{L}^-(y)(3) = 0$ *since not* yQz,
$\tilde{L}^-(z)(1) = \pi_1(x, z)$ *since* zQx, $\tilde{L}^-(z)(2) = \pi_2(y, z)$ *since* zQy,
$\tilde{L}^-(z)(3) = 0$ *since not* zQz,

and

$\tilde{L}^+(x)(1) = 0$ *since not* xQx, $\tilde{L}^+(x)(2) = \pi_2(y, x)$ *since* yQx,
$\tilde{L}^+(x)(3) = \pi_3(z, x)$ *since* zQx, $\tilde{L}^+(y)(1) = 0$ *since not* xQy,
$\tilde{L}^+(y)(2) = 0$ *since not* yQy, $\tilde{L}^+(y)(3) = \pi_3(z, y)$ *since* zQy,
$\tilde{L}^+(z)(1) = 0$ *since not* xQz, $\tilde{L}^+(z)(2) = 0$ *since not* yQz,
$\tilde{L}^+(z)(3) = 0$ *since not* zQz.

Let $\tilde{f}$ *be a simple majority rule. Let* $\lambda \in \mathcal{L}(\tilde{f})$. *Then* $\forall \rho \in \mathcal{FR}^n$, $\pi_i(x, y) > 0$ $\forall i \in Supp(\lambda) \Rightarrow \pi(x, y) > 0$. *Thus* $\{Supp(\lambda) \mid \lambda \in \mathcal{L}(\tilde{f})\} = \{\{1, 2\}, \{1, 3\},$

$\{2,3\}$, $\{1,2,3\}\}$. *Since*

$$Supp(\widetilde{L}^-(x)) = \emptyset, \ Supp(\widetilde{L}^-(y)) = \{1\}, \ Supp(\widetilde{L}^-(z)) = \{1,2\},$$
$$Supp(\widetilde{L}^+(x)) = \{2,3\}, \ Supp(\widetilde{L}^+(y)) = \{3\}, \ Supp(\widetilde{L}^+(z)) = \emptyset,$$

it follows that $Supp(\widetilde{L}^-(y)) \notin \{Supp(\lambda) \mid \in \mathcal{L}(\widetilde{f})\}$ *and* $Supp(\widetilde{L}^+(y)) \notin \{Supp(\lambda) \mid \lambda \in \mathcal{L}(\widetilde{f})\}$. *Hence* y *is an* $\widetilde{f}$*-median. Thus* $\mu_{\widetilde{f}}(\rho; Q)(y) = \widetilde{f}(\rho)(y,x) \wedge \widetilde{f}(\rho)(y,z)$. *Now* $M(\widetilde{f}(\rho), 1_X)(y) = \vee\{t \in [0,1] \mid \widetilde{f}(\rho)(y,w) \geq t \ \forall w \in X\} = \widetilde{f}(\rho)(y,x) \wedge \widetilde{f}(\rho)(y,z)$. *It follows that*

$$\widetilde{f}(\rho)(y,x) = \pi_2(y,x) \vee \pi_3(y,x) = 0.2 \vee 0.7 = 0.7,$$
$$\widetilde{f}(\rho)(y,z) = \pi_1(y,z) \vee \pi_2(y,z) = 0.7 \vee 0.9 = 0.9.$$

Hence $\widetilde{f}(\rho)(y,x) \wedge \widetilde{f}(\rho)(y,z) = 0.7$.

The median voter theorem explains things that occur in majoritarian voting systems. In the United States, the Democratic and Republican candidates usually move their platforms to the middle during election campaigns. Politicians tend to adopt similar platforms and rhetoric during campaigns. That is, they tailor platforms to the median voter. The median voter theorem reflects that radical candidates usually are not elected.

The median voter theorem also explains the rise in government programs over the past several decades. See for example Husted and Lawrence [17]. Voters with the median income take advantage of their status as voters by electing politicians who propose taxing those who earn more than the median voter and then redistributing the money. A detailed account can be found in Rice [25] and Husted and Lawrence [17]. Also Holcombe analyzed the Bowen equilibrium level of education expenditure in over 200 Michigan school districts and found that actual expenditures were only about 3% away from the estimated district average. Limitations of the median voter theorem can be found in Krehbiel [18] and Buchanan and Tollisin [7]. A large problem for median voter theory is that government systems are composed of individuals who are self-interested. To continue to obtain their own set of goals, individuals focus on their re-election rather than their constituents' wishes.

We now present our main conclusion.

Theorem 6.3.9 *Let* $\widetilde{f}$ *be a fuzzy aggregation rule. Assume that* $F : [0,1]^2 \rightarrow [0,1]$ *is such that* $F(0,1) = 0$. *If* $\widetilde{f}$ *is a partial fuzzy simple rule, then for any* $\rho \in \mathcal{FS}, C_{F,\widetilde{f}}(\rho)(x) = \mu_{F,\widetilde{f}}(\rho,Q)(x)$ *for all* $x \in X$, *where* Q *is the strict order for which* ρ *is single-peaked.*

Proof. Let $\rho \in \mathcal{FS}$ and $z \in X$. Then

$$
\begin{aligned}
M_F(\widetilde{f}(\rho), 1_X)(z) &= \vee\{t \in [0,1] \mid F(\widetilde{f}(\rho)(z,y), 1_X(x)) \geq t \ \forall y \in Supp(1_X)\}. \\
&= \vee\{t \in [0,1] \mid F(\widetilde{f}(\rho)(z,y), 1) \geq t \ \forall y \in X\} \\
&= \wedge\{F(\widetilde{f}(z,y)(\rho), 1) \mid y \in X\}.
\end{aligned}
$$

Thus if z is an $\tilde{f}$-median, $C_{F,\tilde{f}}(\rho)(z) = \mu_{F,\tilde{f}}(\rho)(z)$. That is, $M_F(\tilde{f}(\rho), 1_X) = \mu_{F,\tilde{f}}(\rho)$ on $\text{Supp}(\mu_{F,\tilde{f}}(\rho))$. Hence it suffices to show $\text{Supp}(M_F(\tilde{f}(\rho), 1_X)) = \text{Supp}(\mu_{F,\tilde{f}}(\rho))$. Let $x \in \text{Supp}(M_F(\tilde{f}(\rho), 1_X))$. Then $F(\tilde{f}(\rho)(x, y), 1_X(x)) = F(\tilde{f}(\rho)(x, y), 1) > 0$ for all $y \in X$. Suppose $\mu_{F,\tilde{f}}(\rho) = 0$, i.e., $x \notin \text{Supp}(\mu_{F,\tilde{f}}(\rho))$. Let Q be a strict ordering of X with respect to which ρ is single-peaked. Since X is finite, we are free to label X so that $a_{t+1} Q a_t$ for all $t \geq 1$ (and $t \leq |X|$). Then $x = a_t$ for some $t \geq 1$ and x is not an $\tilde{f}$-median since $x \notin \text{Supp}(\mu_{F,\tilde{f}}(\rho))$. Thus either $\text{Supp}(\tilde{L}^-(x)) \in \{\text{Supp}(\lambda) \mid \lambda \in \mathcal{L}(\tilde{f})\}$ or $\text{Supp}(\tilde{L}^+(x)) \in \{\text{Supp}(\lambda) \mid \lambda \in \mathcal{L}(\tilde{f})\}$. Assume the former. Then $\text{Supp}(\tilde{L}^-(x)) = \text{Supp}(\lambda)$ for some $\lambda \in \mathcal{L}(\tilde{f})$. Hence $\forall u, v \in X$, $\pi_i(u, v) > 0 \ \forall i \in \tilde{L}^-(x)$ implies $\pi_i(u, v) > 0$. Since ρ is single-peaked, $\pi_i(a_{t-1}, a_t) > 0 \ \forall i \in \text{Supp}(\tilde{L}^-(x))$. Thus since $\mathcal{L}(\tilde{f})$ is decisive, $\tilde{f}(\rho)(a_{t-1}, a_t) > 0$, i.e., $\tilde{f}(\rho)(a_{t-1}, x) > 0$. But then $\tilde{f}(\rho)(x, a_{t-1}) = 0$ and so from the assumption that $F(0, 1) = 0$, we have $F(\tilde{f}(\rho)(x, a_{t-1}), 1) = 0$, a contradiction. It therefore follows that $x \in \text{Supp}(\mu_{F,\tilde{f}}(\rho))$. Hence

$$\text{Supp}(M_F(\tilde{f}(\rho), 1_X)) \subseteq \text{Supp}(\mu_{F,\tilde{f}}(\rho)).$$

Since $\text{Supp}(\mu_{F,\tilde{f}}(\rho)) \subseteq \text{Supp}(M_F(\tilde{f}(\rho), 1_X))$ we have that $\text{Supp}(M_F(\tilde{f}(\rho), 1_X)) = \text{Supp}(\mu_{F,\tilde{f}}(\rho))$. Hence $M_F(\tilde{f}(\rho), 1_X) = \mu_{F,\tilde{f}}(\rho)$. Now assume that $\text{Supp}(\tilde{L}^+(x)) \in \{\text{Supp}(\lambda) \mid \lambda \in \mathcal{L}(\tilde{f})\}$. Then $\text{Supp}(\tilde{L}^+(x)) = \text{Supp}(\lambda)$ for some $\lambda \in \mathcal{L}(\tilde{f})$. Thus $\forall u, v \in X$, $\pi_i(u, v) > 0 \forall i \in \tilde{L}^+(x)$ gives $\pi(u, v) > 0$. Since ρ is single-peaked, $\pi_i(a_{t+1}, a_t) > 0 \forall i \in \text{Supp}(\tilde{L}^+(x))$. Hence since $\mathcal{L}(\tilde{f})$ is decisive, $\tilde{f}(\rho)(a_{t+1}, a_t) > 0$, i.e., $\tilde{f}(\rho)(a_{t+1}, x) > 0$. But then as before, $F(\tilde{f}(\rho)(x, a_{t+1}) = 0$, a contradiction. Thus $x \in \text{Supp}(\mu_{F,\tilde{f}}(\rho))$. Hence

$$\text{Supp}(M_F(\tilde{f}(\rho), 1_X)) \subseteq \text{Supp}(\mu_{F,\tilde{f}}(\rho)).$$

Since $\text{Supp}(\mu_{F,\tilde{f}}(\rho)) \subseteq \text{Supp}(M_F(\tilde{f}(\rho), 1_X))$, we have $\text{Supp}(M_F(\tilde{f}(\rho), 1_X)) = \text{Supp}(\mu_{F,\tilde{f}}(\rho))$. Hence $M_F(\tilde{f}(\rho), 1_X) = \mu_{F,\tilde{f}}(\rho)$. $\blacksquare$

We see from Theorem 6.3.9 that a partial fuzzy simple rule results in a fuzzy core (fuzzy maximal set) in single-dimensional spatial models whenever the policy preferences of actors are single-peaked. Single dimensional fuzzy spatial models of decision making under majority, unanimity, and supermajority rules will result in stable predictions under assumptions of single-peakedness. Hence single-dimensional fuzzy spatial models are no less stable than the conventional single-dimensional spatial models.

Corollary 6.3.10 Let $F = \wedge$. Let $\tilde{f}$ be a fuzzy aggregation rule. If $\tilde{f}$ is a partial fuzzy simple rule, then $\forall \rho \in \mathcal{FS}, C_{\tilde{f}}(\rho) = \mu_{\tilde{f}}(\rho)$.

6.4 The Core in Fuzzy Spatial Models

We rely on the results in [23] for this section. In crisp spatial models, predictions concerning voting outcomes rely heavily on the existence of a core. Individuals choosing among a set of alternatives by majority rule will not be able to arrive at a stable choice in the absence of a core. Most voting rules will permit another option to defeat the previously chosen one. Such problems are particularly problematic to majority rule spatial models at dimensionalities greater than one. We feel that fuzzy spatial models might offer a partial solution to this problem. We present the notion of a fuzzy core. We find that a fuzzy core is more likely in two or more dimensions as the number of players increases.

We make use of fuzzy set theory due to the disconnect between predictions and empirical reality. We consider the basic rationale underlying the notion of a core. A core in conventional spatial models comprises the set of predictable outcomes. Without a highly improbable set of conditions or the imposition of severe restrictions on models, the core is empty under sincere voting whenever the dimensionality of models exceeds one, [20].

We present conditions under which the fuzzy core equals the fuzzy subset of $\widetilde{f}$-medians, Theorem 6.4.11, where $\widetilde{f}$ is a fuzzy aggregation rule. We also give conditions under which the fuzzy core is nontrivial, Theorem 6.4.32.

Recall that $\mathcal{FR}^*(X) = \mathcal{FP}^*(X \times X)$ and $\mathbb{N}$ denotes the set of positive integers. Let $*$ denote a t-norm on $[0,1]$. A fuzzy subset μ of a set S is said to have the **inf property** if for all subsets S' of S, there exists $x \in S'$ such that $\mu(x) = \wedge\{\mu(y) \mid y \in S'\}$. In this section, we define a class of fuzzy maximal subsets which contain these previously defined fuzzy maximal subsets. The strict ordering on X taken in the finite case, [20], is now taken to be the strict inequality $>$ of $\mathbb{R}$, the set of real numbers.

As previously mentioned, strict preferences are of type $\pi_{(0)}$.

Recall that a fuzzy weak order on X is a fuzzy binary relation ρ on X that is reflexive, complete, and (max-min)-transitive and that $\mathcal{FTR}$ denotes the set of all fuzzy weak orders on X.

Let $\rho \in \mathcal{FTR}^n$. $\forall x, y \in X$, recall that $R(x, y; \rho) = \{i \in N \mid \rho_i(x,y) \geq 0\}$ and $P(x, y; \rho) = \{i \in N \mid \pi_i(x,y) > 0\}$.

We use the following definition of a maximal fuzzy subset. Let $\mathcal{C}$ be a function from $X \times X$ into $[0,1]$. Define $M : \mathcal{FR}^*(X) \times \mathcal{FP}^*(X) \to \mathcal{FP}^*(X)$ by $\forall \rho \in \mathcal{FR}^*(X), \forall \mu \in \mathcal{FP}^*(X)$,

$$M(\rho, \mu)(x) = \mu(x) * (\wedge(\vee\{t \in [0,1] \mid \mathcal{C}(x,w) * t \leq \rho(x,w) \; \forall w \in \text{Supp}(\mu)\}))$$

for all $x \in X$.

Players will select alternatives with the highest set inclusion values with respect to $M(\rho, \mu)$. If $M(\rho, \mu) = \theta$, then no alternative will be selected.

If $\mathcal{C}(x, w) = \mu(w) * \rho(w, x) \ \forall x, w \in X$, then M is the maximal fuzzy subset defined in [11] and Chapter 1. Recall that we denote this maximal fuzzy subset by M_G. If $\mathcal{C}(x, w) = 1 \ \forall x, w \in X$, then M is the maximal fuzzy subset defined in [21] and Chapter 1. Recall that we denote this maximal fuzzy subset by M_M.

Proposition 6.4.1 *Let $\mathcal{C}$ be a function from $X \times X$ into $[0, 1]$. Let $\mu \in \mathcal{FP}^*(X)$ and $\rho \in \mathcal{FR}^*(X)$. Then $\wedge \{\vee \{t \in [0, 1] \mid \mathcal{C}(x, w) * t \le \rho(x, w)\} \mid w \in Supp(\mu)\} = \vee \{t \in [0, 1] \mid \mathcal{C}(x, w) * t \le \rho(x, w) \ \forall w \in Supp(\mu)\}$.*

Proof. We have that $\vee \{t \in [0, 1] \mid \mathcal{C}(x, w) * t \le \rho(x, w) \ \forall w \in \mathrm{Supp}(\mu)\} = t_0 \Leftrightarrow \mathcal{C}(x, w) * t_0 \le \rho(x, w) \ \forall w \in \mathrm{Supp}(\mu)$ and $\nexists t_1 \in [0, 1]$ such that $\mathcal{C}(x, w) * t_1 \le \rho(x, w) \ \forall w \in \mathrm{Supp}(\mu)$ and $t_1 > t_0$. Let $w \in \mathrm{Supp}(\mu)$ and let $t_w = \vee \{t \in [0, 1] \mid \mathcal{C}(x, w) * t \le \rho(x, w)\}$. Then

$$\wedge \{t_w \mid w \in \mathrm{Supp}(\mu)\} = \wedge \{\vee \{t \in [0, 1] \mid \mathcal{C}(x, w) * t \le \rho(x, w)\} \mid w \in \mathrm{Supp}(\mu)\}.$$

Let $t_1 = \wedge \{t_w \mid w \in \mathrm{Supp}(\mu)\}$. Then $\mathcal{C}(x, w) * t_1 \le \rho(x, w) \ \forall w \in \mathrm{Supp}(\mu)$ and so $t_1 \le t_0$. Since $t_w \ge t_0 \ \forall w \in \mathrm{Supp}(\mu), t_1 \ge t_0$. Thus $t_0 = t_1$. ∎

Let $x \in X$ and $\rho \in \mathcal{FR}^*(X)$. Let $S_x = \{w \in X \mid \mathcal{C}(x, w) > \rho(x, w)\}$. If $S_x \neq \emptyset$, let $t_{x,w} = \vee \{t \in [0, 1] \mid \mathcal{C}(x, w) * t \le \rho(x, w)\}, w \in S_x$.

Proposition 6.4.2 *Let $\rho \in \mathcal{FR}^*(X)$ and $x \in X$. Then*

$$M(\rho, 1_X)(x) = \begin{cases} 1 & \text{if } S_x = \emptyset, \\ \wedge \{t_{x,w} \mid w \in S_x\} & \text{if } S_x \neq \emptyset. \end{cases}$$

The fuzzy core has substantial utility in k-dimensional spatial models. Our domain of interest is the set of real numbers $\mathbb{R}$. We next develop the notion of fuzzy a core and connect it to a nonempty maximal fuzzy subset. We begin with the concept of single-peakedness. We then show that the maximal fuzzy subset of players whose preferences are single-peaked lies at the median. Thus the players will jointly prefer the median alternative to all others. We conclude the section by showing that the median preference under single-peakedness is not only the fuzzy maximal subset, but it is also the fuzzy core.

Definition 6.4.3 *Let $X \subseteq \mathbb{R}$, $x \in X$, and $\rho \in \mathcal{FTR}^n$. Then ρ is said to be **single-peaked** on X if $\forall i \in N$, $\exists x_i \in X$ such that*
    (1) $\forall y \in X \backslash \{x_i\}$, $\pi_i(x_i, y) > 0$;
    (2) $\forall y, z \in X$, $y < z < x_i$ implies $\pi_i(z, y) > 0$;
    (3) $\forall y, z \in X$, $x_i < z < y$ implies $\pi_i(z, y) > 0$.

Throughout this section, we assume $X \subseteq \mathbb{R}$ unless otherwise specified. Let $\mathcal{FS}$ denote the set of all single-peaked fuzzy preference profiles on X.

Definition 6.4.4 Let $\rho \in \mathcal{FS}$. Let x_i be that element of X such that $\pi_i(x_i, y) > 0 \; \forall y \in X\backslash\{x_i\}$, $i \in N$. Define $\tilde{L}^-$, $\tilde{L}^+ : X \to \mathcal{FP}(N)$ by $\forall z \in X$, $\forall i \in N$,

$$\tilde{L}^-(z)(i) = \begin{cases} \pi_i(x_i, z) & \text{if } x_i < z, \\ 0 & \text{otherwise}, \end{cases} \qquad \tilde{L}^+(z)(i) = \begin{cases} \pi_i(x_i, z) & \text{if } x_i > z, \\ 0 & \text{otherwise}. \end{cases}$$

$\tilde{L}^-(z)(i)$ represents the degree to which player i prefers a position to the left of alternative z and $\tilde{L}^+(z)(i)$ represents the degree to which player i prefers a position to the right of alternative z.

Definition 6.4.5 Let $\tilde{f}$ be a fuzzy preference aggregation rule. Let $\rho \in \mathcal{FS}$. Then an element z of X is called an $\tilde{f}$-**median** if

$$Supp(\tilde{L}^-(z)) \notin \{Supp(\lambda) \mid \lambda \in \mathcal{L}(\tilde{f})\} \quad and$$
$$Supp(\tilde{L}^+(z)) \notin \{Supp(\lambda) \mid \lambda \in \mathcal{L}(\tilde{f})\}.$$

An $\tilde{f}$-median passes through any alternative for which there is no plausible alternative either to its right or left that is even minimally preferred by some majority coalition.

We now develop the concept of a fuzzy core and connect it to a nonempty fuzzy maximal subset.

Definition 6.4.6 Let $\tilde{f}$ be a fuzzy aggregation rule. Define $C_{\tilde{f}} : \mathcal{FTR}^n \to \mathcal{FP}(X)$ by $\forall \rho \in \mathcal{FTR}^n$, $C_{\tilde{f}}(\rho) = M(\tilde{f}(\rho), 1_X)$. Then $C_{\tilde{f}}$ is called the **fuzzy core** of ρ.

In previous chapters, the set of alternatives X did not possess any structure such as an ordering of its elements. With single peaked preferences this is no longer true. The maximal outcomes are located with respect to the ordering Q yielding preferences single-peaked. Thus the core of a preference aggregation rule $\tilde{f}$ at a profile ρ is the set of maximal elements in X under the binary relation $\tilde{f}(\rho)$.

Definition 6.4.7 Let $\tilde{f}$ be a fuzzy aggregation rule. Define $\mu_{\tilde{f}} : \mathcal{FTR}^n \to \mathcal{FP}(X)$ by $\forall \rho \in \mathcal{FTR}^n$, $\forall z \in X$,

$$\mu_{\tilde{f}}(\rho)(z) = \begin{cases} 1 & \text{if } S_z = \emptyset \text{ and } z \text{ is an } \tilde{f}\text{-median}, \\ \wedge\{t_{z,w} \mid w \in S_z\} & \text{if } S_z \neq \emptyset \text{ and } z \text{ is an } \tilde{f}\text{- median}, \\ 0 & \text{otherwise}. \end{cases}$$

If $\mathcal{C}(x, w) = 1 \; \forall x, w \in X$, then $\mu_{\tilde{f}}$ becomes

$$\mu_{\tilde{f}}(\rho)(z) = \begin{cases} \wedge\{\tilde{f}(\rho)(z, y) \mid y \in X\} & \text{if } z \text{ is an } \tilde{f}\text{-median}, \\ 0 & \text{otherwise}. \end{cases}$$

The following examples illustrate these concepts.

Example 6.4.8 Let $C(x, w) = 1 \; \forall x, w \in X$, where $X = [0, 1]$. Let $N = \{1, 2, 3\}$. Let $\tilde{f}$ be a partial fuzzy simple rule, i.e., $\tilde{f}(\rho)(x) > 0 \Leftrightarrow \tilde{f}_{\mathcal{L}(\tilde{f})}(\rho)(x) > 0$. Let $\mathcal{L}(\tilde{f}) = \{\{1, 2\}, \{1, 3\}, \{2, 3\}, N\}$. Let $x_i \in X$ be player i's preferred alternative, $i = 1, 2, 3$, where $x_1 < x_2 < x_3$. For all $i \in N$, define $\rho_i \in \mathcal{FR}^n$ by $\forall y, z \in X$,

$$\rho_i(z, y) = \begin{cases} 1 & \text{if } |z - x_i| = |y - x_i|, \\ 1 - |z - x_i| & \text{if } |z - x_i| < |y - x_i|, \\ 0 & \text{if } |z - x_i| > |y - x_i|. \end{cases}$$

It follows that ρ_i is reflexive and complete, $i = 1, 2, 3$. We next show ρ_i is max-min transitive $\forall i \in N$. Let $u, v, w \in X$ and $i \in N$. If either $\rho_i(u, v) = 0$ or $\rho_i(v, w) = 0$, then clearly $\rho_i(u, v) \wedge \rho_i(v, w) \leq \rho_i(u, w)$. Suppose $\rho_i(u, v) > 0$ and $\rho_i(v, w) > 0$. Then $|u - x_i| \leq |v - x_i|$ and $|v - x_i| \leq |w - x_i|$. Thus $|u - x_i| \leq |w - x_i|$. Hence it follows that $\rho_i(u, v) \wedge \rho_i(v, w) \leq \rho_i(u, w)$.

Now

$$\begin{aligned}
Supp(\tilde{L}^-(x)) &= \emptyset \text{ if } x \leq x_1, \\
Supp(\tilde{L}^-(x)) &= \{1\} \text{ if } x_1 < x \leq x_2, \\
Supp(\tilde{L}^-(x)) &= \{1, 2\} \text{ if } x_2 < x \leq x_3, \\
Supp(\tilde{L}^-(x)) &= \{1, 2, 3\} \text{ if } x_3 < x, \\
Supp(\tilde{L}^+(x)) &= \{1, 2, 3\} \text{ if } x < x_1, \\
Supp(\tilde{L}^+(x)) &= \{2, 3\} \text{ if } x_1 \leq x < x_2, \\
Supp(\tilde{L}^+(x)) &= \{3\} \text{ if } x_2 \leq x < x_3, \\
Supp(\tilde{L}^+(x)) &= \emptyset \text{ if } x_3 \leq x.
\end{aligned}$$

Also

$$\begin{aligned}
Supp(\tilde{L}^-(x_1)) &= \emptyset, \;\; Supp(\tilde{L}^+(x_1)) = \{2, 3\}, \\
Supp(\tilde{L}^-(x_2)) &= \{1\}, Supp(\tilde{L}^+(x_2)) = \{3\}, \\
Supp(\tilde{L}^-(x_3)) &= \{1, 2\}, Supp(\tilde{L}^+(x_3)) = \emptyset.
\end{aligned}$$

Thus $\{x_2\}$ is the set of all $\tilde{f}$-medians.

Example 6.4.9 Let $X = [0, 1]$ and $x_2 \in (0, 1)$. Let $\tilde{f}$ be a fuzzy aggregation rule such that $\forall \rho \in \mathcal{FR}^n$ and $\forall x, y \in X$,

$$\tilde{f}(\rho)(x, y) = \begin{cases} 1 & \text{if } x = y, \\ \frac{1}{x_2} y & \text{if } x = x_2 \text{ and } 0 < y \leq x_2, \\ \frac{1}{x_2 - 1}(y - 1) & \text{if } x = x_2 \text{ and } 1 > y \geq x_2, \\ 0 & \text{if } x \neq x_2 \text{ and } y = x_2, \\ .5 & \text{if } x_2 \neq x \neq y \neq x_2 \text{ or } y = 0 \text{ or } y = 1. \end{cases}$$

Then $\tilde{f}(\rho)$ is reflexive and complete. (Note that if we let $\tilde{f}(\rho)(x_2, 0) = 0$ or $\tilde{f}(\rho)(x_2, 1) = 0$, then $\tilde{f}(\rho)$ is not complete since $\tilde{f}(\rho)(0, x_2) = 0$ and $\tilde{f}(\rho)(1, x_2) = 0$.) However, $\wedge\{\tilde{f}(\rho)(x, y) \mid y \in X\} = 0$ for all $x \in X$. Thus $\mu_{\tilde{f}}(\rho)(x) = 0$ for all $x \in X$, where $C(x, w) = 1$ for all $x, w \in X$.

Definition 6.4.10 *Let* $\rho \in \mathcal{FR}(X)$. *Then* $\mathcal{C}$ *and* ρ *are called* **compatible** *if* $\forall x \in X$ *there does not exist* $w \in X$ *such that* $\mathcal{C}(x, w) = \rho(x, w) = 0$.

If $\mathcal{C}(x, w) = 1 \;\forall x, w \in X$, then $\mathcal{C}$ and any ρ are compatible. If $\mathcal{C}(x, w) = \rho(w, x) \;\forall w, x \in X$ and ρ is complete, then $\mathcal{C}$ and ρ are compatible. Hence the next result holds for M_M and M_G.

The following theorem is the main result of this section.

Theorem 6.4.11 *Let* $\widetilde{f}$ *be a partial fuzzy simple rule and* $X \subseteq \mathbb{R}$. *Then* $\forall \rho \in \mathcal{FS}$, *if* $\widetilde{f}(\rho)(z, \_)$ *is bounded away from 0 for each* $\widetilde{f}$-*median* z *and* $\mathcal{C}$ *and* $\widetilde{f}(\rho)$ *are compatible, then* $C_{\widetilde{f}}(\rho) \neq \theta$ *and* $C_{\widetilde{f}}(\rho) = \mu_{\widetilde{f}}(\widetilde{f}(\rho))$.

Proof. Let $\rho \in \mathcal{FS}$ and $z \in X$. Then by Proposition 6.4.2,

$$M(\widetilde{f}(\rho), 1_X)(z) = \begin{cases} 1 \text{ if } S_z = \emptyset, \\ \wedge\{t_{z,w} \mid w \in S_z\} \text{ if } S_z \neq \emptyset. \end{cases}$$

Hence if z is an $\widetilde{f}$-median, $C_{\widetilde{f}}(\rho)(z) = \mu_{\widetilde{f}}(\rho)(z)$. Thus it suffices to show that $\text{Supp}(M(\widetilde{f}(\rho), 1_X)) = \text{Supp}(\mu_{\widetilde{f}}(\rho)$. Let $x \in \text{Supp}(M(\widetilde{f}(\rho), 1_X))$. Then $M(\widetilde{f}(\rho), 1_X)(x) > 0$ and so $\vee\{t \in [0, 1] \mid \mathcal{C}(x, w) * t \leq \widetilde{f}(\rho)(x, w) \;\forall w \in X\} > 0$. Since $\mathcal{C}$ and $\widetilde{f}(\rho)$ are compatible, $\widetilde{f}(\rho)(x, y) > 0 \;\forall y \in X$. Suppose $\mu_{\widetilde{f}}(\rho)(x) = 0$, i.e., $x \notin \text{Supp}(\mu_{\widetilde{f}}(\rho))$. Then x is not an $\widetilde{f}$-median. Hence either $\text{Supp}(\widetilde{L}^-(x)) \in \{\text{Supp}(\lambda) \mid \lambda \in \mathcal{L}(\widetilde{f})\}$ or $\text{Supp}(\widetilde{L}^+(x)) \in \{\text{Supp}(\lambda) \mid \lambda \in \mathcal{L}(\widetilde{f})\}$. Assume the former. Then $\text{Supp}(\widetilde{L}^-(x)) = \text{Supp}(\lambda)$ for some $\lambda \in \mathcal{L}(\widetilde{f})$. Thus $\forall u, v \in X$, $\pi_i(u, v) > 0 \;\forall i \in \text{Supp}(\widetilde{L}^-(x))$ implies $\pi(u, v) > 0$. Since ρ is single-peaked, $\pi_i(y, x) > 0 \;\forall i \in \text{Supp}(\widetilde{L}^-(x))$ for $x_i < y < x$. Hence since $\mathcal{L}(\widetilde{f})$ is decisive, $\widetilde{f}(\rho)(y, x) > 0$ for $x_i < y < x$. But then $\widetilde{f}(\rho)(x, y) = 0$ which contradicts the fact that $\widetilde{f}(\rho)(x, y) > 0 \;\forall y \in X$. It thus follows that $x \in \text{Supp}(\mu_{\widetilde{f}}(\rho))$. Hence $\text{Supp}(M(\widetilde{f}(\rho), 1_X)) \subseteq \text{Supp}(\mu_{\widetilde{f}}(\rho))$. Now assume the latter. Then $\text{Supp}(\widetilde{L}^+(x)) = \text{Supp}(\lambda)$ for some $\lambda \in \mathcal{L}(\widetilde{f})$. Since ρ is single-peaked, $\pi_i(y, x) > 0 \;\forall i \in \text{Supp}(\widetilde{L}^-(x))$ for $x_i > y > x$. Hence since $\mathcal{L}(\widetilde{f})$ is decisive, $\widetilde{f}(\rho)(y, x) > 0$ for $x_i > y > x$. But then $\widetilde{f}(\rho)(x, y) = 0$ which contradicts the fact that $\widetilde{f}(\rho)(x, y) > 0 \;\forall y \in X$. It thus follows that $x \in \text{Supp}(\mu_{\widetilde{f}}(\rho))$. Hence $\text{Supp}(M(\widetilde{f}(\rho), 1_X)) \subseteq \text{Supp}(\mu_{\widetilde{f}}(\rho))$. Since clearly $\text{Supp}(\mu_{\widetilde{f}}(\rho) \subseteq \text{Supp}(M(\widetilde{f}(\rho), 1_X))$, $\text{Supp}(M(\widetilde{f}(\rho), 1_X)) = \text{Supp}(\mu_{\widetilde{f}}(\rho))$. Since $\widetilde{f}(\rho)(z, \_)$ is bounded away from 0, $C_{\widetilde{f}}(\rho) \neq \theta$. $\blacksquare$

The previous theorem states that any alternative at the median position for players with single peaked preference profiles is an element in the fuzzy maximal subset and constitutes the fuzzy core. An element in the fuzzy core will be chosen by players by any partial fuzzy simple rule.

Corollary 6.4.12 *Let $\widetilde{f}$ be a partial fuzzy simple rule and $X \subseteq R$. Then $\forall \rho \in \mathcal{FS}$, if $\widetilde{f}(\rho)(z, \_)$ is bounded away from 0 for some $\widetilde{f}$-median z and C and $\widetilde{f}(\rho)$ are compatible, then $C_{\widetilde{f}}(\rho) \neq \theta$.*

We now consider the implications of the core for k-dimensional spatial models, $\mathbb{R}^k$. These models are widely used in political science. We use the usual topology in the Euclidean space $\mathbb{R}^k$. Suppose X is a subset of $\mathbb{R}^k$. A subset Z of X is called open relative to X (or simply open) if $Z = Y \cap X$ for some subset Y that is open in $\mathbb{R}^k$.

We assume throughout that ρ is reflexive and a complete fuzzy binary relation on X.

Definition 6.4.13 *Let ρ be a fuzzy binary relation on X and let $x \in X$. Define $\rho^{-1}(x) = \{y \in X \mid \pi(x, y) > 0\}$ and $\rho(x) = \{y \in X \mid \pi(y, x) > 0\}$.*

Definition 6.4.14 *Let ρ be a fuzzy binary relation on X. Then ρ is called **lower continuous** at $x \in X$ if $\rho^{-1}(x)$ is open and ρ is called **upper continuous** if $\rho(x)$ is open for all $x \in X$.*

Definition 6.4.15 *Let ρ be a fuzzy binary relation on X. The ρ is said to satisfy **condition** F if $\forall$ finite subsets Y of $X, \exists x \in X$ such that $\rho(x, y) > 0$ $\forall y \in Y$.*

Note that condition F does not require that x be a member of Y. It follows that if ρ is acyclic on X, then condition F holds.

Lemma 6.4.16 *Suppose X is compact and ρ is lower continuous. Suppose that C and ρ are compatible. Suppose ρ satisfies condition F and that $M(Supp(\rho), X) \neq \emptyset$, implies there exists $x \in M(Supp(\rho), X)$ such that $\rho(x, \_)$ is bounded away from 0. Then $M(\rho, 1_X) \neq \theta$. Conversely if $M(\rho, 1_X) \neq \theta$, then ρ satisfies condition F and if $M(Supp(\rho), X) \neq \emptyset$, there exists $x \in M(Supp(\rho), X)$ such that $\rho(x, w) > 0 \forall w \in X$.*

Proof. Suppose ρ satisfies condition F and $M(Supp(\rho), X) \neq \emptyset$ implies $\exists x \in M(Supp(\rho), X)$ such that $\rho(x, \_)$ is bounded away from 0. Then for all finite subsets Y of X, there exists $x \in X$ such that $(x, y) \in Supp(\rho)$ $\forall y \in Y$. Thus $Supp(\rho)$ satisfies condition F, [2]. Since ρ is lower continuous, $\{y \in X \mid (x, y) \in Supp(\pi)\} = \{y \in X \mid \pi(x, y) > 0\}$ is open for all $x \in X$. Thus $Supp(\rho)$ is lower continuous since if $R = Supp(\rho)$, then $P = Supp(\pi)$. Hence $M(Supp(\rho), X) = \{x \in X \mid (x, y) \in Supp(\rho) \ \forall y \in X\} \neq \emptyset$ by [2, Lemma 5.1, p. 124]. Let $x \in M(Supp(\rho), X)$ be such that $\rho(x, \_)$ is bounded away from 0. Then $\{t \in (0, 1] \mid (x, y) \in \rho_t \ \forall y \in X\}$ is not empty. Thus there exists

$t_0 \in (0,1]$ such that $\rho(x,y) \geq t_0 \ \forall y \in X$. Hence $C(x,y) * t_0 \leq \rho(x,y) \ \forall y \in X$. Thus $M(\rho,1_X)(x) > 0$. Hence $M(\rho,1_X) \neq \theta$.

Conversely, suppose $M(\rho,1_X) \neq \theta$. Then there exists $x \in X$ such that $M(\rho,1_X)(x) = \vee \{t \in [0,1] \mid C(x,w) * t \leq \rho(x,w) \ \forall w \in X\} > 0$. Since C and ρ are compatible, $\rho(x,w) > 0 \ \forall w \in X$. It follows that $x \in M(\text{Supp}(\rho), X)$. Let Y be a finite subset of X. Then $\vee \{t \in [0,1] \mid (x,y) \in \rho_t \ \forall y \in Y\} > 0$. Hence there exists $x \in X$ such that $\rho(x,y) > 0 \ \forall y \in Y$. That is, ρ satisfies condition F. ∎

Corollary 6.4.17 *Suppose X is compact and ρ is lower continuous. Suppose C and ρ are compatible. Then $\text{Supp}(M_M(\rho,1_X)) = M(\text{Supp}(\rho), X) \neq \emptyset$ if and only if ρ satisfies condition F and for all $x \in M(\text{Supp}(\rho), X)$, $\rho(x,\_)$ is bounded away from 0.*

Example 6.4.18 *Let ρ be the fuzzy binary relation on $X = [0,1]$ defined as follows: $\forall (x,y) \in X \times X$,*

$$\rho(x,y) = \begin{cases} x & \text{if } x > y, \\ 1 & \text{if } x = y = 1, \\ 0 & \text{otherwise.} \end{cases}$$

Let $x \in (0,1)$. Then $\rho^{-1}(x) = \{y \in X \mid \pi(x,y) > 0\} = [0,x)$ which is open in X. Also $\rho^{-1}(0) = \emptyset$ which is open and $\rho^{-1}(1) = [0,1]$ which is open in X. Thus ρ is lower continuous. Let Y be a finite subset of X. If $1 \notin Y$, then $\exists x \in X$ such that $\pi(x,y) > 0 \ \forall y \in Y$. Suppose $1 \in Y$. Then $\pi(1,y) > 0 \ \forall y \in Y$. Thus ρ satisfies condition F. Now $\text{Supp}(\rho) = \{(1,1)\} \cup (\{(x,y) \in X \times X \mid x > y\}$. There does not exist $t \in [0,1]$ such that $\text{Supp}(\rho) = \rho_t$ since $\rho(x,y)$ approaches 0 as x approaches 0. Note however that $M_M(\rho,1_X)(1) = \vee \{t \in [0,1] \mid (1,y) \in \rho_t \ \forall y \in X\} = 1$ and $M_M(\rho,1_X)(x) = 0 \ \forall x \in [0,1)$ since it is not the case $(x,y) \in \rho_t \ \forall y \in X$; only for those $y < x$ does $(x,y) \in \rho_t$. Also, $M_G(\rho,1_X)(1) = \vee \{t \in [0,1] \mid \rho(y,1) * t \leq \rho(1,y) \ \forall y \in X\} = 1$ and $M_G(\rho,1_X)(x) = \vee \{t \in [0,1] \mid \rho(y,x) * t \leq \rho(x,y) \ \forall y \in X\} = 0 \ \forall x \in [0,1)$ since $\rho(x,y) = 0$ if $y > x$. Let $x \in X$. Let $t \in [0,1]$. Then

$$\rho_t^{-1}(x) = \begin{cases} \emptyset & \text{if } x < t, \\ [0,x) & \text{if } x \in [t,1), t < 1, \\ [0,1] & \text{if } t = 1 \text{ and } x = 1. \end{cases}$$

Since $\emptyset$, $[0,x)$, and $[0,1)$ are open in X, ρ_t is lower continuous $\forall t \in [0,1)$. Let $t \in [0,1]$. Then $(1,y) \in \rho_t \ \forall t \in [0,1]$ and $\forall y \in X$. Hence $1 \in M(\rho_t, X)$ and so $M(\rho_t, X) \neq \emptyset \ \forall t \in [0,1]$. In fact, $M(\rho_t, X) = \{1\} \ \forall t \in [0,1]$.

Definition 6.4.19 *Let $X \subseteq \mathbb{R}^k$ be a convex set. Let ρ be a fuzzy preference ordering on X. Then ρ is said to be **strictly convex** if $\forall x, y \in X$, $\rho(x,y) > 0$ and $x \neq y$ imply $\pi(ax + (1-a)y, y) > 0 \ \forall a \in (0,1)$. If $\rho \in \mathbb{R}^n$, then ρ is called **strictly convex** if ρ_i is strictly convex for all $i \in N$.*

Definition 6.4.20 *Let $Y \subseteq X$. The **convex hull** of Y, written $Con(Y)$, is defined to be the set*

$$Con(Y) = \left\{ \sum_{i=1}^{m} a_i y_i \mid m \in \mathbb{N}, \; a_i \in [0,1], \; y_i \in Y, i = 1, \ldots, m, \; \sum_{i=1}^{m} a_i = 1 \right\}.$$

Definition 6.4.21 *Let ρ be a fuzzy binary relation on X. Then ρ is called **semi-convex** if $\forall x \in X, x \notin Con(\rho(x))$.*

KKM [6] Let $A = \{a_0, a_1, \ldots, a_m\} \subseteq \mathcal{R}^k$ and let $\{S_0, S_1, \ldots, S_m\}$ be a collection of closed sets. Let $M = \{0, 1, \ldots, m\}$. If $\forall L \subseteq M$, $Con(\{a_i \mid i \in L\}) \subseteq \cup_{i \in L} S_i$. Then $\cap_{j \in L} S_j \neq \emptyset$.

Lemma 6.4.22 *Suppose X is convex. Let ρ be a fuzzy binary relation on X. If ρ is lower continuous and semi-convex on X, then ρ satisfies condition F.*

Proof. Let $Y = \{y_0, y_1, \ldots, y_m\}$ be a finite subset of X. Let $M = \{0, 1, \ldots, m\}$. Since $\rho^{-1}(x)$ is open $\forall x \in X$, $X \backslash \rho^{-1}(x)$ is closed $\forall x \in X$. Suppose there exists $L \subseteq M$ such that

$$Con(\{y_i \mid i \in L\}) \nsubseteq \cup_{i \in L}(X \backslash \rho^{-1}(y_i)) \cap Con(Y).$$

Then there exists $z \in Con(\{y_i \mid i \in L\})$ such that $z \notin \cup_{i \in L}(X \backslash \rho^{-1}(y_i)) \cap Con(Y)$. Hence $z \notin \cup_{i \in L}(X \backslash \rho^{-1}(y_i))$ since $z \in Con(Y)$. Thus $z \notin X \backslash \rho^{-1}(y_i) \; \forall i \in L$. Hence $z \in \rho^{-1}(y_i) \; \forall i \in L$ and so $y_i \in \rho(z)$ $\forall i \in L$. Thus $z \in Con(\{y_i \mid i \in L\}) \subseteq Con(\rho(z))$. However, this contradicts the hypothesis that ρ is semi-convex. Thus

$$Con(\{y_i \mid i \in L\}) \subseteq \cup_{i \in L}(X \backslash \rho^{-1}(y_i)) \cap Con(Y).$$

Hence by KKM, $\cap_{i \in L}(X \backslash \rho^{-1}(y_i)) \cap Con(Y) \neq \emptyset$. Let $x \in \cap_{i \in M}(X \backslash \rho^{-1}(y_i)) \cap Con(Y)$. Then $x \in X \backslash \rho^{-1}(y_i) \forall i \in M$ and so $x \notin \rho^{-1}(y_i) \; \forall i \in M$. That is, $\forall i \in M$, it is not the case that $\rho(y_i, x) > 0$ and $\rho(x, y_i) = 0$. Hence $\forall i \in M$, either $\rho(y_i, x) \geq 0$ and $\rho(x, y_i) > 0$ or $\rho(y_i, x) = 0$ and $\rho(x, y_i) = 0$. In the former case, we have that ρ satisfies condition F. The latter case is impossible since ρ is complete. ∎

We now consider the characteristics of collective preferences over a set of alternatives X when a maximal fuzzy subset is nonzero.

Theorem 6.4.23 *Suppose X is compact and convex. Let ρ be a fuzzy binary relation on X. Suppose C and ρ are compatible. If ρ is lower continuous and semi-convex and $M(Supp(\rho), X) \neq \emptyset$ implies there exists $x \in M(Supp(\rho), X)$ such that $\rho(x, \_)$ is bounded away from 0, then $M(\rho, 1_X) \neq \theta$.*

Proof. The result follows by Lemmas 6.4.16 and 6.4.22. ∎

We next consider the degree to which a fuzzy core is likely to be nonempty as the dimensionality of spatial models increases.

Let $\rho \in \mathcal{FR}^n$. Recall that $\forall x, y \in X, P(x, y; \rho) = \{i \in N \mid \pi_i(x, y) > 0\}$.

Suppose $X \subseteq \mathbb{R}^k$. Then k is the dimension of the policy space. Assume in what follows that X is convex and compact.

Theorem 6.4.24 *Let $\widetilde{f}$ be a partial fuzzy voting rule. If $\rho \in \mathcal{FTR}^n$ is strictly convex and $\widetilde{f}(\rho)(x, \_)$ satisfies the inf property$\forall x \in X$ and $\widetilde{f}_{\mathcal{L}(\widetilde{f})}(\rho)(x, \_)$ is bounded away from 0 $\forall x \in X$ and $\mathcal{C}$ and $\widetilde{f}(\rho)$ are compatible, then $Supp(C_{\widetilde{f}}(\rho)) = Supp(C_{\widetilde{f}_{\mathcal{L}(\widetilde{f})}}(\rho))$.*

Proof. Let $x \in Supp(C_{\widetilde{f}}(\rho))$. Since $\mathcal{C}$ and $\widetilde{f}(\rho)$ are compatible, $\widetilde{f}(\rho)(x, y) > 0 \,\forall y \in X$. Hence $\forall y \in X$, either $\pi(x, y) > 0$ or $(\widetilde{f}(\rho)(x, y) > 0$ and $\widetilde{f}(\rho)(y, x) > 0)$. Thus $\forall y \in X$, either $(x, y) \in \widetilde{\mathcal{P}}(\rho)$ or $(x, y) \notin Symm\,(\widetilde{\mathcal{P}}(\rho))$. Hence $\widetilde{f}_{\mathcal{L}(\widetilde{f})}(\rho)(x, y) > 0 \,\forall y \in X$. Since $\widetilde{f}_{\mathcal{L}(\widetilde{f})}(\rho)(x, \_)$ is bounded away from 0, $C_{\widetilde{f}_{\mathcal{L}(\widetilde{f})}}(\rho)(x) > 0$. Thus $Supp(C_{\widetilde{f}}(\rho)) \subseteq Supp(C_{\widetilde{f}_{\mathcal{L}(\widetilde{f})}}(\rho))$. Suppose $x \notin Supp(C_{\widetilde{f}}(\rho))$. Then $M(\widetilde{f}(\rho), 1_X)(x) = 0$ and so $\wedge\{\widetilde{f}(\rho)(x, y) \mid y \in X\} = 0$. Since $\widetilde{f}(\rho)(x, \_)$ has the inf property, there exists $y \in X$ such that $\widetilde{f}(\rho)(x, y) = 0$. Thus $\pi_{\widetilde{f}}(y, x) > 0$ since $\widetilde{f}(\rho)$ is complete. Since $\widetilde{f}$ is a partial fuzzy voting rule, $\widetilde{f}$ is monotonic and neutral by Theorem 3.14, [24]. Hence $(P(y, x; \rho), R(y, x; \rho)) \in \mathcal{D}(\widetilde{f})$. Since ρ is strictly convex, we have $\forall i \in R(y, x; \rho)$ that $\pi_i(z, x) > 0$, where $z = ax + (1 - a)y$ for some $a \in (0, 1)$. Since $R(y, x; \rho)$ is a decisive coalition, i.e., $R(y, x; \rho) \in \mathcal{L}(\widetilde{f}), \pi_{\widetilde{f}_{\mathcal{L}(\widetilde{f})}}(z, x) > 0$ and so $\widetilde{f}_{\mathcal{L}(\widetilde{f})}(\rho)(x, z) = 0$. Thus $x \notin Supp(C_{\widetilde{f}_{\mathcal{L}(\widetilde{f})}}(\rho))$. ∎

For the crisp case, Theorem 6.4.24 corresponds to [2, Theorem 4.3, p. 100] for voting rules. With single-peaked preferences, the core outcomes of f and $f_{\mathcal{L}(f)}$ are the same.

Let $\mathcal{L} \subseteq \mathcal{FP}(N)$. Let $\mathcal{L}_{Supp} = \{Supp(\lambda) \mid \lambda \in \mathcal{L}\}$. Let $\mathcal{L}_{Supp}(\widetilde{f}) = \{Supp(\lambda) \mid \lambda \in \mathcal{L}(\widetilde{f})\}$. Define $K(\mathcal{L}) = \cap_{\lambda \in \mathcal{L}} Supp(\lambda)$.

Definition 6.4.25 *Let $\widetilde{f}$ be a fuzzy aggregation rule. If $K(\mathcal{L}(\widetilde{f})) \neq \emptyset$, then $\widetilde{f}$ is called* **collegial**.

Definition 6.4.26 *Let $\widetilde{f}$ be a fuzzy aggregation rule. The* **Nakamura number** *associated with a simple rule $\widetilde{f}$, written $s(\widetilde{f})$, is defined as follows:*

$$s(\widetilde{f}) = \begin{cases} \infty & \text{if } \widetilde{f} \text{ is collegial,} \\ min\{|\mathcal{L}_{Supp}| \mid \mathcal{L} \subseteq \mathcal{L}(\widetilde{f}) \text{ and } K(\mathcal{L}) = \emptyset\} & \text{otherwise.} \end{cases}$$

Definition 6.4.27 *Let $\mathcal{L} \subseteq \mathcal{FP}(N)$. Then $\mathcal{L}$ is called*
  (1) **monotonic** *if $\forall \lambda \in \mathcal{L}$, $\forall \lambda' \in \mathcal{FP}(N)$, $Supp(\lambda') \supseteq Supp(\lambda)$ implies $\lambda' \in \mathcal{L}$;*
  (2) **proper** *if $\forall \lambda \in \mathcal{L}$, $\forall \lambda' \in \mathcal{FP}(N)$, $N \backslash Supp(\lambda) = Supp(\lambda')$ implies $\lambda' \notin \mathcal{L}$.*

Lemma 6.4.28 *Let $\widetilde{f}$ be a fuzzy aggregation rule. Then $\mathcal{L}(\widetilde{f})$ is monotonic and proper.*

  Proof. Let $\lambda \in \mathcal{L}(\widetilde{f})$ and $\lambda' \in \mathcal{FP}(N)$. Suppose $Supp(\lambda') \supseteq Supp(\lambda)$. Let $x, y \in X$. If $\pi_i(x, y) > 0$ $\forall i \in Supp(\lambda')$, then $\pi_i(x, y) > 0$ $\forall i \in Supp(\lambda)$. Hence $\pi(x, y) > 0$ and so $\lambda' \in \mathcal{L}(\widetilde{f})$. Thus $\mathcal{L}(\widetilde{f})$ is monotonic.
  Now suppose $Supp(\lambda') = N \backslash Supp(\lambda)$. There exists $\rho \in \mathcal{R}^n$ such that $\pi_i(x, y) > 0$ $\forall i \in Supp(\lambda)$ and $\pi_i(y, x) > 0$ $\forall i \in Supp(\lambda')$. Thus if $\lambda' \in \mathcal{L}(\widetilde{f})$, then $\pi(y, x) > 0$. However, this is impossible since $\lambda \in \mathcal{L}(\widetilde{f})$ and so $\pi(x, y) > 0$. Thus $\lambda' \notin \mathcal{L}(\widetilde{f})$. ∎

Lemma 6.4.29 *Suppose $\widetilde{f}$ is a partial fuzzy simple rule. Then $s(\widetilde{f}) \geq 3$. If $\widetilde{f}$ is noncollegial, then $s(\widetilde{f}) \leq n$.*

  Proof. Suppose $|\mathcal{L}_{\text{Supp}}(\widetilde{f})| = 1$. Then $\widetilde{f}$ is collegial and so $s(\widetilde{f}) = \infty$. Suppose $|\mathcal{L}_{\text{Supp}}(\widetilde{f})| \geq 2$. Let $\lambda_1, \lambda_2 \in \mathcal{L}(\widetilde{f})$. By Lemma 6.4.28, $\mathcal{L}(\widetilde{f})$ is proper. Hence $Supp(\lambda_1) \cap Supp(\lambda_2) \neq \emptyset$. Thus $K(\mathcal{L}) \neq \emptyset$ if $|\mathcal{L}_{\text{Supp}}| = 2$ and $\mathcal{L} \subseteq \mathcal{L}(\widetilde{f})$. Hence $s(\widetilde{f}) \geq 3$.
  Suppose $\widetilde{f}$ is noncollegial. Then $\forall i \in N$, there exists $\lambda \in \mathcal{L}(\widetilde{f})$ such that $i \notin Supp(\lambda)$. Since $\mathcal{L}(\widetilde{f})$ is monotonic, it follows $\forall i \in N$ that $1_{N \backslash \{i\}} \in \mathcal{L}(\widetilde{f})$. Let $\mathcal{L} = \{1_{N \backslash \{i\}} \mid i \in N\}$. Then $\mathcal{L} \subseteq \mathcal{L}(\widetilde{f})$, $K(\mathcal{L}) = \emptyset$ and $|\mathcal{L}_{\text{Supp}}| = n$. Thus $s(\widetilde{f}) \leq n$. ∎

  Let $i \in N$ and $x \in X$. Define $P_i(x) = \{y \in X \mid \pi_i(y, x) > 0\}$ and $P_i^{-1}(x) = \{y \in X \mid \pi_i(x, y) > 0\}$. Let $\lambda \in \mathcal{FP}(N)$ be a coalition. Define $P_\lambda(x) = \cap_{i \in Supp(\lambda)} P_i(x)$ and $P_\lambda^{-1}(x) = \cap_{i \in Supp(\lambda)} P_i^{-1}(x)$. Let $\widetilde{f}$ be a fuzzy aggregation rule. Define $P_{\widetilde{f}}(x) = \cup_{\lambda \in \mathcal{L}(\widetilde{f})} P_\lambda(x)$ and $P_{\widetilde{f}}^{-1}(x) = \cup_{\lambda \in \mathcal{L}(\widetilde{f})} P_\lambda^{-1}(x)$.

Lemma 6.4.30 *Suppose $\rho \in \mathcal{R}^n$ is lower continuous. Then for all partial fuzzy simple rules $\widetilde{f}$, $\widetilde{f}(\rho)$ is lower continuous.*

  Proof. Let $x \in X$ and $\lambda \in \mathcal{L}(\widetilde{f})$. Then $P_\lambda^{-1}(x)$ is a finite intersection of open sets and hence is open. Thus $P_{\widetilde{f}}^{-1}(x)$ is a union of open sets and hence is open. Since $P_{\widetilde{f}}^{-1}(x) = \cup_{\lambda \in \mathcal{L}(\widetilde{f})} P_\lambda^{-1}(x)$ and $\lambda \in \mathcal{L}(\widetilde{f})$ in the definition of $P_{\widetilde{f}}^{-1}(x)$, it follows that $P_{\widetilde{f}}^{-1}(x) = \rho^{-1}(x) = \{y \in X \mid \pi(x, y) > 0\}$. That is, $\widetilde{f}(\rho)^{-1}(x) = \rho^{-1}(x)$ is open. ∎

Lemma 6.4.31 *Let $\widetilde{f}$ be a partial fuzzy simple rule. Suppose $\rho \in \mathcal{R}^n$ is semiconvex. If $k \leq s(\widetilde{f}) - 2$, then $\widetilde{f}(\rho)$ is semi-convex.*

Proof. Suppose $\widetilde{f}(\rho)$ is not semi-convex. Then there exists $z \in X$ such that $z \in \mathrm{Con}(P_{\widetilde{f}}(z)) = \mathrm{Con}(\cup_{\lambda \in \mathcal{L}(\widetilde{f})} P_\lambda(z)) = \mathrm{Con}(\cup_{\lambda \in \mathcal{L}(\widetilde{f})} \cap_{i \in \mathrm{Supp}(\lambda)} P_i(z)) = \mathrm{Con}(\cup_{\lambda \in \mathcal{L}(\widetilde{f})} \cap_{i \in \mathrm{Supp}(\lambda)} \{y \in X \mid \pi_i(y, z) > 0\})$. By Caratheodory's Theorem [6], there exist $y_1, \ldots, y_{k+1} \in P_{\widetilde{f}}(z)$ such that $z \in \mathrm{Con}(\{y_1, \ldots y_{k+1}\})$. Since $y_j \in P_{\widetilde{f}}(z)$ for $j = 1, \ldots, k+1$ and $\widetilde{f}$ is simple, there exist decisive coalitions $\lambda_1, \ldots, \lambda_{k+1}$ such that $y_j \in P_{\lambda_j}(z)$ for $j = 1, \ldots, k+1$. Let $\mathcal{L} = \{\lambda_1, \ldots, \lambda_{k+1}\}$. By hypothesis, $k \leq s(\widetilde{f}) - 2$. Hence $|\mathcal{L}_{\mathrm{Supp}}| \leq |\mathcal{L}| \leq k+1 \leq s(\widetilde{f}) - 1$. Thus $s(\widetilde{f}) \neq \min\{|\mathcal{L}_{\mathrm{Supp}}| \mid \mathcal{L} \subseteq \mathcal{L}(\widetilde{f})$ and $K(\mathcal{L}) = \emptyset\}$. Hence $K(\mathcal{L}) = \cap_{\lambda \in \mathcal{L}} \mathrm{Supp}(\lambda) \neq \emptyset$ by the definition of s. Thus there exists $i \in N$ such that $i \in \mathrm{Supp}(\lambda_j)$ for $j = 1, \ldots, k+1$. Hence $y_j \in P_i(z)$ for $j = 1, \ldots, k+1$. Thus $z \in \mathrm{Con}(\{y_1, \ldots, y_{k+1}\}) \subseteq \mathrm{Con}(P_i(z))$ which contradicts the fact that ρ_i is semi-convex. ∎

We conclude with our main result concerning k-dimensional policy space.

Theorem 6.4.32 *Let $\widetilde{f}$ be a partial fuzzy simple rule. Let $\rho \in \mathcal{R}^n$. Suppose $\mathcal{C}$ and $\widetilde{f}(\rho)$ are compatible. Suppose ρ is lower continuous and semi-convex. Suppose $M(\mathrm{Supp}(\widetilde{f}(\rho)), X) \neq \emptyset$ implies there exists $x \in M(\mathrm{Supp}(\widetilde{f}(\rho)), X)$ such that $\widetilde{f}(\rho)(x, \_)$ is bounded away from 0. If $k \leq s(\widetilde{f}) - 2$, then $C_{\widetilde{f}}(\rho) \neq \theta$.*

Proof. By Lemmas 6.4.30 and 6.4.31, $\widetilde{f}(\rho)$ is lower continuous and semi-convex. The desired result is now immediate from Theorem 6.4.23. (Recall that X is assumed to be convex and compact.) ∎

Lemma 6.4.29 shows that for a partial fuzzy simple rule $\widetilde{f}$, the Nakamura number $s(\widetilde{f})$ is greater than or equal to 3. The Nakamura number is the number of winning coalitions in a noncollegial subfamily. Hence as the number of these coalitions increases past three, a fuzzy core will exist in two-dimensional spatial models. To the extent that the number of players potentially increases the number of winning coalitions, an increase in their numbers will result in a greater likelihood of a core.

We conclude the chapter with a short history of the median voter theorem. Majoritarian voting is an ancient method of group decision making. There is no clear statement of the median voter theorem until around 1950. Aristotle's analysis of political decision making written in 330 B.C. made no mention of a pivotal or decisive voter. Condorcet (1785) discovered the idea of a pivotal voter and noted how the accuracy of decisions can be improved by majority decisions, but made no clear statement of the median voter theorem. The median voter theorem was first stated by Black (1948) and extended by Downs [9] to representative democracy (1957).

6.5   Exercises

1. [14] Let ρ be an FWPR. Then π, the strict preference relation strict preference relation with respect to ρ, is said to be **partial** if, for all $x, y \in X$, $\pi(x, y) > 0 \iff \rho(y, x) = 0$. An FWPR ρ on X is **regularly acyclic** if for all $\{x_1, x_2, x_3, \ldots, x_{n-1}, x_n\} \in X$, $\pi(x_1, x_2) \wedge \pi(x_2, x_3) \wedge \ldots \wedge \pi(x_{n-1}, x_n) > 0$ implies $\pi(x_n, x_1) = 0$. Let ρ be an FWPR. If π is regular, then prove that the following properties hold:

   (1) ρ is max-min transitive implies ρ is partially quasi-transitive.

   (2) ρ is weakly transitive implies ρ is partially quasi-transitive.

   (3) ρ is partially quasi-transitive implies ρ is regularly acyclic.

2. [14] Let $\mathcal{L} \subseteq \mathcal{F}(N)$. Let $\rho \in \mathcal{FR}^n$ and set $\tilde{\mathcal{P}}(\rho) = \{(x, y) \in X \times X \mid \exists \lambda \in \mathcal{L}, \pi(x, y) > 0, \forall i \in Supp(\lambda)\}$. Define $\tilde{g}_{\mathcal{L}} : \mathcal{FR}^n \to \mathcal{FR}$ by for all $\rho \in \mathcal{FR}^n$ and all $x, y \in X$,

$$\tilde{g}_{\mathcal{L}}(\rho)(x, y) = \begin{cases} 1 \text{ if } (x, y) \notin \text{Symm}(\tilde{\mathcal{P}}(\rho)) \\ \vee\{\wedge\{\rho_i(x, y) \mid i \in Supp(\lambda)\} \mid \lambda \in \mathcal{L} \text{ and } \pi_i(x, y) > 0, \\ \quad \forall i \in Supp(\lambda)\} \quad \text{if } (x, y) \in \tilde{\mathcal{P}}(\rho) \\ \wedge\{\wedge\{\rho_i(x, y) \mid i \in Supp(\lambda)\} \mid \lambda \in \mathcal{L} \text{ and } \pi_i(y, x) > 0, \\ \quad \forall i \in Supp(\lambda)\} \quad \text{if } (x, y) \in \text{Symm}(\tilde{\mathcal{P}}(\rho)) \backslash \tilde{\mathcal{P}}(\rho). \end{cases}$$

   If $(x, y) \in \tilde{\mathcal{P}}(\rho)$, then prove that $g_{\mathcal{L}}(\rho)(x, y) > g_{\mathcal{L}}(\rho)(y, x)$. If strict preferences are partial (regular), then $\pi_{\mathcal{L}}$ is partial (regular).

3. [14] Let $\tilde{f}$ be a fuzzy aggregation rule. Then
   (1) $\tilde{f}$ is called a **regular fuzzy simple rule** if for all $\rho \in \mathcal{FR}^n$ and all $x, y \in X$,

$$\tilde{f}(\rho)(x, y) > \tilde{f}(\rho)(y, x) \iff \tilde{g}_{\mathcal{L}(\tilde{f})}(\rho)(x, y) > \tilde{g}_{\mathcal{L}(\tilde{f})}(\rho)(y, x).;$$

   (2) $\tilde{f}$ is called **regularly neutral** if $\forall \rho, \rho' \in \mathcal{FR}^n$ and $\forall x, y, z, w \in X$, $[P(x, y; \rho) = P(z, w; \rho')$ and $P(y, x; \rho) = P(w, z; \rho')]$ imply $\tilde{f}(\rho)(x, y) > \tilde{f}(\rho)(y, x) \Leftrightarrow \tilde{f}(\rho')(z, w) > \tilde{f}(\rho')(w, z)$.

   Let $\tilde{f}$ be an FPAR and suppose π is regular. Then prove that $\tilde{f}$ is a regular fuzzy simple rule if and only if $\tilde{f}$ is decisive, monotonic, and regularly neutral.

4. [14] Let $\mathcal{D} \subseteq \mathcal{F}(N) \times \mathcal{F}(N)$ be such that $Supp(\sigma) \subseteq Supp(\omega)$ for all $(\sigma, \omega) \in \mathcal{D}$. Let $\rho \in \mathcal{FR}^n$ and set $\tilde{\mathcal{R}}(\rho) = \{(x, y) \in X \times X \mid \exists (\sigma, \omega) \in$

$\mathcal{D}, \pi_i(x,y) > 0 \; \forall i \in \mathrm{Supp}(\sigma)$ and $\pi_j(y,x) = 0 \; \forall j \in \mathrm{Supp}(\omega)\}$. Define $\tilde{g}_{\mathcal{D}} : \mathcal{FR}^n \to \mathcal{FR}$ by for all $\rho \in \mathcal{FR}^n$ and all $x, y \in X$,

$$\tilde{g}_{\mathcal{D}}(\rho)(x,y) = \begin{cases} 1 & \text{if } (x,y) \notin \mathrm{Symm}(\tilde{\mathcal{R}}(\rho)) \\[4pt] \vee\{\wedge\{\rho_j(x,y) \mid j \in \mathrm{Supp}(\omega)\} \mid (\sigma,\omega) \in \mathcal{D}, \; \pi_i(x,y) > 0 \\ \quad \forall i \in \mathrm{Supp}(\sigma), \text{ and } \pi_j(y,x) = 0 \; \forall j \in \mathrm{Supp}(\omega)\} \\ \quad\quad\quad \text{if } (x,y) \in \tilde{\mathcal{R}}(\rho) \\[4pt] \wedge\{\wedge\{\rho_j(x,y) \mid j \in \mathrm{Supp}(\omega)\} \mid (\sigma,\omega) \in \mathcal{D}, \; \pi_i(y,x) > 0 \\ \quad \forall i \in \mathrm{Supp}(\sigma), \text{ and } \pi_j(x,y) = 0 \; \forall j \in \mathrm{Supp}(\omega)\} \\ \quad\quad\quad \text{if } (x,y) \in \mathrm{Symm}(\tilde{\mathcal{R}}(\rho))\backslash\tilde{\mathcal{R}}(\rho). \end{cases}$$

Prove the following statements. If $(x,y) \in \tilde{\mathcal{P}}(\rho)$, then $g_{\mathcal{D}}(\rho)(x,y) > g_{\mathcal{D}}(\rho)(y,x)$. If strict preferences are partial (regular), then $\pi_{\mathcal{D}}$ is partial (regular).

5. [14] Let $\tilde{f}$ be a fuzzy aggregation rule. Then

(1) $\tilde{f}$ is called a **regular fuzzy voting rule** if for all $\rho \in \mathcal{FR}^n$ and all $x, y \in X$,

$$\tilde{f}(\rho)(x,y) > \tilde{f}(\rho)(y,x) \iff \tilde{g}_{\mathcal{D}(\tilde{f})}(\rho)(x,y) > \tilde{g}_{\mathcal{D}(\tilde{f})}(\rho)(y,x),$$

and $\tilde{f}$ is called a **partial fuzzy voting rule** if for all $\bar{\rho} \in \mathcal{FR}^n$ and all $x, y \in X$,

$$\tilde{f}(\rho)(x,y) > 0 \iff \tilde{g}_{\mathcal{D}(\tilde{f})}(\rho)(x,y) > 0.$$

(2) $\tilde{f}$ is called a **regular fuzzy simple rule** if for all $\rho \in \mathcal{FR}^n$ and all $x, y \in X$,

$$\tilde{f}(\rho)(x,y) > \tilde{f}(\rho)(y,x) \iff \tilde{g}_{\mathcal{L}(\tilde{f})}(\rho)(x,y) > \tilde{g}_{\mathcal{L}(\tilde{f})}(\rho)(y,x).$$

(3) $\tilde{f}$ is called **regularly neutral** if $\forall \rho, \rho' \in \mathcal{FR}^n$ and $\forall x, y, z, w \in X, [P(x,y;\rho) = P(z,w;\rho')$ and $P(y,x;\rho) = P(w,z;\rho')]$ imply $\tilde{f}(\rho)(x,y) > \tilde{f}(\rho)(y,x) \Leftrightarrow \tilde{f}(\rho')(z,w) > \tilde{f}(\rho')(w,z)$.

Prove the following statements. Let $\tilde{f}$ be an FPAR and suppose π is regular. Then $\tilde{f}$ is a regular fuzzy simple rule if and only if $\tilde{f}$ is decisive, monotonic, and regularly neutral.

6. [14] Let $\mathcal{D} \subseteq \mathcal{F}(N) \times \mathcal{F}(N)$ be such that $\mathrm{Supp}(\sigma) \subseteq \mathrm{Supp}(\omega)$ for all $(\sigma, \omega) \in \mathcal{D}$. Let $\rho \in \mathcal{FR}^n$ and set $\tilde{\mathcal{R}}(\rho) = \{(x,y) \in X \times X \mid \exists(\sigma,\omega) \in$

$\mathcal{D}, \pi_i(x, y) > 0 \; \forall i \in \mathrm{Supp}(\sigma)$ and $\pi_j(y, x) = 0 \; \forall j \in \mathrm{Supp}(\omega)\}$. Define $\tilde{g}_{\mathcal{D}} : \mathcal{FR}^n \rightarrow \mathcal{FR}$ by for all $\rho \in \mathcal{FR}^n$ and all $x, y \in X$,

$$
\tilde{g}_{\mathcal{D}}(\rho)(x, y) = \begin{cases} 1 & \text{if } (x, y) \notin \mathrm{Symm}(\tilde{\mathcal{R}}(\rho)) \\ \vee\{\wedge\{\rho_j(x, y) \mid j \in \mathrm{Supp}(\omega)\} \mid (\sigma, \omega) \in \mathcal{D}, \; \pi_i(x, y) > 0 \\ \qquad \forall i \in \mathrm{Supp}(\sigma), \text{ and } \pi_j(y, x) = 0 \; \forall j \in \mathrm{Supp}(\omega)\} \\ \qquad \text{if } (x, y) \in \tilde{\mathcal{R}}(\rho) \\ \wedge\{\wedge\{\rho_j(x, y) \mid j \in \mathrm{Supp}(\omega)\} \mid (\sigma, \omega) \in \mathcal{D}, \; \pi_i(y, x) > 0 \\ \qquad \forall i \in \mathrm{Supp}(\sigma), \text{ and } \pi_j(x, y) = 0 \; \forall j \in \mathrm{Supp}(\omega)\} \\ \qquad \text{if } (x, y) \in \mathrm{Symm}(\tilde{\mathcal{R}}(\rho)) \backslash \tilde{\mathcal{R}}(\rho). \end{cases}
$$

Prove the following statements. If $(x, y) \in \tilde{\mathcal{P}}(\rho)$, then $\tilde{g}_{\mathcal{D}}(\rho)(x, y) > \tilde{g}_{\mathcal{D}}(\rho)(y, x)$. If strict preferences are partial (regular), then $\pi_{\mathcal{D}}$ is partial (regular).

7. Let $\tilde{f}$ be a fuzzy aggregation rule. Then $\tilde{f}$ is called a **regular fuzzy voting rule** if for all $\rho \in \mathcal{FR}^n$ and all $x, y \in X$,

$$
\tilde{f}(\rho)(x, y) > \tilde{f}(\rho)(y, x) \iff \tilde{g}_{\mathcal{D}(\tilde{f})}(\rho)(x, y) > \tilde{g}_{\mathcal{D}(\tilde{f})}(\rho)(y, x),
$$

and $\tilde{f}$ is called a **partial fuzzy voting rule** if for all $\rho \in \mathcal{FR}^n$ and all $x, y \in X$,

$$
\tilde{f}(\rho)(x, y) > 0 \iff \tilde{g}_{\mathcal{D}(\tilde{f})}(\rho)(x, y) > 0.
$$

Let $\tilde{f}$ be an FPAR and suppose π is regular. Prove that $\tilde{f}$ is a regular fuzzy voting rule if and only if $\tilde{f}$ is monotonic and regularly neutral.

8. [14] Show that the following fuzzy preference aggregation rule is monotonic and decisive, but not neutral. Let $X = \{w, x, y, z\}$ and $N = \{1, 2\}$. Suppose that π is regular. Let $\succ$ be the lexicographical (alphabetical) order of X. Then $\succ$ is irreflexive, complete, and asymmetric. Define the fuzzy aggregation rule $\tilde{f} : \mathcal{FR}^n \rightarrow \mathcal{FR}$ by for all $a, b \in X$ and for all $\rho \in \mathcal{FR}^n$,

$$
\tilde{f}(\rho)(a, b)) = \begin{cases} 1 \text{ if } a = b \text{ or } \pi_i(a, b) > 0 \; \forall i \in N, \\ \beta \text{ if } a \succ b, \text{ not } [\pi_i(a, b) > 0 \; \forall i \in N] \\ \qquad \text{and not } [\pi_i(b, a) > 0 \; \forall i \in N], \\ \gamma \text{ if } a \prec b, \text{ not } [\pi_i(a, b) > 0 \; \forall i \in N] \\ \qquad \text{and not } [\pi_i(b, a) > 0 \; \forall i \in N], \\ 0 \text{ if } \pi_i(b, a) > 0 \; \forall i \in N, \end{cases}
$$

where $0 < \gamma < \beta < 1$.

9. [14] Show that the following fuzzy preference aggregation rule is neutral and monotonic, but not decisive. Let $X = \{x, y, z\}$ and $N = \{1, 2\}$. Suppose that π is regular. Define the fuzzy aggregation rule $\tilde{f} : \mathcal{FR}^n \to \mathcal{FR}$ by for all $a, b \in X$ and for all $\rho \in \mathcal{FR}^n$,

$$\tilde{f}(\rho)(a, b)) = \begin{cases} 1 \text{ if } a = b \text{ or } \pi_i(a, b) > 0 \ \forall i \in N, \\ \beta \text{ if } P(b, a; \rho) = \emptyset, \text{ not } [\pi_i(a, b) > 0 \ \forall i \in N] \\ \qquad \text{and not } [\pi_i(b, a) > 0 \ \forall i \in N], \\ \gamma \text{ if } P(b, a; \rho) \neq \emptyset, \text{ not } [\pi_i(a, b) > 0 \ \forall i \in N] \\ \qquad \text{and not } [\pi_i(b, a) > 0 \ \forall i \in N], \\ 0 \text{ if } \pi_i(b, a) > 0 \ \forall i \in N, \end{cases}$$

where $0 < \gamma < \beta < 1$.

10. [14] Show that the following fuzzy preference aggregation rule is decisive and neutral, but not monotonic. Let $X = \{x, y, z\}$ and $N = \{1, 2, 3, 4, 5\}$. Suppose that π is regular. Define the fuzzy aggregation rule $\tilde{f} : \mathcal{FR}^n \to \mathcal{FR}$ by for all $a, b \in X$ and for all $\rho \in \mathcal{FR}^n$,

$$\tilde{f}(\rho)(a, b)) = \begin{cases} 1 \text{ if } a = b \text{ or } \pi_i(a, b) > 0 \ \forall i \in N, \\ \beta \text{ if } |P(b, a; \rho)| \in (\frac{n}{2}, n - 1), \text{ not } [\pi_i(a, b) > 0 \ \forall i \in N] \\ \qquad \text{and not } [\pi_i(b, a) > 0 \ \forall i \in N], \\ \gamma \text{ if } |P(b, a; \rho)| \notin (\frac{n}{2}, n - 1), \text{ not } [\pi_i(a, b) > 0 \ \forall i \in N] \\ \qquad \text{and not } [\pi_i(b, a) > 0 \ \forall i \in N], \\ 0 \text{ if } \pi_i(b, a) > 0 \ \forall i \in N. \end{cases}$$

where $0 < \gamma < \beta < 1$.

6.6   References

1. K. J. Arrow, *Social Choice and Individual Values*, John Wiley and Sons, New York, 1951.

2. D. Austen-Smith and J. S. Banks, *Positive Political Theory I: Collective Preference*, Michigan Studies in Political Science, The University of Michigan Press, 2000.

3. D. Austen-Smith and J. S. Banks, *Positive Political Theory II: Strategy and Structure*, Michigan Studies in Political Science, The University of Michigan Press, 2005.

4. D. Black, *The Theory of Committees and Elections*, London, Cambridge University Press, 1958.

5. D. Black, On Arrow's impossibility theorem, *Journal of Law and Economics*, 12 (1969) 227–248.

6. K. C. Border, *Fixed Point Theorems with Applications to Economics and Game Theory*, London Cambridge University Press 1985.

7. J. M. Buchanan and R. D. Tollison, *The Theory of Public Choice-II*, University of Michigan Press, 1984.

8. T. D. Clark and J. N. Mordeson, Applying fuzzy mathematics to political science: What is to be done? *Critical Review* 2 (2008) 13–19.

9. A. Downs, *An Economic Theory of Democracy*, New York, Harper and Row Publishers, 1957.

10. L. A. Fono and N. G. Andjiga, Fuzzy strict preference and social choice, *Fuzzy Sets and Systems,* 155 (2005) 372–389.

11. I. Georgescu, *Fuzzy Choice Functions: A Revealed Preference Approach*, Springer, Studies in Fuzziness and Soft Computing Vol. 214 2007.

12. I. Georgescu, Similarity of fuzzy choice functions, *Fuzzy Sets and Systems,* 158 (2007) 1314–1326.

13. A. Gibbard, Manipulation of voting schemes: A general result, *Econometrica,* 41 (1973) 587–601.

14. M. B. Gibilisco, A. M. Gowen, K. E. Albert, J. N. Mordeson, M. J. Wierman, and T. D. Clark, *Fuzzy Social Choice*, Springer, Studies in Fuzziness and Soft Computing 315 Springer 2014.

15. M. B. Gibilisco, J. N. Mordeson, and T. D. Clark, Fuzzy Black's median voter theorem: Examining the structure of fuzzy preference rules and strict preference, *New Mathematics and Natural Computation,* 8 (2012) 195–217.

16. R. G. Holcomb, An empirical test of the median voter model, *Economic Inquiry,* 18 (1980) 260–275.

17. T. A. Husted and A. W. Kenny, The effect of the expansion of the voting franchise on the size of government model, *Journal of Political Economy,* 105 (1997) 54–82.

18. K. Krehbiel, Legislative organization, *Journal of Economic Perspectives,* 18 (2004) 113–128.

19. D. R. Kiewit and M. D. McCubbins, Presidential influence on congressional appropriations decisions, *American Journal of Political Science,* 32 (1988) 713–736.

20. R. D. McKelvey, Intransitivities in multidimensional voting models and some implications for agenda control, *Journal of Economic Theory,* 12 (1976) 472–482.

21. J. N. Mordeson, K. R. Bhutani, and T. D. Clark, The rationality of fuzzy choice functions, *New Mathematics and Natural Computation,* 4 (2008) 309–327.

22. J. N. Mordeson and T. D. Clark, Fuzzy Arrow's Theorem, *New Mathematics and Natural Computation,* 5 (2009) 371–383.

23. J. N. Mordeson and T. D. Clark, The core in fuzzy spatial models, *New Mathematics and Natural Computation,* 6 (2010) 17–29.

24. J. N. Mordeson, L. Nielsen, and T. D. Clark, Single peaked fuzzy preferences in one-dimensional models: Does Black's median voter theorem hold?, *New Mathematics and Natural Computation* **6** (2010) 1–16.

25. T. W. Rice, An examination of the median voter hypothesis, *The Western Political Quarterly,* 38 (1985) 211–223.

26. T. Romer and H. Rosenthal, The elusive median voter, *Journal of Public Economics,* 12 (1979) 143–170.

27. M. A. Satterthwaite, Strategy-proofness and Arrows conditions, existence and correspondence theorems for voting procedures and social welfare functions, *Journal of Economic Theory,* 10 (1975) 187–217.

28. A. K. Sen, *Collective Choice and Social Welfare,* North Holland Publishing Company, 1979.

Chapter 7

Rationality

7.1 Fuzzy Preference and Preference-Based Choice Functions

The results in this section are based on Barrett, Pattanaik, and Salles [9]. We consider the question as to how unambiguous or exact choices are generated by fuzzy preferences and whether the exact choices induced by fuzzy preferences satisfy certain plausible rationality properties. Also, see [7, 11, 12, 15, 21, 23, 30]). Several alternative rules for generating exact choice sets from fuzzy weak preference relations are presented. To what extent the exact choice sets generated by these alternative rules satisfy certain fairly weak rationality conditions is also examined.

Given fuzzy weak preference relations, we present the nine rules given in [9] for inducing exact choice sets. Different sets of rationality properties are considered in terms of which the performance of the nine preference-based choice functions is measured. The first set of properties gives sufficient conditions for choosing an alternative, given a set of feasible alternatives, and given a fuzzy weak preference relation. The second set of properties gives sufficient conditions for rejecting an alternative, and the third set of properties states that if an exact best alternative exists, then the choice set induced by a preference based choice function should possess certain relations to the set of exactly best alternatives. It is shown that if the fuzzy preferences are not assumed to satisfy max-min transitivity, then all the preference-based choice functions which are examined here perform rather badly in terms of some of these plausible criteria. However, if the admissible fuzzy-preferences are assumed to satisfy max-min transitivity, then one of the preference-based choice functions satisfies all the rationality conditions under consideration. Let X be a given finite set of alternatives such that $|X| \geq 4$.

Definition 7.1.1 *Let $\rho \in \mathcal{FR}(X)$. Then*

(1) ρ is called a **fuzzy ordering** if ρ is reflexive, strongly complete, and max-min transitive;

(2) ρ is called **exact** if $\rho(X \times X) \subseteq \{0,1\}$.

We let $\mathcal{FC}(R)$ denote the set of all fuzzy relations on X that are reflexive and strongly complete. The set of all fuzzy orderings will be denoted by $\mathcal{FT}(R)$. Throughout, we let $\mathcal{FC}(R)'$ denote a nonempty subset of $\mathcal{FC}(R)$. Recall $\mathcal{P}^*(X)$ denotes $\mathcal{P}(X)\backslash\{\emptyset\}$.

Definition 7.1.2 *A function* $C : \mathcal{P}^*(X) \times \mathcal{FC}(R)' \rightarrow \mathcal{P}^*(X)$ *is called a* **preference-based choice function** **(PCF)** *if for all* $B \in \mathcal{P}^*(X)$ *and for all* $\rho \in \mathcal{FC}(R)'$, $\emptyset \neq C(B, \rho) \subseteq B$.

We think of $\mathcal{FC}(R)'$ as those fuzzy relations which are "admissible." Given $\rho \in \mathcal{FC}(R)'$ and a crisp set B of available alternatives, then $C(B, \rho)$ is the crisp set of alternatives chosen from B with respect to ρ. Often an individual's preferences are fuzzy, but his actual choices are not fuzzy, i.e., they are crisp.

We next introduce nine preference-based choice functions.

Let $B \in \mathcal{P}^*(X)$ and $\rho \in \mathcal{FC}(R)$. Using the notation of [9], we define the following real-valued functions of B as follows: $\forall x \in B$,

$$MF(B, \rho)(x) = \vee\{\rho(x,y) \mid y \in B\backslash\{x\}\},$$

$$mF(B, \rho)(x) = \wedge\{\rho(x,y) \mid y \in B\backslash\{x\}\},$$

$$MA(B, \rho)(x) = \vee\{\rho(y,x) \mid y \in B\backslash\{x\}\},$$

$$mA(B, \rho)(x) = \wedge\{\rho(y,x) \mid y \in B\backslash\{x\}\},$$

The notation can be interpreted as follows: $MF(B,\rho)(x)$ as "max for x in B, with respect to ρ" and $mF(B,\rho)(x)$ as "min for x in B, with respect to ρ." Similarly MA and mA can be interpreted as "max against" and "min against," respectively. These four functions can be considered to be fuzzy subsets of B with respect to ρ. Define the remaining five functions as follows: $\forall x \in B$,

$$SF(B, \rho)(x) = \sum_{y \in B\backslash\{x\}} \rho(x,y),$$

$$SA(B, \rho)(x) = \sum_{y \in B\backslash\{x\}} \rho(y,x),$$

$$MD(B, \rho)(x) = \vee\{\rho(x,y) - \rho(y,x) \mid y \in B\backslash\{x\}\},$$

$$mD(B, \rho)(x) = \wedge\{\rho(x,y) - \rho(y,x) \mid y \in B\backslash\{x\}\},$$

and

$$MD(B, \rho)(x) = \sum_{y \in B\backslash\{x\}} [\rho(x,y) - \rho(y,x)].$$

The notation SF, SA, MD, mD, and SD can be interpreted as "sum for," "sum against," "max difference," "min difference," and "sum difference," respectively.

We present some different possible types of preference-based choice functions.

Definition 7.1.3 [9] *Let* $C : \mathcal{P}^*(X) \times \mathcal{FC}(R)' \to \mathcal{P}^*(X)$ *be a PCF. Then* C *is called*

(1) **max-MF** *if for all* $B \in \mathcal{P}^*(X)$ *and for all* $\rho \in \mathcal{FC}(R)'$,

$$C(B, \rho) = \{x \in B \mid MF(B, \rho)(x) \geq MF(B, \rho)(y) \quad \text{for all } y \in B\backslash\{x\}\}.$$

(2) **max-mF** *if for all* $B \in \mathcal{P}^*(X)$ *and for all* $R \in \mathcal{FC}(R)'$,

$$C(B, \rho) = \{x \in B \mid mF(B, \rho)(x) \geq mF(B, \rho)(y) \quad \text{for all } y \in B\backslash\{x\}\}.$$

(3) **max-SF** *if for all* $B \in \mathcal{P}^*(X)$ *and for all* $\rho \in \mathcal{FC}(R)'$,

$$C(B, \rho) = \{x \in B \mid SF(B, \rho)(x) \geq SF(B, \rho)(y) \quad \text{for all } y \in B\backslash\{x\}\}.$$

(4) **min-MA** *if for all* $B \in \mathcal{P}^*(X)$ *and for all* $\rho \in \mathcal{FC}(R)'$,

$$C(B, \rho) = \{x \in B \mid MA(B, \rho)(x) \geq MA(B, \rho)(y) \quad \text{for all } y \in B\backslash\{x\}\}.$$

(5) **min-mA** *if for all* $B \in \mathcal{P}^*(X)$ *and for all* $\rho \in \mathcal{FC}(R)'$,

$$C(B, \rho) = \{x \in B \mid mA(B, \rho)(x) \geq mA(B, \rho)(y) \quad \text{for all } y \in B\backslash\{x\}\}.$$

(6) **min-SA** *if for all* $B \in \mathcal{P}^*(X)$ *and for all* $\rho \in \mathcal{FC}(R)'$,

$$C(B, \rho) = \{x \in B \mid SA(B, \rho)(x) \geq SA(B, \rho)(y) \quad \text{for all } y \in B\backslash\{x\}\}.$$

(7) **max-MD** *if for all* $B \in \mathcal{P}^*(X)$ *and for all* $\rho \in \mathcal{FC}(R)'$,

$$C(B, \rho) = \{x \in B \mid MD(B, \rho)(x) \geq MD(B, \rho)(y) \quad \text{for all } y \in B\backslash\{x\}\}.$$

(8) **max-mD** *if for all* $B \in \mathcal{P}^*(X)$ *and for all* $\rho \in \mathcal{FC}(R)'$,

$$C(B, \rho) = \{x \in B \mid mD(B, \rho)(x) \geq mD(B, \rho)(y) \quad \text{for all } y \in B\backslash\{x\}\}.$$

(9) **max-SD** *if for all* $B \in \mathcal{P}^*(X)$ *and for all* $\rho \in \mathcal{FC}(R)'$,

$$C(B, \rho) = \{x \in B \mid SD(B, \rho)(x) \geq SD(B, \rho)(y) \quad \text{for all } y \in B\backslash\{x\}\}.$$

Let $\mathcal{C}^{(i)}$ denote the set of all PCFs satisfying condition (i) of Definition 7.1.3, $i = 1, \ldots, 9$.

An equivalent way of defining some of the PCFs introduced in Definition 7.1.3 is by using the following approach. Let ρ be a fuzzy relation. Recall that $\pi_{(2)}(x, y) = 1 - \rho(y, x)$ for all $x, y \in X$. Given $B \in \mathcal{P}^*(X)$, $C(B, \rho)$ under the max-MF rule is the same as $C(B, \pi_{(2)})$ under the min-mA rule. Also, $C(B, \rho)$ under the max-mF rule is the same as $C(B, \pi_{(2)})$ under the min-MA rule.

The class of max-mF PCFs was introduced by Dutta, Panda, and Pattanaik [15], and Switaiski [30]. The min-MA PCFs are the same as Nurmi's [20],

$$C(B, \rho) = \{x \in X \mid \wedge\{1 - \rho(y, x) \mid y \in B\backslash\{x\}\}$$
$$\geq \wedge\{1 - \rho(y, x') \mid y \in B\backslash\{x\}\} \text{ for all } x' \in B\backslash\{x\}\}.$$

If ρ is strongly complete, then the max-mF rule here is the same as a choice rule discussed by Switalski [29],

$$C(B, \rho) = \{x \in B \mid \wedge\{\rho(x, y) \vee (1 - \rho(y, x)) \mid y \in B\backslash\{x\}\}$$
$$\geq \wedge\{\rho(x', y) \vee (1 - \rho(y, x')) \mid y \in B\backslash\{x\}\} \text{ for all } x' \in B\backslash\{x\}\}.$$

Proposition 7.1.5 below shows that max-mD PCFs are the same as certain PCFs introduced by Orlovsky [21]. An interesting discussion of the intuition underlying max-mF PCFs and Orlovsky's PCFs can be found in [29].

Let $B \in \mathcal{P}^*(X)$ and $\rho \in \mathcal{FC}(R)$. Define the fuzzy subset $OV(B, \rho)$ of B by for all $x \in B$,

$$OV(B, \rho)(x) = \wedge\{(1 - \rho(y, x) + \rho(x, y)) \wedge 1 \mid y \in B\backslash\{x\}\}.$$

Definition 7.1.4 *Let* $C : \mathcal{P}^*(X) \times \mathcal{FC}(R)' \to \mathcal{P}^*(X)$ *be a PCF. C is **max-OV** if for all $B \in \mathcal{P}^*(X)$ and for all $\rho \in \mathcal{FC}(R)'$,*

$$C(B, \rho) = \{x \in B \mid OV(x, B, \rho) \geq OV(y, B, \rho) \text{ for all } y \in y \in B\backslash\{x\}\}.$$

max-OV PCFs were introduced by Orlovsky [21]. Related ideas can be found in Ovchinnikov and Ozernoy [23].

For all $B \in \mathcal{P}^*(X)$ and for all $\rho \in \mathcal{FT}(R)$, define the set $T(B, \rho)$ as follows:

$$T(B, \rho) = \{x \in B \mid \rho(x, y) \geq \rho(y, x) \text{ for all } y \in B\backslash\{x\}\}.$$

Proposition 7.1.5 *(Barrett, Pattanaik, and Salles [9]) Let* $C : \mathcal{P}^*(X) \times \mathcal{FC}(R)' \to \mathcal{P}^*(X)$ *be a PCF. Then C is max-mD if and only if it is max-OV.*

Proof. Let $B \in \mathcal{P}^*(X)$ and $\rho \in \mathcal{FC}(R)'$.

Suppose $T(B, \rho) \neq \emptyset$. Then for all $x \in T(B, \rho)$, $OV(B, \rho)(x) = 1$, and for all $x \in B\backslash T(B, \rho)$, $OV(B, \rho)(x) < 1$. Thus

$$T(B, \rho) = \{x \in B \mid OV(B, \rho)(x) \geq OV(B, \rho)(y) \text{ for all } y \in B\backslash\{x\}\}. \quad (7.1)$$

We also have for all $x \in T(B, \rho)$, $mD(B, \rho)(x) \geq 0$ and for all $x \in B \backslash T(B, \rho)$, $mD(B, \rho)(x) < 0$. If $|T(B, \rho)| \geq 2$, then for all $x \in T(B, \rho)$, $mD(B, \rho)(x) = 0$. Hence

$$T(B, \rho) = \{x \in B \mid mD(B, \rho)(x) \geq mD(B, \rho)(y) \text{ for all } y \in B \backslash \{x\}\}. \quad (7.2)$$

By (7.1) and (7.2), it follows that when $T(B, \rho) \neq \emptyset$, a PCF is max-mD if and only if it is max-OV.

Suppose that $T(B, \rho) = \emptyset$. Then it can be shown that for all $x \in B$,

$$OV(B, \rho)(x) = \wedge\{1 - \rho(y, x) + \rho(x, y) \mid y \in B \backslash \{x\}\}.$$

Thus for all $x, z \in B$, $OV(x, B, \rho) \geq OV(z, B, \rho)$ if and only if

$$\wedge\{\rho(x, y) - \rho(y, x) \mid y \in B \backslash \{x\}\} \geq \wedge\{\rho(z, y) - \rho(y, z) \mid y \in B \backslash \{x\}\}.$$

Hence

$$\{x \in B \mid OV(B, \rho)(x) \geq OV(B, \rho)(y) \text{ for all } y \in B \backslash \{x\}\}$$
$$= \{x \in B \mid mD(B, \rho)(x) \geq mD(y, B, \rho)(y) \text{ for all } y \in B \backslash \{x\}\}.$$

Thus when $T(B, \rho) = \emptyset$, a PCF is max-mD if and only if it is max-OV. ∎

We now consider some basic restrictions on preference-based choice functions, which represent different aspects of "rationality." See [2, 28, 29], for counterparts on rational choice and exact preference to some of these properties. We examine the performance of the different PCFs in terms of these properties.

The following rationality conditions are considered: conditions for choosing an alternative; conditions for rejecting an alternative; and conditions stipulating certain relations between the set of chosen alternatives and the set of alternatives which, in an exact sense, weakly dominate all feasible alternatives. Each set contains two different, but related conditions.

Throughout the rest of this section, C denotes a PCF as defined in Definition 7.1.2.

The following definition provides some conditions for choosing alternatives.

Definition 7.1.6 [9] *Let C be a preference-based choice function. Then*

(1) *C is said to satisfy the **reward for pairwise weak dominance** (**RPWD**) if for all $B \in \mathcal{P}^*(X)$, all $\rho \in \mathcal{FC}(R)$, and all $x \in B$, $[\rho(x, y) \geq \rho(y, x)$ for all $y \in B \backslash \{x\}]$ implies $x \in C(B, \rho)$;*

(2) *C is said to satisfy the **reward for pairwise strict dominance** (**RPSD**) if for all $B \in \mathcal{P}^*(X)$, all $\rho \in \mathcal{FC}(R)$, and all $x \in B$, $[\rho(x, y) > \rho(y, x)$ for all $y \in B \backslash \{x\}]$ implies $x \in C(B, \rho)$.*

If an available alternative x strictly dominates every other available alternative in a pairwise comparison, then RPSD requires that x should be chosen. RPWD requires x to be chosen if it weakly dominates every other alternative in a pairwise comparison. If C satisfies the property that for all $\rho \in \mathcal{FC}(R)'$ and for all $x, y \in X$, $\rho(x, y) \geq \rho(y, x)$ (resp. $\rho(y, x) \geq \rho(x, y)$) implies $x \in C(\{x, y\})$, then RPWD (resp. RPSD) is implied by Sen's [12] Condition γ (or Aizerman and Malishevski's [1] Concordance). All of the PCFs introduced in Definition 7.1.3 satisfy this property.

Example 7.1.7 *We show (1) and (2) in this example.*
(1) *If* $\mathcal{FT}(R) \subseteq \mathcal{FC}(R)'$ *and* $C \in \mathcal{C}^{(1)} \cup \mathcal{C}^{(3)} \cup \mathcal{C}^{(4)} \cup \mathcal{C}^{(6)} \cup \mathcal{C}^{(7)} \cup \mathcal{C}^{(9)}$, *then* C *satisfies neither RPWD nor RSPD.*
(2) *If* $\mathcal{FC}(R)' = \mathcal{FC}(R)$ *and* $C \in \cup_{i=1}^{7} \mathcal{C}^{(i)} \cup \mathcal{C}^{(9)}$, *then* C *satisfies neither RPWD nor RPSD.*
Consider (1). Let $\mathcal{FT}(R) \subseteq \mathcal{FC}(R)'$ *and let* $B = \{x, y, z\}$. *Let* $\rho, \rho' \in \mathcal{FT}(R)$ *be such that*

$$\rho(x, x) = \rho(y, y) = \rho(z, z) = 1,$$
$$\rho(x, y) = s, \ \rho(x, z) = s,$$
$$\rho(y, x) = r, \ \rho(y, z) = t,$$
$$\rho(z, x) = r, \ \rho(z, y) = r$$

and

$$\rho'(x, x) = \rho'(y, y) = \rho'(z, z) = 1,$$
$$\rho'(x, y) = s, \ \rho'(x, z) = t,$$
$$\rho'(y, x) = r, \ \rho'(y, z) = r,$$
$$\rho'(z, x) = q, \ \rho'(z, y) = s.$$

where $1 > t > q > s > r > 0$. *Then* $\rho(x, y) > \rho(y, x)$, $\rho(x, z) > \rho(z, x)$, $\rho'(x, y) > \rho'(y, x)$, *and* $\rho'(x, z) > \rho'(z, x)$. *However, if* C *is max-MF, or max-SF, or max-MD, or max-SD, then* $x \notin C(B, \rho)$. *Thus RPWD and RPSD do not hold. If* C *is min-MA or min-SA, then* $x \notin C(B, \rho')$. *Thus neither RPWD nor RPSD hold.*
Consider (2). Let $\mathcal{FC}(R)' = \mathcal{FC}(R)$. *Suppose* C *is max-MF, or max-SF, or min-SA, or max-MD, or max-SD. Since* $\mathcal{FT}(R) \subseteq \mathcal{FC}(R)$, *it follows from (1) that both RPWD and RPSD do not hold. Therefore, we have only to consider the case where* C *is either max-mF or min-mA. Let* $B = \{x, y, z\}$. *Let* $\rho \in \mathcal{FC}(R)$ *be such that*

$$\rho(x, x) = \rho(y, y) = \rho(z, z) = 1,$$
$$\rho(x, y) = t_1, \ \rho(x, z) = t_3,$$
$$\rho(y, x) = t_2, \ \rho(y, z) = t_2,$$
$$\rho(z, x) = t_4, \ \rho(z, y) = t_5$$

where $1 > t_1 > t_2 > t_3 > t_4 > t_5 > 0$. *Then* $\rho(x, y) > \rho(y, x)$ *and* $\rho(x, z) > \rho(z, x)$. *However, if* C *is either max-mF or min-mA, then* $x \notin C(B, \rho)$. *Thus RPWD and RPSD do not hold.*

Theorem 7.1.8 *(Barrett, Pattanaik, and Salles* [9]*)* (1) *If $C \in \mathcal{C}^{(8)}$, then C satisfies both RPWD and RPSD.*

(2) *If $\mathcal{FC}(R) \subseteq \mathcal{FT}(R)$ and $C \in \mathcal{C}^{(2)} \cup \mathcal{C}^{(5)} \cup \mathcal{C}^{(8)}$, then C satisfies both RPWD and RPSD.*

Proof. (1) The proof here is clear.

(2) Let $\mathcal{FC}(R)' \subseteq \mathcal{FT}(R)$. We show that if $C \in \mathcal{C}^{(2)} \cup \mathcal{C}^{(5)}$, then C satisfies RPWD; the result will then follow from the fact that RPWD implies RPSD and part (i).

Let $B \in \mathcal{P}^*(X)$, $\rho \in \mathcal{FC}(R)'$, and $x \in B$ be such that

$$\rho(x, y) \geq \rho(y, x) \text{ for all } y \in B \backslash \{x\}. \tag{7.3}$$

Suppose C is max-mF. Then we must show that $mF(B, \rho)(x) \geq mF(B, \rho)(z)$ for all $z \in B \backslash \{x\}$. It is sufficient to show that for all distinct $z, w \in B \backslash \{x\}$, $\rho(x, w) \geq mF(B, \rho)(z)$. Let $z, w \in B \backslash \{x\}, z \neq w$. Since ρ is a fuzzy ordering, $\rho(x, w) \geq \rho(x, z) \wedge \rho(z, w)$. If $\rho(x, z) = \rho(x, z) \wedge \rho(z, w)$, then $\rho(x, w) \geq \rho(x, z)$, so that by (7.3), we have $\rho(x, w) \geq \rho(x, z) \geq \rho(z, x) \geq mF(B, \rho)(z)$. If $\rho(z, x) = \rho(x, z) \wedge \rho(z, w)$, then $\rho(x, w) \geq \rho(z, w) \geq mF(B, \rho)(z)$. This completes the case where C is max-mF.

Suppose that C is min-mA. Then we must prove that $mA(B, \rho)(x) \leq mA(B, \rho)(z)$ for all $z \in B \backslash \{x\}$. It is sufficient to show that for all distinct $z, w \in B \backslash \{x\}$, $\rho(w, z) \geq mA(B, \rho)(x)$. Let $z, w \in B \backslash \{x\}, z \neq w$. Since ρ is max-min transitive, $\rho(w, z) \geq \rho(w, x) \wedge \rho(x, z)$. If $\rho(w, x) = \rho(w, x) \wedge \rho(x, z)$, then $\rho(w, z) \geq \rho(w, x) \geq mA(B, \rho)(x)$. If $\rho(x, z) = \rho(w, x) \wedge \rho(x, z)$, then by (7.3), it follows that $\rho(w, z) \geq \rho(x, z) \geq \rho(z, x) \geq mA(B, \rho)(x)$. ∎

Even though RPSD is a weaker condition than RPWD, the performance of the nine types of PCFs does not improve when RPSD rather than RPWD is assumed. Note also that only three (max-mF, min-mA, max-mD) of our type of PCFs satisfy RPSD.

We now introduce "rationality" conditions which specify when an alternative should be rejected.

Definition 7.1.9 [9] *Let C be a preference-based choice function. Then*

(1) *C is said to satisfy **strict rejection** (**SREJ**) if $\forall\ B \in \mathcal{P}^*(X), \forall$ $\rho \in \mathcal{FC}(R)'$, and $\forall\ x \in B$, $(\rho(y, x) \geq \rho(x, y)$ for all $y \in B \backslash \{x\}$ and $\rho(y, x) > \rho(x, y)$ for some $y \in B \backslash \{x\})$ implies $x \notin C(B, \rho)$;*

(2) *C is said to satisfy **weak rejection** (**WREJ**) if for all $B \in \mathcal{P}^*(X)$, all $\rho \in \mathcal{FC}(R)'$, and all $x \in B$, $\rho(y, x) > \rho(x, y)$ for all $y \in B \backslash \{x\}$ implies $x \notin C(B, \rho)$.*

SREJ implies WREJ, but not conversely. SREJ says that if an alternative is weakly dominated in every relevant pairwise comparison in which it is involved, and if it is strictly dominated in at least one of these pairwise comparisons, then it should not be chosen. WREJ states that if an alternative is

strictly dominated in every relevant pairwise comparison, then it should not
be chosen.

Suppose C satisfies the property that for all $\rho \in \mathcal{FC}(R)'$ and for all
$x, y \in X$, $\rho(x, y) > \rho(y, x)$ implies $C(\{x, y\}, \rho) = \{x\}$. Then both SREJ
and WREJ are weaker restrictions than Sen's [27] Condition α (or Aizerman
and Malishevski's [1]). Note that all the PCFs introduced in Definition 7.1.3
satisfy this property.

We next determine which of the nine types of PCFs satisfy SREJ and
WREJ.

Lemma 7.1.10 *Suppose that ρ is a fuzzy ordering and $B \in \mathcal{P}^*(X)$. Then
$T(B, \rho) \neq \emptyset$.*

Proof. Suppose that $T(B, \rho) = \emptyset$. Then there exist $x_1, x_2, \ldots, x_n \in B$ such
that for all i, $1 \leq i \leq n-1$, $\rho(x_i, x_{i+1}) > \rho(x_{i+1}, x_i)$ and $\rho(x_n, x_1) > \rho(x_1, x_n)$.
We assume without loss of generality that $\rho(r_n, x_1) = \rho(x_1, x_2) \wedge \rho(x_2, x_3) \wedge
\cdots \wedge \rho(x_n, x_{n-1}) \wedge \rho(x_n, x_1)$. Since ρ is max-min transitive, we have that
$\rho(x_1, x_n) \geq \rho(x_1, x_2) \wedge \rho(x_2, x_n)$. Suppose $\rho(x_1, x_2) \geq \rho(x_n, x_1) > \rho(x_1, x_n)$.
Then it follows that

$$\rho(x_1, x_n) \geq \rho(x_2, x_n). \tag{7.4}$$

Since ρ is transitive, $\rho(x_2, x_n) \geq \rho(x_2, x_3) \wedge \rho(x_3, x_n)$. Given $\rho(x_2, x_3) \geq
\rho(x_n, x_1) > \rho(x_1, x_n)$, and (7.4), it follows that $\rho(x_2, x_n) \geq \rho(x_3, x_n)$. Hence
by (7.4), we have that

$$\rho(x_1, x_n) \geq \rho(x_3, x_n). \tag{7.5}$$

Continuing this process, it follows that

$$\rho(x_1, x_n) \geq \rho(x_{n-1}, x_n). \tag{7.6}$$

Given $\rho(x_{n-1}, x_n) \geq \rho(x_n, x_1) > \rho(x_1, x_n)$, we have a contradiction. ∎

Example 7.1.11 *We show conditions (1) and (2) hold.*
*(1) If $\mathcal{FT}(R) \subseteq \mathcal{FC}(R)'$ and $C \in \mathcal{C}^{(2)} \cup \mathcal{C}^{(4)} \cup \mathcal{C}^{(6)}$, then C does not satisfy
WREJ. Also, if $\mathcal{FT}(R) \subseteq \mathcal{FC}(R)'$ and $C \in \cup_{i=1}^6 \mathcal{C}^{(i)}$, then C does not satisfy
SREJ.*
*(2) If $\mathcal{FC}(R)' = \mathcal{FC}(R)$ and $C \in \mathcal{C}^{(2)} \cup \mathcal{C}^{(3)} \cup \mathcal{C}^{(4)} \cup \mathcal{C}^{(6)} \cup \mathcal{C}^{(8)}$, then C
does not satisfy WREJ. Also, if $\mathcal{FC}(R)' = \mathcal{FC}(R)$ and $C \in \mathcal{C}^{(1)} \cup \mathcal{C}^{(2)} \cup \mathcal{C}^{(3)} \cup
\mathcal{C}^{(4)} \cup \mathcal{C}^{(5)} \cup \mathcal{C}^{(6)} \cup \mathcal{C}^{(8)}$, then C does not satisfy SREJ.*
(1) Let $B = \{x, y, z\}$.
(a) Let $\rho \in \mathcal{FT}(R) \subseteq \mathcal{FC}(R)'$ be such that

$$\begin{aligned}
\rho(x, x) &= \rho(y, y) = \rho(z, z) = 1, \\
\rho(x, y) &= s, \quad \rho(x, z) = s, \\
\rho(y, x) &= t, \quad \rho(y, z) = s, \\
\rho(z, x) &= t, \quad \rho(z, y) = s.
\end{aligned}$$

where $1 > t > s > 0$. If C is max-mF, then $x \in C(B, \rho)$ and so WREJ and hence SREJ do not hold.

(b) Let $\rho' \in \mathcal{FT}(R) \subseteq \mathcal{FC}(R)'$ be such that

$$\begin{aligned}
\rho'(x, x) &= \rho'(y, y) = \rho'(z, z) = 1, \\
\rho'(x, y) &= r, \ \rho'(x, z) = r, \\
\rho'(y, x) &= s, \ \rho'(y, z) = t, \\
\rho'(z, x) &= s, \ \rho'(z, y) = t.
\end{aligned}$$

where $1 > t > s > r > 0$. If C is min-MA or min-SA, then $x \in C(B, \rho')$ and so WREJ and hence SREJ do not hold.

(c) Let $\rho'' \in \mathcal{FT}(R) \subseteq \mathcal{FC}(R)'$ be such that

$$\begin{aligned}
\rho''(x, x) &= \rho''(y, y) = \rho''(z, z) = 1, \\
\rho''(x, y) &= r, \ \rho''(x, z) = t, \\
\rho''(y, x) &= s, \ \rho''(y, z) = s, \\
\rho''(z, x) &= t, \ \rho''(z, y) = r.
\end{aligned}$$

where $1 > t > s > r > 0$. If C is max-MF or max-SF, then $x \in C(B, \rho'')$. Therefore, SREJ does not hold.

(d) Let $\rho^* \in \mathcal{FT}(R) \subseteq \mathcal{FC}(R)'$ be such that

$$\begin{aligned}
\rho^*(x, x) &= \rho^*(y, y) = \rho^*(z, z) = 1, \\
\rho^*(x, y) &= s, \ \rho^*(x, z) = s, \\
\rho^*(y, x) &= t, \ \rho^*(y, z) = t, \\
\rho^*(z, x) &= s, \ \rho^*(z, y) = s.
\end{aligned}$$

where $1 > s > r > 0$. If C is min-mA, then $x \in C(B, \rho^*)$ and so SREJ fails to hold. This completes the proof of (i).

(2) Let $\mathcal{FC}(R)' = \mathcal{FC}(R)$. Since $\mathcal{FT}(R) \subseteq \mathcal{FC}(R)$, from part (1), it follows that if C is max-mF, or min-mF, or min-MA, or min-SA, then WREJ does not hold for C. We now show that if C is max-SF or max-mD, then WREJ does not hold. Let $B = \{x, y, z, w\}$. Let $\rho \in \mathcal{FC}(R)'$ be such that its restriction to $B \times B$ is given by

ρ	x	y	z	w
x	1	s	s	s
y	t	1	r	s
z	t	s	1	r
w	t	r	s	1

where $1 > t > s > r > 0$. If C is max-SF or max-mD, then $x \in C(B, R)$ and so WREJ fails to hold

Given part (i) and given that $\mathcal{FC}(R)' = \mathcal{FC}(R)$ and C is max-mD, then C violates WREJ, it follows that if $\mathcal{FC}(R)' = \mathcal{FC}(R)$ and $C \in K(\text{max-MF, max-mF, max-SF, min-MA, min-mA, min-SA, max-mD})$, then C violates SREJ.

Theorem 7.1.12 (*Barrett, Pattanaik, and Salles* [9]) (1) *If* $C \in \mathcal{C}^{(7)} \cup \mathcal{C}^{(9)}$, *then* C *satisfies SREJ.*

If $C \in \mathcal{C}^{(1)} \cup \mathcal{C}^{(5)} \cup \mathcal{C}^{(7)} \cup \mathcal{C}^{(9)}$, *then* C *satisfies WREJ.*

(2) *If* $\mathcal{FC}(R)' \subseteq \mathcal{FT}(R)$ *and* $C \in \mathcal{C}^{(7)} \cup \mathcal{C}^{(8)} \cup \mathcal{C}^{(9)}$, *then* C *satisfies SREJ.*

If $\mathcal{FC}(R)' \subseteq \mathcal{FT}(R)$ *and* $C \in \mathcal{C}^{(1)} \cup \mathcal{C}^{(3)} \cup \mathcal{C}^{(5)} \cup \mathcal{C}^{(7)} \cup \mathcal{C}^{(8)} \cup \mathcal{C}^{(9)}$, *then* C *satisfies WREJ.*

Proof. (1) If C is max-MD, then clearly C satisfies SREJ. We now show that if C is max-SD, then C satisfies SREJ.

Suppose that C is max-SD. Let $\rho \in \mathcal{FC}(R)'$, $B \in \mathcal{P}^*(X)$, and $x \in B$ be such that for all $y \in B \backslash \{x\}$, $\rho(x,y) \leq \rho(y,x)$. Suppose there exists $y \in B \backslash \{x\}$ such that $\rho(x,y) < \rho(y,x)$. Suppose that $x \in C(B,\rho)$. Since $x \in C(B,\rho)$ and $SD(B,\rho)(x)$ is negative, it follows that $SD(B,\rho)(y) < 0$ for all $y \in B$. Let $B = \{z_1, z_2, \ldots, z_n\}$ and for all distinct $i, j \in \{1, 2, \ldots, n\}$, let $d_{ij} = \rho(x_i, x_j) - \rho(x_j, x_i)$. Then for all distinct $i, j \in \{1, 2, \ldots, n\}$, $d_{ij} + d_{ji} = 0$. Hence

$$
\begin{aligned}
&(d_{12} + d_{13} + \cdots + d_{1n}) + (d_{21} + d_{23} + \cdots + d_{2n}) + \\
&\cdots + (d_{n1} + d_{n2} + \cdots + d_{n,n-1}) \\
= \ &0.
\end{aligned}
$$

$$(d_{12} + d_{13} + \cdots + d_{1n}) + (d_{21} + d_{23} + \cdots + d_{2n}) + \cdots + (d_{n1} + d_{n2} + \cdots + d_{n,n-1}) = 0.$$

Since $SD(B,\rho)(y) < 0$ for all $y \in B$, we have

$$
\begin{aligned}
&(d_{12} + d_{13} + \cdots + d_{1n}) + (d_{21} + d_{23} + \cdots + d_{2n}) + \\
&\cdots + (d_{n1} + d_{n2} + \cdots + d_{n,n-1}) \\
= \ &[SD(B,\rho)(z_1) + SD(B,\rho)(z_2) + \cdots + SD(B,\rho)(z_n)] \\
< \ &0,
\end{aligned}
$$

which is a contradiction. Thus that if C is max-SD, then C satisfies SREJ.

Suppose C is max-MD or max-SD. Then C satisfies SREJ and it follows that if C is max-MD or max-SD, then C satisfies WREJ. Also if C is max-MF or min-mA, then C satisfies WREJ.

(2) By part (1), if $\mathcal{FC}(R)' \subseteq \mathcal{FT}(R)$ and C is max-MD or max-SD, then C satisfies SREJ. We show that if $\mathcal{FC}(R)' \subseteq \mathcal{FT}(R)$ and C is max-mD, then C satisfies SREJ.

Let $B \in \mathcal{P}^*(X)$, $\rho \in \mathcal{FC}(R)' \subseteq \mathcal{FT}(R)$, and $x \in B$ be such that for all $y \in B \backslash \{x\}$, $\rho(x,y) \leq \rho(y,x)$, and for some $y \in B \backslash \{x\}$, $\rho(x,y) < \rho(y,x)$. Suppose C is max-mD. By Lemma 7.1.10, $T(B,\rho) \neq \emptyset$. Let $x_0 \in T(B,\rho)$. Then $mD(B,\rho)(x_0) \geq 0 > mD(B,\rho)(x)$. Hence $x \notin C(B,\rho)$.

By part (1), and given that if $\mathcal{FC}(R)' \subseteq \mathcal{FT}(R)$ and C is max-mD, then C satisfies SREJ, it follows that if $\mathcal{FC}(R)' \subseteq \mathcal{FT}(R)$ and $C \in \mathcal{C}^{(1)} \cup \mathcal{C}^{(5)} \cup \mathcal{C}^{(7)} \cup \mathcal{C}^{(8)} \cup \mathcal{C}^{(9)}$ then C satisfies WREJ. Let $B \in \mathcal{P}^*(X)$. Let $\rho \in \mathcal{FC}(R)' \subseteq \mathcal{FT}(R)$ and $x \in B$ be such that for all $y \in B \backslash \{x, z\}$, $\rho(x,y) < \rho(y,x)$. Let $z \in B \backslash \{x\}$ be such that for all $y \in B \backslash \{x\}$, $\rho(x,z) \geq \rho(y,x)$. Let $w \in B \backslash \{x, z\}$. Then by the max-min transitivity of ρ, $\rho(z,w) \geq \rho(z,x) \wedge \rho(x,w)$. Since by our assumptions $\rho(z,x) > \rho(x,z) \geq \rho(z,w)$, it follows that $\rho(z,w) \geq \rho(x,w)$. Thus for all $w \in B\backslash\{x,z\}$, $\rho(z,w) \geq \rho(x,w)$. Given $\rho(z,x) > \rho(x,z)$, it follows that $SF(B,\rho)(z) > SF(B,\rho)(x)$ and thus $x \in C(B,\rho)$. ∎

The properties in Definition 7.1.13 below require that if a best alternative exists under the FBPR concerned in the exact sense, then the set of alternatives actually chosen must be faithful to the set of exactly best alternatives.

For all $B \in \mathcal{P}^*(X)$ and for all $\rho \in \mathcal{FC}(R)$, define the set $D(B,\rho)$ as follows:

$$D(B,\rho) = \{x \in B \mid \text{for all } y \in B, \ \rho(x,y) = 1\}.$$

Definition 7.1.13 [9] *Let C be a preference-based choice function. Then*
(1) C *is said to satisfy* **upper faithfulness** *(UF) if for all $B \in \mathcal{P}^*(X)$ and for all $\rho \in \mathcal{FC}(R)'$, $D(B,\rho) \neq \emptyset$ implies $C(B,\rho) \subseteq D(B,\rho)$;.*
(2) C *is said to satisfy* **lower faithfulness** *(LF) if for all $B \in \mathcal{P}^*(X)$ and for all $\rho \in \mathcal{FC}(R)'$, $D(B,\rho) \neq \emptyset$ implies $D(B,\rho) \subseteq C(B,\rho)$.*

Example 7.1.14 *We show that conditions (1) and (2) hold.*
(1) *If $\mathcal{FT}(R) \subseteq \mathcal{FC}(R)'$ and $C \in \mathcal{C}^{(1)} \cup \mathcal{C}^{(4)} \cup \mathcal{C}^{(5)} \cup \mathcal{C}^{(7)}$, then C does not satisfy UF.*
(2) *If $\mathcal{FC}(R)' = \mathcal{FC}(R)$ and $C \in \mathcal{C}^{(1)} \cup \mathcal{C}^{(4)} \cup \mathcal{C}^{(5)} \cup \mathcal{C}^{(6)} \cup \mathcal{C}^{(7)} \cup \mathcal{C}^{(8)}$, then C does not satisfy UF. Also, if $\mathcal{FC}(R)' = \mathcal{FC}(R)$ and $C \in K($min-mA, min-SA, max-MD, max-SD$)$, then C does not satisfy LF.*
Consider (1). Let $B = \{x,y,z\}$. Let $\rho' \in \mathcal{FT}(R)$ be such that

$$\rho'(x,x) = \rho'(y,y) = \rho'(z,z) = \rho'(x,y) = \rho'(x,z) = \rho'(y,z) = 1,$$
$$\rho'(y,x) = \rho'(z,x) = \rho'(z,y) = 0.$$

If C is max-MF, or min-MA, or max-MD, then $x \in C(B,\rho')$ and so UF does not hold.
Let $\rho'' \in \mathcal{FT}(R) \subseteq \mathcal{FC}(R)'$ be such that $\rho''(y,x) = 1$ and $\rho''(u,v) = \rho'(u,v)$ for all $(u,v) \in X \times X\backslash\{(y,x)\}$. If C is min-MA, then $z \in C(B,\rho'')$ and so UF does not hold.
(2) *Since $\mathcal{FT}(R) \subseteq \mathcal{FC}(R)$, from part (1), it follows that $\mathcal{FC}(R)' = \mathcal{FC}(R)$ and if $C \in \mathcal{C}^{(1)} \cup \mathcal{C}^{(4)} \cup \mathcal{C}^{(5)} \cup \mathcal{C}^{(7)}$, then UF does not hold for C. We now show that if $\mathcal{FC}(R) = \mathcal{FC}(R)'$ and $C \in \mathcal{C}^{(6)} \cup \mathcal{C}^{(8)} \cup \mathcal{C}^{(9)}$, then UF does not hold for C. Let $\rho''' \in \mathcal{FC}(R)$ be such that*

$$\rho'''(y,z) = t, \rho'''(z,y) = s, \rho'''(u,v) = 1 \ \forall (u,v) \in X \times X\backslash\{(y,z),(z,y)\}$$

where $1 > t > s > 0$. If C is min-SA, or max-mD, or max-SD, then $y \in C(B,\rho''')$ and so UF does not hold.

We now consider LF. If C is min-mA, min-SA, max-MD, or max-SD, then $x \notin C(B,\rho''')$ and thus LF does not hold.

Theorem 7.1.15 (*Barrett, Pattanaik, and Salles* [9]) (1) *If* $C \in \mathcal{C}^{(1)} \cup \mathcal{C}^{(3)}$, *then* C *satisfies UF.*

(2) *If* $C \in \mathcal{C}^{(1)} \cup \mathcal{C}^{(2)} \cup \mathcal{C}^{(3)} \cup \mathcal{C}^{(4)} \cup \mathcal{C}^{(7)}$, *then* C *satisfies LF.*

(3) *If* $\mathcal{FC}(R)' \subseteq \mathcal{FT}(R)$ *and* $C \in \mathcal{C}^{(2)} \cup \mathcal{C}^{(3)} \cup \mathcal{C}^{(6)} \cup \mathcal{C}^{(8)} \cup \mathcal{C}^{(9)}$, *then* C *satisfies UF.*

(4) *If* $\mathcal{FC}(R)' \subseteq \mathcal{FT}(R)$ *and* $C \in \cup_{i=1}^{9} \mathcal{C}^{(i)}$, *then* C *satisfies LF.*

Proof. That (1) and (2) hold is clear.

Consider (3) and (4). From part (1), it follows that if $\mathcal{FC}(R)' \subseteq \mathcal{FT}(R)$ and C is max-mF or max-SF, then C satisfies UF. We next show that if $\mathcal{FC}(R)' \subseteq \mathcal{FT}(R)$ and C is min-SA, or max-mD, or max-SD, then C satisfies UF.

Let $B \in \mathcal{P}^*(X)$, $\rho \in \mathcal{FC}(R)' \subseteq \mathcal{FT}(R)$ and $x \in D(B, \rho)$, and suppose $z \in C(B, \rho) \setminus \{x\}$. Suppose C is min-SA. By the max-min transitivity of ρ, we have since $\rho(x, z) = 1$ that

$$\rho(w, z) \geq \rho(w, x) \text{ for all } w \in B \setminus \{x, z\}. \tag{7.7}$$

For $z \in C(B, \rho)$, we have

$$\sum_{w \in B \setminus \{z\}} \rho(w, z) \leq \sum_{w \in B \setminus \{x\}} \rho(w, x). \tag{7.8}$$

From (7.7) and (7.8), and since $\rho(x, z) = 1$, it follows that $\rho(z, x) = 1$. Hence by the max-min transitivity of ρ and the fact that $\rho(x, w) = 1$ for all $w \in B$, it follows that $\rho(z, x) = 1$ for all $w \in B$. Thus $z \in D(B, \rho)$.

Let C be max-mD. For $x \in D(B, \rho)$, it is clear that $mD(B, \rho)(x) \geq 0$. Since $z \in C(B, \rho)$, it follows that $mD(B, \rho)(z) \geq 0$ and hence $\rho(z, x) \geq \rho(x, z) = 1$. Therefore, by the max-min transitivity of ρ, we have $\rho(z, w) = 1$ for all $w \in B$ since $\rho(x, w) = 1$ for all $w \in B$. Thus $z \in D(B, \rho)$.

Let C be max-SD. By the max-min transitivity of ρ and the fact that $\rho(x, z) = 1$, we have

$$\rho(w, z) \geq \rho(w, x) \text{ for all } w \in B. \tag{7.9}$$

Also,

$$\rho(z, w) \leq \rho(x, w) = 1 \text{ for all } w \in B. \tag{7.10}$$

From (7.9) and (7.10), it is clear that if $\rho(z, w) < 1$ for some $w \in B$, then $SD(B, \rho)(z) < SD(B, \rho)(x)$, which contradicts $z \in C(B, \rho)$. Therefore, $\rho(z, w) = 1$ for all $w \in B$, i.e., $z \in D(B, \rho)$.

Now consider LF. Given $\mathcal{FC}(R)' \subseteq \mathcal{FT}(R)$, by part ($i$), if C is max-MF, or max-mF, or max-SF, or min-MA, or max-mD, then C satisfies *LF*. We have only to consider the cases where $C \in \mathcal{C}^{(5)} \cup \mathcal{C}^{(6)} \cup \mathcal{C}^{(7)} \cup \mathcal{C}^{(9)}$.

Let $B \in \mathcal{P}^*(X)$ and $\rho \in \mathcal{FC}(R)' \subseteq \mathcal{FT}(R)$, and let $x \in D(B, \rho)$ and $z \in C(B, \rho)$. Then by the max-min transitivity of ρ, we have since $\rho(x, z) = 1$ that

$$\rho(w, z) \geq \rho(w, x) \text{ for all } w \in B. \tag{7.11}$$

Also,

$$\rho(z, w) \le \rho(x, w) = 1 \text{ for all } w \in B. \tag{7.12}$$

From (7.11) and (7.12), it follows that

$$\rho(z, w) - \rho(w, z) \le \rho(x, w) - \rho(w, x) \text{ for all } w \in B. \tag{7.13}$$

From (7.11) and since $z \in C(B, \rho)$, it follows that if C is min-mA or min-SA, then $x \in C(B, \rho)$. From (7.13) and since $z \in C(B, \rho)$, it follows that if C is max-MD or max-SD, then $x \in C(B, \rho)$. ∎

In [9], other preference-based choice functions were considered. We provide some of these in the exercises.

If the fuzzy preferences are not constrained to satisfy transitivity, then all the preference-based choice functions discussed in this section violate some properties which have often been considered to be basic properties of rational choice. However, if the admissible fuzzy preferences are constrained to be fuzzy orderings, then the preference-based choice function, max-mD, satisfies all the rationality properties of Barrett, Pattanaik, and Salles that we have presented in this section. We summarize Theorems 7.1.8, 7.1.12, and 7.1.15 and Examples 7.1.7, 7.1.11, and 7.1.14. The symbol $*$ indicates that the PCF satisfies the property under consideration when $\mathcal{F}(R)' = \mathcal{F}(R)$ and where the symbols $*$ and $\circledast$ indicate the PFC satisfies the property under consideration when $\mathcal{F}(R)' = \mathcal{F}\mathcal{T}(R)$.

	$RPWD$	$RPSD$	$WREJ$	$SREJ$	UF	LF
max-MF			$*$			$*$
max-mF	$\circledast$	$\circledast$			$*$	$*$
max-SF			$\circledast$		$*$	$*$
min-MA						$*$
min-mA	$\circledast$	$\circledast$	$*$			$\circledast$
min-SA					$\circledast$	$\circledast$
max-MD			$*$	$*$		$\circledast$
max-mD	$*$	$*$	$\circledast$	$\circledast$	$\circledast$	$*$
max-SD			$*$	$*$	$\circledast$	$\circledast$

In [16], the structure of the nine preference-based choice functions of this section were examined. It was shown that for crisp total pre-orders, first and last alternatives exist in a finite set of alternatives that has a strongly complete fuzzy pre-order. This result is used to characterize each of those crisp choice functions for crisp total pre-orders and strongly complete fuzzy pre-orders. When preferences are strongly complete fuzzy pre-orders, the six consistency conditions of Sen of those preference-based choice functions are determined by means of those characterizations.

7.2   Fuzzy Choice Functions, Revealed Preference and Rationality

This section is concerned with fuzzy revealed preference theory and with the problem of rationalizing fuzzy choice functions. The work is based on that presented in Banerjee [5]. We assume the domain of the choice function consists of all (crisp) finite subsets of the universal set of alternatives. However, we also assume that the choice is fuzzy in the sense that the decision maker can state the degree of his choice of an alternative. This is in contrast to the previous section, where the choice was crisp and the preferences were fuzzy. Banerjee's two types of weak and strong axioms of fuzzy revealed preferences (WAFRP and SAFRP) are stated. We show that WAFRP is not equivalent to SAFRP and that SAFRP does not characterize rationality. In the special case where every set of available alternatives has at least one element which is unambiguously chosen, we show that WAFRP is equivalent to SAFRP, but SAFRP still does not characterize rationality. In [5], a fuzzy congruence condition stronger than SAFRP was proposed and is presented here. It is shown to be necessary and sufficient for rationality. The interested reader may examine [8-11, 14, 19, 22] for results concerning choice with fuzzy preferences.

In crisp revealed preference theory, the congruence condition was introduced by Richter [26]. A weaker version was proposed by Sen [28]. In crisp revealed preference theory, the case where the finite subsets of the universal set are included in the domain of the choice function is discussed in Arrow [2] and Sen [28]. In traditional theory, it is known that WARP and SARP are equivalent conditions and that WARP is a characterization of rationality of the choice functions. It is also known that in the case of the competitive consumer, WARP is not equivalent to SARP, but SARP characterizes rationality. (See [17] and [24].)

We now consider fuzzy choice functions and rationality.

Let X denote the (crisp) universal set of alternatives. Let $\mathcal{E}$ denote the set of all non-empty, finite, and crisp subsets of X. Let $\mathcal{FP}_f^*(X)$ denote the set of all nonempty fuzzy subsets of X with finite supports.

The following definition is the third definition of a fuzzy choice function in this book.

Definition 7.2.1 A ***fuzzy choice function*** *is a function $C : \mathcal{E} \to \mathcal{FP}_f^*(X)$ such that $\forall S \in \mathcal{E}$, $Supp(C(S)) \subseteq S$.*

For all $S \in \mathcal{E}$, $C(S)(x)$ represents the degree to which x belongs to the set of chosen alternatives when the available set of alternatives is S.

In the previous definition of a fuzzy choice function, every nonempty available set S has an element which is chosen to a positive extent. A natural question arises as to whether it should be assumed that every such available set of alternatives has an element which is unambiguously chosen. We discuss this question later.

Throughout the section, we assume that

$$C(\{x\})(x) = C(\{y\})(y) \text{ for all } x, y \in X.$$

Definition 7.2.2 *For all $S \in \mathcal{E}$, an element $x \in S$ is called **dominant** in S if $\forall y \in S, C(S)(x) \geq C(S)(y)$.*

Recall that a fuzzy preference relation ρ (a function from X^2 into $[0, 1]$) is weakly transitive if $\forall x, y, z \in X$, $\rho(x, y) \geq \rho(y, x)$ and $\rho(y, z) \geq \rho(z, y)$ implies $\rho(x, z) \geq \rho(z, x)$.

Definition 7.2.3 *[5] Let ρ be a fuzzy preference relation on X. Then*
   *(1) ρ is called **weakly reflexive** if $\forall x \in X$, $0 < \rho(x, x) \geq \rho(x, y)$ $\forall y \in S$,*
   *(2) ρ is called **intensely transitive** if it is max-min transitive and weakly transitive,*
   *(3) ρ is called a **weak fuzzy ordering** if it is weakly reflexive, complete, and intensely transitive.*

The notion of what has been called weak transitivity appeared in Ponsard [25].

Definition 7.2.4 *A fuzzy choice function C is said to **reveal a fuzzy preference relation** ρ if $\forall x, y \in X$,*

$$\rho(x, y) = \vee \{C(S)(x) \mid S \in \mathcal{E}, x, y \in S\}.$$

It follows from Definition 7.2.4 that $\forall x, y \in X$, if $\exists S \in \mathcal{E}$ such that $C(S)(x) > 0$, then $\rho(x, y) > 0$ and also that $\forall S \in \mathcal{E}$, $C(S)(x) > 0$ implies $\rho(x, y) > 0$ $\forall y \in S$. In the crisp case, Definition 7.2.4 reduces to the most frequently used definition of a revealed (weak) preference relation.

The fuzzy strict preference relation π and the fuzzy indifference relation ι corresponding to ρ for this section is defined as follows: $\forall x, y \in X$,

$$\pi(x, y) = 0 \vee (\rho(x, y) - \rho(y, x))$$

and

$$\iota(x, y) = \rho(x, y) \wedge \rho(y, x).$$

The definition of a fuzzy strict preference relation given here follows the ideas presented in Orlovsky [21]. It can be shown that the major results of this section are not sensitive to the choice between alternative derivation rules, [14]. Recall that this choice of π is $\pi_{(3)}$ of Chapter 1.

Definition 7.2.5 *Let C be a fuzzy choice function. Define the fuzzy relation $\widetilde{\pi}$ on X by for all $x, y \in X$,*

$$\widetilde{\pi}(x,y) = 0 \vee (\vee\{C(S)(x) - C(S)(y) \mid S \in \mathcal{E}, x, y \in S\}).$$

In the crisp case, Definition 7.2.5 reduces to Arrow's [2] definition of revealed preference. The results of this section can be established without explicitly obtaining the weak preference relation corresponding to $\widetilde{\pi}$.

Definition 7.2.6 *Let ρ be the fuzzy preference relation revealed by the choice function C. The **image** of C is defined to be the function $\widehat{C} : \mathcal{E} \to \mathcal{FP}_f^*(X)$ such that $\forall S \in \mathcal{E}$ and $\forall x \in S$,*

$$\widehat{C}(S)(x) = \wedge\{\rho(x,y) \mid y \in S\}.$$

*C is called **normal** if $\forall S \in \mathcal{E}$ and $\forall x \in S$,*

$$C(S)(x) = \widehat{C}(S)(x).$$

From the definition of the revealed preference relation ρ, it can be shown that normality implies that $C(\{x,y\})(x) = \rho(x,y)$ for all $x, y \in X$. That is, normality of a choice function means that it can be recovered by its own revealed preference relation.

In the crisp case, Sen [28] introduced the concepts of image and normality. In the crisp case, there is a natural way of constructing a choice function from a weak reference relation. This is as follows: $\forall S \in \mathcal{E}$, let $\widehat{C}(S) = \{x \in S \mid x\rho y \ \forall y \in S\}$, where $\forall x, y \in X$ and $x\rho y$ can be interpreted to be $\rho(x,y) = 1$. In the fuzzy case, however, there are many alternatives. Definition 7.2.6 is a reasonable one for the situation when choice is fuzzy.

Definition 7.2.7 *Let ρ be a fuzzy preference relation on X. Let $S \in \mathcal{E}$ and $x \in S$. Then x is said to be **relation dominant** in S in terms of ρ if $\rho(x,y) \geq \rho(y,x) \ \forall y \in S$.*

The next two conditions are additional regularity conditions.

Relation dominant revealed preference (RDRP): A choice function is said to satisfy RDRP if $\forall S \in \mathcal{E}$ and $\forall x \in S$, x is dominant in S if and only if it is relation dominant in S in terms of the fuzzy revealed preference relation ρ.

Condition of revealed preference dominance (CRPD): A fuzzy choice function is said to satisfy CRPD if the fuzzy revealed preference relation ρ is such that $\forall S \in \mathcal{E}$ and for all distinct $x, y, z \in S$, if x is relation dominant in S, then $\rho(x,z) \geq \rho(y,x)$.

Condition CRPD says that if x "dominates" all elements of S, then the extent to which it is preferred to another element $z \in S$ cannot be less than the extent to which a third element y of that set is preferred to x. It is easily verified that in the crisp case RDRP is implied by normality and CRPD is trivially satisfied.

Definition 7.2.8 *A fuzzy choice function is called* **rational** *if it is normal, satisfies RDRP and CRPD, and the fuzzy revealed preference relation is a weak fuzzy ordering.*

We next consider weak and strong axioms of fuzzy revealed preference and congruence.

The weak axiom of revealed preference (WARP) was proposed by Samuelson [26] and the strong axiom of revealed preference (SARP) was proposed by Houthakker [18], Ville [31], and von Neumann and Morgenstern [32]. In Arrow [2] and Sen [28], these axioms were adapted to the case of crisp set-valued choice functions. We now present fuzzy versions of these axioms.

Definition 7.2.9 *Let C be a fuzzy choice function. Define a fuzzy preference relation $\widetilde{\pi}^*(x, y)$ on X so that for all $x, y \in X$, $\widetilde{\pi}^*(x, y) > 0$ if and only if there exists $x_0, x_1, \ldots, x_n \in X$ such that $x_0 = x$, $x_n = y$, and $\widetilde{\pi}(x_{i-1}, x_i) > 0$ for $i = 1, 2, \ldots, n$.*

The reader is asked in the exercises to show that for all $x, y \in X$, $\widetilde{\pi}^*(x, y) > 0$ implies $\widetilde{\pi}^*(y, x) = 0$. $\widetilde{\pi}^*$ can be thought of as providing a degree of indirect strict preference.

Definition 7.2.10 *Let C be a fuzzy choice function. Then C is said to satisfy the*

(1) **weak axiom of fuzzy revealed preference** 1 (**WAFRP**1) *if $\forall x, y \in X$, $\widetilde{\pi}(x, y) > 0$ implies $\rho(y, x) < 1$.*

(2) **strong axiom of fuzzy revealed preference** 1 (**SAFRP**1) *if $\forall x, y \in X$, $\widetilde{\pi}^*(x, y) > 0$ implies $\rho(y, x) < 1$.*

WAFRP1 and SAFRP1 are consistency conditions. WAFRP1 says that if x is directly strictly preferred to y to a positive extent, then y cannot be definitely weakly preferred to x. SAFRP1 says that this conclusion is valid if x is indirectly strictly preferred to y to a positive extent under the relation $\widetilde{\pi}^*$.

Definition 7.2.11 *Let C be a fuzzy choice function. Define a fuzzy preference relation π^* on X so that for all $x, y \in X$ and for all $t \in (0, 1]$, $\pi^*(x, y) \geq t$ if and only if there exists $x_0, \ldots, x_n \in X$ such that $x_0 = x$, $x_n = y$, and $\widetilde{\pi}(x_{i-1}, x_i) \geq t$ for $i = 1, 2, \ldots, n$.*

Under the relation π^*, x is indirectly strictly preferred to y at least to the extent t if and only if there is a sequence of elements connecting x to y so that each alternative in the sequence is directly strictly preferred to its immediate successor to the extent t.

Definition 7.2.12 *Let C be a fuzzy choice function. Then C is said to satisfy the*

(1) **weak axiom of fuzzy revealed preference** 2 (**WAFRP2**) *if for all* $x, y \in X$ *and for all t such that* $0 < t \leq 1$, $\tilde{\pi}(x, y) \geq t$ *implies* $\rho(y, x) \leq 1 - t$;

(2) **strong axiom of fuzzy revealed preference** 2 (*SAFRP2*) *if for all* $x, y \in X$ *and for all t such that* $0 < t \leq 1$, $\pi^*(x, y) \geq t$ *implies* $\rho(y, x) \leq 1 - t$.

WAFRP2 and SAFRP2 are also consistency conditions. Clearly, SAFRP1 (SAFRP2) implies WAFRP1 (WAFRP2). Also, SAFRP2 (WAFRP2) implies SAFRP1 (WAFRP1). Both WAFRP1 and WAFRP2 reduce to WARP in the case crisp. Similarly, both SAFRP1 and SAFRP2 reduce to SARP. Thus both WAFRP1 and WAFRP2 are valid fuzzifications of the crisp axiom WARP. Also, SAFRP1 and SAFRP2 are valid fuzzifications of SARP.

Definition 7.2.13 (*Banerjee* [5]) *Let C be a fuzzy choice function. Then*

(1) *C is called* **fuzzy congruent** 1 (**FC1**) *if* $\forall S \in \mathcal{E}$ *and* $\forall x \in S, y$ *dominant in* $S \Rightarrow C(S)(x) = \rho(x, y)$;

(2) *C is called* **fuzzy congruent** 2 (**FC2**) *if* $\forall S \in \mathcal{E}$ *and* $\forall x \in S, y$ *dominant in S and* $\rho(x, y) \Rightarrow x$ *is dominant in S*;

(3) *C is called* **fuzzy congruent** 3 (**FC3**) *if* $\forall S \in \mathcal{E}$ *and* $\forall x \in S$, *and all t such that* $0 < t \leq 1$, $\{C(S)(y) \geq t$ *and* $\rho(x, y) \geq t\} \Rightarrow C(S)(x) \geq t$;

(4) *C is called* **fuzzy congruent** *if it satisfies FC1, FC2, and FC3.*

It follows that in the crisp case each of the three conditions FC1, FC2, and FC3 reduces to the weak congruence condition defined in Sen [27]. However, in the fuzzy case these conditions are independent.

Banerjee states the following in [5]: A congruence condition was introduced in a crisp revealed preference theory by Richter [26]. Attention was focussed there on the case of the competitive consumer. It was based on the concept of "indirect revealed preference in the wide sense" which used the idea of a transitive closure of the crisp revealed preference relation. Richter's congruence condition is, in general, independent of SARP. In the case of non-satiety of the consumer, it is weaker. In the case studied by Sen (i.e., where the choice function is set valued and its domain includes all finite subsets of the universal set of alternatives), the weak congruence condition is equivalent to Richter's condition and both are equivalent to WARP as well as SARP.

The fuzzy congruence condition defined here is stronger than SAFRP2 and hence is stronger than SAFRP1, WAFRP2 and WAFRP2.

We next characterize rationality.

It is known in crisp revealed preference theory that SARP characterizes rationality. This provides the motivation for determining whether a fuzzy version of this proposition is true.

Example 7.2.14 *We show that SAFRP1 does not imply normality: Let* $X = \{x, y, z\}$ *and* r, s, t *real numbers such that* $0 < r < s < t < 1$. *Define the fuzzy choice function* C *as follows:*

$$C(\{x\})(x) = C(\{y\})(y) = C(\{z\})(z) = 1,$$
$$C(\{x, y\})(x) = 1, \quad C(\{y, z\})(y) = 1, \quad C(\{x, z\})(z) = 1,$$
$$C(\{x, y\})(y) = t, \quad C(\{y, z\})(z) = s, \quad C(\{x, z\})(z) = 0,$$
$$C(\{x, y, z\})(x) = 1, \quad C(\{x, y, z\})(y) = r, \quad C(\{x, y, z\})(z) = 0.$$

By Definition 7.2.4, we get the following values for ρ : $\rho(x, x) = \rho(y, y) = \rho(z, z) = 1$ *and* $\rho(x, y) = 1$, $\rho(y, x) = t$, $\rho(y, z) = 1$, $\rho(z, y) = s$, $\rho(x, z) = 1$, $\rho(z, x) = 0$. *By Definition 7.2.5,*

$$\begin{aligned}
\widetilde{\pi}(x, y) &= 0 \vee ((C(\{x, y\})(x) - C(\{x, y\})(y)) \\
&\quad \vee (C(\{x, y, z\})(x) - C(\{x, y, z\})(y))) \\
&= (1 - s) \vee (1 - r) = 1 - r.
\end{aligned}$$

We also have

$$\widetilde{\pi}(y, z) = s, \quad \widetilde{\pi}(x, z) = 1, \quad \widetilde{\pi}(y, x) = 0, \quad \widetilde{\pi}(z, y) = 0, \quad \widetilde{\pi}(z, x) = 0.$$

It follows for all $x, y \in X$ *such that* $\widetilde{\pi}(x, y) > 0$ *that* $\rho(y, x) < 1$. *It also follows that there does not exist* (u, v) *such that* $\widetilde{\pi}(u, v) = 0$, *but* $\widetilde{\pi}^*(u, v) > 0$. *Thus SAFRP1 holds. However,* $\widehat{C}(\{x, y, z\})(y) = \wedge\{\rho(y, x), \rho(y, y), \rho(y, z)\} = t \neq r = C(\{x, y, z\})(y)$. *Hence* C *is not normal.*

It follows that WAFRP1 does not imply normality. However, it can be shown that WAFRP1 and hence SAFRP1 imply a weak form of normality in that $\forall S \in \mathcal{E}$ and $\forall x \in S$, $C(S)(x) > 0 \iff \widehat{C}(S)(x) > 0$. However, this is insufficient for generating a fuzzy choice function back from its own fuzzy revealed preference relation.

We establish the following lemmas in order to study the rationality implications of the other fuzzy revealed preference conditions stated previously.

Lemma 7.2.15 *If* C *is fuzzy congruent 1, then it is normal.*

Proof. It follows that $\forall S \in \mathcal{E}$ and $\forall x \in S$, $C(S)(x) \leq \widehat{C}(S)(x)$. Suppose C is not normal. Then $\exists S \in \mathcal{E}$ and $x \in S$ such that $C(S)(x) < \widehat{C}(S)(x) = \wedge\{\rho(x, y) \mid S \in \mathcal{E}, y \in S\}$. Thus $C(S)(x) < \rho(x, y) \; \forall y \in S$. This a contradiction of fuzzy congruence 1 since the inequality does not hold for y which is dominant in S. ∎

The aspect of rationality provided by FC1 guarantees that a fuzzy choice function can be generated back from the preference relation revealed by it.

Lemma 7.2.16 *Suppose* C *is fuzzy congruence 1. Then* $\forall S \in \mathcal{E}$ *and* $\forall x \in S$, *if* x *is dominant in* S, *then* x *is relation dominant in* S *in terms of* ρ.

Proof. Let C be FC1 and x be dominant in $S \in \mathcal{E}$. Then $C(S)(y) = \rho(y, x) \forall y \in S$. However, $C(S)(x) \geq C(S)(y)$. Hence if $\exists y \in S$ such that $\rho(x, y) < \rho(y, x)$, then $\rho(y, x) > \rho(x, y) \geq C(S)(x) \geq C(S)(y) = \rho(y, x)$, a contradiction. ∎

We ask the reader to prove the following result.

Lemma 7.2.17 *Suppose C is fuzzy congruent 2. Then $\forall S \in \mathcal{E}$ and $\forall x \in S$, if x is relation dominant in S in terms of ρ, then x is dominant in S.*

Lemmas 7.2.16 and 7.2.17 show that under FC1 and FC2 the two concepts of dominance and relation dominance become equivalent.

Lemma 7.2.18 *Suppose C is fuzzy congruent 1 and fuzzy congruent 2. Then ρ is weakly transitivity.*

Proof. Suppose that there exist $x, y, z \in X$ such that $\rho(x, y) \geq \rho(y, x)$ and $\rho(y, z) \geq \rho(z, y)$. Let $S = \{x, y, z\}$. If y is dominant in S, $\rho(x, y) \geq \rho(y, x)$ implies that x is dominant in S by FC2. If z is dominant in S, $\rho(y, z) \geq \rho(z, y)$ implies that y is dominant in S. Hence by x is dominant in S. Thus x is dominant and so relation dominant in S in all cases by FC1. Hence $\rho(x, z) \geq \rho(z, x)$. ∎

Lemma 7.2.19 *Suppose C is fuzzy congruence 1 and fuzzy congruence 2. Then C satisfies CRPD.*

Proof. Assume FC1 and FC2. Suppose there exist $S \in \mathcal{E}$ and $x, y \in S$ such that x is relation dominant in S, but $\rho(x, z) < \rho(y, x)$. Then x is dominant in S since it is relation dominant by Lemma 7.2.18. Now $\rho(y, x) > \rho(x, z) \geq C(S)(x) \geq C(S)(y)$. However, this contradicts FC1 since x is dominant. ∎

Lemma 7.2.20 *If C is fuzzy congruent 1, then ρ is weakly reflexive.*

Proof. Since FC1 implies normality, we have if for any $x \in X$, $\rho(x, x) = 0$, then $C(\{x\})(x) = \hat{C}(\{x\})(x) = \rho(x, x) = 0$, a contradiction. Thus $\rho(x, x) > 0$ $\forall x \in X$. Suppose that there exist $x, y \in X$ such that $\rho(x, x) < \rho(x, y)$. Let $S = \{x, y\}$. Since FC1 implies normality, $C(S)(x) = \rho(x, y)$. Similarly, $C(S)(y) = \rho(y, x)$. If $\rho(x, y) \geq \rho(y, x)$, then $C(S)(x) \geq C(S)(y)$ and so x is dominant in S. Hence $C(S)(x) = \rho(x, x)$ by FC1. Thus $\rho(x, x) = \rho(x, y)$, a contradiction. If $\rho(y, x) > \rho(x, y)$, then $C(S)(y) > C(S)(x)$ and so y is dominant in S. Hence by FC1, $C(S)(y) = \rho(y, y)$. Thus $\rho(y, x) = \rho(y, y)$. The assumption that $C(\{x\})(x) = C(\{y\})(y)$ implies $\rho(x, x) = \rho(y, y)$. Hence $\rho(y, x) = \rho(y, y) = \rho(x, x) < \rho(x, y)$, a contradiction. Thus ρ is weakly reflexive. ∎

The above lemmas show that FC1 and FC2 imply a large part of rationality as defined in Definition 7.2.8. However, they do not imply full rationality which includes Condition A as well as weak completeness and max-min transitivity of the revealed preference relation.

Lemma 7.2.21 *Suppose C is fuzzy congruent 1. Then $\forall S \in \mathcal{E}$ and $\forall x, y, z \in X$, if x is dominant in S, then $\rho(x, z) \geq \rho(y, z)$.*

Proof. Suppose there exist $S \in \mathcal{E}$ and $x, y, z \in X$ such that x is dominant in S, but $\rho(x, z) < \rho(y, z)$. Then by normality and FC1, $\rho(y, z) > \rho(x, z) \geq \widehat{C}(S)(x) = C(S)(x) = \rho(x, x) = \rho(y, y)$. This contradicts the fact that ρ is weakly reflexive by Lemma 7.2.20. ∎

Theorem 7.2.22 (*Banerjee* [5]) *Let C be a fuzzy choice function. Then C satisfies FC if and only if C is rational.*

Proof. Suppose C satisfies FC. By Lemmas 7.2.15-7.2.20, we have that C is normal, C satisfies RDRP and CRPD and that the fuzzy revealed preference relation ρ is weakly reflexive and weakly transitive.

Suppose $\exists x, y \in X$ such that $\rho(x, y) = \rho(y, x) = 0$. Let $S = \{x, y\}$. By normality, we have $C(S) = \emptyset$, which is impossible. Thus ρ is complete. Let $x, y, z \in X$ be such that $\rho(x, z) < \rho(x, y) \wedge \rho(y, z) = t$, say. Then $\rho(x, y) \geq t$ and $\rho(y, z) \geq t$. Therefore,

$$\rho(x, y) \geq t > \rho(x, z). \tag{7.14}$$

Also,

$$\rho(x, z) < \rho(y, z). \tag{7.15}$$

Let $S = \{x, y, z\}$. Then

$$\widehat{C}(S)(x) = \rho(x, y) \wedge \rho(x, z) = \rho(x, z), \text{ by } (7.14).$$

Hence $C(S)(x) = \rho(x, z)$ by normality.

Now if y is dominant in S, FC1 implies $C(S)(x) = \rho(x, y) \geq t$. Since $\rho(x, y) \geq t$ and by FC1, $C(S)(x) = \rho(x, z) < t$ and so FC3 is contradicted. Thus x is dominant in S. This contradicts Lemma 7.2.21 by (7.15).

Conversely, assume C is rational. Assume that C is normal and ρ is a weak fuzzy ordering satisfying RDRP and CRPD. Suppose FC1 doesn't hold. Then there exists $S \in \mathcal{E}$ and $x, y \in S$ such that y is dominant in S, but $C(S)(x) < \rho(x, y)$. Since C is normal, $C(S)(x) = \widehat{C}(S)(x)$. Let $\widehat{C}(S)(x) = \wedge\{\rho(x, y) \mid y \in S\} = \rho(x, z)$ for some $z \in S$. Then $C(S)(x) = \rho(x, z) < \rho(x, y)$. If $\rho(x, y) \leq \rho(y, z)$, then max-min transitivity is violated. Hence $\rho(x, y) > \rho(y, z)$. But then CRPD doesn't hold since y is dominant in S by RDRP.

Let $S \in E$ and $x, y \in S$ be such that y is dominant in S and $\rho(x, y) \geq \rho(y, x)$. Since y is dominant in S, rationality implies FC1 which in turn implies $\rho(y, z) \geq \rho(z, y)$ by Lemma 7.2.16. Now weak transitivity implies $\rho(x, z) \geq$

$\rho(z, x)$ and so x is relation dominant in S. Hence x is dominant in S. Thus rationality implies FC2.

Let $S \in \mathcal{E}$ and $x, y \in S$ be such that

$$C(S)(y) \geq t, \tag{7.16}$$

and

$$\rho(x, y) \geq t. \tag{7.17}$$

By (7.16), $\rho(y, z) \geq t \; \forall z \in S$. Hence by the max-min transitivity of ρ, (7.17) implies $\rho(x, z) \geq t \; \forall z \in S$. Thus $\widehat{C}(S)(x) \geq t$. Hence by normality $C(S)(x) \geq t$. Consequently, FC3 by rationality. $\blacksquare$

Theorem 7.2.22 shows that fuzzy congruence constitutes a full characterization of rationality of a fuzzy choice function.

We next examine the relationship between fuzzy congruence and other fuzzy revealed preference conditions.

Theorem 7.2.23 (*Banerjee* [5]) *Let C be a fuzzy choice function. If C is rational, then WAFRP2 holds.*

Proof. Suppose WAFRP2 does not hold. Then $\exists x, y \in X$ and $t, 0 < t \leq 1$, such that

$$\widetilde{\pi}(x, y) \geq t, \tag{7.18}$$

$$\rho(y, x) > 1 - t. \tag{7.19}$$

(7.18) implies $\exists S \in \mathcal{E}$ such that $C(S)(x) - C(S)(y) \geq t$. Now S has a dominant element, say z. Thus

$$\begin{aligned} C(S)(y) &= C(S)(x) - t \leq R(x, y) - t \\ &\leq 1 - t \\ &< \rho(y, x) \text{ by (7.19).} \end{aligned}$$

If $\rho(y, x) \leq \rho(y, z)$, $C(S)(y) < \rho(y, z)$ so that FC1 is violated and consequently so is rationality. Thus $\rho(x, y) > \rho(y, z)$. If $\rho(y, x) \leq \rho(x, z)$, then ρ fails to be intensely transitive. Hence, $\rho(y, x) > \rho(x, z)$. If $\rho(y, z) < \rho(x, z)$, then max-min transitivity is violated. Hence $\rho(y, z) \geq \rho(x, z)$. Thus

$$\begin{aligned} C(S)(y) &\leq C(S)(x) - t < C(S)(x) \leq \rho(x, y) \\ &\leq \rho(y, z). \end{aligned}$$

Since z is dominant in S, the strict inequality violates FC1 and thus rationality. $\blacksquare$

We see from Theorem 7.2.22 that FC $\Rightarrow$ WAFRP2.

Lemma 7.2.24 *Let C be a fuzzy choice function If C satisfies FC1, then$\forall x, y, z \in X$ and $\forall t, 0 < t \leq 1$, $(\rho(x, y) \leq 1 - t$ and $\rho(y, z) \leq 1 - t) \Rightarrow \rho(x, z) \leq 1 - t$.*

Proof. Suppose that there exist $x, y, z \in X$ such that $\rho(x, y) \leq 1 - t$ and $\rho(y, z) \leq 1 - t$, but $\rho(x, z) > 1 - t$.

Let $S = \{x, y, z\}$. If z is dominant in S, $C(S)(x) = \rho(x, z)$ since C satisfies FC1. FC1 $\Rightarrow C(S)(x) = \rho(x, z)$. But $\rho(x, z) > 1 - t \geq \rho(x, y)$. Since FC1 implies normality, $C(S)(x) = \widehat{C}(S)(x) = \rho(x, y)$. Hence $\rho(x, z) = \rho(x, y)$, contradiction.

If x is dominant in S, $C(S)(x) = \rho(x, x)$ by FC1. Thus $\rho(x, x) = \rho(x, y) < \rho(x, z)$ as above. However, this contradicts the weak reflexivity of ρ which holds by FC1.

Suppose y is dominant in S. Then by Lemma 7.2.21, $\rho(y, z) \geq \rho(x, z) > 1 - t$. This contradicts the hypothesis $\rho(y, z) \leq 1 - t$. However, $C(S) \neq \emptyset$. Hence it must be the case that $\rho(x, z) \leq 1 - t$. ∎

Proposition 7.2.25 *Let C be a fuzzy choice function. If C satisfies FC1, then WAFRP2 and SAFRP2 are equivalent.*

Proof. It suffices to show that SAFRP2 follows from FC1 and WAFRP2. Suppose that $x, y \in X$ are such that $\pi^*(x, y) \geq t$ for some $t, 0 < t \leq 1$. Then there exists $x_0, x_1, \ldots, x_n \in X$, $x_0 = x$, $x_n = y$ and $\widetilde{\pi}(x_{i-1}, x_i) \geq t$ for all $i = 1, 2, \ldots n$. Since C satisfies FC1, repeated application of Lemma 7.2.24 7.2.24 yields $\rho(y, x) \leq 1 - t$. ∎

It follows that rationality implies SAFRP2 since rationality implies FC1. However, WAFRP2 and SAFRP2 are not equivalent. We show in the following example that neither of these axioms guarantees rationality.

Example 7.2.26 [5] *We show that WAFRP2 does not imply SAFRP2. Let $X = \{x, y, z\}$ and s be a real number such that $\frac{1}{2} < s < 1$. Define the fuzzy choice function C as follows:*

$$C(\{x\})(x) = C(\{y\})(y) = C(\{z\})(z) = 1,$$
$$C(\{x, y\})(x) = 1, \; C(\{y, z\})(y) = 1, \; C(\{x, z\})(x) = 1,$$
$$C(\{x, y\})(y) = s, \; C(\{y, z\})(z) = s, \; C(\{x, z\})(z) = 1,$$
$$C(\{x, y, z\})(x) = s, \; C(\{x, y, z\})(y) = s, \; C(\{x, y, z\})(z) = s.$$

The fuzzy revealed preference relation ρ is as follows: $\rho(x, x) = \rho(y, y) = \rho(z, z) = 1$. $\rho(x, y) = \rho(y, z) = \rho(x, z) = \rho(z, x) = 1$, $\rho(y, x) = \rho(z, y) = s$.
Thus $\widetilde{\pi}(x, y) = \widetilde{\pi}(y, z) = 1 - s$, $\widetilde{\pi}(x, z) = \widetilde{\pi}(y, x) = \widetilde{\pi}(z, y) = \widetilde{\pi}(z, x) = 0$.
For $t > 1 - s$, there does not exist (u, v) such that $\widetilde{\pi}(u, v) \geq t$. If t is such that $0 < t \leq 1 - s$, $\widetilde{\pi}(x, y) \geq t$, $\rho(y, x) \leq 1 - t$; $\widetilde{\pi}(y, z) \geq t$, $\rho(z, y) \leq 1 - t$. There does not exist any other (u, v) such that $\widetilde{\pi}(u, v) \geq t$. Thus WAFRP2 holds. However, $\pi^*(x, z) \geq t$ and $\rho(z, x) = 1 > 1 - t$. Hence SAFRP2 does not hold.

Example 7.2.27 [5] *We show that SAFRP2 does not imply normality of the choice function. We modify Example 7.2.26 by letting* $C(\{y, z\})(y) = s$. *The revealed preference relation* ρ *is now as follows:* $\rho(x, x) = \rho(y, y) = \rho(z, z) = 1$; $\rho(x, y) = \rho(x, z) = \rho(z, x) = 1 - s$; $\rho(y, x) = \rho(y, z) = \rho(z, y) = s$.

The $\widetilde{\pi}$ *relation is now such that* $\widetilde{\pi}(x, y) = 1 - s$, $\widetilde{\pi}(x, z) = \widetilde{\pi}(z, x) = \widetilde{\pi}(y, z) = \widetilde{\pi}(y, x) = \widetilde{\pi}(z, y) = 0$.

It follows that WAFRP2 holds. Since there is no ordered triple (x, y, z) *and no* t, $0 < t \leq 1$, *such that* $\widetilde{\pi}(x, z) < t$, *but* $\widetilde{\pi}(x, y) \geq t$ *and* $\widetilde{\pi}(y, z) \geq t$, *SAFRP2 also holds. The choice function* C *is not normal since for* $S = \{x, y, z\}$, $\widehat{C}(S)(x) = \rho(x, y) \wedge \rho(x, z) = 1 \wedge 1 = 1$, *but* $C(S)(x) = s$.

We have demonstrated that FC $\Leftrightarrow$ Rationality $\Rightarrow$ SAFRP2 $\Rightarrow$ WAFRP2 and neither of the last two arrows can be reversed.

We now consider the case where there is at least one unambiguous choice, i.e., $\forall S \in \mathcal{E}$, $\exists x \in S$ such that $C(S)(x) = 1$. Note that the choice function continues to be fuzzy.

We note that FC3 implies FC1. This holds since if x is dominant in $S \in \mathcal{E}$, but $C(S)(y) < \rho(y, x) = t$, then $C(S)(x) = 1 \geq t$, $\rho(y, x) = t$. However, $C(S)(y) < t$ so that FC3 is violated. Similarly, it can be shown that FC3 implies FC2. Thus FC and FC3 are now equivalent.

Conditions A and B are now trivially satisfied by every well-defined fuzzy choice function since if x is dominant in $S \in \mathcal{E}$, $C(S)(x) = 1$ so that by definition, $\rho(x, y) = 1 \; \forall y \in S$.

A fuzzy preference relation is called a **fuzzy ordering** if it is reflexive, strongly complete, and intensely transitive.

Definition 7.2.28 *Let* C *be a choice function. Suppose* $\forall S \in \mathcal{E}$, $\exists x \in S$ *such that* $C(S)(x) = 1$. *Then* C *is called* **rational** *if* C *is normal and the fuzzy revealed preference relation* ρ *is a fuzzy ordering.*

Proposition 7.2.29 *Let* C *be a choice function. Suppose* $\forall S \in \mathcal{E}$, $\exists x \in S$ *such that* $C(S)(x) = 1$. *Then* C *satisfies FC if and only if* C *is rational.*

There is now a significant change in the result regarding WAFRP2 and SAFR2.

The proof of the next result is similar to the crisp case, Arrow [2] and Sen [28].

Proposition 7.2.30 *Let* C *be a choice function. Suppose* $\forall S \in \mathcal{E}$, $\exists x \in S$ *such that* $C(S)(x) = 1$. *Then WAFRP2 and SAFRP2 are equivalent.*

Theorem 7.2.31 *(Banerjee* [5]*) Let* C *be a choice function. Suppose* $\forall S \in \mathcal{E}$, $\exists x \in S$ *such that* $C(S)(x) = 1$. *Then WAFRP2 implies normality of the choice function and weak transitivity of the fuzzy revealed preference relation.*

Proof. By Lemma 7.2.15, it suffices to show for normality that WAFRP2 implies FC1. If FC1 does not hold for C, $\exists S \in \mathcal{E}$ and $x, y \in S$ such that x is dominant in S and $C(S)(y) < \rho(x, y)$. Since $C(S)(x) = 1$,

$$C(S)(x) - C(S)(y) > C(S)(x) - \rho(y, x) = 1 - \rho(y, x).$$

Thus $\widetilde{\pi}(x, y) > 1 - \rho(y, x)$. Let $\widetilde{\pi}(x, y) = t$. If WAFRP2 holds, then $\rho(y, x) \leq 1 - t < \rho(y, x)$, a contradiction.

For the weak transitivity of ρ, it suffices in view of Lemma 7.2.16 to show in addition that WAFRP2 implies FC2. If FC2 does not hold, $\exists S \in \mathcal{E}$ and $x, y \in S$ such that y is dominant in S, $\rho(x, y) \geq \rho(y, x)$, but x is not dominant in S. Therefore, $C(S)(y) = 1 > C(S)(x)$. Hence $C(S)(y) - C(S)(x) = 1 - C(S)(x) = t > 0$. Thus $\widetilde{\pi}(y, x) \geq t$. If WAFRP2 holds, then $\rho(x, y) \leq 1 - t < 1$. However, since $C(S)(y) = 1$, $\rho(y, x) = 1$. Hence $\rho(x, y) \geq \rho(y, x) = 1$, a contradiction. Thus WAFRP2 does not hold. ∎

By Proposition 7.2.30, the Theorem 7.2.31 holds for SAFRP2. Since a well-defined choice function guarantees that ρ is reflexive and complete, Theorem 7.2.31 shows that WAFRP2 and SAFRP2 come close to fulfilling the requirements of rationality. However, Example 7.2.32 shows that a gap remains.

Example 7.2.32 *We show that SAFTRP2 (hence, WAFRP2) does not imply max-min transitivity of ρ. We modify Example 7.2.14 by letting $C(\{x, y\})(y) = 0.4$, leaving other specifications unchanged. Then $\rho(y, x) = 0.4$, but for all other ordered pairs $(u, v) \in X \times X$, $\rho(u, v)$ remains the same as in Example 7.2.14. It follows easily that $\widetilde{\pi}(x, y) = 1 - \rho(y, x) \; \forall x, y \in X$, which is a sufficient condition for WAFRP2 and hence SAFRP2. However, $0 = \rho(z, x) < \rho(z, y) \wedge \rho(y, x) = 0.5 \wedge 0.4 = 0.4$. Hence ρ is not max-min transitive.*

When at least one unambiguous choice is assumed, we have FC ⇔ Rationality ⇒ SAFRP2 ⇔ WAFRP2.

7.3 Exercises

1. Show that all the PCFs introduced in Definition 7.1.3 satisfy the property that for all $\rho \in \mathcal{F}(R)'$ and for all $x, y \in X$, $\rho(x, y) \geq \rho(y, x)$ (resp. $\rho(y, x) \geq \rho(x, y)$) implies $x \in C(\{x, y\})$.

   Let $C : \mathcal{P}^*(X) \times \mathcal{F}(R)' \to \mathcal{P}^*(X)$ be a PCF. Define J^1 and J^2 as follows:

   (i) $C \in J^1$ if and only if there exists $\alpha \geq 0.5$ such that for all $B \in P^*(X)$ and all $\rho \in \mathcal{F}(R)'$,

   $$C(B, \rho) = \{x \in B \mid \rho(x, y) \geq \alpha \text{ for all } y \in B\};$$

   (ii) $C \in J^2$ if and only if for all $B \in \mathcal{P}^*(X)$ and all $\rho \in \mathcal{F}(R)'$, $C(B, \rho) = T(B, \rho)$.

$(J^1$ is a slightly generalized version of some PCFs presented in [11] and [15]. J^2 has been discussed by Dutta, Panda, Pattanaik [15].)

2. Let $C : \mathcal{P}^*(X) \times \mathcal{F}(R)' \to \mathcal{P}^*(X)$ be a PCF. If $C \in J^1 \cup J^2$, then prove that for all $\rho \in \mathcal{F}(R)'$, there exists an exact $\rho' \in \mathcal{F}(R)$ such that for all $B \in \mathcal{P}^*(X)$, $C(B, \rho) = D(B, \rho')$.

3. Let $C : \mathcal{P}^*(X) \times \mathcal{F}(R)' \to \mathcal{P}^*(X)$ be a max-mD PCF, where $\mathcal{F}(R)' \subseteq \mathcal{FT}(R)$. Prove that $C \in J^2$.

4. Use Definition 7.2.5 to show that for all $x, y \in X$, $\tilde{\pi}^*(x, y) > 0$ implies $\tilde{\pi}^*(y, x) = 0$.

5. [4] Use numerical examples to show that max-min transitivity and weak transitivity are independent conditions in the set inclusion sense.

6. Show that conditions FC1, FC2, and FC3 are independent.

7. Prove Lemma 7.2.17.

8. Suppose $\forall S \in \mathcal{E}$, $\exists x \in S$ such that $C(S)(x) = 1$. Prove Proposition 7.2.29.

The following exercises are based on [33]. In the following exercises, a fuzzy choice function is a function $C : \mathcal{E} \to \mathcal{FP}^*(X)$

Let C be a fuzzy choice function and ρ the revealed preference. Consider the statements:

(1) ρ is a *-regular revealed preference relation and C is normal;

(2) $\overline{\rho}$ is a *-regular generated preference relation and C is normal;

(3) C satisfies WFCA;

(4) C satisfies SFCA;

(5) C satisfies WAFRP;

(6) C satisfies SAFRP;

(7) $\rho = \tilde{\rho}$;

(8) $\overline{\rho} = \tilde{\rho}$ and C is normal.

9. Prove that

(*i*) (1), (2), (3), and (4) are equivalent.

(*ii*) (5), (6), (7), and (8) are equivalent.

(*iii*) If (T, S, N) is a strong DeMorgan triple, then (3) $\Leftrightarrow$ (5).

10. Let N denote a strong negation.

Condition $F\alpha$: If $S_1 \subseteq S_2$, then $C(S_2)(x) \leq C(S_1)(x)$ for any $x \in S_1$.

Condition $F\alpha_2$: For all $x, y \in S$, $C(S)(x) \leq C(\{x, y\})(x)$.

Condition $F\beta$: If $S_1 \subseteq S_2$ and $x, y \in S_1$, then $C(S_1)(x) * C(S_1)(y) * C(S_2)(x) \leq C(S_2)(y)$.

Condition $F\beta^+$: If $S_1 \subseteq S_2$ and $x, y \in S_1$, then $C(S_1)(x) * C(S_2)(y) \leq C(S_2)(y)$.

Condition $F\delta$: If $S_1 \subseteq S_2, x, y \in S_1$, and $x \neq y$, then $C(S_1)(x) * C(S_1)(y) \leq N(C(S_2)(x) * (*\{N(C(S_2)(z)) \mid z \neq x\}))$.

Condition $F\beta_2$: For all $x, y \in S, C(\{x,y\})(x) * C(\{x,y\})(y) * C(S)(x) \leq C(S)(y)$.

Condition $F\beta'$: For all $x, y \in S_1 \cap S_2, C(S_1)(x) * C(S_1)(y) * C(S_2)(x) \leq C(S_2)(y)$.

Condition $F\gamma$: For all $x \in S_1 \cap S_2, C(S_1)(x) * C(S_2)(x) \leq C(S_1 \cup S_2)(y)$.

Prove that if a fuzzy choice function C satisfies $F\beta$, then it satisfies $F\delta$.

Prove that if C is a fuzzy choice function that salsifies $F\beta^+$ and $*$ is continuous, then C satisfies $F\beta, F\gamma$, ρ is $*$-transitive.

11. If C is a fuzzy choice function that satisfies $F\alpha_2$ and $F\beta_2$, prove that C satisfies $F\alpha, F\beta', F\beta, F\gamma$, and $F\delta$.

12. Prove that a fuzzy choice function C satisfies WFCA if and only if it satisfies $F\alpha$ and $F\beta^+$.

13. Prove that if a fuzzy choice function satisfies $F\alpha$ and $F\gamma$, then C is normal under the Gödel t-norm.

14. Prove that if a fuzzy choice function C satisfies $F\alpha$ and $F\delta$, then ρ is $*$-quasi-transitive.

7.4 References

1. M. M. Aizerman and A. V. Malishevski, General theory of best-variant choices. Some aspects, *IEEE Trans. Automat. Control*, 26 (1981), 1030–1040.

2. K. J. Arrow, Rational choice functions and ordering, *Economica*, NS 26 (1959) 121–127.

3. A. K. Banerjee, On deriving strict preference and indifference from a fuzzy weak preference relation, Mimeo, Dept. of Economics, Calcutta Univ. (1991).

4. A. K. Banerjee, Rational choice under fuzzy preferences: The Orlovsky choice function, *Fuzzy Sets and Systems*, 54 (1993) 295–299.

5. A. K. Banerjee, Fuzzy choice functions, revealed preference and rationality, *Fuzzy Sets and Systems*, 70 (1995) 31–43.

6. C. R. Barrett, Fuzzy preferences and choice: A progress report, *Fuzzy Sets and Systems*, 21 (1987) 127–129.

7. C. R. Barratt and P. K. Pattanik, On vague preferences, in: H. von G. Enderle, Ed., *Ethik and Wirtschafwswissenschaftz*, Dunker and Humbolt, Berlin 1985.

8. C. R. Barrett, P. K. Pattanaik, and M. Salles, On the structure of fuzzy social welfare functions, *Fuzzy Sets and Systems*, 19 (1986) 1–11.

9. C. R. Barrett, P. K. Pattanaik, and M. Salles, On choosing rationally when preferences are fuzzy, *Fuzzy Sets and Systems*, 34 (1990) 197–212.

10. C. R. Barrett, P. K. Pattanaik, and M. Salles, Rationality and aggregation of preferences in an ordinally fuzzy framework, *Fuzzy Sets and Systems*, 49 (1992), 9–13.

11. K. Basu, Fuzzy revealed preference theory, *Journal Econom. Theory*, 32 (1984) 212–227.

12. M. Dasgupta and R. Deb, Rational choice with fuzzy preferences, Mimeograph (1987).

13. M. Dasgupta and R. Deb, Fuzzy Choice Functions, Mimeo, Dept. of Economics, Southern Methodist Univ., 1988.

14. B. Dutta, Fuzzy preferences and social choices, *Math. Social Sci.*, 13 (1987) 215–229.

15. B. Dutta, S. Panda, and P. R. Pattanaik, Exact choice and fuzzy preferences, *Math. Social Sci.*, 11 (1986) 53–68.

16. S. Fotso and L. A. Fono, On the consistency of some crisp choice functions based on a strongly complete fuzzy pre-order, *New Mathematics and Natural Computation*, 8 (2012) 257–272.

17. D. G. Gale, A note on revealed preferences, *Economica*, 27 (1960) 971–978.

18. H. S. Houthakker, Revealed preference and utility function, *Economica*, NS 17 (1950) 159–174.

19. J. Kacprzyck and M. Roubens, Eds., *Nonconventional Preference Relations in Decison-Making*, Lecture notes in Economics and Mathematical Systems, Springer, Berlin, 1988.

20. H. Nurumi, Approaches to collective decision-making with fuzzy preference relations, *Fuzzy Sets and Systems*, 6 (1981) 249–259.

21. S. A. Orlovsky, Decision-making with a fuzzy preference relation, *Fuzzy Sets and Systems*, 1 (1978) 155–167.

22. S. V. Ovchnnikov, Structure of fuzzy binary relations, *Fuzzy Sets and Systems*, 6 (1981) 169–195.

23. S. V. Ovchinnikov and V. R. Ozernoy, Using fuzzy binary relations for identifying noninferior decision alternatives, *Fuzzy Sets and Systems*, 25 (1988) 21–32.

24. H. Peters and P. Wakker, Independence of irrelevant alternatives and revealed group preference, *Econometrica*, 59 (1992) 1787–1801.

25. C. Ponsard, An application of fuzzy subsets theory to the analysis of the consumer's spatial preference, *Fuzzy Sets and Systems*, 5 (1981) 235–244.

26. M. Richter, Revealed preference theory, *Econometrica*, 34 (1996) 635–645.

27. P. A. Samuelson, A note on the pure theory of consumer's behavior, *Economica*, 5 (1938) 61–67, 353–354.

28. A. K. Sen, Choice functions and revealed preference, *Rev Economica Studied*, 38 (1971) 307–317.

29. A. K. Sen, Social choice theory: A re-examination, *Econometrica*, 45 (1977) 53–89.

30. Z. Switalski, Choice functions associated with fuzzy preference relations, in [19] 106–118.

31. J. Ville, Sur les conditions d'existence of d'une ophelimite totale et d'un indice du niveau des pix, *Annales de l' Universite de Lyon,* 9 (1946) 32–29.

32. J. von Neumann and O. Morgenstern, *Theory of Games and Economic Behavior*, Princeton University Press, Princeton 1957.

33. C. Wu, X. Wang, and Y. Hao, A further study on rationality conditions of fuzzy choice functions, *Fuzzy Sets and Systems,* 176 (2011) 1–19.

Chapter 8

Arrow-Type Results under Intuitionistic Fuzzy Preferences

8.1 Fuzzy Preference Profiles and Fuzzy Aggregation Rules

The proofs of many factorization results for an intuitionistic fuzzy binary relation $\langle \rho_\mu, \rho_\nu \rangle$ involve dual proofs, one for ρ_μ with respect to a t-conorm $\oplus$ and one for ρ_ν with respect to a t-norm $\otimes$. We show that one proof can be obtained from the other by considering $\oplus$ and $\otimes$ dual under an involutive fuzzy complement. This section follows the development of that in [23].

Recall that an intuitionistic fuzzy set consists of a set X and two fuzzy subsets μ and ν of X such that $\mu(x) + \nu(x) \leq 1$. In [16], a factorization of an intuitionistic fuzzy binary relation into a unique indifference component and a family of regular strict components was provided. This result generalizes a result in [10] with the (max, min) intuitionistic fuzzy t-conorm. In [16], a characterization of C-transitivity for a continuous t-representable intuitionistic fuzzy t-conorm C was established. This enabled the authors to determine necessary and sufficient conditions on a C-transitive intuitionistic fuzzy binary relation ρ under which a component of ρ satisfies pos-transitivity and negative transitivity.

In [9], some results on t-representable intuitionistic fuzzy t-norms (T, S) were established, where T is a fuzzy t-norm and S is a fuzzy t-conorm such that $T(a, b) = 1 - S(1 - a, 1 - b)$ for all $a, b \in [0, 1]$.

We extend some known Arrowian results involving fuzzy set theory to results involving intuitionistic fuzzy sets. Many of the proofs for intuitionistic fuzzy sets can be obtained from existing results for fuzzy sets in a dual manner.

We point out how the use of an involutive fuzzy complement can be used to obtain the results.

It was shown in Chapter 5 that Arrow's theorem, [2, 30, 5, 32], could be avoided by using fuzzy preferences and by using a weak version of transitivity. In [6], it is shown that the choice of definitions for indifference and strict preference associated with a given fuzzy preference relation also have Arrowian implications. In [26], it is shown that the dictatorship theorem in [6] can be strengthened to cover any version of transitivity for fuzzy preferences no matter how weak, and that this dictatorship result holds for any regular strict preference including the one used in [6].

In [16], the notion of intuitionistic fuzzy relations was introduced to study Arrowian like conditions. In Chapter 5, other types of independence of irrelevant alternative conditions were introduced and it was shown that they can be profitably used in the examination of Arrow's theorem. In this chapter, we extend some known Arrowian results involving fuzzy set theory to results involving intuitionistic fuzzy sets. The interested reader can find results dealing with vague preference relations to study Arrow's Impossibility Theorem and invariants in [1, 17, 12, 13, 14, 19, 21, 26, 31].

In this section, we lay the groundwork for our study of fuzzy Arrow results.

An **intuitionistic fuzzy binary relation** on X is an ordered pair $\langle \rho_\mu, \rho_\nu \rangle$ of fuzzy binary relations on X such that $\forall x, y \in X$, $\rho_\mu(x, y) + \rho_\nu(x, y) \leq 1$. Let c be a fuzzy complement on $[0, 1]$. Recall that c is called involutive if $c(c(a)) = a$ for all $a \in [0, 1]$. It follows that an involutive fuzzy complement is a one-to-one continuous function of $[0, 1]$ onto $[0, 1]$. Let c be an involutive fuzzy complement of $[0, 1]$. Let ρ be a fuzzy binary relation on X. Define the fuzzy relation ρ^c on X by $\forall (x, y) \in X \times X$, $\rho^c(x, y) = c(\rho(x, y))$. If ρ is a fuzzy relation on X, then $\langle \rho, \rho^c \rangle$ is an intuitionistic fuzzy relation on X. Let $\mathcal{FR}$ denote the set of all fuzzy relations on X and define the function $f : \mathcal{FR} \to \mathcal{FR}$ by $\forall \rho \in \mathcal{FR}$, $f(\rho) = \rho^c$. Then f maps $\mathcal{FR}$ one-to-one onto $\mathcal{FR}$. We showed in Chapter 3 how a result concerning ρ^c can thus be determined from a result concerning ρ as long as the definitions involving ρ^c are dual to those of ρ with respect to c. This follows since a result concerning ρ^c is a general result for fuzzy preference relation due to the fact that f is one-to-one onto. We do not use this approach in this chapter. The proofs of results in Sections 8.1-8.4, can be compared to those in Chapter 3 to see their dual nature. We explain the situation further in Section 8.5.

Definition 8.1.1 *Let $\langle \rho_\mu, \rho_\nu \rangle$ be an intuitionistic fuzzy binary relation on X. Then*

(1) $\langle \rho_\mu, \rho_\nu \rangle$ *is called **reflexive** if $\forall x \in X, \rho_\mu(x, x) = 1$ and $\rho_\nu(x, x) = 0$;*

(2) $\langle \rho_\mu, \rho_\nu \rangle$ *is called **complete** if $\forall x, y \in X$, either $\rho_\mu(x, y) > 0$ or $\rho_\mu(y, x) > 0$ and $\rho_\nu(x, y) < 1$ or $\rho_\nu(y, x) < 1$;*

(3) $\langle \rho_\mu, \rho_\nu \rangle$ *is called **min-max transitive** or simply **transitive** if $\forall x, y, z \in X, \rho_\mu(x, y) \wedge \rho_\mu(y, x) \leq \rho_\mu(x, z)$ and $\rho_\nu(x, y) \vee \rho_\nu(y, x) \geq \rho_\nu(x, z)$.*

(4) $\langle \rho_\mu, \rho_\nu \rangle$ *is called* **partially transitive** *if* $\forall x, y, z \in X$, $\rho_\mu(x, y) > 0$ *and* $\rho_\mu(y, z) > 0$ *implies* $\rho_\mu(x, z) > 0$ *and* $\rho_\nu(x, y) < 1$ *and* $\rho_\nu(y, z) < 1$ *implies* $\rho_\nu(x, z) < 1$.

Definition 8.1.2 *Let* $\langle \rho_\mu, \rho_\nu \rangle$ *be an intuitionistic fuzzy binary relation on* X. *Then*

(1) $\langle \rho_\mu, \rho_\nu \rangle$ *is called* **symmetric** *if* $\forall x, y \in X$, $\rho_\mu(x, y) = \rho_\mu(y, x)$ *and* $\rho_\nu(x, y) = \rho_\nu(y, x)$.

(2) $\langle \rho_\mu, \rho_\nu \rangle$ *is called* **asymmetric** *if* $\forall x, y \in X$, $\rho_\mu(x, y) > 0$ *implies* $\rho_\mu(y, x) = 0$ *and* $\rho_\nu(x, y) < 1$ *implies* $\rho_\nu(y, x) = 1$.

Let T denote a set of intuitionistic fuzzy binary relations on X. An **intuitionistic fuzzy preference profile** on X is a n-tuple of intuitionistic fuzzy binary relations $\rho = (\langle \rho_{\mu 1}, \rho_{\nu 1} \rangle, \ldots, \langle \rho_{\mu n}, \rho_{\nu n} \rangle)$ describing the intuitionistic fuzzy preferences of all individuals. Let $T^n = \{ \rho = (\langle \rho_{\mu 1}, \rho_{\nu 1} \rangle, \ldots, \langle \rho_{\mu n}, \rho_{\nu n} \rangle) \mid \langle \rho_{\mu i}, \rho_{\nu i} \rangle \in T, i = 1, \ldots, n \}$. Let $\mathcal{P}(X)$ denote the power set of X. For any $\rho \in T^n$ and for all $S \in \mathcal{P}(X)$, let $\rho]_S = (\langle \rho_{\mu 1}]_S, \rho_{\nu 1}]_S \rangle, \ldots, \langle \rho_{\mu n}]_S, \rho_{\nu n}]_S \rangle)$, where $\rho_{\mu i}]_S = \rho_{\mu i}]_{S \times S}$, the restriction of $\rho_{\mu i}$ to $S \times S$, $i = 1, \ldots, n$ and similarly for $\rho_{\nu i}]_S$. For $\langle \rho_\mu, \rho_\nu \rangle \in T$, we associate an intuitionistic asymmetric fuzzy relation $\langle \pi_\mu, \pi_\nu \rangle$ with $\langle \rho_\mu, \rho_\nu \rangle$. For all $\rho \in T^n$ and $\forall x, y \in X$, let

$$R(x, y; \rho) = \{ i \in N \mid \rho_{\mu i}(x, y) > 0 \text{ and } \rho_{\nu i}(x, y) < 1 \}$$
$$\text{and } P(x, y; \rho) = \{ i \in N \mid \pi_{\mu i}(x, y) > 0 \text{ and } \pi_{\nu i}(x, y) < 1 \},$$

where $< \pi_{\mu i}, \pi_{\nu i} >$ is an intuitionistic asymmetric fuzzy binary relation associated with $< \rho_{\mu i}, \rho_{\nu i} >, i = 1, \ldots, n$.

Definition 8.1.3 *Let* $\langle \rho_\mu, \rho_\nu \rangle \in T$ *and* $\langle \pi_\mu, \pi_\nu \rangle$ *be an asymmetric intuitionistic fuzzy preference relation associated with* $\langle \rho_\mu, \rho_\nu \rangle$.

(1) $\langle \pi_\mu, \pi_\nu \rangle$ *is called* **simple** *if* $\forall x, y \in X$, $\rho_\mu(x, y) = \rho_\mu(y, x)$ *implies* $\pi_\mu(x, y) = \pi_\mu(y, x)$ *and* $\rho_\nu(x, y) = \rho_\nu(y, x)$ *implies* $\pi_\nu(x, y) = \pi_\nu(y, x)$.

(2) $\langle \pi_\mu, \pi_\nu \rangle$ *is called* **regular** *if* $\forall x, y \in X$, $\rho_\mu(x, y) > \rho_\mu(y, x) \Leftrightarrow \pi_\mu(x, y) > 0$ *and* $\rho_\nu(x, y) < \rho_\nu(y, x) \Leftrightarrow \pi_\nu(x, y) < 1$.

It follows easily that if π_μ is regular, then π_μ is simple as is the case for π_ν.

Definition 8.1.4 *Let* $\langle \rho_\mu, \rho_\nu \rangle \in T$. *Define* $\langle \pi_{(i)_\mu}, \pi_{(i)_\nu} \rangle \in T$, $i = 0, 1, 2, 3, 4, 5$, *as follows:* $\forall x, y \in X$,

(1) $\pi_{(0)_\mu}(x, y) = \rho_\mu(x, y)$ *if* $\rho_\mu(y, x) = 0$ *and* $\pi_{(0)_\mu}(x, y) = 0$ *otherwise;*

$\pi_{(0)_\nu}(x, y) = \rho_\nu(x, y)$ *if* $\rho_\nu(y, x) = 1$ *and* $\pi_{(0)_\nu}(x, y) = 1$ *otherwise;*

(2) $\pi_{(1)_\mu}(x, y) = \rho_\mu(x, y)$ *if* $\rho_\mu(x, y) > \rho_\mu(y, x)$ *and* $\pi_{(1)_\mu}(x, y) = 0$ *otherwise;*

$\pi_{(1)_\nu}(x, y) = \rho_\nu(x, y)$ *if* $\rho_\nu(x, y) < \rho_\nu(y, x)$ *and* $\pi_{(1)_\nu}(x, y) = 1$ *otherwise;*

(3) $\pi_{(2)_\mu}(x,y) = 1 - \rho_\mu(y,x)$;

$\quad \pi_{(2)_\nu}(x,y) = \rho_\nu(y,x)$;

(4) $\pi_{(3)_\mu}(x,y) = 0 \vee (\rho_\mu(x,y) - \rho_\mu(y,x))$.

$\quad \pi_{(3)_\nu}(x,y) = 1 \wedge (1 + \rho_\nu(x,y) - \rho_\nu(y,x))$.

(5) $\pi_{(4)\mu}(x,y) = (\rho_\mu(x,y) - \rho_\mu(y,x))/(1 - \rho_\mu(y,x))$ if $\rho_\mu(x,y) > \rho_\mu(y,x)$ and $\pi_{(4)_\mu}(x,y) = 0$ otherwise;

$\quad \pi_{(4)\nu}(x,y) = \rho_\nu(x,y)/\rho_\nu(y,x))$ if $\rho_\nu(x,y) < \rho_\nu(y,x)$ and $\pi_{(4)_\nu}(x,y) = 1$ otherwise;

(6) $\pi_{(5)\mu}(x,y) = (\rho_{\mu,}(x,y)^w - \rho_\mu(y,x)^w)^{1/w}$ if $\rho_\mu(x,y) > \rho_\mu(y,x)$ and $\pi_{(5)_\mu}(x,y) = 0$ otherwise, $\omega > 0$;

$\quad \pi_{(5)\nu}(x,y) = 1 - [(1 - \rho_\nu(x,y))^w - (1 - \rho_\nu(y,x))^w]^{1/w}$ if $\rho_\nu(x,y) < \rho_\nu(y,x)$ and $\pi_{(5)_\nu}(x,y) = 1$ otherwise, $\omega > 0$.

Definition 8.1.5 *A function* $\widetilde{f} : T^n \to T$ *is called a **fuzzy aggregation rule**.*

Let $\widetilde{f}$ be a fuzzy aggregation rule. Let $\rho = (\langle \rho_{\mu 1}, \rho_{\nu 1}\rangle, \ldots, \langle \rho_{\mu n}, \rho_{\nu n}\rangle) \in T^n$. Let $\mathcal{FR}_\mu^n(X) = \{(\rho_{\mu 1}, \ldots, \rho_{\mu n}) \mid (\langle \rho_{\mu 1}, \rho_{\nu 1}\rangle, \ldots, \langle \rho_{\mu n}, \rho_{\nu n}\rangle) \in T^n\}$ and $\mathcal{FR}_\nu^n(X) = \{(\rho_{\nu 1}, \ldots, \rho_{\nu n}) \mid (\langle \rho_{\mu 1}, \rho_{\nu 1}\rangle, \ldots, \langle \rho_{\mu n}, \rho_{\nu n}\rangle) \in T^n\}$. Let $\widetilde{f}_\mu : \mathcal{FR}_\mu^n(X) \to \mathcal{FR}(X)$ and $\widetilde{f}_\nu : \mathcal{FR}_\nu^n(X) \to \mathcal{FR}(X)$. Then $\widetilde{f}_\mu$ and $\widetilde{f}_\nu$ induce a fuzzy aggregation rule $\widetilde{f}$ on T. That is $\widetilde{f}(\rho) = \langle \widetilde{f}_\mu(\rho_\mu), \widetilde{f}_\nu(\rho_\nu)\rangle$, where $\rho_\mu = (\rho_{1_\mu}, \ldots, \rho_{n_\mu})$ and $\rho_\nu = (\rho_{1_\nu}, \ldots, \rho_{n_\nu})$. We sometimes write $\widetilde{f} = (\widetilde{f}_\mu, \widetilde{f}_\nu)$. We sometimes suppress the notation $\langle \widetilde{f}_\mu(\rho_\mu), \widetilde{f}_\nu(\rho_\nu)\rangle$ and write $\langle \rho_\mu, \rho_\nu \rangle$. In this case, we write $\langle \pi_\mu, \pi_\nu \rangle$ for the asymmetric fuzzy preference relation associated with $\langle \rho_\mu, \rho_\nu \rangle$.

Definition 8.1.6 *Let* $\widetilde{f} : T^n \to T$ *be a fuzzy aggregation rule. Then* $\widetilde{f}$ *satisfies*

(1) ***Independence of irrelevant alternatives (IIA1)*** *if* $\forall(\langle \rho_{\mu 1}, \rho_{\nu 1}\rangle, \ldots, \langle \rho_{\mu n}, \rho_{\nu n}\rangle), (\langle \rho'_{\mu 1}, \rho'_{\nu 1}\rangle, \ldots, \langle \rho'_{\mu n}, \rho'_{\nu n}\rangle) \in T^n, \forall x, y \in X,$

$$[\forall i \in N, \rho_{\mu i}(x,y) = \rho'_{\mu i}(x,y) \Rightarrow \rho_\mu(x,y) = \rho'_\mu(x,y)],$$
$$[\forall i \in N, \rho_{\nu i}(x,y) = \rho'_{\nu i}(x,y) \Rightarrow \rho_\nu(x,y) = \rho'_\nu(x,y)].$$

(2) ***Pareto condition (PC)*** *with respect to* $\langle \pi_\mu, \pi_\nu \rangle$ *if* $\forall(\langle \rho_{\mu 1}, \rho_{\nu 1}\rangle, \ldots, \langle \rho_{\mu n}, \rho_{\nu n}\rangle) \in T^n, \forall x, y \in X,$

$$\pi_\mu(x,y) \geq \wedge\{\pi_{\mu i}(x,y) \mid i \in N\},$$
$$\pi_\nu(x,y) \leq \vee\{\pi_{\nu i}(x,y) \mid i \in N\}.$$

(3) ***Positive responsiveness (PR)*** *with respect to* $\langle \pi_\mu, \pi_\nu \rangle$ *if* $\forall(\langle \rho_{\mu 1}, \rho_{\nu 1}\rangle, \ldots, \langle \rho_{\mu n}, \rho_{\nu n}\rangle), (\langle \rho'_{\mu 1}, \rho'_{\nu 1}\rangle, \ldots, \langle \rho'_{\mu n}, \rho'_{\nu n}\rangle) \in T^n, \forall x, y \in X,$

$$\rho_{\mu i} = \rho'_{\nu i} \; \forall i \neq j, \; \rho_\mu(x,y) = \rho_\mu(y,x), \; \text{and}$$
$$(\pi_{\mu j}(x,y) = 0 \; \text{and} \; \pi'_{\mu j}(x,y) > 0$$
$$\text{or} \; \pi_{\mu j}(y,x) > 0 \; \text{and} \; \pi'_{\mu j}(y,x) = 0) \Rightarrow \pi'_\mu(x,y) > 0.$$

$$\rho_{\nu i} \;=\; \rho'_{\nu i} \; \forall i \neq j, \; \rho_\nu(x,y) = \rho_\nu(y,x), \quad and$$
$$(\pi_{\nu j}(x,y) = 1 \;\; and \;\; \pi'_{\nu j}(x,y) < 1$$
$$or \; \pi_{\nu j}(y,x) < 1 \;\; and \;\; \pi'_{\nu j}(y,x) = 1) \Rightarrow \pi'_\nu(x,y) < 1.$$

Let $\rho = (\langle \rho_{\mu 1}, \rho_{\nu 1}\rangle, \ldots, \langle \rho_{\mu n}, \rho_{\nu n}\rangle) \in T^n$. Let $\widetilde{f}$ be a fuzzy aggregation rule. In the following, when an asymmetric fuzzy preference of a particular type for $\langle \rho_\mu, \rho_\mu \rangle = \widetilde{f}(\rho)$ is assumed, then it is assumed that the associated asymmetric fuzzy preference for $\langle \rho_{\mu i}, \rho_{\nu i}\rangle$ is of the same type, $i = 1, \ldots, n$.

The proof of the next result follows easily.

Proposition 8.1.7 *Let $\langle \rho_\mu, \rho_\nu \rangle \in T$. Let $\langle \pi_\mu, \pi_\nu \rangle, \langle \tau_\mu, \tau_\nu \rangle$ be two different types of asymmetric fuzzy preference relations associated with $\langle \rho_\mu, \rho_\nu \rangle$. Suppose $\forall x, y \in X, \pi_\mu(x,y) > 0$ if and only if $\tau_\mu(x,y) > 0$ and $\pi_\nu(x,y) < 1$ if and only if $\tau_\nu(x,y) < 1$. Let $\widetilde{f} : T^n \to T$ be a fuzzy aggregation rule. Let $\rho \in T^n$. Then $\widetilde{f}$ satisfies PR with respect to $\langle \pi_\mu, \pi_\nu \rangle$ if and only if $\widetilde{f}$ satisfies PR with respect to $\langle \tau_\mu, \tau_\nu \rangle$.*

Corollary 8.1.8 *Let $\widetilde{f} : T^n \to T$ be a fuzzy aggregation rule. Then $\widetilde{f}$ satisfies PR with respect to $\langle \pi_{(3)_\mu}, \pi_{(3)_\nu}\rangle$ if and only if $\widetilde{f}$ satisfies PR with respect to $\langle \pi_{(1)_\mu}, \pi_{(1)_\nu}\rangle$.*

Proof. It suffices to show that $\forall x, y \in X, \pi_{(3)_\mu}(x,y) > 0$ if and only if $\pi_{(1)_\mu}(x,y) > 0$ and $\pi_{(3)_\nu}(x,y) < 1$ if and only if $\pi_{(1)_\nu}(x,y) < 1$. However, this follows immediately from the definitions since $\rho_\mu(x,y) > \rho_\mu(y,x)$ if and only if $\rho_\mu(x,y) - \rho_\mu(y,x) > 0$ and $\rho_\nu(x,y) < \rho_\nu(y,x)$ if and only if $1 + \rho_\nu(x,y) - \rho_\nu(y,x) < 1$. ∎

Definition 8.1.9 *Define the fuzzy aggregation rule $\widetilde{f} : T^n \to T$ as follows:*
$$\forall \rho = (\langle \rho_{\mu 1}, \rho_{\nu 1}\rangle, \ldots, \langle \rho_{\mu n}, \rho_{\nu n}\rangle) \in T^n, \forall x, y \in X,$$

$$\widetilde{f}(\rho)(x,y) = \left\langle \sum_{i=1}^{n} w_i \rho_{\mu i}(x,y), \sum_{i=1}^{n} w_i \rho_{\nu i}(x,y) \right\rangle,$$

where $\sum_{i=1}^{n} w_i = 1$ and $w_i > 0, i = 1, \ldots, n$.

It is clear that if $\rho_{\mu i}$ and $\rho_{\nu i}$ are reflexive and complete for all $i \in N$, then $\widetilde{f}(\rho)$ in Definition 8.1.9 is reflexive and complete.

Proposition 8.1.10 *Let $\widetilde{f}$ be defined by Definition 8.1.9. Then $\widetilde{f}$ satisfies PR with respect to $\pi_{(3)}$ and $\pi_{(1)}$.*

Proof. Let $\rho, \rho' \in T^n$. Suppose $\rho_{\mu i} = \rho'_{\mu i}, i = 1, \ldots, n, i \neq j$. Let $x, y \in X$. Suppose $\rho_\mu(x,y) = \rho_\mu(y,x)$. Suppose also that either (1) $(\pi_{\mu j}(x,y) = 0$ and

$\pi'_{\mu j}(x, y) > 0$) or (2) ($\pi_{\mu j}(y, x) > 0$ and $\pi'_{\mu j}(y, x) = 0$), where strict preference is of type $\pi_{(3)}$. Then $\pi'_{\mu}(x, y) = 0 \vee (\rho'_{\mu}(x, y) - \rho'_{\mu}(y, x))$ and

$$\rho'_{\mu}(x, y) - \rho'_{\mu}(y, x)$$

$$= \sum_{i=1}^{n} (w_i \rho'_{\mu i}(x, y) - w_i \rho'_{\mu i}(y, x))$$

$$= \sum_{i=1, i \neq j}^{n} (w_i \rho_{\mu i}(x, y) - w_i \rho_{\mu i}(y, x)) + w_j \rho'_{\mu j}(x, y) - w_j \rho'_{\mu j}(y, x)$$

$$= \sum_{i=1}^{n} (w_i \rho_{\mu i}(x, y) - w_i \rho_{\mu i}(y, x)) - w_j(\rho_{\mu j}(x, y) - \rho_{\mu j}(y, x))$$

$$+ w_j(\rho'_{\mu j}(x, y) - \rho'_{\mu j}(y, x))$$

$$= -w_j(\rho_{\mu j}(x, y) - \rho_{\mu j}(y, x)) + w_j(\rho'_{\mu j}(x, y) - \rho'_{\mu j}(y, x))$$

$$> 0,$$

where the inequality holds if either (1) or (2) hold. Hence $\widetilde{f}$ satisfies PR with respect to $\pi_{(3)}$. The desired result for $\pi_{(1)}$ follows from Corollary 8.1.8.

We now consider the case for $\pi'_{(3)_{\nu}}$. Suppose that either (1) ($\pi_{\nu j}(x, y) = 1$ and $\pi'_{\nu j}(x, y) < 1$) or (2) ($\pi_{\nu j}(y, x) < 1$ and $\pi'_{\nu j}(y, x) = 1$), where strict preference is of type $\pi_{(3)}$. Then $\pi'_{(3)_{\nu}}(x, y) = 1 \wedge (1 - \rho'_{\nu}(x, y) + \rho'_{\nu}(y, x))$ and as above

$$1 - \rho'_{\nu}(x, y) + \rho'_{\nu}(y, x)$$

$$= 1 - [-w_j(\rho_{\nu j}(x, y) - \rho_{\nu j}(y, x)) + w_j(\rho'_{\nu j}(x, y) - \rho'_{\nu j}(y, x))].$$

For case (1), we write

$$1 - \rho'_{\nu}(x, y) + \rho'_{\nu}(y, x)$$

$$= 1 - [-w_j[-1 + \rho_{\nu j}(x, y) - \rho_{\nu j}(y, x)] + w_j(-1 + \rho'_{\nu j}(x, y) - \rho'_{\nu j}(y, x))]$$

$$= 1 - [w_j \pi_{\nu j}(x, y) - w_j \pi'_{\nu j}(x, y)] < 1.$$

For case (2), we write

$$1 - \rho'_{\nu}(x, y) + \rho'_{\nu}(y, x)$$

$$= 1 - [-w_j[1 + \rho_{\nu j}(x, y) - \rho_{\nu j}(y, x)] + w_j(1 + \rho'_{\nu j}(x, y) - \rho'_{\nu j}(y, x))]$$

$$= 1 - [-w_j \pi_{\nu j}(y, x) + w_j \pi'_{\nu j}(y, x)] < 1.$$

Hence $\widetilde{f}$ satisfies PR with respect to $\pi_{(3)_{\nu}}$. The desired result for $\pi_{(1)_{\nu}}$ follows from Corollary 8.1.8. ∎

Proposition 8.1.11 *Let $\widetilde{f}$ be the fuzzy aggregation rule defined in Definition 8.1.9. Then $\forall x, y \in X$, $\pi_{(2)_{\mu}}(x, y) = \sum_{i=1}^{n} w_i \pi_{\mu i}(x, y)$ and $\pi_{(2)_{\nu}}(x, y) = \sum_{i=1}^{n} w_i \pi_{\nu i}(x, y)$.*

Proof. Let $x, y \in X$. Then

$$
\begin{aligned}
\pi_{(2)_\mu}(x, y) &= 1 - \rho_\mu(y, x) = 1 - \sum_{i=1}^{n} w_i \rho_{\mu i}(y, x) = 1 - \sum_{i=1}^{n} w_i(1 - \pi_{\mu i}(x, y)) \\
&= 1 - \sum_{i=1}^{n} w_i + \sum_{i=1}^{n} w_i \pi_{\mu i}(x, y) = \sum_{i=1}^{n} w_i \pi_{\mu i}(x, y).
\end{aligned}
$$

and

$$
\pi_{(2)_\nu}(x, y) = \rho_\nu(y, x) = \sum_{i=1}^{n} w_i \rho_{\nu i}(y, x) = \sum_{i=1}^{n} w_i \pi_{\nu i}(y, x).
$$

∎

Example 8.1.12 Let $\widetilde{f}$ be the fuzzy aggregation rule defined in Definition 8.1.9. That $\widetilde{f}$ does not satisfy PR with respect to $\pi_{(2)}$ is shown in Example 4.1.8.

Proposition 8.1.13 Let $\widetilde{f}$ be the fuzzy aggregation rule defined in Definition 8.1.9. Then $\widetilde{f}$ satisfies PC with respect to $\langle \pi_{(3)_\mu}, \pi_{(3)_\nu} \rangle$.

Proof. Let $x, y \in X$ and $\rho \in T^n$. Let $m = m_{x,y} = \wedge \{\pi_{\mu i}(x, y) \mid i = 1, \ldots, n\}$. There is no loss in generality in assuming $m = \pi_{\mu 1}(x, y)$. Suppose $m > 0$. Then

$$
\begin{aligned}
1 &\le w_1 + w_2 + \ldots + w_n + w_2 \left(\frac{\pi_{\mu 2}(x, y)}{m} - 1 \right) + \ldots + w_n \left(\frac{\pi_{\mu n}(x, y)}{m} - 1 \right) \\
&= w_1 \frac{\pi_{\mu 1}(x, y)}{m} + \ldots + w_n \frac{\pi_{\mu n}(x, y)}{m}.
\end{aligned}
$$

Thus

$$
\pi_{1_\mu}(x, y) = m \le w_1 \pi_{1_\mu}(x, y) + \ldots + w_n \pi_{n_\mu}(x). \tag{8.1}
$$

Since $m > 0$, $\pi_{\mu i}(x, y) > 0 \; \forall i \in N$. Thus $\rho_{\mu i}(x, y) > \rho_{\mu i}(y, x) \; \forall i \in N$. Hence $\sum_{i=1}^{n} w_i \rho_{\mu i}(x, y) > \sum_{i=1}^{n} w_i \rho_{\mu i}(y, x)$. Thus $\sum_{i=1}^{n} w_i [\rho_{\mu i}(x, y) - \rho_{\mu i}(y, x)] > 0$. But $\rho_{\mu i}(x, y) - \rho_{\mu i}(y, x) = \pi_i(x, y), i = 1, \ldots, n$. Hence by (8.1),

$$
\begin{aligned}
0 &< \rho_{\mu 1}(x, y) - \rho_{\mu 1}(y, x) \le \sum_{i=1}^{n} w_i [\rho_{\mu i}(x, y) - \rho_{\mu i}(y, x)] \\
&= \sum_{i=1}^{n} w_i \rho_{\mu i}(x, y) - \sum_{i=1}^{n} w_i \rho_{\mu i}(y, x)],
\end{aligned}
$$

or $\wedge \{\pi_{\mu i}(x, y) \mid i \in N\} \le \rho_\mu(x, y) - \rho_\mu(y, x) = \pi_{(3)_\mu}(x, y)$. Hence $\widetilde{f}$ satisfies PC with respect to $\pi_{(3)_\mu}$. If $m = 0$, then clearly $\widetilde{f}$ satisfies PC with respect to $\pi_{(3)_\mu}$.

We now show $\widetilde{f}$ satisfies PC with respect to $\pi_{(3)_\nu}$. Let $x, y \in X$. Let $m = m_{x,y} = \vee\{\pi_{\nu i}(x,y) \mid i \in N\}$. Suppose $m < 1$. Then Thus $\pi_\nu(x,y) \le \vee\{\pi_{\nu i}(x,y) \mid i \in N\}$. If $m = 1$, then the result is immediate. ∎

Definition 8.1.14 Let $\langle \rho_\mu, \rho_\nu \rangle \in T$. Then $\langle \rho_\mu, \rho_\nu \rangle$ is said to satisfy type T_2-**transitivity** if $\forall x, y, z \in X$, $\rho_\mu(x,z) \ge \rho_\mu(x,y) + \rho_\mu(y,z) - 1$ and $\rho_\nu(x,z) \le \rho_\nu(x,y) + \rho_\nu(y,z)$.

Definition 8.1.15 Let $\widetilde{f}$ be a fuzzy aggregation rule. Then $\widetilde{f}$ is said to be T_2-**transitive** if $\forall \rho \in T^n$, $\widetilde{f}(\rho)$ is T_2-transitive.

Proposition 8.1.16 Let $\widetilde{f}$ be the fuzzy aggregation rule defined in Definition 8.1.9. Let $\rho = (\langle \rho_{\mu 1}, \rho_{\nu 1} \rangle, \ldots, \langle \rho_{\mu n}, \rho_{\nu n} \rangle) \in T^n$. If $\langle \rho_{\mu i}, \rho_{\nu i} \rangle$ is T_2-transitive $\forall i \in N$, then $\widetilde{f}(\rho)$ is T_2-transitive.

Proof. Let $x, y, z \in X$. Then $\rho_{\mu i}(x,z) \ge \rho_{\mu i}(x,y) + \rho_{\mu i}(y,z) - 1$ and $\rho_{i_\nu}(x,z) \le \rho_{\nu i}(x,y) + \rho_{\nu i}(y,z) \ \forall i \in N$. Thus

$$
\begin{aligned}
\rho_\mu(x,z) &= \sum_{i=1}^n w_i \rho_{\mu i}(x,z) \ge \sum_{i=1}^n w_i \rho_{\mu i}(x,y) + \sum_{i=1}^n w_i \rho_{\mu i}(y,z) - \sum_{i=1}^n w_i \\
&= \rho_\mu(x,y) + \rho_\mu(y,z) - 1
\end{aligned}
$$

and

$$
\begin{aligned}
\rho_\nu(x,z) &= \sum_{i=1}^n w_i \rho_{\nu i}(x,z) \le \sum_{i=1}^n w_i \rho_{\nu i}(x,y) + \sum_{i=1}^n w_i \rho_{\nu i}(y,z) \\
&= \rho_\nu(x,y) + \rho_\nu(y,z).
\end{aligned}
$$

∎

8.2 IIA3 and Nondictatorial Fuzzy Aggregation Rules

We now consider alternative definitions of IIA.

Let $H_2 = \{\langle \rho_\mu, \rho_\nu \rangle \in T \mid \langle \rho_\mu, \rho_\nu \rangle \text{ is } T_2\text{-transitive}\}$.

Let C be a nonempty subset of N. Then C is called a **coalition**.

Definition 8.2.1 Let $\widetilde{f}$ be a fuzzy aggregation rule. An individual $j \in N$ is called a **dictator** if for all distinct $x, y \in X$ and all $\rho = (\langle \rho_{\mu 1}, \rho_{\nu 1} \rangle, \ldots, \langle \rho_{\mu n}, \rho_{\nu n} \rangle) \in T^n$, $\pi_{\mu j}(x,y) > 0 \Rightarrow \pi_\mu(x,y) > 0$ and $\pi_{\nu j}(x,y) < 1 \Rightarrow \pi_\nu(x,y) < 1$.

A fuzzy aggregation rule $\widetilde{f}$ is said to be **nondictatorial** if it is not dictatorial.

Theorem 8.2.2 *Let strict preference be defined by $\pi_{(3)}$. Then there exists a nondictatorial fuzzy aggregation rule $\widetilde{f} : H_2^n \to H_2$ satisfying IIA1, PR, and PC.*

Proof. Let $\widetilde{f}$ be the fuzzy aggregation rule defined in Definition 8.1.9. Clearly, $\widetilde{f}$ is not dictatorial. By Propositions 8.1.10, 8.1.13, and 8.1.16, it only remains to show that $\widetilde{f}$ satisfies IIA1. However, this is immediate. ∎

Definition 8.2.3 *Let $\langle \rho_\mu, \rho_\nu \rangle, \langle \rho'_\mu, \rho'_\nu \rangle \in T$. Let $Im(\rho_\mu) = \{s_1, \ldots, s_k\}$ and $Im(\rho'_\mu) = \{t_1, \ldots, t_m\}$ be such that $s_1 < \ldots < s_k$ and $t_1 < \ldots < t_m$. Then ρ_μ and ρ'_μ are said to be **equivalent**, written $\rho_\mu \sim \rho'_\mu$, if $s_1 = 0 \Leftrightarrow t_1 = 0, k = m$, and $(\rho_\mu)_{s_i} = (\rho'_\mu)_{t_i}$ (level sets) for $i = 1, \ldots, k$.*

*Let $Im(\rho_\nu) = \{u_1, \ldots, u_h\}$ and $Im(\rho'_\nu) = \{v_1, \ldots, v_n\}$ be such that $u_1 > \ldots > u_h$ and $v_1 > \ldots > v_n$. Then ρ_ν and ρ'_ν are said to be **equivalent**, written $\rho_\nu \sim \rho'_\nu$, if $s_1 = 1 \Leftrightarrow t_1 = 1, h = n$, and $(\rho_\mu)_{s_i} = (\rho'_\mu)_{t_i}$ (level sets) for $i = 1, \ldots, h$.*

*Let $\langle \rho_\mu, \rho_\nu \rangle, \langle \rho'_\mu, \rho'_\nu \rangle \in T$. Then $\langle \rho_\mu, \rho_\nu \rangle$ and $\langle \rho'_\mu, \rho'_\nu \rangle$ are said to be **equivalent**, written $\langle \rho_\mu, \rho_\nu \rangle \sim \langle \rho'_\mu, \rho'_\nu \rangle$, if ρ_μ and ρ'_μ are equivalent and ρ_ν and ρ'_ν are equivalent.*

If the intuitionistic fuzzy binary relations in Definition 8.2.3 are reflexive, then $\vee \, Im(\rho_\mu) = 1 = \vee \, Im(\rho'_\mu)$ and $\wedge \, Im(\rho_\nu) = 0 = \vee \, Im(\rho'_\nu)$.

Proposition 8.2.4 *Let $\langle \rho_\mu, \rho_\nu \rangle, \langle \rho'_\mu, \rho'_\nu \rangle \in T$. Let $x, y \in X$. Suppose $\langle \rho_\mu, \rho_\nu \rangle \sim \langle \rho'_\mu, \rho'_\nu \rangle$. Then $\rho_\mu(x, y) > \rho_\mu(y, x)$ if and only if $\rho'_\mu(x, y) > \rho'_\mu(y, x)$ and $\rho_\nu(x, y) > \rho_\nu(y, x)$ if and only if $\rho'_\nu(x, y) > \rho'_\nu(y, x)$.*

Proof. Suppose $\rho_\mu(x, y) > \rho_\mu(y, x)$. Let $\rho_\mu(x, y) = s_i$. Then $s_i > \rho_\mu(y, x)$. Hence $(y, x) \notin (\rho_\mu)_{s_i}$. Thus $(y, x) \notin (\rho'_\mu)_{t_i}$. Now $(x, y) \in (\rho_\mu)_{s_i}$ so $(x, y) \in (\rho'_\mu)_{t_i}$. Hence $\rho'_\mu(x, y) \geq t_i > \rho'_\mu(y, x)$. A similar argument holds for ρ_ν and ρ'_ν. ∎

Definition 8.2.5 *Let $(\widetilde{f_\mu}, \widetilde{f_\nu})$ be a fuzzy aggregation operator. Then $(\widetilde{f_\mu}, \widetilde{f_\nu})$ is said to be **independent of irrelevant alternatives (IIA3)** if $\forall x, y \in X, \forall (\langle \rho_{\mu 1}, \rho_{\nu 1} \rangle, \ldots, \langle \rho_{\mu n}, \rho_{\nu n} \rangle), (\langle \rho'_{\mu 1}, \rho'_{\nu 1} \rangle, \ldots, \langle \rho'_{\mu n}, \rho'_{\nu n} \rangle) \in T^n, \rho_{\mu i}]\restriction_{\{x,y\}} \sim \rho'_{\mu i}]\restriction_{\{x,y\}} \forall i \in N \Rightarrow \widetilde{f_\mu}(\rho)]\restriction_{\{x,y\}} \sim \widetilde{f_\mu}(\rho')]\restriction_{\{x,y\}}]$ and $[\rho_{\nu i}]\restriction_{\{x,y\}} \sim \rho'_{\nu i}]\restriction_{\{x,y\}} \forall i \in N \Rightarrow \widetilde{f_\nu}(\rho)]\restriction_{\{x,y\}} \sim \widetilde{f_\nu}(\rho')]\restriction_{\{x,y\}}]$.*

In the following, we let t_μ and t_ν denote functions from $\mathcal{FR}^n$ into $(0, 1]$. Some examples of t_μ are $t_\mu(\rho) = \vee\{\sum_{i=1}^n w_i \rho_{\mu i}(x, y) \mid x, y \in X, x \neq y\}$ and $t_\mu(\rho) = \wedge\{\sum_{i=1}^n w_i \rho_{\mu i}(x, y) > 0 \mid x, y \in X\}$, where $\sum_{i=1}^n w_i = 1$ and $w_i > 0$, $i = 1, \ldots, n$.

Definition 8.2.6 *Let the fuzzy aggregation rule* $\widetilde{f} : T^n \to T$ *be induced by the fuzzy aggregation rules* $\widetilde{f_\mu}$ *and* $\widetilde{f_\nu}$ *as follows* $\forall \rho = (\langle \rho_{\mu 1}, \rho_{\nu 1} \rangle, \ldots, \langle \rho_{\mu n}, \rho_{\nu n} \rangle) \in T^n$, $\forall x, y \in X$,

$$\widetilde{f_\mu}(\rho_\mu)(x, y) = \begin{cases} 1 & \text{if } x = y, \\ 1 & \text{if } \pi_{\mu i}(x, y) > 0 \ \forall \ i \in N, \\ t_\mu(\rho) & \text{otherwise.} \end{cases}$$

$$\widetilde{f_\nu}(\rho_\nu)(x, y) = \begin{cases} 0 & \text{if } x = y, \\ 0 & \text{if } \pi_{\nu i}(x, y) < 1 \ \forall \ i \in N, \\ t_\nu(\rho) & \text{otherwise.} \end{cases}$$

Clearly, $\widetilde{f}(\rho) = (\widetilde{f_\mu}, \widetilde{f_\nu})(\rho)$ *in Definition 8.2.6 is reflexive and complete.*

Proposition 8.2.7 *Let strict preference be defined by* $\pi_{(1)}$ *or* $\pi_{(3)}$. *Let* $(\widetilde{f_\mu}, \widetilde{f_\nu})$ *be the fuzzy aggregation rule defined in Definition 8.2.6. Suppose* $\text{Im}(t_\mu)$, $\text{Im}(t_\nu) \subseteq (0, 1)$. *Then* $(\widetilde{f_\mu}, \widetilde{f_\nu})$ *is independent of irrelevant alternatives IIA3.*

Proof. Let $\langle \rho_\mu, \rho_\nu \rangle, \langle \rho'_\mu, \rho'_\nu \rangle \in T^n$ and $x, y \in X$. Suppose $\rho_{\mu i}\rceil_{\{x,y\}} \sim \rho'_{\mu i}\rceil_{\{x,y\}} \ \forall i \in N$. Let $i \in N$. Then $\rho_{\mu i}\rceil_{\{x,y\}}(x, y) > \rho_{\mu i}\rceil_{\{x,y\}}(y, x) \Leftrightarrow \rho'_{\mu i}\rceil_{\{x,y\}}(x, y) > \rho'_{\mu i}\rceil_{\{x,y\}}(y, x)$ by Proposition 8.2.4. Thus $\rho_{\mu i}(x, y) > \rho_{\mu i}(y, x) \Leftrightarrow \rho'_{\mu i}(x, y) > \rho'_{\mu i}(y, x)$. Hence by definition of $\widetilde{f_\mu}$, $\widetilde{f_\mu}(\rho_\mu)(x, y) = 1 \Leftrightarrow \widetilde{f_\mu}(\rho'_\mu)(x, y) = 1$ and so $\widetilde{f_\mu}(\rho_\mu)(x, y) = t_\mu(\rho_\mu) \Leftrightarrow \widetilde{f}(\rho'_\mu)(x, y) = t_\mu(\rho')$. Hence $\widetilde{f_\mu}(\rho_\mu)\rceil_{\{x,y\}} \sim \widetilde{f_\mu}(\rho'_\mu)\rceil_{\{x,y\}}$. Similarly, $\widetilde{f_\nu}(\rho_\nu)(x, y) = 0 \Leftrightarrow \widetilde{f_\nu}(\rho'_\nu)(x, y) = 0$. Hence $\widetilde{f_\nu}(\rho_\nu)\rceil_{\{x,y\}} \sim \widetilde{f_\nu}(\rho'_\nu)\rceil_{\{x,y\}}$. ∎

If t is a constant function in Proposition 8.2.7, then we can conclude $\widetilde{f}(\rho)\rceil_{\{x,y\}} = \widetilde{f}(\rho')\rceil_{\{x,y\}}$ in the proof. If $t(\rho) \geq 1/2$, then we can conclude that $\widetilde{f}(\rho)$ is strongly complete, i.e., $\widetilde{f}(\rho)(x, y) + \widetilde{f}(\rho)(y, x) \geq 1$.

Definition 8.2.8 *Let* $(\widetilde{f_\mu}, \widetilde{f_\nu}) : T^n \to T$ *be a fuzzy aggregation rule. Then* $(\widetilde{f_\mu}, \widetilde{f_\nu})$ *is said to be **weakly Paretian** with respect to a given strict preference if* $\forall x, y \in X$, $\forall (\langle \rho_{\mu 1}, \rho_{\nu 1} \rangle, \ldots, \langle \rho_{\mu n}, \rho_{\nu n} \rangle) \in T^n$, $\pi_{\mu i}(x, y) > 0 \ \forall i \in N$ *implies* $\pi_\mu(x, y) > 0$ *and* $\pi_{\nu i}(x, y) < 1 \ \forall i \in N$ *implies* $\pi_\nu(x, y) < 1$.

Proposition 8.2.9 *Let strict preference be defined by* $\pi_{(1)}$ *or* $\pi_{(3)}$. *Suppose* $\text{Im}(t) \subseteq (0, 1)$. *Let* $(\widetilde{f_\mu}, \widetilde{f_\nu})$ *be the fuzzy aggregation rule defined in Definition 8.2.6. Then* $\widetilde{f}$ *satisfies PC with respect to* $\pi_{(1)}$ *and is weakly Paretian with respect to* $\pi_{(3)}$.

Proof. Let $x, y \in X$ and $(\langle \rho_{\mu 1}, \rho_{\nu 1} \rangle, \ldots, \langle \rho_{\mu n}, \rho_{\nu n} \rangle) \in T^n$. If $\wedge_{i \in N} \pi_{\mu i}(x, y) = 0$, then clearly $\pi_\mu(x, y) \geq \wedge_{i \in N} \pi_{\mu i}(x, y)$. Suppose $\wedge_{i \in N} \pi_{\mu i}(x, y) > 0$. Then $\pi_{\mu i}(x, y) > 0 \ \forall i \in N$. Hence $\rho_{\mu i}(x, y) > \rho_{\mu i}(y, x) \ \forall i \in N$. Thus $\widetilde{f_\mu}(\rho_\mu)(x, y) = 1$ and $\widetilde{f_\mu}(\rho_\mu)(y, x) = t_\mu(\rho_\mu) < 1$. Hence $\pi_\mu(x, y) \geq \wedge_{i \in N} \pi_{\mu i}(x, y)$ for $\pi_\mu = \pi_{(1)_\mu}$ and $\pi_\mu(x, y) > 0$ for $\pi_\mu = \pi_{(3)_\mu}$.

If $\vee_{i\in N}\pi_{\nu i}(x,y) = 1$, then clearly $\pi_\nu(x,y) \leq \vee_{i\in N}\pi_{\nu i}(x,y)$. Suppose $\vee_{i\in N}\pi_{\nu i}(x,y) < 1$. Then $\pi_{\nu i}(x,y) < 1$ $\forall i \in N$. Hence $\rho_{\nu i}(x,y) < \rho_{\nu i}(y,x)$ $\forall i \in N$. Thus $\widetilde{f}_\nu(\rho_\nu)(x,y) = 0$ and $\widetilde{f}_\nu(\rho_\nu)(y,x) = t_\nu(\rho_\nu) > 0$. Hence $\pi_\nu(x,y) \leq \vee_{i\in N}\pi_{\nu i}(x,y)$ for $\pi_\nu = \pi_{(1)_\nu}$ and $\pi_\nu(x,y) < 1$ for $\pi_\nu = \pi_{(3)_\nu}$.

∎

Definition 8.2.10 *Let $\langle \rho_\mu, \rho_\nu \rangle \in T$. Then $\langle \rho_\mu, \rho_\nu \rangle$ is called (**max-min**)* **quasi-transitive** *if $\forall x, y, z \in X$, $\pi_\mu(x,z) \geq \pi_\mu(x,y) \wedge \pi_\mu(y,z)$ and $\pi_\nu(x,z) \leq \pi_\nu(x,y) \wedge \pi_\nu(y,z)$.*

Proposition 8.2.11 *Let π be defined by $\pi_{(1)}$. Let $\langle \rho_\mu, \rho_\nu \rangle \in T$. If $\langle \rho_\mu, \rho_\nu \rangle$ is max-min transitive, then $\langle \rho_\mu, \rho_\nu \rangle$ is max-min quasi-transitive.*

Proof. The case for ρ_μ is known, Proposition 5.4.4 or [12, Proposition 2.9 (2), p. 220]. Suppose ρ_ν satisfies (max-min) transitivity, but not (max-min) quasi-transitivity. Then there exists $x, y, z \in X$ such that

$$\rho_\nu(x,z) \leq \rho_\nu(x,y) \vee \rho_\nu(y,z), \pi_\nu(x,z) > \pi_\nu(x,y) \vee \pi_\nu(y,z).$$

Thus

$$1 > \pi_\nu(x,y) = \rho_\nu(x,y) < \rho_\nu(y,x), \qquad (8.2)$$

$$1 > \pi_\nu(y,z) = \rho_\nu(y,z) < \rho_\nu(z,y) \qquad (8.3)$$

and so

$$1 = \pi_\nu(x,z) \Rightarrow \rho_\nu(x,z) \geq \rho_\nu(z,x) \qquad (8.4)$$

Suppose

$$\rho_\nu(x,y) \vee \rho_\nu(y,z) = \rho_\nu(x,y) \qquad (8.5)$$

Then

$$\rho_\nu(x,z) \leq \rho_\nu(x,y). \qquad (8.6)$$

Now

$$\rho_\nu(y,x) \leq \rho_\nu(y,z) \vee \rho_\nu(z,x). \qquad (8.7)$$

But since $\rho_\nu(x,y) \vee \rho_\nu(y,z) = \rho_\nu(x,y)$ and $\rho_\nu(y,x) \leq \rho_\nu(y,z)$, we have $\rho_\nu(y,x) \leq \rho_\nu(x,y)$, contradicting (8.2). Similarly, (8.4), (8.6) and $\rho_\nu(y,x) \leq \rho_\nu(z,x)$ imply $\rho_\nu(y,x) \leq \rho_\nu(x,y)$. Hence $\rho_\nu(y,z) \vee \rho_\nu(z,x) < \rho_\nu(y,x)$ contradicting (8.7).

Suppose

$$\rho_\nu(x,y) \vee \rho_\nu(y,z) = \rho_\nu(y,z). \qquad (8.8)$$

Then

$$\rho_\nu(x,z) \leq \rho_\nu(y,z). \qquad (8.9)$$

By the max-min transitivity of ρ,

$$\rho_\nu(z,y) \leq \rho_\nu(z,x) \qquad (8.10)$$

or

$$\rho_\nu(z, y) \le \rho_\nu(x, y). \tag{8.11}$$

Now (8.10), (8.4), and (8.9) imply $\rho_\nu(z, y) \le \rho_\nu(y, z)$. Similarly, (8.11) and imply $\rho_\nu(z, y) \le \rho_\nu(y, z)$. However, this contradicts (8.3). ∎

Let $\mathcal{FR}$ denote the set of all intuitionistic fuzzy binary relations on X which are reflexive and complete. Let $\mathcal{FR}_{Tr} = \{\langle \rho_\mu, \rho_\nu \rangle \in \mathcal{FR} \mid \langle \rho_\mu, \rho_\nu \rangle$ is transitive$\}$. We say that a fuzzy aggregation rule $(\widetilde{f}_\mu, \widetilde{f}_\nu)$ is **max-min transitive** if $\widetilde{f}_\mu(\rho_\mu)$ and $\widetilde{f}_\nu(\rho_\nu)$ are max-min transitive for every intuitionistic fuzzy preference profile $(\langle \rho_{\mu 1}, \rho_{\nu 1} \rangle, \ldots, \langle \rho_{\mu n}, \rho_{\nu n} \rangle) \in T^n$, where $\rho_\mu = (\rho_{\mu 1}, \ldots, \rho_{\mu n})$ and $\rho_\nu = (\rho_{\nu 1}, \ldots, \rho_{\nu n})$.

Theorem 8.2.12 *Let π be defined by $\pi_{(1)}$. Then there exists a nondictatorial fuzzy aggregation rule $\widetilde{f} = (\widetilde{f}_\mu, \widetilde{f}_\nu) : \mathcal{FR}^n_{Tr} \to \mathcal{FR}$ that is max-min transitive, independent of irrelevant alternatives IIA3, and PC.*

Proof. Let $\widetilde{f}$ be defined as in Definition 8.2.6. Clearly, $\widetilde{f}$ is reflexive and complete. By Propositions 8.2.7 and 8.2.9, $\widetilde{f}$ is independent of irrelevant alternatives IIA3 and PC. By definition, $\widetilde{f}((\langle \rho_{\mu 1}, \rho_{\nu 1} \rangle, \ldots, \langle \rho_{\mu n}, \rho_{\nu n} \rangle))$ is reflexive $\forall (\langle \rho_{\mu 1}, \rho_{\nu 1} \rangle, \ldots, \langle \rho_{\mu n}, \rho_{\nu n} \rangle) \in \mathcal{FR}^n_{Tr}$. It remains to be shown that $\widetilde{f}$ is transitive. Let $(\langle \rho_{\mu 1}, \rho_{\nu 1} \rangle, \ldots, \langle \rho_{\mu n}, \rho_{\nu n} \rangle) \in \mathcal{FR}^n_{Tr}$. Let $x, y, z \in X$. Suppose $\widetilde{f}_\mu(\rho_\mu)(x, y) \wedge \widetilde{f}_\mu(\rho_\mu)(y, z) = 1$. Assume $x \ne y \ne z$. Then $\forall i \in N, \pi_{\mu i}(x, y) > 0$ and $\pi_{\mu i}(y, z) > 0$. By Proposition 8.2.11, we have $\forall i \in N$ that $\pi_{\mu i}(x, z) \ge \pi_{\mu i}(x, y) \wedge \pi_{\mu i}(y, z)$ since $\rho_{\mu i}$ is transitive. Thus $\pi_{\mu i}(x, z) > 0 \,\forall i \in N$. Hence $\rho_{\mu i}(x, z) > \rho_{\mu i}(z, x) \,\forall i \in N$. Thus $\widetilde{f}_\mu(\rho_\mu)(x, z) = 1$. Assume $x = y \ne z$. Then $\forall i \in N, \pi_{\mu i}(y, z) > 0$. Now $\forall i \in N, \pi_{\mu i}(x, z) = \pi_{\mu i}(y, z) > 0$. Thus $\widetilde{f}_\mu(\rho_\mu)(x, z) = 1$. A similar argument holds for $x \ne y = z$. The desired result when $x = y = z$ is immediate. If $\widetilde{f}_\mu(\rho_\mu)(x, y) \wedge \widetilde{f}_\mu(\rho_\mu)(y, z) = t_\mu(\rho_\mu)$, then clearly $\widetilde{f}_\mu(\rho_\mu)(x, z) \ge \widetilde{f}_\mu(\rho_\mu)(x, y) \wedge \widetilde{f}_\mu(\rho_\mu)(y, z)$.

Suppose $\widetilde{f}_\nu(\rho_\nu)(x, y) \vee \widetilde{f}_\nu(\rho_\nu)(y, z) = 0$. Assume $x \ne y \ne z$. Then $\forall i \in N, \pi_{\nu i}(x, y) < 1$ and $\pi_{\nu i}(y, z) < 1$. By Proposition 8.2.11, we have $\forall i \in N$ that $\pi_{\nu i}(x, z) \le \pi_{\nu i}(x, y) \vee \pi_{\nu i}(y, z)$ since ρ_{i_ν} is transitive. Thus $\pi_{\nu i}(x, z) < 1 \,\forall i \in N$. Hence $\rho_{\nu i}(x, z) < \rho_{\nu i}(z, x) \,\forall i \in N$. Thus $\widetilde{f}_\nu(\rho_\nu)(x, z) = 0$. Assume $x = y \ne z$. Then $\forall i \in N, \pi_{\nu i}(y, z) < 1$. Now $\forall i \in N, \pi_{\nu i}(x, z) = \pi_{\nu i}(y, z) < 1$. Thus $\widetilde{f}_\nu(\rho_\nu)(x, z) = 0$. A similar argument holds for $x \ne y = z$. The desired result when $x = y = z$ is immediate. If $\widetilde{f}_\nu(\rho_\nu)(x, y) \vee \widetilde{f}_\nu(\rho_\nu)(y, z) = t_\nu(\rho_\nu)$, then clearly $\widetilde{f}_\nu(\rho_\nu)(x, z) \le \widetilde{f}_\nu(\rho_\nu)(x, y) \vee \widetilde{f}_\nu(\rho_\nu)(y, z)$. ∎

Example 8.2.13 *An example of $\pi_{(2)}$ that is T_2-transitive and partially transitive, but not max-min quasi-transitive can be found in Example 4.1.24.*

Theorem 8.2.14 *There exists a weakly Paretian (with respect to $\pi_{(2)}$) fuzzy aggregation rule $\widetilde{f} = (\widetilde{f_\mu}, \widetilde{f_\nu}) : T^n \to T$ that is reflexive, complete, max-min transitive, satisfies IIA3 and is such that every member of N is a dictator.*

Proof. Define $\widetilde{f} : T^n \to T$ by $\forall(\langle \rho_{\mu 1}, \rho_{\nu 1} \rangle, \ldots, \langle \rho_{\mu n}, \rho_{\nu n} \rangle) \in T^n$, $\forall x, y \in X$,

$$\widetilde{f_\mu}(\rho_\mu)(x,y) = \begin{cases} 1 \text{ if } x = y, \\ t_\mu(\rho_\mu) \text{ if } x \neq y, \end{cases}$$

where $0 < t_\mu(\rho_\mu) < 1$ and

$$\widetilde{f_\nu}(\rho_\nu)(x,y) = \begin{cases} 0 \text{ if } x = y, \\ t_\nu(\rho_\nu) \text{ if } x \neq y, \end{cases}$$

where $0 < t_\nu(\rho_\nu) < 1$. Let $\pi = \pi_{(2)}$. Suppose $\pi_{i_\mu}(x,y) > 0 \; \forall i \in N$. Then $x \neq y$. Hence $\pi_\mu(x,y) = 1 - \widetilde{f_\mu}(\rho_\mu)(y,x) = 1 - t_\mu(\rho_\mu) > 0$. Thus $\widetilde{f_\mu}$ is weakly Paretian with respect $\pi_{(2)}$. Clearly, $\widetilde{f_\mu}$ satisfies IIA3 by a similar argument as in Proposition 8.2.7. Now suppose $\pi_{i_\nu}(x,y) < 1 \; \forall i \in N$. Then $x \neq y$. Hence $\pi_\nu(x,y) = \widetilde{f_\nu}(\rho_\nu)(y,x) = t_\nu(\rho_\nu) < 1$. Thus $\widetilde{f_\nu}$ is weakly Paretian with respect to $\pi_{(2)}$. Clearly, $\widetilde{f_\nu}$ satisfies IIA3 by a similar argument as in Proposition 8.2.7. That every member of N is a dictator follows by definition since $\forall x, y \in N$, $x \neq y, \pi_\mu(x,y) > 0$ and $\pi_\nu(x,y) < 1$. ∎

8.3 IIA2, IIA4, and Fuzzy Arrow's Theorem

In this section, we give consideration to two new definitions of independence of irrelevant alternatives, IIA. Recall that we sometimes suppress the underlying fuzzy preference aggregation rule and write $\rho(x,y)$ for $\widetilde{f}(\rho)(x,y)$ and $\pi(x,y) > 0$ for $\widetilde{f}(\rho)(x,y) > 0$ and $\widetilde{f}(\rho)(y,x) = 0$.

Let $(\langle \rho_{\mu 1}, \rho_{\nu 1} \rangle, \ldots, \langle \rho_{\mu n}, \rho_{\nu n} \rangle) \in T^n$. Recall that we sometimes write $\langle \rho_\mu, \rho_\nu \rangle$ for $(\langle \rho_{\mu 1}, \rho_{\nu 1} \rangle, \ldots, \langle \rho_{\mu n}, \rho_{\nu n} \rangle)$, where $\rho_\mu = (\rho_{\mu 1}, \ldots, \rho_{\mu n})$ and $\rho_\nu = (\rho_{\nu 1}, \ldots, \rho_{\nu n})$.

Definition 8.3.1 *Let $\widetilde{f} = (\widetilde{f_\mu}, \widetilde{f_\nu})$ be a fuzzy preference aggregation rule. Then $\widetilde{f}$ is said to be* **independent of irrelevant alternatives IIA2** *if $\forall(\langle \rho_{\mu 1}, \rho_{\nu 1} \rangle, \ldots, \langle \rho_{\mu n}, \rho_{\nu n} \rangle), (\langle \rho'_{\mu 1}, \rho'_{\nu 1} \rangle, \ldots, \langle \rho'_{\mu n}, \rho'_{\nu n} \rangle) \in T^n$, $\forall x, y \in X$, $Supp(\rho_{\mu i} \rceil_{\{x,y\}}) = Supp(\rho'_{\mu i} \rceil_{\{x,y\}}) \; \forall i \in N$ implies $Supp(\widetilde{f_\mu}(\rho_\mu) \rceil_{\{x,y\}}) = Supp(\widetilde{f_\mu}(\rho'_\mu) \rceil_{\{x,y\}})$ and $Cosupp(\rho_{\nu i} \rceil_{\{x,y\}}) = Cosupp(\rho'_{\nu i} \rceil_{\{x,y\}}) \; \forall i \in N$ implies $Cosupp(\widetilde{f_\nu}(\rho_\nu) \rceil_{\{x,y\}}) = Cosupp(\widetilde{f_\nu}(\rho'_\nu) \rceil_{\{x,y\}})$.*

Definition 8.3.2 *Let $\widetilde{f}$ be a fuzzy preference aggregation rule. Let $(x,y) \in X \times X$. Let λ be a fuzzy subset of N. Then*

(1) λ is called **semidecisive** for x against y, written $x\tilde{D}_\lambda y$, if $\forall(\langle\rho_{\mu 1},\rho_{\nu 1}\rangle,$ $\ldots,\ \langle\rho_{\mu n},\rho_{\nu n}\rangle) \in T^n,\ [(\pi_{\mu i}(x,y) > 0\ \forall i \in Supp(\lambda)$ and $\pi_{\mu j}(y,x) > 0$ $\forall j \notin Supp(\lambda)]$ implies $\pi_\mu(x,y) > 0)$ and $(\pi_{\nu i}(x,y) < 1\ \forall i \in Cosupp(\lambda)$ and $\pi_{\nu j}(y,x) < 1\ \forall j \notin Cosupp(\lambda))$ implies $\pi_\nu(x,y) > 0]$.

(2) λ is called **decisive** for x against y, written $xD_\lambda y$, if $\forall\langle\rho_\mu,\rho_\nu\rangle \in \mathcal{FR}^n$, $[(\pi_{\mu i}(x,y) > 0\ \forall i \in Supp(\lambda)]$ implies $\pi_\mu(x,y) > 0)$ and $(\pi_{\nu i}(x,y) < 1\ \forall i \in Cosupp(\lambda)]$ implies $\pi_\nu(x,y) < 1.]$

Definition 8.3.3 Let $\tilde{f}$ be a fuzzy preference aggregation rule. Let λ be a fuzzy subset of N. Then λ is called **semidecisive (decisive)** if $\forall(x,y) \in X \times X$, λ is semidecisive (decisive) for x against y.

Proposition 8.3.4 Let $\tilde{f}$ be a fuzzy preference aggregation rule. Suppose $\tilde{f}$ is dictatorial with dictator i. Let λ be a fuzzy subset of N. Suppose $Supp(\lambda)$ is a coalition. Then λ is decisive if and only if $i \in Supp(\lambda) \cap Cosupp(\lambda)$.

Proof. Suppose λ is decisive. Suppose $i \notin Supp(\lambda) \cap Cosupp(\lambda)$. Let $x, y \in X$ be such that $x \neq y$. Then $\exists(\langle\rho_{\mu 1},\rho_{\nu 1}\rangle,\ldots,\langle\rho_{\mu n},\rho_{\nu n}\rangle) \in T^n$ such that $\pi_{\mu j}(x,y) > 0\ \forall j \in Supp(\lambda)$ and $\pi_{\mu i}(y,x) > 0$ and so $\pi_\mu(x,y) > 0$ and $\pi_\mu(y,x) > 0$, a contradiction, or $\pi_{\nu j}(x,y) < 1\ \forall j \in Cosupp(\lambda)$ and $\pi_{\nu i}(y,x) < 1$. Thus $\pi_\nu(x,y) < 1$ and $\pi_\nu(y,x) < 1$, a contradiction. Hence $i \in Supp(\lambda) \cap Cosupp(\lambda)$. The converse is immediate. ∎

Definition 8.3.5 Let $\langle\rho_\mu,\rho_\nu\rangle \in \mathcal{FR}$. Then

(1) $\langle\rho_\mu,\rho_\nu\rangle$ is said to be **partially quasi-transitive** if $\forall x,y,z \in X$, $\pi_\mu(x,y) \wedge \pi_\mu(y,z) > 0$ implies $\pi_\mu(x,z) > 0$ and $\forall x,y,z \in X$, $\pi_\nu(x,y) \vee \pi_\nu(y,z) < 1$ implies $\pi_\nu(x,z) < 1$.

(2) $\langle\rho_\mu,\rho_\nu\rangle$ is said to be **acyclic** if $\forall x_1,\ldots,x_n \in X$, $\pi_\mu(x_1,x_2) \wedge \ldots \wedge \pi_\mu(x_{n-1},x_n) \leq \pi_\mu(x_1,x_n)$ and $\forall x_1,\ldots,x_n \in X$, $\pi_\nu(x_1,x_2) \vee \ldots \vee \pi_\nu(x_{n-1},x_n) \geq \pi_\nu(x_1,x_n)$

(3) $\langle\rho_\mu,\rho_\nu\rangle$ is said to be **partially acyclic** if $\forall x_1,\ldots,x_n \in X$, $\pi_\mu(x_1,x_2) \wedge \ldots \wedge \pi_\mu(x_{n-1},x_n) > 0$ implies $\rho_\mu(x_1,x_n) > 0$ and $\forall x_1,\ldots,x_n \in X$, $\pi_\nu(x_1,x_2) \vee \ldots \vee \pi_\nu(x_{n-1},x_n) < 1$ implies $\rho_\nu(x_1,x_n) < 1$.

Definition 8.3.6 Let $\tilde{f}$ be a fuzzy preference aggregation rule. Then

(1) $\tilde{f}$ is said to be **(partially) transitive** if $\forall\rho \in \mathcal{FR}^n$, $\tilde{f}(\rho)$ is (partially) transitive;

(2) $\tilde{f}$ is said to be **(partially) quasi-transitive** if $\forall\rho \in \mathcal{FR}^n$, $\tilde{f}(\rho)$ is (partially) quasi-transitive;

(3) $\tilde{f}$ is said to be **(partially) acyclic** if $\forall\rho \in \mathcal{FR}^n$, $\tilde{f}(\rho)$ is (partially) acyclic.

In the remainder of this and the next section, we assume that $|X| \geq 3$ unless otherwise specified.

Proposition 8.3.7 *Let $\langle \rho_\mu, \rho_\nu \rangle$ be an intuitionistic fuzzy binary relation on X. If $\langle \rho_\mu, \rho_\nu \rangle$ is partially transitive, then $\langle \rho_\mu, \rho_\nu \rangle$ is partially quasi-transitive with respect to $\pi = \pi_{(0)}$.*

Proof. Let $x, y, z \in X$. Suppose $\pi_\mu(x, y) > 0$ and $\pi_\mu(y, z) > 0$. Then $\rho_\mu(x, y) > 0, \rho_\mu(y, x) = 0$ and $\rho_\mu(y, z) > 0, \rho_\mu(z, y) = 0$. Hence $\rho_\mu(x, z) > 0$. Suppose $\pi_\mu(x, z) = 0$. Then $\rho_\mu(z, x) > 0$. Thus since $\rho_\mu(x, y) > 0$, $\rho_\mu(z, y) > 0$ by partial transitivity, a contradiction. Thus $\pi_\mu(x, z) > 0$.

Suppose $\pi_\nu(x, y) < 1$ and $\pi_\nu(y, z) < 1$. Then $\rho_\nu(x, y) < 1$, $\rho_\nu(y, x) < 1$ and $\rho_\nu(y, z) < 1, \rho_\nu(z, y) = 1$. Hence $\rho_\nu(x, z) < 1$. Suppose $\pi_\nu(x, z) = 1$. Then $\rho_\nu(z, x) < 1$. Thus since $\rho_\nu(x, y) < 1$, $\rho_\nu(z, y) < 1$ by partial transitivity, a contradiction. Hence $\pi_\nu(x, z) < 1$. ∎

Example 8.3.8 *An example such that partial transitivity $\nRightarrow$ partial quasi-transitivity for $\pi = \pi_{(1)}$ or $\pi = \pi_{(3)}$ can be found in Example 5.1.12.*

Definition 8.3.9 *Let $\widetilde{f}$ be a fuzzy aggregation rule. Then $\widetilde{f}$ is said to be **independent of irrelevant alternatives IIA4** if $\forall x, y \in X$, $\forall (\langle \rho_{\mu 1}, \rho_{\nu 1} \rangle,$ $\ldots, \langle \rho_{\mu n}, \rho_{\nu n} \rangle), (\langle \rho'_{\mu 1}, \rho'_{\nu 1} \rangle, \ldots, \langle \rho'_{\mu n}, \rho'_{\nu n} \rangle) \in T^n$ $[Cosupp(\rho_{\mu i}]_{\{x,y\}}) = Cosupp(\rho'_{\mu i}]_{\{x,y\}}) \; \forall i \in N \Rightarrow Cosupp(\widetilde{f}(\rho_\mu)]_{\{x,y\}}) = Cosupp(\widetilde{f}(\rho'_\mu)]_{\{x,y\}})]$ and $Supp(\rho_{\nu i}]_{\{x,y\}}) = Supp(\rho'_{\nu i}]_{\{x,y\}}) \; \forall i \in N \Rightarrow Supp(\widetilde{f}(\rho_\nu)]_{\{x,y\}}) = Supp(\widetilde{f}(\rho'_\nu)]_{\{x,y\}})]$.*

Proposition 8.3.10 *Let $\langle \rho_\mu, \rho_\nu \rangle, \langle \rho'_\mu, \rho'_\nu \rangle \in \mathcal{FR}$. Let $x, y \in X$. Then*

(1) $Cosupp(\rho_\mu]_{\{x,y\}}) = Cosupp(\rho'_\mu]_{\{x,y\}})$ if and only if $\forall u, v \in \{x, y\}$, $\rho_\mu(u, v) < 1 \Leftrightarrow \rho'_\mu(u, v) < 1$;

(2) $Supp(\rho_\nu]_{\{x,y\}}) = Supp(\rho'_\nu]_{\{x,y\}})$ if and only if $(\forall u, v \in \{x, y\}, \rho_\nu(u, v) > 0 \Leftrightarrow \rho'_\nu(u, v) > 0$.

Proof. (1) Suppose $Cosupp(\rho_\mu]_{\{x,y\}}) = Cosupp(\rho'_\mu]_{\{x,y\}})$. Let $u, v \in \{x, y\}$. Then $\rho_\mu(u, v) < 1 \Leftrightarrow (u, v) \in Cosupp(\rho_\mu]_{\{x,y\}}) \Leftrightarrow (u, v) \in Cosupp(\rho'_\mu]_{\{x,y\}}) \Leftrightarrow \rho'_\mu(u, v) < 1$. Conversely, suppose $\forall u, v \in \{x, y\}, \rho_\mu(u, v) < 1 \Leftrightarrow \rho'_\mu(u, v) < 1$. Then $(u, v) \in Cosupp(\rho_\mu]_{\{x,y\}}) \Leftrightarrow \rho_\mu(u, v) < 1 \Leftrightarrow \rho'_\mu(u, v) < 1 \Leftrightarrow Cosupp(\rho'_\mu]_{\{x,y\}})$.

(2) Suppose $Supp(\rho_\nu]_{\{x,y\}}) = Supp(\rho'_\nu]_{\{x,y\}})$. Let $u, v \in \{x, y\}$. Then $\rho_\nu(u, v) > 0 \Leftrightarrow (u, v) \in Supp(\rho_\nu]_{\{x,y\}}) \Leftrightarrow (u, v) \in Supp(\rho'_\nu]_{\{x,y\}}) \Leftrightarrow \rho'_\nu(u, v) > 0$. Conversely, suppose $\forall u, v \in \{x, y\}, \rho_\nu(u, v) > 0 \Leftrightarrow \rho'_\nu(u, v) > 0$. Then $(u, v) \in Supp(\rho_\nu]_{\{x,y\}}) \Leftrightarrow \rho_\nu(u, v) > 0 \Leftrightarrow \rho'_\nu(u, v) > 0 \Leftrightarrow Supp(\rho'_\nu]_{\{x,y\}})$. ∎

For $\pi = \pi_{(2)}$ and $x, y \in X, \pi_\mu(x, y) > 0$ if and only if $\rho_\mu(y, x) < 1$ and $\pi_\nu(x, y) < 1$ if and only if $\rho_\nu(y, x) > 0$. The strict preferences $\pi_{(1)}, \pi_{(3)}, \pi_{(4)},$ and $\pi_{(5)}$ are regular. These properties play a key role in the proof of the next result.

Lemma 8.3.11 *Let λ be a fuzzy subset of N. Let $\widetilde{f}$ be a partially quasi-transitive fuzzy preference aggregation rule that is weakly Paretian and either independent of irrelevant alternatives IIA2 with strict preferences of type $\pi_{(0)}$ or independent of irrelevant alternatives IIA3 with regular strict preferences or is independent of irrelevant alternatives IIA4 with strict preferences of type $\pi_{(2)}$. If λ is semidecisive for x against y w.r.t π, then $\forall (v, w) \in X \times X$, λ is decisive for v against w.*

Proof. The proof for ρ_μ can be found in Lemma 5.1.15. Suppose λ is semidecisive for x against y. Let ρ_ν be any fuzzy preference profile such that $\pi_{\nu i}(x, z) < 1 \ \forall i \in \text{Supp}(\lambda)$, where $z \notin \{x, y\}$. Let ρ'_ν be a fuzzy preference profile such that

$$
\begin{aligned}
\rho'_{\nu i}(x, z) &= \rho_{\nu i}(x, z) \text{ and } \rho'_{\nu i}(z, x) = \rho_{\nu i}(z, x) \ \forall i \in N, \\
\pi'_{\nu i}(x, y) &> 0 \ \forall i \in \text{Supp}(\lambda) \text{ and } \pi'_{\nu j}(y, x) > 0 \ \forall j \in N \backslash \text{Supp}(\lambda), \\
\pi'_{\nu i}(y, z) &> 0 \ \forall i \in N.
\end{aligned}
$$

Since $\pi_{\nu i}(x, z) < 1 \ \forall i \in \text{Supp}(\lambda), \pi'_{\nu i}(x, z) < 1 \ \forall i \in \text{Supp}(\lambda)$ by the definition of ρ'_ν. Since $x \widetilde{D}_\lambda y, \pi'_\nu(x, y) < 1$. Since $\widetilde{f}$ is weakly Paretian, $\pi'_\nu(y, z) < 1$. Since $\widetilde{f}$ is partially quasi-transitive, $\pi'_\nu(x, z) < 1$. For IIA3 and π_ν regular, $\rho'_\nu(x, z) < \rho'_\nu(z, x)$. Since $\rho_{\nu i}]_{\{x, z\}} = \rho'_{\nu i}]_{\{x, z\}} \ \forall i \in N$ and $\widetilde{f}$ is IIA3, $\rho_\nu]_{\{x, z\}} \sim \rho'_\nu]_{\{x, z\}}$. Thus $\rho_\nu(x, z) < \rho_\nu(z, x)$. (For strict preferences of type $\pi_{(0)}$ and the crisp case, we write: Hence $\rho'_\nu(x, z) < 1$ and $\rho'_\nu(z, x) = 1$. Since $\text{Cosupp}(\rho_{\nu i}]_{\{x, z\}}) = \text{Cosupp}(\rho'_{\nu i}]_{\{x, z\}}) \ \forall i \in N$ and $\widetilde{f}$ is IIA2, $\text{Cosupp}(\rho_\nu]_{\{x, z\}}) = \text{Cosupp}(\rho'_\nu]_{\{x, z\}})$. Thus $\rho_\nu(x, z) < 1$ and $\rho_\nu(z, x) = 1$.) (For strict preferences of type $\pi_{(2)}$, we write: Hence $\rho'_\nu(z, x) > 0$. Since $\text{Supp}(\rho_{i_\nu i}]_{\{x, z\}}) = \text{Supp}(\rho'_{\nu i}]_{\{x, z\}}) \ \forall i \in N$ and $\widetilde{f}$ is IIA4, $\text{Supp}(\rho_\nu]_{\{x, z\}}) = \text{Supp}(\rho'_\nu]_{\{x, z\}})$. Thus $\rho_\nu(z, x) > 0$.) Hence $\pi_\nu(x, z) < 1$. Since ρ was arbitrary, $x D_\lambda z$. Thus since z was arbitrary in $X \backslash \{x, y\}$,

$$\forall z \in X \backslash \{x, y\}, x \widetilde{D}_\lambda y \Rightarrow x D_\lambda z. \tag{8.12}$$

Since λ is decisive for x against z implies λ is semidecisive for x against z, interchanging y and z in the above argument implies λ is decisive for x against y.

Let ρ''_ν be any fuzzy preference profile with $\pi''_{vi}(y, z) < 1 \ \forall i \in \text{Supp}(\lambda)$ and let ρ^+_ν be a fuzzy preference profile such that

$$
\begin{aligned}
\rho^+_{\nu i}(y, z) &= \rho''_{vi}(y, z) \text{ and } \rho^+_{vi}(z, y) = \rho''_{vi}(z, y) \ \forall i \in N, \\
\pi^+_{vi}(y, x) &< 1 \ \forall i \in N, \\
\pi^+_{vi}(x, z) &< 1 \ \forall i \in \text{Supp}(\lambda) \text{ and } \pi^+_{\nu j}(z, x) < 1 \ \forall j \in N \backslash \text{Supp}(\lambda).
\end{aligned}
$$

Then $\pi^+_{vi}(y, z) < 1 \ \forall i \in \text{Supp}(\lambda)$. Since $x D_\lambda z, \pi^+_\nu(x, z) < 1$. Since $\widetilde{f}$ is weakly Paretian, $\pi^+_\nu(y, x) < 1$. Since $\widetilde{f}$ is partially quasi-transitive, $\pi^+_\nu(y, z) < 1$. For

IIA3 and π_ν regular, we have since $\rho_{vi}^+\rceil\{y,z\} = \rho_{vi}''\rceil\{y,z\}$ $\forall i \in N$ and $\widetilde{f}$ is IIA3, $\rho_\nu''\rceil\{y,z\} \sim \rho_\nu''\rceil\{y,z\}$. Since $\pi_\nu^+(y,z) < 1$, $\rho_\nu^+(y,z) < \rho_\nu^+(z,y)$. Hence $\rho_\nu''(y,z) < \rho_\nu''(z,y)$. (For $\pi = \pi_{(0)}$ and the crisp case, we write: Since $\rho_{vi}^+\rceil\{y,z\} = \rho_\nu''\rceil\{y,z\}$ $\forall i \in N$ and $\widetilde{f}$ is IIA2, $\mathrm{Cosupp}(\rho_\nu^+\rceil\{y,z\}) = \mathrm{Cosupp}(\rho_\nu''\rceil\{y,z\})$. Since $\pi_\nu^+(y,z) < 1$, $\rho_\nu^+(y,z) < 1$ and $\rho_\nu^+(z,y) = 1$. Hence $\rho_\nu''(y,z) < 1$ and $\rho_\nu''(z,y) = 1$.) (For $\pi = \pi_{(2)}$, we write: Since $\rho_{vi}^+\rceil\{y,z\} = \rho_\nu''\rceil\{y,z\}$ $\forall i \in N$ and $\widetilde{f}$ is IIA4, $\mathrm{Supp}(\rho_\nu^+\rceil\{y,z\}) = \mathrm{Supp}(\rho_\nu''\rceil\{y,z\})$. Since $\pi_\nu^+(y,z) < 1$, $\rho_\nu^+(z,y) > 0$. Hence $\rho_\nu''(z,y) > 0$.) Thus $\pi_\nu''(y,z) < 1$ and so $yD_\lambda z$. Hence since z is arbitrary in $X\backslash\{x,y\}$, the preceding two steps yield

$$\forall z \notin \{x,y\}, \quad x\widetilde{D}_\lambda y \Rightarrow yD_\lambda z. \tag{8.13}$$

Now λ decisive for y against z implies λ is semidecisive for y against z. Thus by (8.12), λ is decisive for y against x. We have $\forall(v,w) \in X \times X$,

$$x\widetilde{D}_\lambda y \Rightarrow xD_\lambda v \text{ (by (4.1))} \Rightarrow x\widetilde{D}_\lambda v \Rightarrow vD_\lambda w \text{ (by (8.13)) with } v \text{ replacing } y.$$

∎

Let $\mathcal{FR}_T = \{\rho \in \mathcal{FR} \mid \rho \text{ is partially transitive}\}$.

Theorem 8.3.12 (*Fuzzy Arrow's Theorem*) *Let $\widetilde{f}$ be a fuzzy aggregation rule from $\mathcal{FR}_T^n$ into $\mathcal{FR}$. Suppose strict preferences are of type $\pi_{(0)}$. Let $\widetilde{f}$ be weakly Paretian, partially transitive, and independent of irrelevant alternatives IIA2. Then $\widetilde{f}$ is dictatorial.*

Proof. The proof for ρ_μ can be found in Theorem 5.1.16. Since $\widetilde{f}$ is partially transitive, $\widetilde{f}$ is partially quasi-transitive. By Lemma 8.3.11, it suffices to show that $\exists i \in N, \exists x,y \in X$ such that $\{i\}$ is semidecisive for x against y. (That is, $\exists \lambda$ with $\mathrm{Supp}(\lambda) = \{i\}$ such that λ is semidecisive for x against y.) This follows from Lemma 8.3.11 because λ is decisive for x against y for all $x,y \in X$. Hence $\forall \rho_\nu \in \mathcal{FR}^n(X)$, $\forall x,y \in X$, $\pi_{\nu_i}(x,y) < 1$ implies $\pi_\nu(x,y) < 1$.

Since $\widetilde{f}$ is weakly Paretian, $\exists$ a decisive λ for any pair of alternatives, namely $\lambda = 1_N$. For all $(u,v) \in X \times X$, let $m(u,v) = \wedge\{m(u,v) \mid (u,v) \in X^2\}$. Without loss of generality suppose λ is semidecisive for x against y with $|\mathrm{Supp}(\lambda)| = m$. If $m = 1$, the proof is complete. Suppose > 1. Let $i \in \mathrm{Supp}(\lambda)$. Consider any fuzzy profile $\rho_\nu \in \mathcal{FR}^n$ such that

$$\pi_{\nu i}(x,y) < 1, \ \pi_{\nu i}(y,z) < 1, \ \pi_{\nu i}(x,z) < 1,$$
$$\forall j \in \mathrm{Supp}(\lambda)\backslash\{i\}, \ \pi_{vj}(z,x) < 1, \ \pi_{vj}(x,y) < 1, \ \pi_{vj}(z,y) < 1,$$
$$\forall k \notin \mathrm{Supp}(\lambda), \ \pi_{vk}(y,z) < 1, \ \pi_{vk}(z,x) < 1, \ \pi_{vk}(y,x) < 1.$$

Since λ is semidecisive for x against y and $\pi_{\nu j}(x,y) < 1$ $\forall j \in \mathrm{Supp}(\lambda)$, $\pi_\nu(x,y) < 1$. Since $|\mathrm{Supp}(\lambda)| = m$, it is not the case that $\pi_\nu(z,y) < 1$ otherwise λ' is semidecisive for z against y, where $\lambda'(j) = \lambda(j)$ $\forall j \in \mathrm{Supp}(\lambda)\backslash\{i\}$ and $\lambda'(i) = 0$. (Suppose $\pi_\nu(z,y) < 1$. $\mathrm{Supp}(\rho_{vi}\rceil\{x,y\}) = \mathrm{Supp}(\rho_{vi}'\rceil\{x,y\})$

$\forall i \in N, \forall \rho'_\nu \in \mathcal{FR}^n(X) \Rightarrow \text{Supp}(\widetilde{f}(\rho_\nu]\{x,y\}) = \text{Supp}(\widetilde{f}(\rho'_\nu]\{x,y\}) \; \forall \rho'_\nu \in$
$\mathcal{FR}^n(X)$ by IIA2 $\Rightarrow (z,y) \in \text{Supp}(\widetilde{f}(\rho'_\nu))$, $(y,z) \notin \text{Supp}(\widetilde{f}(\rho'_\nu)) \; \forall (z,y) \in$
$\text{Supp}(\widetilde{f}(\rho'_\nu)) \; \forall \rho'_\nu \in \mathcal{FR}^n(X)$ since this is the case for $\widetilde{f}(\rho_\nu)$ and $\pi_{\nu j}(z,y) < 1$
and $\pi_{\nu j}(y,z) < 1$ otherwise.) However, this contradicts the minimality of
m since $|\text{Supp}(\lambda')| = m - 1$. Thus $\widetilde{f}(\rho_\nu)(y,z) < 1$ since $\widetilde{f}(\rho_\nu)$ is complete.
(Actually since $\pi_\nu(x,y) = 1, \pi_\nu(y,z) < 1$.) Since $\widetilde{f}$ is partially transitive,
$\widetilde{f}(\rho_\nu)(x,z) < 1$. (Actually $\pi_\nu(x,z) < 1$ by partial quasi-transitivity.) By IIA2,
λ^* is semidecisive for x against z, where $\lambda^*(i) = \lambda(i)$ and $\lambda^*(j) = 0$ for
$j \in N\setminus\{i\}$. However, this contradicts the fact that $m > 1$. ∎

Definition 8.3.13 *Let $\langle \rho_\mu, \rho_\nu \rangle$ be an intuitionistic fuzzy relation on X. Then
$\langle \rho_\mu, \rho_\nu \rangle$ is said to be **weakly transitive** if $\forall x, y, z \in X$, (1) $\rho_\mu(x,y) \geq \rho_\mu(y,x)$
and $\rho_\mu(y,z) \geq \rho_\mu(z,y)$ implies $\rho_\mu(x,z) \geq \rho_\mu(z,x)$ and (2) $\rho_\nu(x,y) \leq \rho_\nu(y,x)$
and $\rho_\nu(y,z) \leq \rho_\nu(z,y)$ implies $\rho_\nu(x,z) \leq \rho_\nu(z,x)$. A fuzzy aggregation rule $\widetilde{f}$
is said to be **weakly transitive** if $\widetilde{f}(\rho)$ is weakly transitive for all $\rho \in \mathcal{FR}^n$.*

Example 8.3.14 *Examples such that ρ is max-min transitive, but ρ is not
weakly transitive; ρ is weakly transitive, but not max-min transitive; ρ is weakly
transitive, but not transitive in the crisp sense (and not complete) can be found
in Example 5.1.18.*

Remark 8.3.15 *Suppose $\langle \rho_\mu, \rho_\nu \rangle$ is exact. If $\langle \rho_\mu, \rho_\nu \rangle$ is weakly transitive and
complete, then $\langle \rho_\mu, \rho_\nu \rangle$ is transitive in the crisp sense.*

Proof. Suppose $\rho_\mu(x,y) = 1$ and $\rho_\mu(y,x) = 1$. Since ρ is weakly transitive,
$\rho_\mu(x,z) \geq \rho_\mu(z,x)$. Since ρ is complete, $\rho_\mu(x,z) = 1$. Suppose $\rho_\nu(x,y) = 0$
and $\rho_\nu(y,x) = 0$. Since ρ_ν is weakly transitive, $\rho_\nu(x,z) \leq \rho_\nu(z,x)$. Since ρ_ν
is complete, $\rho(x,z) = 0$. ∎

Proposition 8.3.16 *Let $\langle \rho_\mu, \rho_\nu \rangle$ be an intuitionistic fuzzy relation on X.
Then the following properties are equivalent:*
(1) $\langle \rho_\mu, \rho_\nu \rangle$ is weakly transitive.
*(2) For all $x, y, z \in X$, (i) $\rho_\mu(x,y) \geq \rho_\mu(y,x)$ and $\rho_\mu(y,z) \geq \rho_\mu(z,y)$
with strict inequality holding at least once, then $\rho_\mu(x,z) > \rho_\mu(z,x)$ and (ii)
$\rho_\nu(x,y) \leq \rho_\nu(y,x)$ and $\rho_\nu(y,z) \leq \rho_\nu(z,y)$ with strict inequality holding at
least once, then $\rho_\nu(x,z) < \rho_\nu(z,x)$*

Proof. The proof for ρ_μ can be found in Proposition 5.1.20.
Suppose (1) holds. Assume that $\rho_\nu(x,y) \leq \rho_\nu(y,x)$ and $\rho_\nu(y,z) < \rho_\nu(z,y)$.
Suppose $\rho_\nu(z,x) \leq \rho_\nu(x,z)$. Then $\rho_\nu(z,y) \leq \rho_\nu(y,z)$ by (1), a contradiction.
Thus $\rho_\nu(x,z) < \rho_\nu(z,x)$. A similar argument shows that $\rho_\nu(x,y) < \rho_\nu(y,x)$
and $\rho_\nu(y,z) \leq \rho_\nu(z,y)$ implies $\rho_\nu(x,z) < \rho_\nu(z,x)$.
Suppose (2) holds. Let $x, y, z \in X$. Suppose $\rho_\nu(x,y) \leq \rho_\nu(y,x)$ and
$\rho_\nu(y,z) \leq \rho_\nu(z,y)$. Suppose $\rho_\nu(z,x) < \rho_\nu(x,z)$. Then by (2), $\rho_\nu(z,x) <$

$\rho_\nu(x,z)$ and $\rho_\nu(x,y) \leq \rho_\nu(y,x)$, we have $\rho_\nu(z,y) < \rho_\nu(y,z)$, a contradiction. Hence $\rho_\nu(x,z) \leq \rho_\nu(z,x)$. ∎

Corollary 8.3.17 *Let $\langle \rho_\mu, \rho_\nu \rangle$ be a fuzzy relation on X. Suppose strict preferences are regular. If $\langle \rho_\mu, \rho_\nu \rangle$ is weakly transitive, then $\langle \rho_\mu, \rho_\nu \rangle$ is partially quasi-transitive.*

Let $\mathcal{FCR}$ denote the set of all reflexive and complete intuitionistic fuzzy preference relations that are consistent.

Theorem 8.3.18 *(Fuzzy Arrow's Theorem) Let $\widetilde{f} : \mathcal{FCR}^n \to \mathcal{FR}$ be a fuzzy aggregation rule. Suppose strict preferences are regular. Let $\widetilde{f}$ be weakly transitive, weakly Paretian, and independent of irrelevant alternatives IIA3. Then $\widetilde{f}$ is dictatorial.*

Proof. The proof for ρ_μ can be found in Theorem 5.1.22. Since $\widetilde{f}$ is consistent, $\widetilde{f}$ is partially quasi-transitive by Corollary 8.3.17. By Lemma 8.3.11, it suffices to show that $\exists i \in N$, $\exists x, y \in X$ such that $\{i\}$ is semidecisive for x against y. (That is, $\exists \lambda$ with $\mathrm{Supp}(\lambda) = \{i\}$ such that λ is semidecisive for x against y.) This follows by Lemma 8.3.11 because then λ is decisive for x against y for all $x, y \in X$, where $\mathrm{Supp}(\lambda) = \{i\}$. Hence $\forall \langle \rho_\mu, \rho_\nu \rangle \in \mathcal{FR}^n$, $\forall x, y \in X$, $\pi_{i_\mu}(x,y) > 0$ implies $\pi_\mu(x,y) > 0$ and $\pi_{i_\nu}(x,y) < 1$ implies $\pi_\nu(x,y) < 1$.

Since $\widetilde{f}$ is weakly Paretian, $\exists$ a decisive λ for any pair of alternatives, namely $\lambda = 1_N$. For all $(u,v) \in X \times X$, let $m(u,v)$ denote the size of the smallest $|\mathrm{Supp}(\lambda)|$ for λ semidecisive for u against v. Let $m = \wedge\{m(u,v) \mid (u,v) \in X \times X\}$. Without loss of generality suppose λ is semidecisive for x against y with $|\mathrm{Supp}(\lambda)| = m$. If $m = 1$, the proof is complete. Suppose $m > 1$. Let $i \in \mathrm{Supp}(\lambda)$. Consider any fuzzy profile $(\langle \rho_{\mu 1}, \rho_{\nu 1} \rangle, \ldots, \langle \rho_{\mu n}, \rho_{\nu n} \rangle) \in T^n$ such that

$$\pi_{\mu i}(x,y) > 0, \pi_{\mu i}(y,z) > 0, \pi_{\mu i}(x,z) > 0,$$
$$\forall j \in \mathrm{Supp}(\lambda)\backslash\{i\}, \ \pi_{\mu j}(z,x) > 0, \ \pi_{\mu j}(x,y) > 0, \ \pi_{\mu j}(z,y) > 0,$$
$$\forall k \notin \mathrm{Supp}(\lambda), \ \pi_{\mu k}(y,z) > 0, \ \pi_{\mu k}(z,x) > 0, \ \pi_{\mu k}(y,x) > 0.$$

$$\pi_{\nu i}(x,y) < 1, \pi_{\nu i}(y,z) < 1, \pi_{\nu i}(x,z) < 1,$$
$$\forall j \in \mathrm{Supp}(\lambda)\backslash\{i\}, \ \pi_{\nu j}(z,x) < 1, \ \pi_{\nu j}(x,y) < 1, \ \pi_{\nu j}(z,y) < 1,$$
$$\forall k \notin \mathrm{Supp}(\lambda), \ \pi_{\nu k}(y,z) < 1, \ \pi_{\nu k}(z,x) < 1, \ \pi_{\nu k}(y,x) < 1.$$

Since λ is semidecisive for x against y, $\pi_{\mu i}(x,y) > 0 \ \forall i \in \mathrm{Supp}(\lambda)$, and $\pi_{\mu j}(y,x) > 0 \ \forall j \notin \mathrm{Supp}(\lambda)$, we have that $\pi_\mu(x,y) > 0$. Since $|\mathrm{Supp}(\lambda)| = m$, it is not the case that $\pi_\mu(z,y) > 0$, otherwise λ' is semidecisive for z against y, where $\lambda'(j) = \lambda(j) \ \forall j \in \mathrm{Supp}(\lambda)\backslash\{i\}$ and $\lambda'(i) = 0$. (Suppose $\pi_\mu(z,y) > 0$. Then $\rho_{\mu i}]_{\{z,y\}} \sim \rho'_{\mu i}]_{\{z,y\}}$) $\forall i \in N$, $\forall \rho'_\mu \in \mathcal{FR}^n(X) \Rightarrow \widetilde{f}(\rho_\mu)]_{\{z,y\}} \sim$

$\widetilde{f}(\rho'_\mu)]_{\{z,y\}} \; \forall \rho'_\mu \in \mathcal{FR}^n(X)$ by independence of irrelevant alternatives IIA3
$\Rightarrow \widetilde{f}(\rho'_\mu)(z,y) > \widetilde{f}(\rho'_\mu)(y,z)$ since $\widetilde{f}(\rho_\mu)(z,y) > \widetilde{f}(\rho_\mu)(y,z)$. Thus $\pi'_\mu(z,y) > 0$.) However, this contradicts the minimality of m since $|\mathrm{Supp}(\lambda')| = m-1$.
Since $\pi_\mu(z,y) = 0$, $\widetilde{f}(\rho_\mu)(y,z) \geq \widetilde{f}(\rho_\mu)(z,y)$. Since $\widetilde{f}(\rho_\mu)(x,y) > \widetilde{f}(\rho_\mu)(y,x)$,
$\widetilde{f}(\rho_\mu)(x,z) > \widetilde{f}(\rho_\mu)(z,x)$ by consistency. Hence $\pi_\mu(x,z) > 0$. By independence
of irrelevant alternatives IIA3, λ^* is semidecisive for x against z, where $\lambda^*(i) = \lambda(i)$ and $\lambda^*(j) = 0$ for $j \in N\backslash\{i\}$. However, this contradicts the fact that $m > 1$.

Since λ is semidecisive for x against y, $\pi_{i_\nu}(x,y) < 1 \; \forall i \in \mathrm{Supp}(\lambda)$, and $\pi_{j_\nu}(y,x) < 1 \; \forall j \notin \mathrm{Supp}(\lambda)$, we have that $\pi_\nu(x,y) < 1$. Since $|\mathrm{Supp}(\lambda)| = m$, it is not the case that $\pi_\nu(z,y) < 1$, otherwise λ' is semidecisive for z against y, where $\lambda'(j) = \lambda(j) \; \forall j \in \mathrm{Supp}(\lambda)\backslash\{i\}$ and $\lambda'(i) = 0$. (Suppose $\pi_\nu(z,y) < 1$. Then $\rho_{\nu i}]_{\{z,y\}} \sim \rho'_{\nu i}]_{\{z,y\}}) \; \forall i \in N, \forall \rho'_\nu \in \mathcal{FR}^n(X) \Rightarrow \widetilde{f}(\rho_\nu)]_{\{z,y\}} \sim \widetilde{f}(\rho'_\nu)]_{\{z,y\}} \; \forall \rho'_\nu \in \mathcal{FR}^n(X)$ by independence of irrelevant alternatives IIA3
$\Rightarrow \widetilde{f}(\rho'_\nu)(z,y) < \widetilde{f}(\rho'_\nu)(y,z)$ since $\widetilde{f}(\rho_\nu)(z,y) < \widetilde{f}(\rho_\nu)(y,z)$. Thus $\pi'_\nu(z,y) < 1$.) However, this contradicts the minimality of m since $|\mathrm{Supp}(\lambda')| = m-1$.
Since $\pi_\nu(z,y) = 1$, $\widetilde{f}(\rho_\nu)(y,z) \leq \widetilde{f}(\rho_\nu)(z,y)$. Since $\widetilde{f}(\rho_\nu)(x,y) < \widetilde{f}(\rho_\nu)(y,x)$,
$\widetilde{f}(\rho_\nu)(x,z) < \widetilde{f}(\rho_\nu)(z,x)$ by consistency. Hence $\pi_\nu(x,z) < 1$. By independence of irrelevant alternatives IIA3, λ^* is semidecisive for x against z, where $\lambda^*(i) = \lambda(i)$ and $\lambda^*(j) = 0$ for $j \in N\backslash\{i\}$. However, this contradicts the fact that $m > 1$. ∎

8.4   Representation Rules, Veto Players, and Oligarchies

In the previous section, we demonstrated that fuzzy intuitionistic versions of Arrow's Theorem hold under a certain set of definitions and various IIA conditions. In this section, we lay out some concepts that give an application of the fuzzy version of Arrow's Theorem.

Recall that by ρ, we mean $\langle \rho_\mu, \rho_\nu \rangle = (\langle \rho_{\mu 1}, \rho_{\nu 1} \rangle, \dots, \langle \rho_{\mu n}, \rho_{\nu n} \rangle)$.

Definition 8.4.1 *Let $i \in N$ and $t \in (0,1]$. Define the fuzzy subset i_t of N by $\forall j \in N$, $i_t(j) = t$ if $j = i$ and $i_t(j) = 0$ otherwise. If $g : T^n \to \{i_t \mid i \in N, t \in (0,1]\}$, then g is called a* **representation rule** *for T.*

Let g be a representation rule for T. Then $\forall \rho = \langle \rho_\mu, \rho_\nu \rangle \in T^n$, $\mathrm{Supp}(g(\rho)) = \{i_0\}$ for some $i_0 \in N$. By the notation, $\rho_{\mathrm{Supp}(g(\rho))_\mu}$, we mean $\rho_{\mu i_0}$ and by $\rho_{\mathrm{Supp}(g(\rho))_\nu}$, we mean $\rho_{\nu i_0}$.

Definition 8.4.2 *Let g be a representation rule for T. Then*
    (1) g is called **dictatorial** *for T if and only if $\forall \rho, \rho' \in T^n$, $\mathrm{Supp}(g(\rho)) = \mathrm{Supp}(g(\rho'))$;*

(2) g is called **independent of irrelevant alternatives IIA2** if $\forall \rho, \rho' \in T^n$, $\forall x, y \in X$, $Supp(\rho_{\mu i}]\{_{x,y}\}) = Supp(\rho'_{\mu i}]\{_{x,y}\})$ $\forall i \in N$ implies

$$Supp(\rho_{Supp(g(\rho))_\mu}]\{_{x,y}\}) = Supp(\rho_{Supp(g(\rho')_\mu}]\{_{x,y}\})$$

and

$$Cosupp(\rho_{\nu i}]\{_{x,y}\}) = Cosupp(\rho'_{\nu i}]\{_{x,y}\})$$

$\forall i \in N$ implies $Cosupp(\rho_{Supp(g(\rho))_\nu}]\{_{x,y}\}) = Cosupp(\rho_{Supp(g(\rho')_\nu}]\{_{x,y}\})$.

(3) g is called **independent of irrelevant alternatives IIA3** if $\forall \rho, \rho' \in T^n$, $\forall x, y \in X$, $\rho_{\mu i}]\{_{x,y}\} \sim \rho'_{\mu i}]\{_{x,y}\}$ $\forall i \in N$ implies $\rho_{Supp(g(\rho))_\mu}]\{_{x,y}\} \sim \rho_{Supp(g(\rho')_\mu}]\{_{x,y}\}$ and $\rho_{\nu i}]\{_{x,y}\} \sim \rho'_{\nu i}]\{_{x,y}\}$ $\forall i \in N$ implies $\rho_{Supp(g(\rho))_\nu}]\{_{x,y}\} \sim \rho_{Supp(g(\rho')_\nu}]\{_{x,y}\}$.

It follows that $Cosupp(g(\rho_\nu)) = \{j\}$ for some $j \in N$. By $\rho_{\nu Cosupp(g(\rho_\nu))}$, we mean $\rho_{\nu j}$.

Theorem 8.4.3 *Let strict preferences be of type $\pi_{(0)}$. Let g be a representation rule for $\mathcal{FR}^n_T$. Then g is independent of irrelevant alternatives IIA2 if and only if g is dictatorial.*

Proof. The proof for ρ_μ can be found in Theorem 5.2.3. Let $\widetilde{f}$ be a fuzzy aggregation rule determined by g. Let $\rho_\nu \in \mathcal{FR}^n$. Then $\widetilde{f}(\rho_\nu) = \rho_{\nu Cosupp(g(\rho_\nu))}$. Since $\rho_{\nu i}$ is partially transitive $\forall i \in N$, $\rho_{\nu Cosupp(g(\rho_\nu))}$ is partially transitive. Thus $\widetilde{f}(\rho_\nu)$ is partially transitive. Hence $\widetilde{f}$ is partially transitive, Let $x, y \in X$. Suppose $\pi_{\nu_i}(x,y) < 1$ $\forall i \in N$. Then $\pi_{\nu Cosupp(g(\rho_\nu))}(x,y) < 1$, but $\pi_{\nu Cosupp(g(\rho_\nu))}(x,y) = \widetilde{f}(x,y)$. Thus $\widetilde{f}$ is weakly Paretian. Since $\widetilde{f}(\rho_\nu) = \rho_{\nu Cosupp(g(\rho_\nu))}$ and $\widetilde{f}(\rho'_\nu) = \rho'_{\nu Cosupp(g(\rho'_\nu))}$, $\widetilde{f}$ is IIA2 if and only if g is. Thus if g is IIA2, the $\widetilde{f}$ is dictatorial by Theorem 8.3.12. Conversely, if g is dictatorial, then it follows easily that g is IIA2. ∎

Theorem 8.4.4 *Let strict preferences be regular. Let g be a representation rule for $\mathcal{FCR}$. Then g is independent of irrelevant alternatives IIA3 if and only if g is dictatorial.*

Proof. Let $\widetilde{f}$ be a fuzzy aggregation rule determined by g. Let $\rho \in \mathcal{FCR}^n$. Then $\widetilde{f}(\rho) = \langle \rho_{Supp(g(\rho))_\mu}, \rho_{Supp(g(\rho))_\nu} \rangle$. Since $\rho_{\mu i}$ and $\rho_{\nu i}$ are consistent $\forall i \in N$, $\langle \rho_{Supp(g(\rho))_\mu}, \rho_{Supp(g(\rho))_\nu} \rangle = \langle \rho_{\mu i_0}, \rho_{\nu i_0} \rangle$ is consistent. Thus $\widetilde{f}(\rho)$ is consistent. Hence $\widetilde{f}$ is consistent. Let $x, y \in X$. Suppose $\pi_{\mu i}(x,y) > 0$ and $\pi_{\nu i}(x,y) < 1$ $\forall i \in N$. Then $\pi_{Supp(g(\rho))_\mu}(x,y) > 0$ and $\pi_{Supp(g(\rho))_\nu}(x,y) < 1$. Thus $\rho_{\mu i_0}(x,y) > \rho_{\mu i_0}(y,x)$ and $\rho_{\nu i_0}(x,y) < \rho_{\nu i_0}(y,x)$. Hence $\pi_\mu(x,y) > 0$ and $\pi_\nu(x,y) < 1$. Thus $\widetilde{f}$ is weakly Paretian. Since $\widetilde{f}(\rho) = \langle \rho_{Supp(g(\rho))_\mu}, \rho_{Supp(g(\rho))_\nu} \rangle$ and $\widetilde{f}(\rho') = \langle \rho_{Supp(g(\rho'))_\mu}, \rho_{Supp(g(\rho'))_\nu} \rangle$, $\widetilde{f}$ is independent of irrelevant alternatives IIA3 if and only if g is. Thus if g is independent of irrelevant alternatives IIA3, then $\widetilde{f}$ is dictatorial by Theorem 8.3.18. Thus g

is dictatorial. Conversely, if g is dictatorial, then it follows easily that g is independent of irrelevant alternatives IIA3. ∎

Definition 8.4.5 *Let $x, y \in X$. An individual $i \in N$ is said to have a **veto** for x against y, written xV_iy, if for every $\rho \in \mathcal{FR}^n$, $\pi_{\mu i}(x, y) > 0$ implies not $\pi(y, x) > 0$ and $\pi_{\nu i}(x, y) < 1$ implies not $\pi_{\nu}(y, x) < 1$. An element $i \in N$ is said to have a **veto** if for all $x, y \in X$, i has a veto for x against y.*

Definition 8.4.6 *A fuzzy aggregation rule is called **oligarchic** if there exists $\lambda \in \mathcal{FP}(N)$ (called an **oligarchy**) such that*
(1) *every member of $Supp(\lambda)$ has a veto;*
(2) *λ is decisive.*

Definition 8.4.7 *Let $x, y \in X$. An individual $i \in N$ is said to have a **semiveto** for x against y, written $x\widetilde{V}_iy$, if $\forall \rho \in \mathcal{FR}^n$, $\pi_{\mu i}(x, y) > 0$ and $\pi_{\mu j}(y, x) > 0 \ \forall j \neq i$ implies not $\pi_{\mu}(y, x) > 0$ and $\pi_{\nu i}(x, y) < 1$ and $\pi_{\nu j}(y, x) < 1 \ \forall j \neq i$ implies not $\pi_{\nu}(y, x) < 1$.*

Lemma 8.4.8 *Let strict preferences be of type $\pi_{(0)}$. Suppose ρ is partially quasi-transitive. Then*
(1) *$\pi_{\mu}(x, y) > 0$ and $\rho_{\mu}(y, z) > 0$ implies $\rho_{\mu}(x, z) > 0$,*
    $\pi_{\nu}(x, y) < 1$ and $\rho_{\nu}(y, z) < 1$ implies $\rho_{\nu}(x, z) < 1$.
(2) *$\rho_{\mu}(x, y) > 0$ and $\pi_{\mu}(y, z) > 0$ implies $\rho_{\mu}(x, z) > 0$,*
    $\rho_{\nu}(x, y) < 1$ and $\pi_{\nu}(y, z) < 1$ implies $\rho_{\nu}(x, z) < 1$.

Proof. (1) The case for ρ_{μ} can be found in Proposition 5.1.20 or [21, Lemma 3.8, p. 378]. Suppose it's not the case that $\rho_{\nu}(x, z) < 1$. Since ρ_{ν} is complete, $\pi_{\nu}(z, x) < 1$. Since also $\pi_{\nu}(x, y) < 1$, we have by partial quasi-transitivity that $\pi_{\nu}(z, y) < 1$. However, this is impossible since $\rho_{\nu}(y, z) < 1$.

(2) Suppose it's not the case that $\rho_{\nu}(x, z) < 1$. Since ρ_{ν} is complete, $\rho_{\nu}(z, x) < 1$ and so $\pi_{\nu}(z, x) < 1$. Since also $\pi_{\nu}(y, z) < 1$, we have by partial quasi-transitivity that $\pi_{\nu}(y, x) < 1$. However, this is impossible since $\rho_{\nu}(x, y) < 1$. ∎

Lemma 8.4.9 *Let strict preferences be of type $\pi_{(0)}$. Let $\widetilde{f} : \mathcal{FR}_T^n \to \mathcal{FR}$ be a fuzzy aggregation rule which is partially quasi-transitive, weakly Paretian, and independent of irrelevant alternatives IIA2. If $i \in N$ has a semiveto for x against y for some $x, y \in X$, then i has a veto over all ordered pairs $(v, w) \in X \times X$.*

Proof. The proof for ρ_{μ} can be found in Lemma 5.2.9 or [21, Lemma 3.7, p. 377]. Let ρ_{ν} be such that
$$\pi_{\nu i}(x, y) < 1, \ \pi_{\nu i}(x, z) < 1, \ \pi_{\nu i}(y, z) < 1,$$
$$\forall j \neq i, \ \pi_{\nu j}(y, x) < 1, \ \pi_{\nu j}(y, z) < 1.$$

Since i has a semiveto for x against y, not $\pi_\nu(y, x) < 1$. Since $\widetilde{f}$ is complete, $\rho_\nu(x, y) < 1$ and in fact $\pi_\nu(x, y) < 1$. Since $\widetilde{f}$ is weakly Paretian, $\pi_\nu(y, z) < 1$. Since $\widetilde{f}$ is partially quasi-transitive, $\pi_\nu(x, z) < 1$. By IIA2, i has a veto for x against z. (We have $\pi_\nu(x, z) < 1$ for this ρ_ν, but we need independence to get it for all ρ_ν. Let ρ'_ν be such that $\pi'_{\nu i}(x, z) < 1$. Then $\pi'_\nu(x, z) < 1$ by independence.) Thus we have

$$\forall z \notin \{x, y\}, \ x\widetilde{V}_i y \Rightarrow xV_i z. \tag{8.14}$$

Now since i has a veto for x against z, i has a semiveto for x against z. Thus switching y and z in the above argument implies i has a veto for x against y. Let ρ'' be any fuzzy preference profile such that $\pi''_i(y, z) < 1$. Let ρ^+ be a fuzzy preference profile such that

$$\pi^+_{\nu i}(y, x) < 1, \ \pi^+_{\nu i}(y, z) < 1, \ \pi^+_{\nu i}(x, z) < 1,$$
$$\forall j \neq i, \ \pi^+_{\nu j}(z, x) < 1, \ \pi^+_{\nu j}(y, x) < 1.$$

Since i has a semiveto x against $z, \pi^+_\nu(x, z) < 1$. Since $\widetilde{f}$ is weakly Paretian, $\pi^+_\nu(y, x) < 1$. Since $\widetilde{f}$ is partially quasi-transitive, $\pi^+_\nu(y, z) < 1$. Since only the preferences for i are specified and $\mathrm{Supp}(\rho^+_{\nu i}\rceil_{\{y,z\}}) = \mathrm{Supp}(\rho''_{\nu i}\rceil_{\{y,z\}})$, IIA2 implies $\pi''_\nu(y, z) < 1$ and so i has a veto for y against z. Thus

$$\forall z \notin \{x, y\}, \ x\widetilde{V}_i y \Rightarrow yV_i z. \tag{8.15}$$

Now i has a veto for y against z, i has a semiveto for y against z. Thus by (8.14), i has a veto for y against x. We have $\forall (v, w) \in X \times X$,

$$x\widetilde{V} y \Rightarrow xV_i v \text{ by (8.14)} \Rightarrow x\widetilde{V} v \Rightarrow v\widetilde{V}_i w \text{ by (8.15)}$$

with v replacing y. ∎

Theorem 8.4.10 *Let strict preferences be of type $\pi_{(0)}$. If a fuzzy aggregation rule $\widetilde{f} : \mathcal{FR}^n_T \to \mathcal{FR}$ is partially quasi-transitive, weakly Paretian, and independent of irrelevant alternatives IIA2, then it is oligarchic.*

Proof. The proof for ρ_μ can be found in Theorem 5.2.10. Since $\widetilde{f}$ is weakly Paretian, there exists $\lambda \in \mathcal{FP}(N)$ such that λ is semidecisive, namely $\lambda = 1_N$. Let λ be such that $\mathrm{Supp}(\lambda)$ is smallest for which λ is semidecisive. Let $m = |\mathrm{Supp}(\lambda)|$. By Lemma 8.3.11, λ is decisive for all ordered pairs $(u, v) \in X^2$. If $m = 1$, then the proof is complete. Suppose $m > 1$. Let $x, y, z \in X$ be distinct. Let ρ_ν be a fuzzy preference profile such that for some $i \in \mathrm{Supp}(\lambda)$,

$$\pi_{\nu i}(x, y) < 1, \ \pi_{\nu i}(x, z) < 1, \ \pi_{\nu i}(y, z) < 1,$$
$$\forall j \in \mathrm{Supp}(\lambda)\backslash\{i\}, \ \pi_{\nu j}(z, x) < 1, \ \pi_{\nu j}(z, y) < 1, \ \pi_{\nu j}(x, y) < 1, \tag{8.16}$$
$$\forall k \notin \mathrm{Supp}(\lambda), \ \pi_{\nu k}(y, z) < 1, \ \pi_{\nu k}(y, x) < 1, \ \pi_{\nu k}(z, x) < 1.$$

Since $i \in \text{Supp}(\lambda)$ and λ is decisive, $\pi_\nu(x, y) > 0$. Since $|\text{Supp}(\lambda)| = m$, there is such a ρ_ν such that not $\pi_\nu(z, y) < 1$ else $\text{Supp}(\lambda)\backslash\{i\}$ would be the support of some fuzzy subset of N semidecisive for z over y and thus decisive, contradicting the minimality of m. (Now for $\rho, \rho' \in \mathcal{FR}^n$, satisfying (8.16), $\text{Supp}(\rho_{\nu i}]_{\{x,y,z\}}) = \text{Supp}(\rho'_{\nu i}]_{\{x,y,z\}}), i = 1, \ldots, n$.) Hence for any such ρ_ν, $\widetilde{f}(\rho_\nu)(y, z) < 1$ by completeness. Since $\widetilde{f}$ is partially quasi-transitive, $\widetilde{f}(\rho_\nu)(x, z) < 1$ and hence not $\pi_\nu(z, x) < 1$. Thus i has a semiveto for x against z. Hence i has a veto over all ordered pairs $(u, v) \in X^2$ by Lemma 8.4.9. Thus $\widetilde{f}$ is oligarchic since i was arbitrary. $\blacksquare$

Let $\mathcal{FQR} = \{\langle \rho_\mu, \rho_\nu \rangle \in \mathcal{FR} \mid \langle \rho_\mu, \rho_\nu \rangle$ is partially quasi-transitive$\}$.

Lemma 8.4.11 *Let strict preferences be regular. Let $\widetilde{f} : \mathcal{FQR}^n \to \mathcal{FR}$ be a fuzzy aggregation rule which is partially quasi-transitive, weakly Paretian, and independent of irrelevant alternatives IIA3. If $i \in N$ has a semiveto for x against y for some $x, y \in X$, then i has a veto over all ordered pairs $(v, w) \in X \times X$.*

Proof. The proof for ρ_μ can be found in Lemma 5.2.11. Let $\langle \rho_\mu, \rho_\nu \rangle \in \mathcal{FQR}^n$ and $z \in X\backslash\{x, y\}$ be such that

$$\pi_{\mu i}(x, y) > 0, \ \pi_{\mu i}(x, z) > 0, \ \pi_{\mu i}(y, z) > 0,$$
$$\forall j \neq i, \ \pi_{\mu j}(y, x) > 0, \ \pi_{\mu j}(y, z) > 0.$$

$$\pi_{\nu i}(x, y) < 1, \ \pi_{\nu i}(x, z) < 1, \ \pi_{\nu i}(y, z) < 1,$$
$$\forall j \neq i, \ \pi_{\nu j}(y, x) < 1, \ \pi_{\nu j}(y, z) < 1.$$

Note that $\rho_{\mu j}]_{\{x,z\}}$ and $\rho_{\nu j}]_{\{x,z\}}$ for all $j \neq i$ can be arbitrarily assigned. Since i has a semiveto for x against y, $\pi_\mu(y, x) = 0$ and $\pi_\nu(y, x) = 1$. Thus $\rho_\mu(x, y) \geq \rho_\mu(y, x)$ and $\rho_\nu(x, y) \leq \rho_\nu(y, x)$. Since $\widetilde{f}$ is weakly Paretian, $\pi_\mu(y, z) > 0$ and $\pi_\nu(y, z) < 1$. Now suppose $\pi_\mu(z, x) > 0$ and $\pi_\nu(z, x) < 1$. Since $\widetilde{f}$ is partially quasi-transitive by assumption, $\pi_\mu(y, z) > 0$ and $\pi_\mu(z, x) > 0$ imply $\pi_\mu(y, x) > 0$ and $\pi_\nu(y, z) < 1$ and $\pi_\nu(z, x) < 1$ imply $\pi_\nu(y, x) < 1$, a contradiction. Hence $\pi_\mu(z, x) = 0$ and $\pi_\nu(z, x) = 1$. (We have $\pi_\mu(z, x) = 0$ for this ρ_μ, but not an arbitrary preference profile. Let $\langle \rho'_\mu, \rho'_\nu \rangle \in \mathcal{FQR}^n$ be such that $\pi'_{\mu i}(x, z) > 0$ and $\pi'_{\nu i}(x, z) < 1$. Because $\rho_{\mu j}]_{\{x,z\}}$ for all $j \neq i$ are arbitrarily assigned, IIA3 implies $\pi'_\mu(z, x) = 0$. Also, we have $\pi_\nu(z, x) = 1$ for this ρ_ν, but not an arbitrary preference profile. Because $\rho_{\nu j}]_{\{x,z\}}$ for all $j \neq i$ are arbitrarily assigned, IIA3 implies $\pi'_\nu(z, x) = 1$.). Thus we have

$$\forall z \in X\backslash\{x, y\}, \ x\widetilde{V}_i y \Rightarrow xV_i z. \tag{8.17}$$

Now since i has a veto for x against z, i has a semiveto for x against z. Thus switching y and z in the above argument implies i has a veto for x against y. Let $\langle \rho''_\mu, \rho''_\nu \rangle \in \mathcal{FQR}^n$ be any fuzzy preference profile such that $\pi''_{\mu i}(y, z) > 0$

and $\pi''_{\nu i}(y, z) < 1$. Let $\langle \rho^+_\mu, \rho^+_\nu \rangle \in \mathcal{FQR}^n$ be another fuzzy preference profile such that

$$\pi^+_{\mu i}(y, x) > 0, \ \pi^+_{\mu i}(y, z) > 0, \ \pi^+_{\mu i}(x, z) > 0$$
$$\forall j \neq i, \ \pi^+_{\mu j}(z, x) > 0, \ \pi^+_{\mu j}(y, x) > 0.$$

$$\pi^+_{\nu i}(y, x) < 1, \ \pi^+_{\nu i}(y, z) < 1, \ \pi^+_{\nu i}(x, z) < 1$$
$$\forall j \neq i, \ \pi^+_{\nu j}(z, x) < 1, \ \pi^+_{\nu j}(y, x) < 1.$$

Since i has a semiveto for x against z, $\pi^+_\mu(z, x) = 0$ and $\pi^+_\nu(z, x) = 1$. Thus $\rho^+_\mu(x, z) \geq \rho^+_\mu(z, x)$ and $\rho^+_\nu(x, z) \leq \rho^+_\nu(z, x)$. Since $\tilde{f}$ is weakly Paretian, $\pi^+_\mu(y, x) > 0$ and $\pi^+_\nu(y, x) < 1$. Now suppose $\pi^+_\mu(z, y) > 0$ and $\pi^+_\nu(z, y) < 1$. Since $\tilde{f}$ is partially quasi-transitive, $\pi^+_\mu(z, x) > 0$ and $\pi^+_\nu(z, x) < 1$, a contradiction. Thus $\pi^+_\mu(z, y) = 0$ and $\pi^+_\nu(z, y) < 1$. Since $\rho''_{\mu j}]_{\{y,z\}}$ and $\rho^+_{\mu j}]_{\{y,z\}}$ can be arbitrarily assigned for all $j \neq i$ so that $\rho^+_{\mu i}]_{\{y,z\}} \sim \rho''_{\mu i}]_{\{y,z\}}$ and $\rho''_{\nu j}]_{\{y,z\}}$ and $\rho^+_{\nu j}]_{\{y,z\}}$ can be arbitrarily assigned for all $j \neq i$ so that $\rho^+_{\nu i}]_{\{y,z\}} \sim \rho''_{\nu i}]_{\{y,z\}}$, IIA3 implies $\pi''_\mu(z, y) = 0$ and $\pi''_\nu(z, y) = 1$ and so i has a veto for y against z. Thus

$$\forall z \in X \backslash \{x, y\}, \ x \tilde{V}_i y \Rightarrow y V_i z. \tag{8.18}$$

Now i has a veto for y against z, i has a semiveto for y against z. Thus by (8.17), i has veto for y against x. We have $\forall (v, w) \in X \times X, \ x \tilde{V}_i y \Rightarrow x V_i v$ (by (8.17))$\Rightarrow x \tilde{V}_i v \Rightarrow v V_i w$ (by (8.18)) with v replacing y. $\blacksquare$

Theorem 8.4.12 *Suppose strict preferences are regular. If a fuzzy aggregation rule $\tilde{f} : \mathcal{FQR}^n \to \mathcal{FR}$ is partially quasi-transitive, weakly Paretian, and independent of irrelevant alternatives IIA3, then it is oligarchic.*

Proof. The proof for ρ_μ can be found in Theorem 5.2.12. Since $\tilde{f}$ is weakly Paretian, there exists $\lambda \in \mathcal{FP}(N)$ such that λ is semidecisive, namely $\lambda = 1_N$. Let λ be such that $\text{Supp}(\lambda)$ is smallest for which λ is semidecisive. Let $m = |\text{Supp}(\lambda)|$. By Lemma 8.3.11, λ is decisive for all ordered pairs $(u, v) \in X \times X$. If $m = 1$, then the proof is complete because a dictator is an oligarchy of size 1. Suppose $m > 1$. Let $x, y, z \in X$ be distinct and let $\langle \rho_\mu, \rho_\nu \rangle \in \mathcal{FQR}^n$ be such that for some $i \in \text{Supp}(\lambda)$,

$$\pi_{\mu i}(x, y) > 0, \ \pi_{\mu i}(x, z) > 0, \ \pi_{\mu i}(y, z) > 0,$$
$$\forall j \in \text{Supp}(\lambda) \backslash \{i\}, \ \pi_{\mu j}(z, x) > 0, \ \pi_{\mu j}(z, y) > 0, \ \pi_{\mu j}(x, y) > 0,$$
$$\forall k \notin \text{Supp}(\lambda), \ \pi_{\mu k}(y, z) > 0, \ \pi_{\mu k}(y, x) > 0, \ \pi_{\mu k}(z, x) > 0.$$

$$\pi_{\nu i}(x, y) < 1, \ \pi_{\nu i}(x, z) < 1, \ \pi_{\nu i}(y, z) < 1,$$
$$\forall j \in \text{Supp}(\lambda) \backslash \{i\}, \ \pi_{\nu j}(z, x) < 1, \ \pi_{\nu j}(z, y) < 1, \ \pi_{\nu j}(x, y) < 1,$$
$$\forall k \notin \text{Supp}(\lambda), \ \pi_{\nu k}(y, z) < 1, \ \pi_{\nu k}(y, x) < 1, \ \pi_{\nu k}(z, x) < 1.$$

Since λ is decisive, $\pi_\mu(x, y) > 0$ and so $\rho_\mu(x, y) > \rho_\mu(y, x)$ because π_μ is regular and $\pi_\nu(x, y) < 1$ and so $\rho_\nu(x, y) < \rho_\nu(y, x)$ because π_μ is regular. Since $|\text{Supp}(\lambda)| = m$, there is no $\langle \rho'_\mu, \rho'_\nu \rangle \in \mathcal{FQR}^n$ such that $\rho'_\mu]_{\{y,z\}} \sim \rho_\mu]_{\{y,z\}}$ and $\pi'_\mu(z, y) > 0$ and $\rho'_\nu]_{\{y,z\}} \sim \rho_\nu]_{\{y,z\}}$ and $\pi'_\nu(z, y) < 1$ else $\pi_\mu(z, y) > 0$ and

$\pi_\nu(z, y) < 1$ by IIA3 and so $\text{Supp}(\lambda)\backslash\{i\} = \text{Supp}(\lambda') \in \mathcal{FP}(N)$ would be such that λ' is semidecisive and thus decisive by Lemma 8.3.11, a contradiction of the minimality of m. Hence for any such $\langle \rho_\mu, \rho_\nu \rangle$, $\pi_\mu(z, y) = 0$ and $\pi_\nu(z, y) = 1$, By the regularity of π_μ and π_ν, $\rho_\mu(y, z) \geq \rho_\mu(z, y)$ and $\rho_\nu(y, z) \leq \rho_\nu(z, y)$.

Now suppose $\pi_\mu(z, x) > 0$ and $\pi_\nu(z, y) < 1$. Since $\widetilde{f}$ is partially quasi-transitive, $\pi_\mu(z, x) > 0$ and $\pi_\mu(x, y) > 0$ imply $\pi_\nu(z, y) > 0$; and $\pi_\nu(z, x) < 1$ and $\pi_\nu(x, y) < 1$ imply $\pi_\nu(z, y) < 1$. Then by the regularity of π_μ and π_ν, $\rho_\mu(z, y) > \rho_\mu(y, z)$ and $\rho_\nu(z, y) < \rho_\nu(y, z)$. However, this is a contradiction. Thus $\pi_\mu(z, x) = 0$ and $\pi_\nu(z, x) = 1$. Hence i has a veto for x against z. By Lemma 8.4.11, i has a veto over all ordered pairs $(u, v) \in X \times X$. Because i was an arbitrary member of $\text{Supp}(\lambda)$, $\widetilde{f}$ is oligarchic. ∎

In this section, we extended some known Arrowian results involving fuzzy set theory to results involving intuitionistic fuzzy sets. We pointed out how the use of an involutive fuzzy complement can be used to obtain the results. A future research project might be to consider more aggressively the use of an involutive fuzzy complement in obtaining dual results for ρ_ν by using known results for ρ_μ.

8.5　Fuzzy Preference and Arrowian Results

In this section, we use intuitionistic fuzzy preference relations to examine Arrowian results in the intuitionistic case as developed by Nana and Fono [24]. The concepts of pos-transitivity and the negative transitivity of a strict component of a relation are important properties. They allow for the establishment of Arrowian results for crisp and fuzzy preferences, Arrow [2] and Fono and Andjiga [13]. Motivated by Fono et al. [16], a factorization of an intuitionistic fuzzy relation (IFR) is established in Chapter 3 and previous sections in this chapter. A unique indifference component and a family of regular strict components was obtained. Necessary and sufficient conditions on a (T, S)-transitive IFR were obtained such that each of its strict components is pos-transitive and negative transitive. The purpose of this section is to characterize, by means of the (min, max)-transitivity, the pos-transitivity and the negative transitivity of a regular strict component obtained in [16] and also to establish some new Arrowian type results with (T, S)-transitive IFRs.

In Chapter 3, we presented results on factorization of an IFR $\rho = \langle \rho_\mu, \rho_\nu \rangle$ into a unique indifference and a family of strict components. We present here a result from [16] that provides the characterization of two properties of a strict component of a (T, S)-transitive IFR ρ which stipulates that a strict component of ρ is pos-transitive if and only if satisfies $C_1^{\rho_\mu}$ or $C_1^{\rho_\nu}$ and also a strict component of R is negative transitive if and only if ρ satisfies $(C_1^{\rho_\mu}$ and $C_2^{\rho_\mu})$ or $(C_1^{\nu_R}$ and $C_2^{\nu_R})$. We show that when ρ is a (min, max)-transitive IFR, i.e., $(T, S) = (\min, \max)$), a strict component of ρ is pos-transitive and a strict component of ρ is negative transitive if and only if ρ satisfies one of the new conditions T^{μ_R} or T^{ν_R} which as stated in [24] can be interpreted as transitivity of indifference in the pairwise comparisons of degrees of preference. We present some properties of an IFAR which are intuitionistic fuzzy versions of the four concepts of a crisp aggregation rule, namely, Pareto condition (PC), independence of irrelevant alternatives (IIA), decisive coalition, veto, oligarchy and dictator. By using the conditions ($C_1^{\rho_\mu}$ and $C_2^{\rho_\mu}$) or ($C_1^{\nu_R}$ and $C_2^{\nu_R}$), an intuitionistic fuzzy version of Arrow's impossibility theorem is established. Without these conditions, a non-dictatorial IFAR and an intuitionistic fuzzy version of Gibbard's oligarchy theorem is obtained.

In Sections 8.1-8.4 and Chapter 3, we showed how Arrowian results could be obtained for intuitionistic fuzzy preference relations using a dual argument. In the remainder of the chapter, such results can not be so easily obtained. We next explain the logic. In previous sections, definitions were of the type $(p \rightarrow q) \wedge (r \rightarrow s)$, where $p \rightarrow q$ was an implication for ρ_μ and $r \rightarrow s$ was an implication for ρ_ν. Many definitions in the remainder of this chapter are of the type $(p \wedge r) \rightarrow (q \wedge s)$. These definitions are not in general equivalent. Definitions of the type $(p \rightarrow q) \wedge (r \rightarrow s)$ imply those of the type $(p \wedge r) \rightarrow (q \wedge s)$, but not conversely. However, in many situations $p \rightarrow q$ and $r \rightarrow s$ are equivalent due to the duality involved between ρ_μ and ρ_ν. In this case, the two types of definitions $(p \rightarrow q) \wedge (r \rightarrow s)$ and $(p \wedge r) \rightarrow (q \wedge s)$ are equivalent. Consequently, arguments are dual in nature when this occurs. However, not all definitions in the remainder of the chapter are of this nature and so full proofs are needed in those spots.

Let X be a set of alternatives with $|X| \geq 3$. Let $L^* = \{(a_1, a_2) \in [0, 1]^2, a_1 + a_2 \leq 1\}$. Define the relation $\leq_{L^*}$ on L^* defined by $\forall (a_1, a_2), (b_1, b_2) \in L^*$, $(a_1, a_2) \leq_{L^*} (b_1, b_2) \Leftrightarrow (a_1 \leq b_1 \text{ and } a_2 \geq b_2)$. $(L^*, \leq_{L^*})$ is a complete lattice. $0_{L^*} = (0, 1)$ and $1_{L^*} = (1, 0)$ are the units of L^*.

In the following, we recall some useful definitions and results on fuzzy operators and intuitionistic fuzzy operators (see Cornelis et al. [9], Fono et al. [15, 13, 16], Fono and Salles [17], Fotso and Fono [18], Klement et al. [20] and Njanpong [255]).

Throughout the remainder of the chapter, we assume that T is a left-continuous t-norm and S is right-continuous t-conorm.

The fuzzy ρ-implication I_T associated with T is a binary operation on $[0, 1]$ defined by $\forall a, b \in [0, 1]$, $I_T(a, b) = \vee \{t \in [0, 1], T(a, t) \leq b\}$. The fuzzy coimplicator J_S associated with S is a binary operation on $[0, 1]$ defined by $\forall a, b \in [0, 1]$, $J_S(a, b) = \wedge \{t \in [0, 1], b \leq S(a, t)\}$.

Some usual examples are: $T = T_M = \min$; $S = S_M = \max$; $\forall a, b \in [0; 1]$,

$$I_{T_M}(a, b) = I_{\min}(a, b) = \begin{cases} 1 \text{ if } a \leq b \\ b \text{ if } a > b \end{cases} \text{ and}$$

$$J_{S_M}(a, b) = J_{\max}(a, b) = \begin{cases} b \text{ if } a < b \\ 0 \text{ if } a \geq b \end{cases}.$$

Definition 8.5.1 (*Cornelis et al.* [9]) *A **t-representable intuitionistic fuzzy t-norm** $\mathcal{T}$ (t-conorm $\mathcal{J}$) is an increasing, commutative, associative binary operation on L^* such that there exists a fuzzy t-norm T and a fuzzy t-conorm S (a fuzzy t-conorm S and a fuzzy t-norm T) satisfying $\forall a = (a_1, a_2), b = (b_1, b_2) \in L^*$, $\mathcal{T}(a, b) = (T(a_1, b_1), S(a_2, b_2))$ $(\mathcal{J}(a, b)) = (S(a_1, b_1), T(a_2, b_2))$. T and S (S and T) are called the **representants** of $\mathcal{T}$ ($\mathcal{J}$).*

Cornelis et al. [9, Theorem 2, pp. 60-61] showed that, given a t-norm T and a t-conorm S, (T, S) is a t-representable intuitionistic fuzzy t-norm if $\forall a_1, a_2 \in [0, 1]$, $T(a_1, a_2) \leq 1 - S(1 - a_1, 1 - a_2)$.

It follows that $\mathcal{T} = (\min, \max)$ and $\mathcal{J} = (\max, \min)$ are respectively a t-representable intuitionistic fuzzy t-norm and a t-representable intuitionistic fuzzy t-conorm. Condition G defined below and introduced by Fono et al. [16] is an important condition that holds for a t-representable intuitionistic fuzzy t-conorm $\mathcal{J} = (S, T)$.

Definition 8.5.2 $\mathcal{J} = (S, T)$ *satisfies **condition** G if $\forall (a_1, a_2), (b_1, b_2) \in L^*, a_1 > b_1, a_2 < b_2$, and $a_1 + a_2 = b_1 + b_2$ imply $I_T(b_2, a_2) + J_S(b_1, a_1) \leq 1$.*

It follows that $\mathcal{J} = (\max, \min)$ satisfies condition G. See also Fono et al. [16] for some examples of t-representable intuitionistic fuzzy t-conorms satisfying condition G and those which violate it. We ask the reader to provide the proofs in the Exercises.

We next recall some basic notions on intuitionistic fuzzy binary relations (see Atanassov [3], Bustince and Burillo [8], Dimitrov [11], Fono et al. [16] and Njanpong [24]).

Definition 8.5.3 [16]

(1) *An **intuitionistic fuzzy set** (**IFS**) in X is given $\{\langle x, \mu(x), \nu(x) \rangle \mid x \in X\}$, where the functions $\mu : X \to [0, 1]$ and $\nu : X \to [0, 1]$ satisfy the condition $\forall x \in X, \mu(x) + \nu(x) \leq 1$.*

(2) *An **intuitionistic fuzzy relation** (**IFR**) on X is a function $\rho : X \times X \to L^*$ where $\forall x, y \in X$ $\rho(x, y) = (\rho_\mu(x, y), \rho_\nu(x, y))$.*

(3) *Let $\langle \mu, \nu \rangle$ and $\langle \mu', \nu' \rangle$ be two IFSs and $\mathcal{J} = (S, T)$. The union $\langle \mu, \nu \rangle \cup_{(S,T)} \langle \mu', \nu' \rangle$ associated with $\mathcal{J}$ is an IFS on X defined by $\forall x \in X, (\mu_{\cup_{(S,T)}} \mu')(x) = S(\mu(x), \mu'(x))$ and $(\nu_{\cup_{(S,T)}} \nu')(x) = T(\nu(x), \nu'(x))$.*

$\mu(x)$ and $\nu(x)$ denote, respectively, the degree of membership and the degree of non-membership of the element x in the intuitionistic fuzzy set. The number $\sigma(x) = 1 - \mu(x) - \nu(x)$ is called the **index** of the element x in X. Clearly, if $\forall x \in X, \nu(x) = 1 - \mu(x)$, then the IFS is a fuzzy subset of X.

An intuitionistic fuzzy binary relation ρ represents a (weak) preference which for $x, y \in X$ can be interpreted as saying that $\rho_\mu(x, y)$ is the degree to which x is "at least as good as" y and $\rho_\nu(x, y)$ is the degree to which x is not "at least as good as" y.

Definition 8.5.4 *Let ρ be an IFR and $\mathcal{T} = (T, S)$.*

(1) ρ *is **strongly complete** if $\forall x, y \in X$, $\rho_\mu(x, y) + \rho_\mu(y, x) \geq 1$ and* $\rho_\nu(x, y) + \rho_\nu(y, x) \leq 1$.

(2) ρ *is σ-**symmetric** if $\forall x, y \in X$, $\sigma(x, y) = \sigma(y, x)$, i.e.,*

$$\rho_\mu(x, y) + \rho_\nu(x, y) = \rho_\mu(y, x) + \rho_\nu(y, x).$$

(3) ρ *is **perfect antisymmetric** if $\forall (x, y) \in X \times X$, $x \neq y$,*

$$
\begin{aligned}
\rho_\mu(x, y) &> 0 \ \ or \ [\rho_\mu(x, y) = 0 \ and \ \rho_\nu(x, y) < 1] \Rightarrow \\
\rho_\mu(y, x) &= 0 \ \ and \ \rho_\nu(y, x) = 1.
\end{aligned}
$$

(4) ρ *is **simple** if $\forall (x, y) \in X \times X$,*

$$
\left\{
\begin{array}{l}
\rho_\mu(x, y) = \rho_\mu(y, x) \\
\rho_\nu(x, y) = \rho_\nu(y, x)
\end{array}
\right.
\Rightarrow
\left\{
\begin{array}{l}
\pi_\mu(x, y) = \pi_\mu(y, x) \\
\pi_\nu(x, y) = \pi_\nu(y, x)
\end{array}
\right.
$$

.

(5) ρ *is $\mathcal{T}$-**transitive** if $\forall x, y, z \in X$, $\rho_\mu(x, z) \geq T(\rho_\mu(x, y), \rho_\mu(y, z))$ and* $\rho_\nu(x, z) \leq S(\rho_\nu(x, y), \rho_\nu(y, z))$

If $\mathcal{T} = (\min, \max)$, we simply say that ρ is transitive instead of $(\min, \max)$-transitive.

Throughout the remainder of this chapter, we assume that ρ is reflexive, strongly complete and σ-symmetric.

In the next paragraph, we recall a factorization of an IFR into a unique indifference I and a family of strict components.

Definition 8.5.5 [16] *Let $\mathcal{J} = (S, T)$ satisfy condition G, ρ be an IFR and, ι and π be two IFRs. Then ι and π are the **indifference** of ρ and the **strict component** of ρ associated with $\mathcal{J}$ respectively if the following conditions are satisfied:*

(1) $\rho = \iota \cup_{(S,T)} \pi$,

(2) π *is simple,*

(3) π *is perfect antisymmetric,*

(4) ι *is symmetric.*

Consider the following system:

$$
\left\{
\begin{array}{l}
(i) \ a + b \leq 1, \\
(ii) \ S(\rho_\mu(y, x), a) = \rho_\mu(x, y), \\
(iii) \ T(\rho_\nu(y, x), b) = \rho_\nu(x, y).
\end{array}
\right.
\tag{8.19}
$$

Proposition 8.5.6 [16] *Let* $\mathcal{J} = (S,T)$ *satisfy condition* G, ρ *be an IFR and,* ι *and* π *be two IFRs. Then the following statements are equivalent:*

(1) ι *and* π *are the indifference of* ρ *and the strict component of* ρ *associated with* $\mathcal{J}$, *respectively,*

(2) (i) $\forall x, y \in X$, $\iota_\mu(x,y) = \iota_\mu(y,x) = \min(\rho_\mu(x,y), \rho_\mu(y,x))$ *and* $\iota_\nu(x,y) = \iota_\nu(y,x) = \max(\rho_\nu(x,y), \rho_\nu(y,x))$.

(ii) $\forall x, y \in X$, $\exists (c_{xy}, g_{xy}) \in L^*$ *such that* $c_{xy} > 0, c_{xy}$ *is a solution of* (ii) *of* (8.19)

$g_{xy} < 1$, g_{xy} *is a solution of* (iii) *of* (8.19) *and*

$$\pi_\mu(x,y) = \begin{cases} 0 & \text{if } \rho_\mu(x,y) \le \rho_\mu(y,x) \\ c_{xy} & \text{otherwise} \end{cases} \quad \text{and}$$

$$\pi_\nu(x,y) = \begin{cases} 1 \text{ if } \rho_\nu(x,y) \ge \rho_\nu(y,x) \\ g_{xy} & \text{otherwise} \end{cases}.$$

It was established in [16, p. 14] that each strict component π obtained in the previous result satisfies the following property: π is regular, that is, $\forall x, y \in X$,

$$\begin{cases} \rho_\mu(x,y) \le \rho_\mu(y,x) \\ \rho_\nu(x,y) \ge \rho_\nu(y,x) \end{cases} \Leftrightarrow \begin{cases} \pi_\mu(x,y) = 0 \\ \pi_\nu(x,y) = 1 \end{cases}.$$

Definition 8.5.7 [16] *Let* ρ *be an IFR and* π *be a regular strict component of* ρ.

(1) π *is* **pos-transitive** *if* $\forall x, y, z \in X$,

$$\left(\begin{cases} \pi_\mu(x,y) > 0 \\ \pi_\nu(x,y) < 1 \end{cases} \text{ and } \begin{cases} \pi_\mu(y,z) > 0 \\ \pi_\nu(y,z) < 1 \end{cases} \right) \Rightarrow \begin{cases} \pi_\mu(x,z) > 0 \\ \pi_\nu(x,z) < 1 \end{cases}.$$

(2) π *is* **negative transitive** *if* $\forall x, y, z \in X$,

$$\left(\begin{cases} \pi_\mu(x,y) = 0 \\ \pi_\nu(x,y) = 1 \end{cases} \text{ and } \begin{cases} \pi_\mu(y,z) = 0 \\ \pi_\nu(y,z) = 1 \end{cases} \right) \Rightarrow \begin{cases} \pi_\mu(x,z) = 0 \\ \pi_\nu(x,z) = 1 \end{cases}.$$

The pos-transitivity of π says that for the strict preference relation π associated with ρ and for all $x, y, z \in X$, if x is strictly preferred to y and y is strictly preferred to z, then x is strictly preferred to z. Considering the contrapositive, we see that the negative transitivity of π says that if x is strictly preferred to z, then either x is strictly preferred to y or y is strictly preferred to z.

Proposition 8.5.8 [16] *Let* ρ *be an IFR and* π *its regular strict component. Then*

(1) π *is pos-transitive if* $\forall x, y, z \in X$,

$$\begin{cases} (i)(\rho_\mu(x,y) > \rho_\mu(y,x) \text{ and } \rho_\mu(y,z) > \rho_\mu(z,y)) \Rightarrow \rho_\mu(x,z) > \rho_\mu(z,x) \\ (ii)(\rho_\nu(x,y) < \rho_\nu(y,x) \text{ and } \rho_\nu(y,z) < \rho_\nu(z,y)) \Rightarrow \rho_\nu(x,z) < \rho_\nu(z,x) \end{cases}$$

(2) π *is negative transitive if* $\forall x, y, z \in X$, (*i*) *or* (*ii*) *or* (*iii*) *or* (*iv*) *implies* $\rho_\mu(x, z) \geq \rho_\mu(z, x)$, *where*

$$(i) \begin{cases} \rho_\mu(y, x) < \rho_\mu(x, y) \\ \rho_\mu(z, y) < \rho_\mu(y, z) \end{cases}, \quad (ii) \begin{cases} \rho_\mu(y, x) = \rho_\mu(x, y) \\ \rho_\mu(z, y) < \rho_\mu(y, z) \end{cases},$$

$$(iii) \begin{cases} \rho_\mu(y, x) < \rho_\mu(x, y) \\ \rho_\mu(z, y) = \rho_\mu(y, z) \end{cases}, \quad (iv) \begin{cases} \rho_\mu(x, y) = \rho_\mu(y, x) \\ \rho_\mu(y, z) = \rho_\mu(z, y) \end{cases}$$

and (*i*) *or* (*ii*) *or* (*iii*) *or* (*iv*) *implies* $\rho_\nu(x, z) \leq \rho_\nu(z, x)$, *where*

$$(i) \begin{cases} \rho_\nu(y, x) > \rho_\nu(x, y) \\ \rho_\nu(z, y) < \rho_\nu(y, z) \end{cases}, \quad (ii) \begin{cases} \rho_\nu(y, x) = \rho_\nu(x, y) \\ \rho_\nu(z, y) > \rho_\nu(y, z) \end{cases},$$

$$(iii) \begin{cases} \rho_\nu(y, x) > \rho_\nu(x, y) \\ \rho_\nu(z, y) = \rho_\nu(y, z) \end{cases}, \quad (iv) \begin{cases} \rho_\nu(x, y) = \rho_\nu(y, x) \\ \rho_\nu(y, z) = \rho_\nu(z, y) \end{cases}.$$

In [16, Example 5.3, pp. 21-22], it is shown that there exists an IFR such that pos-transitivity and negative transitivity do not hold for a regular strict component. Necessary and sufficient conditions on an IFR such that π satisfy each of these properties is also established. We now recall these results.

The following four conditions of an IFR are from [16].

For all $x, y, z \in X$,

$$\beta_2(x, y, z) = S(\rho_\nu(x, y), \rho_\nu(y, z)),$$
$$\alpha_2(x, y, z) = T(\rho_\mu(x, y), \rho_\mu(y, z)),$$
$$\beta_3(x, y, z) = J_S(\rho_\nu(y, z), \rho_\nu(y, x)) \vee J_S(\rho_\nu(x, y), \rho_\nu(z, y)),$$
$$\alpha_3(x, y, z) = \wedge \{ I_T(\rho_\mu(x, y), \rho_\mu(z, y)), \ 1 - J_S(\rho_\nu(x, y), \rho_\nu(z, y)),$$
$$I_T(\rho_\mu(y, z), \rho_\mu(y, x)), \ 1 - J_S(\rho_\nu(y, z), \rho_\nu(y, x)) \}.$$

We next consider conditions important in the study of pos-transitivity and negative transitivity.

Definition 8.5.9 [16] *Let* ρ *be an IFR.*
(1) ρ *satisfies* **condition** $C_1^{\rho_\mu}$ *if* $\forall x, y, z \in X$,

$$\begin{cases} \rho_\mu(y, x) < \rho_\mu(x, y) \\ \rho_\mu(z, y) < \rho_\mu(y, z) \end{cases} \Rightarrow$$
$$\left(\begin{cases} \rho_\mu(x, z) \in [\alpha_2(x, y, z), \alpha_3(x, y, z)] \\ \rho_\mu(z, x) \in [\alpha_2(x, y, z), \alpha_3(x, y, z)] \end{cases} \Rightarrow \rho_\mu(z, x) < \rho_\mu(x, z) \right).$$

(2) ρ *satisfies* **condition** $C_1^{\rho_\nu}$ *if* $\forall x, y, z \in X$,

$$\begin{cases} \rho_\nu(y, x) > \rho_\nu(x, y) \\ \rho_\nu(z, y) > \rho_\nu(y, z) \end{cases} \Rightarrow$$
$$\left(\begin{cases} \rho_\nu(x, z) \in [\beta_3(x, y, z), \beta_2(x, y, z)] \\ \rho_\nu(z, x) \in [\beta_3(x, y, z), \beta_2(x, y, z)] \end{cases} \Rightarrow \rho_\nu(z, x) > \rho_\nu(x, z) \right).$$

(3) ρ satisfies **condition** $C_2^{\rho_\mu}$ if $\forall x, y, z \in X$,

$$\left(\left(\begin{cases} \rho_\mu(y,x) = \rho_\mu(x,y) \\ \rho_\mu(z,y) < \rho_\mu(y,z) \end{cases} \quad or \quad \begin{cases} \rho_\mu(y,x) < \rho_\mu(x,y) \\ \rho_\mu(z,y) = \rho_\mu(y,z) \end{cases} \right) \Rightarrow \right.$$
$$\left. \begin{cases} \rho_\mu(x,z) \in [\alpha_2(x,y,z), \alpha_3(x,y,z)] \\ \rho_\mu(z,x) \in [\alpha_2(x,y,z), \alpha_3(x,y,z)] \end{cases} \Rightarrow \rho_\mu(z,x) < \rho_\mu(x,z) \right).$$

(4) ρ satisfies **condition** $C_2^{\rho_\nu}$ if $\forall x, y, z \in X$,

$$\left(\left(\begin{cases} \rho_\nu(y,x) = \rho_\nu(x,y) \\ \rho_\nu(z,y) > \rho_\nu(y,z) \end{cases} \quad or \quad \begin{cases} \rho_\nu(y,x) > \rho_\nu(x,y) \\ \rho_\nu(z,y) = \rho_\nu(y,z) \end{cases} \right) \Rightarrow \right.$$
$$\left. \begin{cases} \rho_\nu(x,z) \in [\beta_3(x,y,z), \beta_2(x,y,z)] \\ \rho_\nu(z,x) \in [\beta_3(x,y,z), \beta_2(x,y,z)] \end{cases} \Rightarrow \rho_\nu(z,x) > \rho_\nu(x,z) \right).$$

Consider condition $C_1^{\rho_\mu}$. Assume that x is strictly preferred to y with respect to ρ and y is strictly preferred to z with respect to ρ. Then in the crisp case, we have ρ is a crisp total pre-order and x is strictly preferred to z with respect to ρ. However, in the intuitionistic fuzzy case, this is not true when the two degrees of comparisons of x and z belong to the particular subset $[\alpha_2(x,y,z), \alpha_3(x,y,z)]$ of $[0,1]$, i.e., there is not a comparison between the two degrees $\rho_\mu(x,z)$ and $\rho_\mu(z,x)$. Therefore, condition $C_1^{\rho_\mu}$ alleviates this difficulty.

In the following, we recall general results on characterizations of the pos-transitivity and the negative transitivity of a given regular strict component of an IFR.

Theorem 8.5.10 [16] *Let* $\mathcal{T} = (T, S)$, ρ *be a* $\mathcal{T}$-*transitive IFR, and* π *be a regular strict component of* ρ. *Then*
(1) π *is pos-transitive if and only if* ρ *satisfies* $C_1^{\rho_\mu}$ *or* $C_1^{\rho_\nu}$.
(2) π *is negative transitive if and only if* ρ *satisfies* $(C_1^{\rho_\mu}$ *and* $C_2^{\rho_\mu})$ *or* $(C_1^{\rho_\nu}$ *and* $C_2^{\rho_\nu})$.

In [16] the conditions $C_1^{\rho_\mu}$, $C_1^{\rho_\nu}$, $C_2^{\rho_\mu}$, and $C_2^{\rho_\nu}$ were used to obtain characterizations of the pos-transitivity and the negative transitivity of a regular strict component of an IFR. These results are valid for each t-representable intuitionistic fuzzy t-norm $\mathcal{T} = (T, S)$. For the sake of simplicity, we examine these conditions in the case where $\mathcal{T} = \mathcal{T}_M = (\min, \max)$.

To obtain the particular cases of the previous characterizations when $\mathcal{T} = (\min, \max)$, we need the two following conditions:

Definition 8.5.11 *Let* ρ *be an IFR on* X.
(1) ρ *satisfies* **condition** T^{ρ_μ} *if*

$$\forall x, y, z \in X, \quad \rho_\mu(x,y) = \rho_\mu(y,x) = \rho_\mu(y,z) = \rho_\mu(z,y)$$
$$implies \quad \rho_\mu(x,z) = \rho_\mu(z,x).$$

(2) ρ satisfies **condition** T^{ρ_ν} if

$$\forall x, y, z \in X, \quad \rho_\nu(x, y) = \rho_\nu(y, x) = \rho_\nu(y, z) = \rho_\nu(z, y)$$
$$implies \quad \rho_\nu(x, z) = \rho_\nu(z, x).$$

These two conditions can be viewed as a type of transitivity of indifference in the pairwise comparisons of degrees of preference. Since ρ is σ-symmetric, T^{ρ_μ} and T^{ρ_ν} are equivalent.

The following result shows that (min, max)-transitive IFRs satisfy conditions $C_1^{\rho_\mu}$ and $C_1^{\rho_\nu}$. Furthermore, when ρ is (min, max)-transitive, then conditions $C_2^{\rho_\mu}$ and $C_2^{\rho_\nu}$ become T^{ρ_μ} and T^{ρ_ν}, respectively.

Lemma 8.5.12 (*Nana and Fono* [24]) *Let* $\mathcal{T} = (T, S)$ *and* ρ *be an IFR.*
(1) *If* $\mathcal{T} = (\min, \max)$, *then* ρ *satisfies* $C_1^{\rho_\mu}$ *and* $C_1^{\rho_\nu}$.
(2) *If* $\mathcal{T} = (\min, \max)$ *and* ρ *is* (min, max)-*transitive, then*
    (*i*) *conditions* $C_2^{\rho_\mu}$ *and* T^{ρ_μ} *are equivalent,*
    (*ii*) *conditions* $C_2^{\rho_\nu}$ *and* T^{ρ_ν} *are equivalent.*

Proof. Suppose that $\mathcal{T} = (\min, \max)$.

(1*i*) We show that ρ satisfies $C_1^{\rho_\mu}$. Let $x, y, z \in X$ be such that $\rho_\mu(y, x) < \rho_\mu(x, y)$ and $\rho_\mu(z, y) < \rho_\mu(y, z)$. We show that $\rho_\mu(x, z) \in [\alpha_2(x, y, z), \alpha_3(x, y, z)]$ and $\rho_\mu(z, x) \in [\alpha_2(x, y, z), \alpha_3(x, y, z)]$ imply $\rho_\mu(z, x) < \rho_\mu(x, z)$. It suffices to show that conjunction $\rho_\mu(x, z) \in [\alpha_2(x, y, z), \alpha_3(x, y, z)]$ and $\rho_\mu(z, x) \in [\alpha_2(x, y, z), \alpha_3(x, y, z)]$ is false. We distinguish the following cases: $\rho_\mu(z, y) \le \rho_\mu(y, x) < \rho_\mu(x, y) \le \rho_\mu(y, z)$ or $\rho_\mu(z, y) \le \rho_\mu(y, x) < \rho_\mu(y, z) < \rho_\mu(x, y)$ or $\rho_\mu(z, y) < \rho_\mu(y, z) \le \rho_\mu(y, x) < \rho_\mu(x, y)$ or $\rho_\mu(y, x) < \rho_\mu(x, y) \le \rho_\mu(z, y) < \rho_\mu(y, z)$ or $\rho_\mu(y, x) < \rho_\mu(z, y) < \rho_\mu(x, y) \le \rho_\mu(y, z)$ or $\rho_\mu(y, x) < \rho_\mu(z, y) < \rho_\mu(y, z) < \rho_\mu(x, y)$. Since $\mathcal{T} = (\min, \max)$, each of these cases yield $\alpha_3(x, y, z) < \alpha_2(x, y, z)$. Thus the interval $[\alpha_2(x, y, z), \alpha_3(x, y, z)]$ is empty and then the above conjunction is false.

(1*ii*) By σ-symmetry, we show that ρ satisfies $C_1^{\rho_\nu}$.

(2*ia*) Assume T^{ρ_μ}. We show that $C_2^{\rho_\mu}$ holds. Let $x, y, z \in X$ be such that

$$\begin{cases} \rho_\mu(y, x) = \rho_\mu(x, y) \\ \rho_\mu(z, y) < \rho_\mu(y, z) \end{cases} \quad or \quad \begin{cases} \rho_\mu(y, x) < \rho_\mu(x, y) \\ \rho_\mu(z, y) = \rho_\mu(y, z) \end{cases} \tag{8.20}$$

It suffices to show that $\rho_\mu(x, z) \in [\alpha_2(x, y, z), \alpha_3(x, y, z)]$ and $\rho_\mu(z, x) \in [\alpha_2(x, y, z), \alpha_3(x, y, z)]$ imply $\rho_\mu(z, x) < \rho_\mu(x, z)$. (8.20) is equivalent to the disjunction

$$[\rho_\mu(z, y) < \rho_\mu(y, x) = \rho_\mu(x, y) < \rho_\mu(y, z)] \ or$$
$$[\rho_\mu(z, y) < \rho_\mu(y, z) \le \rho_\mu(x, y) = \rho_\mu(y, x)] \ or$$
$$[\rho_\mu(z, y) = \rho_\mu(y, z) \le \rho_\mu(y, x) < \rho_\mu(x, y)] \ or$$
$$[\rho_\mu(y, x) = \rho_\mu(x, y) \le \rho_\mu(z, y) < \rho_\mu(y, z)] \ or$$
$$[\rho_\mu(y, x) < \rho_\mu(x, y) \le \rho_\mu(z, y) = \rho_\mu(y, z)] \ or$$
$$[\rho_\mu(y, x) < \rho_\mu(z, y) = \rho_\mu(y, z) < \rho_\mu(x, y)].$$

We consider the following three cases:

(1) If $\rho_\mu(z,y) < \rho_\mu(y,x) = \rho_\mu(x,y) < \rho_\mu(y,z)$ or $\rho_\mu(z,y) < \rho_\mu(y,z) \leq \rho_\mu(x,y) = \rho_\mu(y,x)$ or $\rho_\mu(y,x) < \rho_\mu(x,y) \leq \rho_\mu(z,y) = \rho_\mu(y,z)$ or $\rho_\mu(y,x) < \rho_\mu(z,y) = \rho_\mu(y,z) < \rho_\mu(x,y)$, then interval $[\alpha_2(x,y,z), \alpha_3(x,y,z)]$ is empty since $\alpha_2(x,y,z) > \alpha_3(x,y,z)$. Hence the result.

(2) If $\rho_\mu(z,y) = \rho_\mu(y,z) \leq \rho_\mu(y,x) < \rho_\mu(x,y)$, then $\alpha_2(x,y,z) = \rho_\mu(y,z)$ and $\alpha_3(x,y,z) = \rho_\mu(z,y)$. The assertion $\rho_\mu(x,z), \rho_\mu(z,x) \in [\alpha_2(x,y,z), \alpha_3(x,y,z)]$ becomes $\rho_\mu(x,z) = \rho_\mu(z,x) = \rho_\mu(z,y) = \rho_\mu(y,z)$. These equalities and the inequality $\rho_\mu(y,x) < \rho_\mu(x,y)$ contradict condition T^{ρ_μ}. Thus this case is not possible.

(3) If $\rho_\mu(x,y) = \rho_\mu(y,x) \leq \rho_\mu(z,y) < \rho_\mu(y,z)$, then $\alpha_2(x,y,z) = \rho_\mu(x,y)$ and $\alpha_3(x,y,z) = \rho_\mu(y,x)$. Conjunctions $\rho_\mu(x,z), \rho_\mu(z,x) \in [\alpha_2(x,y,z), \alpha_3(x,y,z)]$ become $\rho_\mu(x,z) = \rho_\mu(z,x) = \rho_\mu(x,y) = \rho_\mu(y,x)$. These equalities and the inequality $\rho_\mu(z,y) < \rho_\mu(y,z)$ contradict condition $T^{\mu R}$. Thus this case is not possible.

(2ib) Suppose that ρ satisfies $C_2^{\rho_\mu}$. We show that ρ satisfies T^{ρ_μ}. Assume to the contrary that $\exists x,y,z \in X$ such that $\rho_\mu(x,y) = \rho_\mu(y,x) = \rho_\mu(y,z) = \rho_\mu(z,y)$ and $\rho_\mu(x,z) \neq \rho_\mu(z,x)$. We show that it is impossible.

If $\rho_\mu(x,z) > \rho_\mu(z,x)$, then the transitivity of ρ implies $\rho_\mu(z,x) \geq \min(\rho_\mu(z,y), \rho_\mu(y,x))$. Then $\rho_\mu(x,y) = \rho_\mu(y,x) = \rho_\mu(y,z) = \rho_\mu(z,y) \leq \rho_\mu(z,x) < \rho_\mu(x,z)$.

Then

$$
\begin{aligned}
\alpha_3(x,z,y) &= \min(I_{\min}(\rho_\mu(z,y),\rho_\mu(z,x)),\ 1 - J_{\max}(\rho_\nu(z,y),\rho_\nu(z,x)), \\
&\qquad I_{\min}(\rho_\mu(x,z),\rho_\mu(y,z)),\ 1 - J_{\max}(\rho_\nu(x,z),\rho_\nu(y,z))) \\
&= \min(1,\rho_\mu(y,z)) = \rho_\mu(y,z) = \rho_\mu(z,y)
\end{aligned}
$$

and $\alpha_2(x,z,y) = \min(\rho_\mu(x,z),\rho_\mu(z,y)) = \rho_\mu(z,y)$. We obtain $\alpha_2(x,z,y) = \alpha_3(x,z,y) = \rho_\mu(z,y)$. Because of hypothesis $\rho_\mu(z,y) = \rho_\mu(x,y) = \rho_\mu(y,x)$, we have $\rho_\mu(y,x), \rho_\mu(x,y) \in [\alpha_2(x,z,y), \alpha_3(x,z,y)]$. Since ρ satisfies $\rho_\mu(z,x) < \rho_\mu(x,z)$ and $\rho_\mu(y,z) = \rho_\mu(z,y)$, $C_2^{\mu R}$ implies $\rho_\mu(y,x) < \rho_\mu(x,y)$. That contradicts the equality $\rho_\mu(y,x) = \rho_\mu(x,y)$. Hence, $\rho_\mu(x,z) > \rho_\mu(z,x)$ is not possible.

If $\rho_\mu(x,z) < \rho_\mu(z,x)$, then the transitivity of ρ implies

$$\rho_\mu(x,z) \geq \min(\rho_\mu(x,y), \rho_\mu(y,z)).$$

Thus

$$\rho_\mu(x,y) = \rho_\mu(y,x) = \rho_\mu(y,z) = \rho_\mu(z,y) \leq \rho_\mu(x,z) < \rho_\mu(z,x).$$

We obtain

$$
\begin{aligned}
\alpha_3(z,x,y) &= \min(I_{\min}(\rho_\mu(x,y),\rho_\mu(x,z)),\ 1 - J_{\max}(\rho_\nu(x,y),\rho_\nu(x,z)), \\
&\qquad I_{\min}(\rho_\mu(z,x),\rho_\mu(y,x)),\ 1 - J_{\max}(\rho_\nu(z,x),\rho_\nu(y,x))) \\
&= \min(1,\rho_\mu(y,x)) = \rho_\mu(y,x)
\end{aligned}
$$

and $\alpha_2(z,x,y) = \min(\rho_\mu(z,x), \rho_\mu(x,y)) = \rho_\mu(x,y)$. Then $\alpha_2(x,z,y) = \alpha_3(x, z, y) = \rho_\mu(x,y)$. Since $\rho_\mu(x,y) = \rho_\mu(z,y) = \rho_\mu(y,z)$, we have $\rho_\mu(y,z), \rho_\mu(z,y) \in [\alpha_2(z,x,y), \alpha_3(z,x,y)]$. Since ρ satisfies $(\rho_\mu(x,z) < \rho_\mu(z,x)$ and $\rho_\mu(y,x) = \rho_\mu(x,y))$, $C_2^{\rho_\mu}$ gives $\rho_\mu(y,z) < \rho_\mu(z,y)$. That contradicts the equality $\rho_\mu(y,z) = \rho_\mu(z,y)$. Hence, $\rho_\mu(z,x) > \rho_\mu(x,z)$ is impossible.

$(2ii)$ Since ρ is σ-symmetric, the proof of the equivalence of conditions $C_2^{\rho_\nu}$ and T^{ρ_ν} follows from the previous result. ∎

In the case where $\mathcal{T} = (\min, \max)$, conditions $C_1^{\rho_\mu}$ and $C_1^{\rho_\nu}$ hold and have become simplified $C_2^{\rho_\mu}$ and $C_2^{\rho_\nu}$. Therefore, it is important to rewrite characterizations of the pos-transitivity and the negative transitivity of a strict component of an IFR in this case.

The following result shows that, when $\mathcal{T} = (\min, \max)$, each regular strict component of a $\mathcal{T}$-transitive IFR is pos-transitive. It also establishes that when $\mathcal{T} = (\min, \max)$, condition T^{ρ_μ} or T^{ρ_ν} is necessary and sufficient on a $\mathcal{T}$-transitive IFR ρ such that each regular strict component of ρ is negative transitive.

Proposition 8.5.13 (*Nana and Fono* [24]) *Let ρ be a* $(\min, \max)$-*transitive IFR and π be a regular strict component of ρ. Then the following statements hold:*

(1) *π is pos-transitive.*
(2) *The following two statements are equivalent:*
   (*i*) *ρ satisfies T^{ρ_μ} or T^{ρ_ν}.*
   (*ii*) *π is negative transitive.*

Proof. (1) Assume that ρ is $(\min, \max)$-transitive. We show that π is pos-transitive. Since ρ is $(\min, \max)$-transitive, the first result of Lemma 8.5.12 implies that ρ satisfies conditions $C_1^{\rho_\mu}$ and $C_1^{\rho_\nu}$. Using the first result of Theorem 8.5.10, it follows that π is pos-transitive.

(2) Suppose (*i*) holds. Assume that ρ satisfies condition T^{ρ_μ}. We show that π is negative transitive. Since ρ is $(\min, \max)$-transitive, the first result of Lemma 8.5.12 implies that ρ satisfies condition $C_1^{\rho_\mu}$. Since ρ satisfies condition T^{ρ_μ} and is $(\min, \max)$-transitive, the second result of Lemma 8.5.12 implies that ρ satisfies condition $C_2^{\rho_\mu}$. Since ρ satisfies conditions $C_1^{\rho_\mu}$ and $C_2^{\rho_\mu}$, the second result of Theorem 8.5.10 implies that π is negative transitive.

Assume that ρ satisfies condition T^{ρ_ν}. By the σ-symmetry of ρ, it follows that π is negative transitive by an analogous prove.

Suppose that (*ii*) holds. Assume that π is negative transitive. We show that ρ satisfies condition T^{ρ_μ} or T^{ρ_ν}. Since ρ is $(\min, \max)$-transitive, it suffices to show that ρ satisfies condition $C_2^{\rho_\mu}$ or $C_2^{\rho_\nu}$ by the second result of Lemma 8.5.12. Since π is negative transitive, the second result of Theorem 8.5.10 implies that ρ satisfies condition $C_2^{\rho_\mu}$ or $C_2^{\rho_\nu}$. ∎

Any strict component of a (min, max)-transitive is pos-transitive and only the strict components of transitive IFRs satisfying conditions T^{μ_R} or T^{ν_R} are negative transitive.

Consequently, we introduce the following notations for a t-norm $\mathcal{T}$.

(1) W^T be the set of reflexive, complete, π-symmetric and $\mathcal{T}$-transitive IFRs.

(2) W_1^T be the set of reflexive, complete, π-symmetric and $\mathcal{T}$-transitive IFRs satisfying condition $C_1^{\rho_\mu}$ or condition $C_1^{\rho_\nu}$.

(3) W_2^T be the set of reflexive, complete, π-symmetric and $\mathcal{T}$-transitive IFRs satisfying conditions ($C_1^{\rho_\mu}$ and $C_2^{\rho_\mu}$) or ($C_1^{\rho_\nu}$ and $C_2^{\rho_\nu}$).

(4) CO be the set of crisp total pre-orders (reflexive, complete and transitive crisp binary relations)

We have $CO \subset W_2^T \subset W_1^T \subset W^T$.

In the case where $\mathcal{T} = (\min, \max)$, we have by Lemma 8.5.12(1) that W^T and W_1^T become the set W of reflexive, strongly complete, π-symmetric and transitive IFRs and by Lemma 8.5.12(2) W_2^T becomes the set W_2 of elements of W satisfying T^{ρ_μ} or T^{ν_R}. Hence the previous inclusions become $CO \subset W_2 \subset W$.

G is the set of admissible IFRs and $G \subseteq W^T$ in the general case. In the particular case, $G \subseteq W$.

8.6   Intuitionistic Arrow's Theorem and Gibbard's Oligarchy Theorem

In this section, we use Theorem 8.5.10 and Proposition 8.5.13 to establish an intuitionistic fuzzy version of the crisp Arrow impossibility theorem and an intuitionistic fuzzy version of the crisp Gibbard's oligarchy theorem.

Let $N = \{1, 2, \ldots, n\}$ with $n \geq 2$ be a finite set of individuals or voters. A coalition C is a non empty subset of N and $\mathcal{P}^*(N)$ is the set of coalitions of N. We recall some notation: $\rho_i = \langle \rho_{\mu i}, \rho_{\nu i} \rangle$, $\pi_i = \langle \pi_{\mu i}, \pi_{\nu i} \rangle$, $i \in N$, and $\rho = (\langle \rho_{\mu 1}, \rho_{\nu 1} \rangle, \ldots, \langle \rho_{\mu n}, \rho_{\nu n} \rangle)$. Let $G^n = \{\rho \mid \rho_i \in G, i \in N\}$. Given $C \in \mathcal{P}^*(N)$ and $x, y \in X$, $\rho_{\mu C}(x, y) > 0$ means that $\forall i \in C$, $\rho_{\mu i}(x, y) > 0$ and, $\rho_{\nu C}(x, y) < 1$ means $\forall i \in C$, $\rho_{\nu i}(x, y) < 1$.

A function $\widetilde{f} : G^n \to G$ is called an intuitionistic fuzzy aggregation rule (**IFAR**). For $\rho \in G^n$, $\widetilde{f}(\rho) = \rho = (\rho_\mu, \rho_\nu) \in G$ is a social preference and $\pi = (\pi_\mu, \pi_\nu)$ is its strict component.

We introduce intuitionistic fuzzy versions of some usual properties of a crisp aggregation rule.

Definition 8.6.1 *Let $\widetilde{f}$ be an IFAR, $u, v \in X$, $i \in N$ and $C \in \mathcal{P}^*(N)$.*

(1) $\tilde{f}$ satisfies **intuitionistic fuzzy Pareto condition** (**IFPC**) if $\forall \rho \in G^n$, $\forall x, y \in X$,

$$\forall i \in N, \quad \left\{ \begin{array}{l} \pi_{\mu i}(x,y) > 0 \\ \pi_{\nu i}(x,y) < 1 \end{array} \right. \Rightarrow \left\{ \begin{array}{l} \pi_{\mu}(x,y) > 0 \\ \pi_{\nu}(x,y) < 1 \end{array} \right. .$$

(2) $\tilde{f}$ satisfies **independence of irrelevant alternatives** (**IIA**) if $\forall \rho, \rho' \in G^n$,

$$\forall x, y \in X, \left(\forall i \in N \left\{ \begin{array}{ll} \left\{ \begin{array}{l} \pi_{\mu i}(x,y) > 0 \\ \pi_{\nu i}(x,y) < 1 \end{array} \right. & \Leftrightarrow \left\{ \begin{array}{l} \pi'_{\mu i}(x,y) > 0 \\ \pi'_{\nu i}(x,y) < 1 \end{array} \right. \\ \left\{ \begin{array}{l} \pi_{\mu i}(y,x) > 0 \\ \pi_{\nu i}(y,x) < 1 \end{array} \right. & \Leftrightarrow \left\{ \begin{array}{l} \pi'_{\mu i}(y,x) > 0 \\ \pi'_{\nu i}(y,x) < 1 \end{array} \right. \end{array} \right. \right)$$

implies

$$\left\{ \begin{array}{ll} \left\{ \begin{array}{l} \pi_{\mu}(x,y) > 0 \\ \pi_{\nu}(x,y) < 1 \end{array} \right. & \Leftrightarrow \left\{ \begin{array}{l} \pi'_{\mu}(x,y) > 0 \\ \pi'_{\nu}(x,y) < 1 \end{array} \right. \\ \left\{ \begin{array}{l} \pi_{\mu}(y,x) > 0 \\ \pi_{\nu}(y,x) < 1 \end{array} \right. & \Leftrightarrow \left\{ \begin{array}{l} \pi'_{\mu}(y,x) > 0 \\ \pi'_{\nu}(y,x) < 1 \end{array} \right. \end{array} \right.$$

(3) C is **decisive** if $\forall \rho \in G^n$, $\forall x, y \in X$,

$$\left\{ \begin{array}{l} \pi_{\mu C}(x,y) > 0 \\ \pi_{\nu C}(x,y) < 1 \end{array} \right. \Rightarrow \left\{ \begin{array}{l} \pi_{\mu}(x,y) > 0 \\ \pi_{\nu}(x,y) < 1 \end{array} \right. .$$

(4) i is **almost semi-decisive** over (u,v) if $\exists \rho \in G^n$,

$$\left\{ \begin{array}{l} \pi_{\mu i}(u,v) > 0 \\ \pi_{\nu i}(u,v) < 1 \end{array} \right. and \left\{ \begin{array}{l} \pi_{\mu N \setminus \{i\}}(v,u) > 0 \\ \pi_{\nu N \setminus \{i\}}(v,u) < 1 \end{array} \right. \Rightarrow \left\{ \begin{array}{l} \pi_{\mu}(v,u) = 0 \\ \pi_{\nu}(v,u) = 1 \end{array} \right. .$$

(5) C is **almost decisive** on (u,v) if $\forall \rho \in G^n$,

$$\left\{ \begin{array}{l} \pi_{\mu C}(u,v) > 0 \\ \pi_{\nu C}(u,v) < 1 \end{array} \right. and \left\{ \begin{array}{l} \pi_{\mu N \setminus C}(v,u) > 0 \\ \pi_{\nu N \setminus C}(v,u) < 1 \end{array} \right. \Rightarrow \left\{ \begin{array}{l} \pi_{\mu}(u,v) > 0 \\ \pi_{\nu}(u,v) < 1 \end{array} \right. .$$

(6) i is a **dictator** if $\forall \rho \in G^n$, $\forall x, y \in X$,

$$\left\{ \begin{array}{l} \pi_{\mu i}(x,y) > 0 \\ \pi_{\nu i}(x,y) < 1 \end{array} \right. \Rightarrow \left\{ \begin{array}{l} \pi_{\mu}(x,y) > 0 \\ \pi_{\nu}(x,y) < 1 \end{array} \right. .$$

(7) i has a **veto** if $\forall \rho \in G^n$, $\forall x, y \in X$,

$$\left\{ \begin{array}{l} \pi_{\mu i}(x,y) > 0 \\ \pi_{\nu i}(x,y) < 1 \end{array} \right. \Rightarrow \left\{ \begin{array}{l} \pi_{\mu}(y,x) = 0 \\ \pi_{\nu}(y,x) = 1 \end{array} \right. .$$

(8) f is **dictatorial** if there exists $i_0 \in N$ such that i_0 is a dictator.

(9) f is an **oligarchy** if there exists a unique decisive coalition S of N such that $\forall i \in S$, i has a veto.

We next establish an analogue of the crisp Arrow impossibility theorem for (T, S)-transitive IFARs.

Theorem 8.6.2 (*Nana and Fono* [24]) *Let* $\mathcal{T} = (T, S)$ *be an intuitionistic fuzzy t-norm and* $\widetilde{f} : (W_1^T)^n \to W_2^T$ *be an IFAR. If* $\widetilde{f}$ *satisfies conditions IIA and IFPC, then* $\widetilde{f}$ *is dictatorial.*

To establish this theorem, we need the two following lemmas. As in crisp case and fuzzy case, we first establish an intuitionistic fuzzy version of the crisp field expansion lemma.

Lemma 8.6.3 (*Nana and Fono*[24]) (*Field Expansion Lemma for an IFAR*) *Let* $\widetilde{f} : (W_1^T)^n \to W_1^T$ *be an IFAR satisfying IIA and IFPC,* $C \in \mathcal{P}^*(N)$ *and* $x, y \in X$. *If* C *is almost decisive on* (x, y), *then* C *is decisive.*

Proof. Suppose that C is an almost decisive coalition on (x, y). We show that C is decisive. Let $\rho' \in (W_1^T)^n$ and $u, v \in X$ be such that

$$\begin{cases} \pi'_{\mu C}(u, v) > 0 \\ \pi'_{\nu C}(u, v) < 1 \end{cases}$$

We show that

$$\begin{cases} \pi'_\mu(u, v) > 0 \\ \pi'_\nu(u, v) < 1. \end{cases} \tag{8.21}$$

Consider the profile $\rho'' \in (W_1^T)^n$ satisfying the following conditions on $\{u, v, x, y\}$:

(1) $\forall i \in C$,

$$\begin{cases} \pi''_{\mu i}(u, x) > 0 \\ \pi''_{\nu i}(u, x) < 1 \end{cases}, \begin{cases} \pi''_{\mu i}(y, v) > 0 \\ \pi''_{\nu i}(y, v) < 1 \end{cases}, \tag{8.22}$$

$$\begin{cases} \pi''_{\mu i}(x, y) > 0 \\ \pi''_{\nu i}(x, y) < 1 \end{cases} \text{ and } \begin{cases} \pi''_{\mu i}(u, v) = \pi'_{\mu i}(u, v) \\ \pi''_{\nu i}(u, v) = \pi'_{\nu i}(u, v) \end{cases}$$

(2) $\forall j \in N \backslash C$,

$$\begin{cases} \pi''_{\mu j}(u, x) > 0 \\ \pi''_{\nu j}(u, x) < 1 \end{cases}, \begin{cases} \pi''_{\mu j}(y, v) > 0 \\ \pi''_{\nu j}(y, v) < 1 \end{cases}, \begin{cases} \pi''_{\mu j}(y, x) > 0 \\ \pi''_{\nu j}(y, x) < 1 \end{cases} \tag{8.23}$$

and π'_j and π''_j coincide on $\{u, v\}$.

We show first that

$$\begin{cases} \pi''_\mu(u, v) > 0 \\ \pi''_\nu(u, v) < 1 \end{cases} \tag{8.24}$$

By the first system of (8.22) and the first system of (8.23), we have that

$$\begin{cases} \pi''_{\mu N}(u, x) > 0 \\ \pi''_{\nu N}(u, x) < 1 \end{cases}$$

Since $\tilde{f}$ satisfies IFPC, we have

$$\begin{cases} \pi''_\mu(u, x) > 0 \\ \pi''_\nu(u, x) < 1 \end{cases} \tag{8.25}$$

Analogously, by the second system of (8.22) and the second system of (8.23), we have that

$$\begin{cases} \pi''_{\mu N}(y, v) > 0 \\ \pi''_{\nu N}(y, v) < 1 \end{cases}$$

Since $\tilde{f}$ satisfies IFPC,

$$\begin{cases} \pi''_\mu(y, v) > 0 \\ \pi''_\nu(y, v) < 1 \end{cases} \tag{8.26}$$

Since C is almost decisive on (x, y), we have by the third system of (8.22)) and the third system of (8.23) that

$$\begin{cases} \pi''_\mu(x, y) > 0 \\ \pi''_\nu(x, y) < 1 \end{cases} \tag{8.27}$$

Since $\rho'' = \tilde{f}(\rho'') \in W_1^T$, the first result of Theorem 8.5.10 implies that π'' is pos-transitive. The pos-transitivity of π'', (8.25) and (8.27) imply

$$\begin{cases} \pi''_\mu(u, y) > 0 \\ \pi''_\nu(u, y) < 1 \end{cases} \tag{8.28}$$

Analogously, by the pos-transitivity of π'' and (8.26) and (8.28), we have (8.24).

We now show (8.21). The last system of (8.22) and the hypothesis yield

$$\begin{cases} \pi''_{\mu C}(u, v) = \pi'_\mu(u, v) > 0 \\ \pi''_{\mu C}(u, v) = \pi'_{\nu C}(u, v) < 1 \end{cases}$$

Since π is perfect antisymmetric,

$$\begin{cases} \pi''_{\mu C}(v, u) = \pi'_{\mu C}(v, u) = 0 \\ \pi''_{\nu C}(v, u) = \pi'_{\nu C}(v, u) = 1 \end{cases}$$

Since $\pi'_{N \setminus C}$ and $\pi''_{N \setminus C}$ coincide on $\{u, v\}$, we have

$$\begin{cases} \pi''_{\mu N \setminus C}(v, u) = \pi'_{\mu N \setminus C}(v, u) \\ \pi''_{\nu N \setminus C}(v, u) = \pi'_{\nu N \setminus C}(v, u) \end{cases}$$

By the two previous systems, we have that

$$\begin{cases} \pi''_{\mu N}(v, u) = \pi'_{\mu N}(v, u) \\ \pi''_{\nu N}(v, u) = \pi'_{\nu N}(v, u) \end{cases} \tag{8.29}$$

The last system of (8.22) and the fact that $\pi'_{N\setminus C}$ and $\pi''_{N\setminus C}$ coincide on $\{u, v\}$ imply

$$\begin{cases} \pi''_{\mu N}(u, v) = \pi'_{\mu N}(u, v) \\ \pi''_{\nu N}(u, v) = \pi'_{\nu N}(u, v) \end{cases} \tag{8.30}$$

By (8.29) and (8.30),

$$\forall i \in N, \begin{cases} \begin{cases} \pi''_{\mu i}(u, v) > 0 \\ \pi''_{\nu i}(u, v) < 1 \end{cases} \Leftrightarrow \begin{cases} \pi'_{\mu i}(u, v) > 0 \\ \pi'_{\nu i}(u, v) < 1 \end{cases} \\ \begin{cases} \pi''_{\mu i}(v, u) > 0 \\ \pi''_{\nu i}(v, u) < 1 \end{cases} \Leftrightarrow \begin{cases} \pi'_{\mu i}(v, u) > 0 \\ \pi'_{\nu i}(v, u) < 1 \end{cases} \end{cases} \tag{8.31}$$

Since $\widetilde{f}$ satisfies IIA, (8.31) and (8.24) imply $\pi'_\mu(u, v) > 0$ and $\pi'_\nu(u, v) < 1$. Thus the desired result holds. ■

The following lemma is an intuitionistic fuzzy version of the Group contraction lemma for crisp aggregation rules.

Lemma 8.6.4 (*Nana and Fono* [24]) (*Group Contraction Lemma for IFARs*) Let $\widetilde{f} : (W_1^T)^n \to W_2^T$ be an IFAR satisfying IIA and IFPC and, $D \in \mathcal{P}^*(N)$ with $|D| \geq 2$.

If D is decisive, then there exists a coalition $K \in \mathcal{P}^*(N)$ such that $K \subset D$ and K is decisive.

Proof. Let $D \in \mathcal{P}^*(N)$ be such that $|D| \geq 2$ and D is decisive. We find a coalition $K \subset D$ which is decisive. Let $\{D_1, D_2\}$ be a partition of D and let the profile $\rho \in (W_1^T)^n$ satisfy the following conditions on $\{x, y, z\}$:

(1) $\forall i \in D_1$,

$$\begin{cases} \pi_{\mu i}(x, y) > 0 \\ \pi_{\nu i}(x, y) < 1 \end{cases} \text{and} \begin{cases} \pi_{\mu i}(y, z) > 0 \\ \pi_{\nu i}(y, z) < 1 \end{cases}, \tag{8.32}$$

(2) $\forall j \in D_2$,

$$\begin{cases} \pi_{\mu j}(x, y) > 0 \\ \pi_{\nu j}(x, y) < 1 \end{cases} \text{and} \begin{cases} \pi_{\mu j}(z, x) > 0 \\ \pi_{\nu j}(z, x) < 1 \end{cases}, \tag{8.33}$$

(3) $\forall k \in N\setminus D$,

$$\begin{cases} \pi_{\mu k}(y, z) > 0 \\ \pi_{\nu k}(y, z) < 1 \end{cases} \text{and} \begin{cases} \pi_{\mu k}(z, x) > 0 \\ \pi_{\nu k}(z, x) < 1 \end{cases}. \tag{8.34}$$

Since $\rho \in (W_1^T)^n$, the first result of Theorem 8.5.10 implies that $\forall i \in N, \pi_i$ is pos-transitive. Hence, (8.32), (8.33) and (8.34) yield

$$\begin{cases} \pi_{\mu D_1}(x, z) > 0 \\ \pi_{\nu D_1}(x, z) < 1 \end{cases}, \begin{cases} \pi_{\mu D_1}(z, y) > 0 \\ \pi_{\nu D_1}(z, y) < 1 \end{cases} \text{and} \begin{cases} \pi_{\mu N\setminus D}(y, x) > 0 \\ \pi_{\nu N\setminus D}(y, x) < 1 \end{cases}. \tag{8.35}$$

Otherwise, by the first system of (8.32) and the first system of (8.33) we have that $\pi_{\mu D}(x, y) > 0$ and $\pi_{\nu D}(x, y) < 1$, by the second system of (8.32) and the first system of (8.34) we have that $\pi_{\mu N \setminus D_2}(y, z) > 0$ and $\pi_{\nu N \setminus D_2}(y, z) < 1$, and by the second system of (8.33) and the second system of (8.34), we have that

$$\begin{cases} \pi_{\mu N \setminus D_1}(z, x) > 0 \\ \pi_{\nu N \setminus D_1}(z, x) < 1 \end{cases} . \tag{8.36}$$

Since $\pi_{\mu D}(x, y) > 0$ and $\pi_{\nu D}(x, y) < 1$ and D is decisive, we have $\pi_\mu(x, y) > 0$ and $\pi_\nu(x, y) < 1$. Moreover, $\widetilde{f}(\rho) = \rho \in W_2^T$ and so by the second result of Theorem 8.5.10, we have that π is negative transitive. Hence by the previous system and the negative transitivity of π, we have that

$$\begin{cases} \pi_\mu(x, z) > 0 \\ \pi_\nu(x, z) < 1 \end{cases} \quad \text{or} \quad \begin{cases} \pi_\mu(z, y) > 0 \\ \pi_\nu(z, y) < 1 \end{cases} .$$

We consider two cases:

(1) If $\pi_\mu(x, z) > 0$ and $\pi_\nu(x, z) < 1$, then this system, the first system of (8.35), (8.36), the fact that D_1 is almost decisive on (x, z) and Lemma 8.6.3 imply that D_1 is decisive. Thus $K = D_1$.

(2) If $\pi_\mu(z, y) > 0$ and $\pi_\nu(z, y) < 1$, then this system, the second system of (8.35), the fact that D_2 is almost decisive on (y, z) and Lemma 8.6.3 imply that D_2 is decisive. Thus $K = D_2$. ∎

We now prove the theorem.

Proof of Theorem 8.6.2 We find $i_0 \in N$ such that $\forall \rho \in (W_1^T)^n$, $\forall x, y \in X$,

$$\begin{cases} \pi_{\mu i_0}(x, y) > 0 \\ \pi_{\nu i_0}(x, y) < 1 \end{cases} \Rightarrow \begin{cases} \pi_\mu(x, y) > 0 \\ \pi_\nu(x, y) < 1 \end{cases} .$$

Since $\widetilde{f}$ satisfies IFPC, N is decisive. Since $|N| \geq 2$, Lemma 8.6.4 (the group contraction lemma) implies that N admits a decisive sub coalition denoted N_1. We consider two cases:

(1) If $|N_1| = 1$ and N_1 is decisive, then i_0 is the unique element of N_1.

(2) If $|N_1| \geq 2$, then we can repeat the previous procedure to obtain a dictator i_0 by applying the group contraction lemma. ∎

We have from Proposition 8.5.13 and Theorem 8.6.2, the intuitionistic fuzzy Arrow Theorem when $\mathcal{T} = (\min, \max)$.

Corollary 8.6.5 *Let $\widetilde{f} : W^n \to W_2$ be an IFAR. If $\widetilde{f}$ satisfies conditions IIA and IFPC, then $\widetilde{f}$ is dictatorial.*

We now show that in the case where $\mathcal{T} = (\min, \max)$ that if we do not assume conditions $T^{\rho \mu}$ and $T^{\rho \nu}$, we obtain a non-dictatorial IFAR.

Proposition 8.6.6 [24] *There exists a non-dictatorial IFAR $\widetilde{f} : W^n \to W$ satisfying simultaneously IIA and IFPC.*

Proof. Let $\beta \in \left(\frac{1}{2}, 1\right)$. Let $\widetilde{f}$ be an IFAR defined by $\forall \rho \in W^n$ with $\widetilde{f}(\rho) = \rho$, $\forall x, y \in X$ (with $x \neq y$), $\rho_\mu(x, x) = 1$, $\rho_\nu(x, x) = 0$

$$\rho_\mu(x, y) = \begin{cases} 1 & if \quad \forall i \in N, \ \rho_{\mu i}(x, y) > \rho_{\mu i}(y, x) \\ \beta & \text{otherwise,} \end{cases}$$

$$\rho_\nu(x, y) = \begin{cases} 0 & if \quad \forall i \in N, \ \rho_{\nu i}(x, y) < \rho_{\nu i}(y, x) \\ 1 - \beta & \text{otherwise.} \end{cases}$$

We first show that $\widetilde{f}$ is well-defined, i.e., $\forall \rho \in W^n$, $\rho = \widetilde{f}(\rho) \in W$.

Let $\rho \in W^n$. We show that $\rho = \widetilde{f}(\rho) \in W$. ρ is reflexive since $\forall x \in X$, $\rho_\mu(x, x) = 1$ and $\rho_\nu(x, x) = 0$.

We show that ρ is strongly complete, i.e., $\forall x, y \in X$, $\rho_\mu(x, y) + \rho_\mu(y, x) \geq 1$ and $\rho_\nu(x, y) + \rho_\nu(y, x) \leq 1$. Let $x, y \in X$. We consider three cases:

(1) If $\forall i \in N$, $\begin{cases} \rho_{\mu i}(x, y) > \rho_{\mu i}(y, x) \\ \rho_{\nu i}(x, y) < \rho_{\nu i}(y, x) \end{cases}$, then

$$\begin{cases} \rho_\mu(x, y) = 1 \\ \rho_\nu(x, y) = 0 \end{cases} \quad \text{and} \quad \begin{cases} \rho_\mu(y, x) = \beta \\ \rho_\nu(y, x) = 1 - \beta \end{cases}.$$

Hence $\rho_\mu(x, y) + \rho_\mu(y, x) = 1 + \beta > 1$ and $\rho_\nu(x, y) + \rho_\nu(y, x) = 0 + (1 - \beta) < 1$.

(2) If $\forall i \in N$, $\begin{cases} \rho_{\mu i}(y, x) > \rho_{\mu i}(x, y) \\ \rho_{\nu i}(y, x) < \rho_{\nu i}(x, y) \end{cases}$, then

$$\begin{cases} \rho_\mu(y, x) = 1 \\ \rho_\nu(y, x) = 0 \end{cases} \quad \text{and} \quad \begin{cases} \rho_\mu(x, y) = \beta \\ \rho_\nu(x, y) = 1 - \beta \end{cases}.$$

Thus $\rho_\mu(y, x) + \rho_\mu(x, y) = 1 + \beta > 1$ and $\rho_\nu(y, x) + \rho_\nu(x, y) = 0 + (1 - \beta) < 1$. If $\exists i \in N$ such that

$$\begin{cases} \rho_{\mu i}(x, y) \leq \rho_{\mu i}(y, x) \\ \rho_{\nu i}(x, y) \geq \rho_{\nu i}(y, x) \end{cases}$$

or $\exists j \in N$ such that

$$\begin{cases} \rho_{\mu j}(y, x) \leq \rho_{\mu j}(x, y) \\ \rho_{\nu j}(y, x) \geq \rho_{\nu j}(y, x) \end{cases},$$

then

$$\begin{cases} \rho_\mu(x, y) = \beta \\ \rho_\nu(x, y) = 1 - \beta \end{cases} \quad \text{and} \quad \begin{cases} \rho_\mu(y, x) = \beta \\ \rho_\nu(y, x) = 1 - \beta \end{cases}.$$

Hence $\rho_\mu(x, y) + \rho_\mu(y, x) = 2\beta > 1$ and $\rho_\nu(x, y) + \rho_\nu(y, x) = (1 - \beta) + (1 - \beta) = 2 - 2\beta < 1$. Thus ρ is strongly complete.

We now show that ρ is transitive. Assume to the contrary that there $\exists x, y, z \in X$ such that

$$\rho_\mu(x, z) < \min(\rho_\mu(x, y), \rho_\mu(y, z))$$

or

$$\rho_\nu(x, z) > \max(\rho_\nu(x, y), \rho_\nu(y, z)).$$

We consider two cases.

(1) If $\rho_\mu(x, z) < \min(\rho_\mu(x, y), \rho_\mu(y, z))$, then $\rho_\mu(x, z) = \beta$ and $\rho_\mu(x, y) = \rho_\mu(y, z) = 1$. By the definition of ρ, we have that

$$\forall i \in N, \quad \rho_{\mu i}(x, y) > \rho_{\mu i}(y, x) \text{ and } \rho_{\mu i}(y, z) > \rho_{\mu i}(z, y). \qquad (8.37)$$

$\forall i \in N, \quad \rho_{\mu i}(x, y) > \rho_{\mu i}(y, x) \text{ and } \rho_{\mu i}(y, z) > \rho_{\mu i}(z, y).$

Otherwise, $\forall i \in N$, π_i is pos-transitive since $\rho_i \in W$. Thus (8.37) and the pos-transitivity of π_i imply $\forall i \in N$, $\rho_i(x, z) > \rho_i(z, x)$. Hence $\rho_\mu(x, z) = 1$. This contradicts the fact that $\rho_\mu(x, z) = \beta \in (\frac{1}{2}, 1)$.

(2) If $\rho_\nu(x, z) > \max(\rho_\nu(x, y), \rho_\nu(y, z))$, then $\rho_\nu(x, z) = 1 - \beta$ and $\rho_\nu(x, y) = \rho_\nu(y, z) = 0$. Thus by the definition of ρ, it follows that

$$\forall i \in N, \quad \rho_{\nu i}(x, y) < \rho_{\nu i}(y, x) \text{ and } \rho_{\nu i}(y, z) < \rho_{\nu i}(z, y). \qquad (8.38)$$

Since $\forall i \in N$, π_i is pos-transitive, (8.34) implies $\forall i \in N$, we have $\rho_{\nu i}(x, z) < \rho_{\nu i}(z, x)$. Hence $\rho_\nu(x, z) = 0$. This contradicts the fact that $\rho_\nu(x, z) = 1 - \beta > 0$.

We show that ρ is σ-symmetric, i.e., $\forall x, y \in X, \rho_\mu(x, y) + \rho_\nu(x, y) = \rho_\mu(y, x) + \rho_\nu(y, x)$. We distinguish two cases:

(i) If $\rho_\mu(x, y) = 1$, then $\rho_\mu(y, x) = \beta$ and $\forall i \in N, \rho_{\mu i}(x, y) > \rho_{\mu i}(y, x)$. By the σ-symmetry of ρ_i, we get $\forall i \in N, \rho_{\nu i}(x, y) < \rho_{\nu i}(y, x)$, i.e., $\rho_\nu(x, y) = 0$ and $\rho_\nu(y, x) = 1 - \beta$. We have $\rho_\mu(x, y) + \rho_\nu(x, y) = 1 = \rho_\mu(y, x) + \rho_\nu(y, x)$.

(ii) If $\rho_\mu(x, y) = \beta$, then $\exists j \in N$ such that $\rho_{\mu j}(x, y) \leq \rho_{\mu j}(y, x)$. By the σ-symmetry of ρ_j, we have $\rho_{\nu j}(x, y) \geq \rho_{\nu j}(y, x)$ and $\rho_\nu(x, y) = 1 - \beta$. Thus, $\rho_\mu(x, y) + \rho_\nu(x, y) = 1$. To compute $\rho_\mu(y, x) + \rho_\nu(y, x)$, we consider three subcases:

Suppose that $\rho_{\mu j}(x, y) = \rho_{\mu j}(y, x)$. We show that $\rho_\mu(y, x) + \rho_\nu(y, x) = 1$. By the σ-symmetry of ρ_j, $\rho_{\nu j}(x, y) = \rho_{\nu j}(y, x)$. Thus $\rho_\mu(y, x) = \beta$ and $\rho_\nu(y, x) = 1 - \beta$. Hence $\rho_\mu(y, x) + \rho_\nu(y, x) = \beta + (1 - \beta) = 1 = \rho_\mu(x, y) + \rho_\nu(x, y)$.

Suppose that $\rho_{\mu j}(x, y) < \rho_{\mu j}(y, x)$ and $\forall k \in N - \{j\}, \rho_{\mu k}(x, y) < \rho_{\mu k}(y, x)$. Let us show that $\rho_\mu(y, x) + \rho_\nu(y, x) = 1$. The σ-symmetry of ρ_k implies $\forall k \in N, \rho_{\nu k}(x, y) > \rho_{\nu k}(y, x)$. Thus $\rho_\mu(y, x) = 1$ and $\rho_\nu(y, x) = 0$. We get $\rho_\mu(y, x) + \rho_\nu(y, x) = 1 + 0 = 1 = \rho_\mu(x, y) + \rho_\nu(x, y)$.

Suppose that $\rho_{\mu j}(x, y) < \rho_{\mu j}(y, x)$ and $\exists k \in N, \rho_{\mu k}(x, y) \geq \rho_{\nu k}(y, x)$. We show that $\rho_\mu(y, x) + \rho_\nu(y, x) = 1$. As in the previous case, the σ-symmetry of ρ_j and ρ_k imply $\rho_{\nu j}(x, y) > \rho_{\nu k}(y, x)$ and $\rho_{\nu k}(x, y) \leq \rho_{\nu k}(y, x)$. The definition of ρ implies $\rho_\mu(y, x) = \beta$ and $\rho_\nu(y, x) = 1 - \beta$. Hence $\rho_\mu(y, x) + \rho_\nu(y, x) = \beta + (1 - \beta) = 1 = \rho_\mu(x, y) + \rho_\nu(x, y)$. Now ρ is σ-symmetric. We conclude that $\rho \in W$.

We next show that $\widetilde{f}$ satisfies condition IFPC. Suppose that $\forall i \in N$, $\pi_{\mu i}(x, y) > 0$ and $\pi_{\nu i}(x, y) < 1$. We show that $\rho_\mu(x, y) > 0$ and $\rho_\nu(x, y) < 1$. By definition of π_i, $\forall i \in N$, $\rho_{\mu i}(x, y) > \rho_{\mu i}(y, x)$ and $\rho_{\nu i}(x, y) < \rho_{\nu i}(y, x)$. By the definition of ρ, we have $\rho_\mu(x, y) = 1 > 0 = \rho_\mu(y, x)$ and $\rho_\nu(x, y) = 0 < 1 - \beta = \rho_\nu(y, x)$. By the definition of π, we have the desired result.

We now show that $\widetilde{f}$ satisfies condition IIA. Let $\rho, \rho' \in W^n$ and $x, y \in X$ be such that

$$\forall i \in N \quad \begin{cases} \begin{cases} \pi_{\mu i}(x, y) > 0 \\ \pi_{\nu i}(x, y) < 1 \end{cases} \Leftrightarrow \begin{cases} \pi'_{\mu i}(x, y) > 0 \\ \pi'_{\nu i}(x, y) < 1 \end{cases} \\ \begin{cases} \pi_{\mu i}(y, x) > 0 \\ \pi_{\nu i}(y, x) < 1 \end{cases} \Leftrightarrow \begin{cases} \pi'_{\mu i}(y, x) > 0 \\ \pi'_{\nu i}(y, x) < 1 \end{cases} \end{cases} \quad (8.39)$$

We show that $\begin{cases} \pi_\mu(x, y) > 0 \\ \pi_\nu(x, y) < 1 \end{cases} \Leftrightarrow \begin{cases} \pi'_\mu(x, y) > 0 \\ \pi'_\nu(x, y) < 1 \end{cases}$ and $\begin{cases} \pi_\mu(y, x) > 0 \\ \pi_\nu(y, x) < 1 \end{cases} \Leftrightarrow$ $\begin{cases} \pi'_\mu(y, x) > 0 \\ \pi'_\nu(y, x) < 1 \end{cases}$.

By definition of π', (8.39) becomes

$$\forall i \in N, \quad \begin{cases} \begin{cases} \rho_{\mu i}(x, y) > \rho_{\mu i}(y, x) \\ \rho_{\nu i}(x, y) < \rho_{\nu i}(y, x) \end{cases} \Leftrightarrow \begin{cases} \rho'_{\mu i}(x, y) > \rho'_{\mu i}(y, x) \\ \rho'_{\nu i}(x, y) < \rho'_{\nu i}(y, x) \end{cases} \\ \begin{cases} \rho_{\mu i}(y, x) > \rho_{\mu i}(x, y) \\ \rho_{\nu i}(y, x) < \rho_{\nu i}(x, y) \end{cases} \Leftrightarrow \begin{cases} \rho'_{\mu i}(y, x) > \rho'_{\mu i}(x, y) \\ \rho'_{\nu i}(y, x) < \rho'_{\nu i}(x, y) \end{cases} \end{cases}$$

By the definition of $\rho = \widetilde{f}(\rho)$ and $\rho' = \widetilde{f}(\rho')$, we have

$$\begin{cases} \rho_\mu(x, y) = 1 > \beta = \rho_\mu(y, x) \\ \rho_\nu(x, y) = 0 < 1 - \beta = \rho_\nu(y, x) \\ \rho_\mu(y, x) = 1 > \beta = \rho_\mu(x, y) \\ \rho_\nu(y, x) = 0 < 1 - \beta = \rho_\nu(x, y) \end{cases} \begin{array}{c} \Leftrightarrow \\ \\ \Leftrightarrow \end{array} \begin{cases} \rho'_\mu(x, y) = 1 > \beta = \rho'_\mu(y, x) \\ \rho'_\nu(x, y) = 0 < 1 - \beta = \rho'_\nu(y, x) \\ \rho'_\mu(y, x) = 1 > \beta = \rho'_\mu(x, y) \\ \rho'_\nu(y, x) = 0 < 1 - \beta = \rho'_\nu(x, y) \end{cases}$$

The result follows upon rewriting this system by the means of π and π'.

We show that $\widetilde{f}$ is non-dictatorial. Assume to the contrary that $\widetilde{f}$ is dictatorial. Then there exists $i_0 \in N$ such that $\forall \rho \in W^n$, $\forall x, y \in X$, $\pi_{\mu i_0}(x, y) > 0$ and $\pi_{\nu i_0}(x, y) < 1$ implies $\pi_\mu(x, y) > 0$ and $\pi_\nu(x, y) < 1$. Consider the profile $\rho \in W^n$ defined on $\{x, y\}$ by $\exists k \in N \setminus \{i_0\}$,

$$\begin{cases} \rho_{\mu k}(x, y) < \rho_{\mu k}(y, x) \\ \rho_{\nu k}(x, y) > \rho_{\nu k}(y, x) \end{cases} \text{ and } \begin{cases} \rho_{\mu N \setminus \{k\}}(x, y) > \rho_{\mu N \setminus \{k\}}(y, x) \\ \rho_{\nu N \setminus \{k\}}(x, y) < \rho_{\nu N \setminus \{k\}}(y, x) \end{cases}$$

Then by definition of π_{i_0}, the previous system implies $\pi_{\mu i_0}(x, y) > 0$ and $\pi_{\nu i_0}(x, y) < 1$. Thus we obtain $\pi_\mu(x, y) > 0$ and $\pi_\nu(x, y) < 1$. Otherwise, the collective preference $\rho = \widetilde{f}(\rho)$ on $\{x, y\}$ satisfies $\rho_\mu(x, y) = \rho_\mu(y, x) = \beta$ and $\rho_\nu(x, y) = \rho_\nu(y, x) = 1 - \beta$. By the definition of π, the previous system gives

$\pi_\mu(x, y) = 0$ and $\pi_\nu(x, y) = 1$ which contradicts $\pi_\mu(x, y) > 0$ and $\pi_\nu(x, y) < 1$. Hence $\widetilde{f}$ is non-dictatorial. ∎

We now show that if we do not assume conditions $(C_1^{\rho_\mu}$ and $C_2^{\rho_\mu})$ or $(C_1^{\rho_\nu}$ and $C_2^{\rho_\nu})$, we obtain an oligarchy IFAR which is an intuitionistic fuzzy version of Gibbard's theorem.

Theorem 8.6.7 (*Nana and Fono* [24]) *Let* $\widetilde{f} : (W_1^T)^n \to W_1^T$ *be an IFAR. If* $\widetilde{f}$ *satisfies conditions IIA and IFPC, then there exists a unique oligarchy.*

To establish this theorem, we prove the following two Lemmas. We first establish an intuitionistic fuzzy version of the Veto-field expansion lemma.

Lemma 8.6.8 (*Nana and Fono* [24]) (*Veto-field Expansion Lemma for IFARs*) *Let* $\widetilde{f} : (W_1^T)^n \to W_1^T$ *be an IFAR satisfying IIA and IFPC,* $i \in N$ *and* $u, v \in X$. *If* i *is almost semi-decisive over* (u, v), *then* i *has a veto.*

Proof. Let $i \in N$ and $u, v \in X$ such that i is almost semi-decisive over (u, v). Let $x, y \in X$ and $\rho' \in (W_1^T)^n$ be such that $\pi'_\mu(x, y) > 0, \pi'_\nu(x, y) < 1$. We show that $\pi'_\mu(y, x) = 0$ and $\pi'_\nu(y, x) = 1$.

Since i is almost semi-decisive over (u, v), $\exists \rho \in (W_1^T)^n$ such that i satisfies (4) of Definition 8.6.1. Let the profile $\rho'' \in (W_1^T)^n$ satisfy the following conditions: $\forall u, v, x, y \in X$,

$$
\begin{cases}
\begin{cases} \pi''_{\mu i}(u, v) = \pi''_{\mu i}(u, v) > 0 \\ \pi''_{\nu i}(u, v) = \pi''_{\nu i}(u, v) < 1 \end{cases} \\
\begin{cases} \pi''_{\mu i}(x, y) = \pi''_{\mu i}(x, y) > 0 \\ \pi''_{\nu i}(x, y) = \pi''_{\nu i}(x, y) < 1 \end{cases} \\
\begin{cases} \pi''_{\mu N\setminus i}(v, u) = \pi''_{\mu N\setminus i}(v, u) > 0 \\ \pi''_{\nu N\setminus i}(v, u) = \pi''_{\nu N\setminus i}(v, u) < 1 \end{cases} \\
\begin{cases} \pi''_{\mu N\setminus i}(y, x) = \pi''_{\mu N\setminus i}(y, x) > 0 \\ \pi''_{\nu N\setminus i}(y, x) = \pi''_{\nu N\setminus i}(y, x) < 1 \end{cases}
\end{cases}
\quad \text{and} \quad
\begin{cases}
\begin{cases} \pi''_{\mu N}(x, u) > 0 \\ \pi''_{\nu N}(x, u) < 1 \end{cases} \\
\begin{cases} \pi''_{\mu N}(v, y) > 0 \\ \pi''_{\nu N}(v, y) < 1 \end{cases}
\end{cases}
. \quad (8.40)
$$

By the two last systems of (8.40) and condition IFPC, we have that

$$
\begin{cases} \pi''_\mu(x, u) > 0 \\ \pi''_\nu(x, u) < 1 \end{cases} \quad \text{and} \quad \begin{cases} \pi''_\mu(v, y) > 0 \\ \pi''_\nu(v, y) < 1 \end{cases} . \quad (8.41)
$$

Since π_i is perfect antisymmetric for all i, we have the first and the third systems of (8.40) that

$$
\pi'' \text{ and } \pi \text{ coincide on } (u, v). . \quad (8.42)
$$

By (4) of Definition 8.6.1, the first and the third systems of (8.40) yield $\pi_\mu(v, u) = 0$ and $\pi_\nu(v, u) = 1$. By this system, condition IIA and (8.41), it follows that

$$
\begin{cases} \pi''_\mu(v, u) = 0 \\ \pi''_\nu(v, u) = 1. \end{cases} \quad (8.43)
$$

We now show that

$$
\begin{cases}
\pi''_\mu(y,x) = 0 \\
\pi''_\nu(y,x) = 1.
\end{cases}
\tag{8.44}
$$

Assume to the contrary that $\pi''_\mu(y,x) > 0$ and $\pi''_\nu(y,x) < 1$. This system, (8.41) and the pos-transitivity of P_L imply $\pi''_\mu(v,u) > 0$ and $\pi''_\nu(v,u) < 1$. This contradicts (8.43).

By the second and the fourth systems of (8.40) and the perfect antisymmetry of π_j, it follows that π'' and π' coincide on (x,y). This assertion, condition IIA and (8.44) imply $\pi'_\mu(y,x) = 0$ and $\pi'_\nu(y,x) = 1$. Hence the desired result holds. ∎

The following Lemma is an intuitionistic fuzzy version of the crisp veto-field expansion lemma for crisp and fuzzy aggregation rules.

Lemma 8.6.9 (*Nana and Fono* [24]) (*Veto-field Expansion Lemma for IFARs*) *Let* $\tilde{f} : (W_1^T)^n \to W_2^T$ *be an IFAR satisfying IIA and IFPC. Then there exists a unique decisive coalition S of N such that $\forall i \in S$, i has a veto.*

Proof. We first determine the unique decisive coalition.
Consider the set $\mathbb{D}_{\tilde{f}}$ of decisive coalitions of N. Condition IFPC implies that N is decisive, that is, $N \in \mathbb{D}_{\tilde{f}}$. Since N is finite and $\mathbb{D}_{\tilde{f}} \neq \emptyset$, $\mathbb{D}_{\tilde{f}}$ has a minimal decisive coalition, say S. Then S is the unique decisive coalition.
We next show that $\forall i \in S$, i has a veto. Let $i \in S$. By the Veto field expansion lemma, it suffices to show that i is almost semi-decisive over a pair of options. Let $x, y, z \in X$. Let $\rho \in (W_1^T)^n$ satisfy the following conditions:

$$
\begin{cases}
(i) & \begin{cases} \pi_{\mu i}(x,y) > 0 \\ \pi_{\nu i}(x,y) < 1 \end{cases} \text{ and } \begin{cases} \pi_{\mu i}(y,z) > 0 \\ \pi_{\nu i}(y,z) < 1 \end{cases} \\[2ex]
(ii) & \begin{cases} \pi_{\mu S\setminus i}(x,y) > 0 \\ \pi_{\nu S\setminus i}(x,y) < 1 \end{cases} \text{ and } \begin{cases} \pi_{\mu S\setminus i}(z,x) > 0 \\ \pi_{\nu S\setminus i}(z,x) < 1 \end{cases} \\[2ex]
(iii) & \begin{cases} \pi_{\mu N\setminus S}(y,z) > 0 \\ \pi_{\nu S\setminus i}(y,z) < 1 \end{cases} \text{ and } \begin{cases} \pi_{\mu N\setminus S}(z,x) > 0 \\ \pi_{\nu N\setminus S}(z,x) < 1 \end{cases}
\end{cases}
\tag{8.45}
$$

Since S is decisive, we have by the first system of (i) of (8.45) and the first system of (ii) of (8.45) that

$$
\begin{cases}
\pi_\mu(x,y) > 0 \\
\pi_\nu(x,y) < 1
\end{cases}
\tag{8.46}
$$

Since $\forall k \in N$, π_k is pos-transitive, we have by (i) of (8.45 8.45) that

$$
\begin{cases}
\pi_{\mu i}(x,z) > 0 \\
\pi_{\nu i}(x,z) < 1
\end{cases}
\tag{8.47}
$$

We consider two cases:

(1) If $\pi_\mu(z, x) = 0$ and $\pi_\nu(z, x) = 1$, then (8.47), the second system of (ii) of (8.45) and the second system of (iii) of (8.45) imply that i is almost semi-decisive over (x, z).

(2) We show that

$$\begin{cases} \pi_\mu(z, x) > 0 \\ \pi_\nu(z, x) < 1 \end{cases}.$$
(8.48)

is not possible.

By the second system of (i) of (8.45), the first system of (iii) of (8.45) and the pos-transitivity of each π_j, it follows that $\pi_{\mu N \setminus (S \setminus i)}(y, z) > 0$ and $\pi_{\nu N \setminus (S \setminus i)}(y, z) < 1$. By (ii) of (8.45) and the pos-transitivity of each π_j, we have that $\pi_{\mu S \setminus i}(z, y) > 0$ and $\pi_{\nu S \setminus i}(z, y) < 1$. The pos-transitivity of π, (8.48) and (8.46) imply $\pi_\mu(z, y) > 0$ and $\pi_\nu(z, y) < 1$. The three previous systems imply that $S \setminus \{i\}$ is almost decisive on (y, z). By Lemma 8.6.3, we have that $S \setminus \{i\}$ is decisive, which contradicts the minimality of S. Thus this case is not possible. ∎

8.7   Exercises

1. Prove Proposition 8.1.7.

2. [22] Let $\widetilde{f}$ be the fuzzy aggregation rule defined in Definition 8.1.9. Show by example that $\widetilde{f}$ does not satisfy PR with respect to $\pi_{(2)}$.

3. [22] Construct an example of $\pi_{(2)}$ that is T_2-transitive and partially transitive, but not quasi-transitive.

4. [22] Show by example that partial transitivity $\not\Rightarrow$ partial quasi-transitivity for $\pi = \pi_{(1)}$ or $\pi = \pi_{(3)}$.

5. [22] Construct examples that show ρ is max-min transitive, but not consistent; ρ is consistent, but not max-min transitive; ρ is consistent, but not transitive in the crisp sense (and not complete).

6. (a) Show that $\mathcal{J} = (\max, \min)$ satisfies condition G.

   (b) For all $l \in [0, \infty]$ show that $\mathcal{I}_l^F = (S_l^F, T_l^F)$ satisfies condition G, where $\mathcal{I}_l^F$ is the class of Frank t-conorms, t-norms.

   (c) If S and T are dual, then the restriction of $\mathcal{I} = (S, T)$ on $L_1^* = \{(x_1, x_2) \mid x_1 + x_2 = 1\}$ satisfies condition G.

7. Prove Proposition 8.5.6.

8. Prove Theorem 8.5.10.

8.8   References

1. F. B. Abdelaziz, J. R. Figueira, and O. Meddeb, On the manipulability of the fuzzy social choice functions, *Fuzzy Sets and Systems*, 159 (2008) 177–184.

2. K. J. Arrow, *Social Choice and Individual Values*, New York, Wiley 1951.

3. K. T. Atanassov, Intuitionistic fuzzy sets, *Fuzzy Sets and Systems*, 20 (1986) 87–96.

4. K. T. Atanassov, *Intuitionistic Fuzzy Sets*, Fuzziness and Soft Computing Physica-Verlag, Heidelberg 1999.

5. D. Austen-Smith and J. S. Banks, *Positive Political Theory I: Collective Preference*, Michigan Studies in Political Science, The University of Michigan Press 2000.

6. A. Banerjee, Fuzzy preferences and Arrow-type problems in social choice, *Social Choice and Welfare*, 11 (1994) 121–130.

7. C. R. Barrett, P. K. Pattanaik, and Maurice Salles, On the structure of fuzzy social welfare functions, *Fuzzy Sets and Systems*, 19 (1980) 1–10.

8. H. Bustince and P. Burillo. Structures on intuitionistic fuzzy relations. *Fuzzy Sets and Systems*, 78 (1996) 293–303.

9. C. Cornelis, E. Kerre and G. Deschrijver. Implication in intuitionistic fuzzy and interval-valued fuzzy set theory: Construction, classification, application. *International Journal of Approximate Reasoning*, 35 (2004) 55–95.

10. D. Dimitrov. Intuitionistic fuzzy preferences: A factorization. *Advanced Studies Contemporary Mathematics*, 5 (2002) 93–104.

11. D. Dimitrov. The Paretian liberal with intuitionistic fuzzy preferences: A result. *Social Choice and Welfare*, 23 (2004) 149–156.

12. B. Dutta, Fuzzy preferences and social choice, *Mathematical Social Sciences*, 13 (1987) 215–229.

13. L. A. Fono and N. G. Andjiga, Fuzzy strict preference and social choice, *Fuzzy Sets and Systems*, 155 (2005) 372–389.

14. L. A. Fono, V. Donfack-Kommogne, and N. G. Andjiga, Fuzzy arrow-type results without the Pareto principle based on fuzzy pre-orders, *Fuzzy Sets and Systems*, 160 (2009) 2658–2672.

15. L.A. Fono, H. Gwet and S. Fotso. On strict lower and upper sections of weakly complete fuzzy pre-orders based on co-implication. *Fuzzy Sets and Systems*, 159 (2008) 2240–2255.

16. L. A. Fono, G. N. Nana, M. Salles, and H. Gwet, A binary intuitionistic fuzzy relation: some new results, a general factorization, and two properties of strict components, *International Journal of Mathematics and Mathematical Sciences*, Hindaw Publishing Corporation Volume 2009, Article ID 580918, 38 pages dos: 10.1155/2009/580918.

17. L.A. Fono and M. Salles. Continuity of utility functions representing fuzzy preferences. *Social Choice and Welfare*, 37 (2011) 669–682, doi: 10.1007/s00355-011-0571-0.

18. S. Fotso and L.A. Fono. On the consistency of some crisp choice functions based on a strongly complete fuzzy pre–order. *New Mathematics and Natural Computation*, 8 (2012) 257–272.

19. Garcia-Lapresesta, J. Luis, and B. Llamazares, Aggregation of fuzzy preferences: Some rules of the mean, *Social Choice and Welfare,* 17 (2000) 673–690.

20. E. Klement, R. Mesiar and E. Pap. *Triangular norms*, Kluwer Academic Publishers, Dordrecht, 2000.

21. J. N. Mordeson and T. D. Clark, Fuzzy Arrow's theorem, *New Mathematics and Natural Computation,* 5 (2009) 371–383.

22. J. N. Mordeson, M. B. Gibilisco, and T. D. Clark, Independence of irrelevant alternatives and fuzzy Arrow's theorem, *New Mathematics and Natural Computation,* 8 (2012) 219–237.

23. J. N. Mordeson, T. D. Clark. M. J. Wierman, K. K. Nelson-Pook, and K. Albert, Intuitionistic fuzzy preference relations and independence of irrelevant alternatives, *Journal Fuzzy Mathematics*, 22 (2014) to appear.

24. G. N. Nana and L. A. Fono, Arrow-type results under intuitionistic fuzzy preferences, *New Mathematics and Natural Computation*, 9 (2013) 1–27.

25. G. N. Njanpong. Relations binaires floues intuitionistes et application au choix social. These de Doctorat PhD, Département de Mathématiques, Faculté des Sciences, Université de Yaoundé I-Cameroun, Janvier 2011.

26. S. V. Ovchinnikov, Social choice and Lukasiewicz logic, *Fuzzy Sets and Systems,* 43 (1991) 275–289.

27. G. Richardson, The structure of fuzzy preferences: Social choice implications, *Social Choice and Welfare,* 15 (1998) 359–369.

28. M. Salles. Fuzzy utility. Volume 1 of Handbook of Utility Theory, pages 321–344. Principles. *Kluwer Academic Publishers*, Boston, 1998.

29. M. Salles, S. Geslin and A. Ziad. Fuzzy aggregation in economic environments: Quantitative fuzziness, public goods and monotonicity assumptions. *Mathematical Social Sciences*, 45 (2003) 155–166.

30. A. K. Sen, *Collective Choice and Social Welfare*, Advanced Textbooks in Economics, Vol. 11, Editors C. J. Bliss and M. D. Intriligator, North Holland 1979.

31. 31. A. K. Sen, *Social Choice Theory*, Handbook of Mathematical Economics, Vol. 3, Elsevier Sciences Publishers B. V., North Holland, 1986, 1073–1181, Chapter 22.

32. H. J. Skala, Arrow's impossibility theorem: Some new aspects, Decision Theory and Social Ethics: Issues in Social Choice, Eds.: H,. W. Gottinger and W. Leinfellner, Vol. 17, 1978, 215–225.

33. Zeshui Xu. Intuitionistic preference relations and their application in group decision-making. *Information Sciences*, 177 (2007) 2363–2379.

Chapter 9

Manipulability of Fuzzy Social Choice Functions

Sections 9.1-9.3 are based on the work of Abdelaziz, Figueria, and Meddeb [1]. In social decision-making contexts, a manipulator often has incentive to change the social choice in his/her favor by strategically misrepresenting his/her preference. The Gibbard-Satterthwaite theorem [19, 26] states that a social choice function over three or more alternatives that does not provide incentive to individuals to misrepresent their sincere preferences must be dictatorial. This chapter extends their result to the case of fuzzy weak preference relations. Here we use the approach given in [1] by defining the best alternative set in three ways and thus provide three generalizations of the Gibbard-Satterthwaite theorem in the fuzzy context.

A voting choice procedure is subject to strategic manipulation when an individual reveals a insincere preference relation with the purpose of changing the social choice in his favor. Attempts to avoid the negative result of Gibbard and Satterthwaite have been made by relaxing its assumptions [3, 29]. This occurs when individuals have difficulty expressing their preference over the set of alternatives but they can, however, specify a preference degree between 0 and 1 for each ordered pair of alternatives [5, 16, 17, 30]. Thus the choice of a single alternative can involve the use of a fuzzy social choice function.

How alternatives are to be compared and how the social choice is to be interpreted must be defined in order to introduce fuzzy manipulation. The concept of best alternative set, as proposed by Dutta, Panda, and Pattanaik [15], is used to provide three possible alternative sets. Hence three new definitions of fuzzy manipulability were presented in [1]. In each case, the strategic manipulation of a fuzzy social choice function by an individual is possible when two conditions are satisfied. The first condition is that the sincere social choice does not belong to the best alternative set of the manipulator. The second condition is that there exists a fuzzy relation assuring the manipulator an outcome at least as good as the sincere social choice. A dictator for a fuzzy social choice function is defined to the individual who secures the social choice in his/her best alternative set in all situations. Under these new definitions, it is shown that the Gibbard-Satterthwaite theorem continues to hold.

9.1   Preliminaries

In this section, we present three fuzzy preference relations and their best alternative sets.

The social choice problem consists of finding the best alternative in accordance with the preferences of all individuals. The best alternative is also called the social choice. Fuzzy rather than crisp preferences are considered. Fuzzy binary relations on the set of alternatives X are used to model the fuzzy preferences.

Definition 9.1.1 A **crisp binary relation** (**CBR**) ρ on X is a function $\rho: X^2 \to \{0,1\}$. A CBR ρ on X is called
(1) **complete** if for all $x, y \in X$, $\rho(x, y) = 1$ or $\rho(y, x) = 1$;
(2) **transitive** if for all $x, y, z \in X$, $\rho(x, y) = 1$ and $\rho(y, x) = 1$ implies $\rho(x, z) = 1$;
(3) **antisymmetric** if for all $x, y \in X$, $\rho(x, y) = 1$ and $\rho(y, x) = 1$ implies $x = y$.

Definition 9.1.2 A **fuzzy binary relation** (**FBR**) ρ on X is a function $\rho: X^2 \to [0,1]$. A FBR ρ on X is called **frail max-min transitive** if for all $x, y, z \in X$, $\rho(x, y) \geq \rho(y, x) \wedge \rho(y, z) \geq \rho(z, y)$ implies $\rho(x, z) \geq \rho(x, y) \wedge \rho(y, z)$.

Definition 9.1.3 A **fuzzy weak preference relation** (**FWPR**) ρ on X is a reflexive FBR.

Definition 9.1.4 Let ρ be a CBR. Then ρ is called a **crisp linear order** (**CLO**) if it is complete, anti-symmetric, and transitive.

Let ρ_i be a CLO, where $i \in N$. The best alternative set for individual i coincides with the most preferred element. It is given by

$$P_0(X, \rho_i) = \{x \in X \mid \forall y \in X, \rho_i(x, y) = 1\}.$$

We now present three different definitions of an alternative to be "at least as good as" another alternative.

Definition 9.1.5 [1] Let $i \in N$. Let $x, y \in X$.
(1) **Simple degree rule** (**SDR**) : $\rho_i(x, y) \geq \rho_i(y, x)$.

The best alternative set is defined as follows:

$$P_1(X, \rho_i) = \{x \in X \mid \forall y \in X, \ \rho_i(x, y) \geq \rho_i(y, x)\}.$$

(2) **Confidence threshold rule (CTR)** : $\rho_i(x, y) \geq t$, *where* $t \in (0, 1/2]$. *The best alternative set is defined as follows:*

$$P_2(X, \rho_i) = \{x \in X \mid \forall y \in X, \rho_i(x, y) \geq t\}.$$

(3) **Dominance degree rule (DDR)** : $d(\rho_i, X)(x) \geq d(\rho_i, X)(y)$, *where* $d(\rho_i, X)(x) = \wedge\{\rho_i(x, z) \mid z \in X\}$, *is a dominance degree of alternative* x *with respect to all the remaining alternatives in* X.

The best alternative set is defined as follows:

$$P_3(X, \rho_i) = \{x \in X \mid d(\rho_i, X)(x) = \vee\{d(\rho_i, X)(y) \mid y \in X\}\}.$$

We use the following notation.

Let D_c denotes the set of all strongly complete FWPRs and let D_0 denotes the set of all CLOs on X.

Let $D_1 = \{\rho \in D_c \mid \rho \text{ satisfies max-min transitivity}\}$ and let $D_2 = \{\rho \in D_c \mid \rho \text{ satisfies frail max-min transitivity}\}$.

D_1 is used in the context of SDR and D_2 is used in the context of CTR.

$D_3 = D_1$ when adopting DDR.

It follows from [13] that $D_0 \subseteq D_1 = D_3 \subseteq D_2$ since any CLO is reflexive and since max-min transitivity implies frail max-min transitivity trivially.

The following result concerning the best alternative sets can be found in [13, 19, 25, 26].

Proposition 9.1.6 *Let A be a nonempty subset of X. Let ρ be a FWPR.*
(1) *If $\rho \in D_j$, then $P_j(A, \rho) \neq \emptyset$ for $j \in \{0, 1, 2, 3\}$.*
(2) *If $\rho \in D_0$, then $P_0(A, \rho) = P_1(A, \rho) = P_2(A, \rho) = P_3(A, \rho)$.*

We let $\rho = (\rho_1, \ldots, \rho_n)$ denote a profile of individual's preference relations.

Example 9.1.7 *Let $X = \{x, y, z\}$ and $N = \{1, 2, 3\}$. Let the fuzzy binary relations $\rho_i, i = 1, 2, 3$, be given by the following table:*

ρ_1	x	y	z
x	1	.6	.64
y	.7	1	.65
z	.65	.6	1

ρ_2	x	y	z
x	1	.7	.9
y	.4	1	.7
z	.4	.4	1

ρ_3	x	y	z
x	1	.7	.6
y	.65	1	.6
z	1	.9	1

Then $\rho_i \in D_1$, $i = 1, 2, 3$. We see that $P_1(X, \rho_1) = \{y\}$ since $\rho_1(y, x) > \rho_1(x, y)$ and $\rho_1(y, z) > \rho_1(z, y)$. Let a confidence threshold be given to be $t = .45$. Then $P_2(X, \rho_2) = \{x\}$ and $P_2(X, \rho_1) = P_2(X, \rho_3) = X$. We also have $P_3(X, \rho_3) = \{z\}$ since $d(\rho_3, X)(x) = d(\rho_3, X)(y) = .6$ and $d(\rho_3)(z) = .9$.

9.2   Fuzzy Social Choice Functions

We now present three types of fuzzy social choice functions and dictatorships together with their corresponding manipulation procedures that were introduced in [1].

Definition 9.2.1 *A **fuzzy social choice function** (**FSCF**) is a function of X into R_n, where R_n is the set of all profiles of individuals' fuzzy weak preference relations.*

Definition 9.2.2 *A j-**fuzzy social choice function** (j-**FSCF**) is a function of D_j^n into R_n, $j = 0, 1, 2, 3$.*

When each individual expresses an FWPR in D_j, a j-FSCF defines a social choice. A 0-FSCF can be viewed as a voting rule.

Example 9.2.3 *Let $\rho_i, i = 1, 2, 3$, be defined as in Example 9.1.7. Then $\rho_i \in D_j$, $i = 1, 2, 3$; $j = 1, 2, 3$. Apply the arithmetic mean to obtain the fuzzy social choice ρ^S and consider $P_1(X, \rho^S) = \{x\}$ as the social choice. Such an FSCF can be viewed as an j-FSCF for $j \in \{1, 2, 3\}$, ρ^S is given by the following table.*

ρ^S	x	y	z
x	1	.67	.71
y	.58	1	.65
z	.68	.63	1

We use the following notation:

$(\rho|\rho_i')$ is the profile of individuals' fuzzy weak preference relations $(\rho_1, \ldots, \rho_{i-1}, \rho_i', \rho_{i+1}, \ldots, \rho_n)$.

$(\rho|\rho_i', \rho_{i+1}')$ is the profile of individuals' preference relations $(\rho_1, \ldots, \rho_{i-1}, \rho_i', \rho_{i+1}', \ldots, \rho_n)$.

$(\rho|\rho_1', \ldots, \rho_k')$ for $k \in \{1, \ldots, n\}$, is the profile of individuals' fuzzy weak preference relations $(\rho_1', \ldots, \rho_k', \rho_{k+1}, \ldots, \rho_n)$.

ρ' is the profile of individuals' fuzzy weak preference relations $(\rho_1', \ldots, \rho_n')$.

Suppose a profile of individuals' fuzzy weak preference relations is in D_j^n and the social choice is obtained by the use of a j-FSCF, v_j, $j \in \{0, 1, 2, 3\}$. If every individual $i \in N$ expresses a fuzzy weak preference relation ρ_i that belongs to D_j, then $P_j(X, \rho_i)$ is the individual's best alternative set. Suppose that an individual $m \in N$ with sincere fuzzy weak preference relation ρ_m knows the $n - 1$ fuzzy weak preference relations of the remaining individuals and the j-FSCF, v_j. Then individual m can anticipate the outcome $v_j(\rho)$. The individual then determines whether or not the outcome $v_j(\rho)$ can be considered as a good social choice on the basis of his sincere preference, ρ_m, that is, does outcome $v_j(\rho)$ belong to individual m's best alternative set $P_j(X, \rho_m)$? If the outcome $v_j(\rho)$ does not belong to the set $P_j(X, \rho_m)$, then individual m seeks a social choice at least as good as the outcome $v_j(\rho_n)$ on the basis of his fuzzy weak preference relation ρ_m. If there exists a binary relation ρ'_m in D_j such that the outcome $v_j(\rho|\rho'_m)$ is better than the outcome $v_j(\rho)$, then individual m can manipulate the j-FSCF, v_j, by revealing ρ'_m. Thus the j-manipulability of a j-FSCF can be defined as follows:

Definition 9.2.4 [1] (j-manipulability) Let v_j be a j-FSCF for $j = 0, 1, 2, 3$.
(1) v_0 is said to be 0-**manipulable** if there exists $m \in N$, $\rho \in D_0^n$ and $\rho'_m \in D_0$ such that $v_0(\rho|\rho'_m) \neq v_0(\rho)$ and $\rho_m(v_0(\rho|\rho'_m), v_0(\rho)) = 1$.
(2) v_1 is said to be 1-**manipulable** if there exists $m \in N$, $\rho \in D_1^n$ and $\rho'_m \in D_1$ such that there exists $x \in X$ such that $\rho_m(x, v_1(\rho)) > \rho_m(v_1(\rho), x)$, $v_1(\rho|\rho'_m) \neq v_1(\rho)$, and $\rho_m(v_1(\rho|\rho'_m), v_1(\rho)) \geq (v_1(\rho), v_1(\rho|\rho'_m))$.
(3) v_2 is said to be 2-**manipulable** if there exists $m \in N$, $\rho \in D_2^n$ and $\rho'_m \in D_2$ such that for a fixed $t \in [0, 1/2]$ there exists $x \in X$ such that $\rho_m(v_2(\rho), x) < t$, $v_2(\rho|\rho'_m) \neq v_2(\rho)$ and $\rho_m(v_2(\rho|\rho'_m), v_2(\rho)) \geq t$.
(4) v_3 is said to be 3-**manipulable** if there exists $m \in N$, $\rho \in D_3^n$ and $\rho'_m \in D_3$, there exists $x \in X$ such that $d(\rho_m, X)(x) > d(\rho_m, X)$ $(v_3(\rho))$, $v_3(\rho|\rho'_m)) \neq v_3(\rho)$ and $d(\rho_m, X)(v_3(\rho|\rho'_m)) \geq d(\rho_m, X)(v_3(\rho))$.

The 0-manipulability of a 0-FSCF coincides with the one of Gibbard-Satterthwaite. The j-manipulability of a j-FSCF implies its 0-manipulability when its domain is restricted to the set $D_0^n, j = 1, 2, 3$.

Definition 9.2.5 (j-strategy proof) A j-FSCF v_j is said to be j-**strategy proof** if it is not j-manipulable, $j = 0, 1, 2, 3$.

Example 9.2.6 [1] Consider Example 9.2.3.
(1) Individual 1 can 1-manipulate the FSCF. The alternative x is not a member of his best alternative set $P_1(X, \rho_1) = \{y\}$. Individual 1 can thus reveal the non-sincere fuzzy relation ρ'_1 to obtain y as the social choice.

ρ'_1	x	y	z
x	1	0	0
y	.95	1	1
z	.9	0	1

The arithmetic mean now yields

	x	y	z
x	1	.47	.5
y	.67	1	.77
z	.77	.43	1

(2) Let $t = .45$ be a confidence threshold for all individuals. Then alternative x belongs to every best alternative set $P_2(X, \rho_i)$ for all $i \in \{1, 2, 3\}$. Hence no individual can 2-manipulate the FSCF in this situation.

(3) *Individual 3 can 3-manipulate the FSCF. Alternative x does not belong to individual 3's best alternative set $P_3(X, \rho_3) = \{z\}$. Individual 3 can thus reveal the non-sincere fuzzy relation ρ_3' to obtain z as the social choice.*

ρ_3'	x	y	z
x	1	0	0
y	.9	1	0
z	.95	1	1

The arithmetic mean here is

	x	y	z
x	1	.43	.51
y	.67	1	.45
z	.67	.67	1

The following definition generalizes the Gibbard-Satterthwaite impossibility result [19, 26] to the fuzzy case. For $j = 0, 1, 2, 3$, an individual $m \in N$ with a fuzzy weak preference relation ρ_m in D_j is considered to be a j-dictator for the j-FSCF v_j if the outcome $v_j(\rho)$ belongs to his best alternative set $P_j(X, \rho_m)$ for all $\rho = (\rho_1, \ldots, \rho_m, \ldots, \rho_n)$. If a j-dictator exists for a j-FSCF v_j, then v_j is considered to be j-dictatorial.

Definition 9.2.7 [1] *(j-dictatorship) Let v_j be a j-FSCF for $j = 0, 1, 2, 3$.*
(1) *v_0 is said to be 0-**dictatorial** if there exists $m \in N$ such that for all $\rho \in D_0^n$, $\forall x \in X$, $\rho_m(v_0(\rho), x) = 1$.*
(2) *v_1 is said to be 1-**dictatorial** if there exists $m \in N$ such that for all $\rho \in D_1^n$, $\forall x \in X$, $\rho_m(v_1(\rho), x) \geq \rho_m(x, v_1(\rho))$.*
(3) *v_2 is said to be 2-**dictatorial** if there exists $m \in N$ such that for all $\rho \in D_2^n$, for a fixed $t \in (0, 1/2]$, $\forall x \in X$, $\rho_m(v_2(\rho), x) \geq t$.*
(4) *v_3 is said to be 3-**dictatorial** if there exists $m \in N$ such that for all $\rho \in D_0^n$, $\forall x \in X$, $d(\rho_m, X)(v_3(\rho)) \geq d(\rho_m, X)(x)$.*

The 0-dictatorship of a 0-FSCF coincides with the G-S dictatorship. Also, for $j \in \{1, 2, 3\}$, the j-dictatorship of a j-FSCF implies its 0-dictatorship when its domain is restricted to D_0^n.

9.3 Impossibility Results

In this section, the strategy-proofness theorem of G-S is generalized to the j-FSCFs for $j = 1, 2, 3$.

Theorem 9.3.1 (*Gibbard* [19], *Satterthwaith* [26]) *Let $v_0 : D_0^n \to X$ be a 0-FSCF such that $v_0(D_0^n) = X$. If v_0 is 0-strategy-proof, then it is 0-dictatorial.*

The Gibbard–Satterthwaite theorem states that for a set N of three or more individuals and set A of at least three alternatives that if the individuals can have any preference ordering of A for his or her preferences, then every nondictatorial social choice procedure is manipulable for some distribution of preferences. In legislative politics for example, there is something known as the *killer amendment*. It is an amendment to a bill that will cause the bill to be defeated. Here a legislator may not like the amendment, but knows the bill will be defeated if the amendment is added. Another example involves plurality voting. Here any number of candidates may compete for a particular office with the candidate receiving the most votes being declared a winner. Individuals do not wish to waste their votes on a candidate they prefer, but who they think has no chance of winning. Consequently, they vote for a different candidate.

Lemma 9.3.2 *Let A be a non-empty subset of X and ρ be a CBR on X. If ρ satisfies the following conditions, then ρ is a CLO on X and $P_0(X, \rho) \subseteq A$.*
  (1) ρ *is a CLO on A;*
  (2) $\rho(x, y) = 1$ *and* $\rho(y, x) = 0$ *for all $x \in A$ and $y \in X \backslash A$;*
  (3) ρ *is a CLO on $X \backslash A$.*

Proof. By (1), $\rho(x, x) = 1$ for all $x \in A$ and by (3) $\rho(x, x) = 1$ for all $x \in X \backslash A$. Thus ρ is reflexive. By (1), $\rho(x, y + \rho(y, x) \geq 1$ for all $x, y \in A$; by (2), $\rho(x, y + \rho(y, x) \geq 1$ for all $x \in A, y \in X \backslash A$; by (3), $\rho(x, y + \rho(y, x) \geq 1$ for all $x, y \in X \backslash A$. Thus ρ is strongly complete. Let $x, y, z \in X$ be such that $\rho(x, y) = 1$ and $\rho(y, z) = 1$. Suppose that $z \in A$ and $x \in X \backslash A$. If $y \in A$, then $\rho(x, y) = 0$, a contradiction. If $y \in X \backslash A$, then $\rho(y, z) = 0$, a contradiction, Thus we have either $x, y \in A$ or $x, z \in X \backslash A$ or ($x \in A$ and $z \in X \backslash A$). In each of the three cases $\rho(x, z) = 1$. Hence ρ is transitive.

Let $x \in P_0(X, \rho)$. Suppose that $x \in X \backslash A$. Then there exists $y \in A$ such that $\rho(x, y) = 0$. However, this contradicts the assumption that $x \in P_0(X, \rho)$. Hence $P_0(X, \rho) \subseteq A$. ∎

We next show that the G–S result can be extended to the fuzzy context. We show that any j-strategy-proof j-FSCF is j-dictatorial. The proof follows along the lines as in [29].

Theorem 9.3.3 (*Abdelaziz, Figueira, and Meddeb* [1]) *Let v_j be a j-FSCF such that $v_j(D_0^n) = X$ and $j \in \{1, 2, 3\}$. If v_j is j-strategy-proof, then it is j-dictatorial.*

Proof. We prove the case for $j = 1$ since the cases for $j = 2, 3$ follow in an entirely similar manner. Consider a 1-strategy-proof 1-FSCF, $v_1 : D_1^n \to X$. Let $v_0 : D_0^n \to X$ be a 0-FSCF such that for all $\rho \in D_0^n$, $v_0(\rho) = v_1(\rho)$. The 0-FSCF is 0-strategy proof since v_1 is 1-strategy-proof. Thus by Theorem 9.3.1, v_0 is 0-dictatorial. Let individual 1 be the 0-dictator for v_0. We show that individual 1 is also a 1-dictator for v_0.

Let $\rho \in D_1^n$. Let x_0 be $v_1(\rho)$. Consider $P_1(X, \rho_1) = \{x \in X \mid \forall y \in X,$ $\rho_1(x, y) \geq \rho_1(y, x)\}$ be the best alternative set of individual 1. We will show that $x_0 \in P_1(X, \rho_1)$.

Let $\rho' \in D_0^n$ be a profile of individuals CLOs such that:
For $i = 1$,

$$\begin{cases} \rho_i' \text{ is a CLO on } P_1(X, \rho_1), \\ \rho_i'(x, y) = 1 \text{ and } \rho_i'(y, x) = 0 \text{ if } x \in P_1(X, \rho_1), y \in X \backslash P_1(X, \rho_1), \\ \rho_i' \text{ is a CLO on } X \backslash P_1(X, \rho_1). \end{cases}$$

For $i \neq 1$,

$$\begin{cases} \rho_i' \text{ is a CLO on } P_1(X, \rho_1), \\ \rho_i'(y, x) = 1 \text{ and } \rho_i'(x, y) = 0 \text{ if } x \in P_1(X, \rho_1), y \in X \backslash P_1(X, \rho_1). \\ \rho_i' \text{ is a CLO on } X \backslash P_1(X, \rho_1). \end{cases}$$

By Lemma 9.3.2, ρ_i' is a CLO for all $i \in N$ and $P_0(X, \rho') \subseteq P_1(X, \rho_1)$.

Suppose that $x_k = v_1(\rho | \rho_1', \ldots, \rho_i', \ldots, \rho_k')$ is the social choice when the k first individuals change their preference relations ρ_i to ρ_i' in order to contradict individual 1.

Note that $k \in \{0, 1, \ldots, n\}$. If $k = n$, then $x_n = v_1(\rho') = v_0(\rho')$. Thus x_n belongs to $P_0(X, \rho_1)$ because of the dictatorship of 1 by v_0. Therefore $x_n \in P_1(X, \rho_1)$.

Let j denote the least k in $\{0, 1, \ldots, n\}$ such that $x_k \in P_1(X, \rho_1)$. For x_0 to be a member of $P_1(x, \rho_1)$, it must be shown that $j = 0$. The proof is by contradiction. Suppose that $j \geq 1$.
If $j = 1$, then

$$x_1 = v_1(\rho | \rho_1') \in P_1(X, \rho_1) \tag{9.1}$$

$$x_0 = v_1(\rho) \notin P_1(X, \rho_1). \tag{9.2}$$

Now (9.1) implies $\forall y \in X$, $\rho_1(x_1, y) \geq \rho_1(y, x_1)$. Thus $\rho_1(x_1, x_0) \geq \rho_1(x_0, x_1)$. Also (9.2) implies $\exists x \in X$ such that $\rho_1(x, x_0) > \rho(x_0, x)$. Therefore, x_1 is 1-manipulable by individual 1 at ρ.
If $j > 1$, then

$$\begin{aligned} x_j &= v(\rho | \rho_1', \rho_2', \ldots, \rho_j') \in P_1(X, \rho_1), \\ x_{j-1} &= v_1(\rho | \rho_1', \rho_2', \ldots, \rho_{j-1}') \notin P_1(X, \rho_1). \end{aligned}$$

Thus $\rho_j'(x_{j-1}, x_j) = 1$ and $\rho_j'(x_j, x_{j-1}) = 0$. Hence

$$\rho_j'(x_{j-1}, x_j) > \rho_j'(x_j, x_{j-1}).$$

Suppose $(\rho | \rho_1', \ldots, \rho_j')$ is the profile of individuals' preference relations. If individual j declares a fuzzy preference ρ_j instead of a crisp relation ρ_j', then $v_1(\rho | \rho_1', \ldots, \rho_j')$ is changed in his favor. Therefore, v_1 is 1-manipulable by individual j at $(\rho | \rho_1', \ldots, \rho_j')$.

Consequently, j must be equal to 0. Hence $x_0 = v_1(\rho) \in P_1(X, \rho_1)$ for all $\rho \in D_1^n$. It follows that individual 1 is also a 1-dictator. ∎

We have shown how an individual could manipulate a social choice even if the preferences of the individuals are ambiguous. Three definitions of strategy-proofness were proposed and it was shown in all the three cases that the G-S's theorem remains valid.

9.4 Non-Manipulable Partitioning

The final two sections are based on the results of Duddy, Perote-Pena, and Piggins [13]. Consider a group of individuals that wishes to classify a set of alternatives as belonging to one of two sets. Since the individuals may disagree, an aggregation rule is employed to determine a compromise outcome. If it is required that the social classification should not be imposed nor be manipulable, then it is shown that the only such rules are dictatorial. A survey concerning the potential for social choices to be manipulated can be found in Barbera [3].

Consider a set N of individuals that wish to partition a finite set X into two subsets, say A and B. Suppose the individual evaluators submit crisp, non-empty, strict subsets of X. The subset submitted by an evaluator contains the alternatives that this evaluator classifies as belonging to set A. The complement of this set is the set of alternatives in X that the evaluator classifies as belonging to set B.

Let X be a finite set of alternatives containing at least three members. Let $N = \{1, \ldots, n\}, n \geq 2$, denote the set of evaluators. Let ρ be a function of X into $\mathcal{P}(N)$. A **profile** is a function ρ of X into $\mathcal{P}(N)$ such that for all $i \in N$, there exists $x, y \in X$ such that $i \in \rho(x)\backslash\rho(y)$. The set $\rho(x)$ can be interpreted as the set of evaluators who have voted for x at profile ρ. We require that for every evaluator i there exists at least one alternative x that the evaluator has voted for and at least one alternative y that the evaluator has not. Let $\mathcal{P}$ denote the set of all profiles.

A **social classification** is a function $v : X \to [0, 1]$ such that (i) if the range of v contains 1 then it also contains a number $< 1/2$ and (ii) if the range of v contains 0, then it also contains a number $> 1/2$. Let V denote the set of all social classifications. In the remainder, of the chapter "greater than" and "less than" mean in the strict sense.

An **aggregation rule** is a function $f : \mathcal{P} \to V$. If ρ and ρ' are distinct profiles and there exists $i \in N$ such that $\rho(x)\backslash\{i\} = \rho'(x)\backslash\{i\}$ for all $x \in X$, then we say that ρ and ρ' differ at evaluator i only.

We now define some properties that aggregation rules may satisfy.

Definition 9.4.1 [13 *Let f be an aggregation rule. Then f is said to be (or satisfy)*

(1) **Non-manipulable (NM)** : $\forall\, x \in X$, $\forall i \in N$, and $\forall \rho, \rho' \in \mathcal{P}$, if ρ and ρ' differ at evaluator i only, then $i \notin \rho(x)$ implies $v(x) \leq v'(x)$;

(2) **Non-imposition (NI)** : $\forall x \in X$, $\exists \rho, \rho' \in \mathcal{P}$ such that $v(x) = 0$ and $v'(x) = 1$;

(3) **Dictatorial:** $\exists i \in N$ such that $\forall x \in X$ and $\forall \rho \in \mathcal{P}$, $i \in \rho(x)$ implies $v(x) > 1/2$ and $i \notin P(x)$ implies $v(x) < 1/2$;

(4) **Monotone independent (MI)** : $\forall x \in X$ and $\forall \rho, \rho' \in \mathcal{P}$, if $\rho(x) \subseteq \rho'(x)$, then $v(x) \leq v'(x)$.

In other words, if every evaluator who votes for x at profile ρ also votes for x at profile ρ', then the social classification of x at ρ must be no greater that it is at ρ'.

Lemma 9.4.2 *If an aggregation rule is non-manipulable, then it satisfies monotone independence.*

Proof. Let f be an aggregation rule that is NM. Assume that f is not MI. Then there exists $\rho, \rho' \in \mathcal{P}$ and $x \in X$ such that $\rho(x) \subseteq \rho'(x)$ and $v(x) > v'(x)$. Let $\rho = \rho^{(0)}$ and $\rho' = \rho^{(n)}$. Let the sequence of profiles $\rho^{(0)}, \rho^{(1)}, \ldots, \rho^{(n)}$ be such that at $\rho^{(1)}$ evaluator 1 votes as he/she does at ρ while evaluators $2, \ldots, n$ vote as they do at profile ρ'; at profile $\rho^{(k)}$ evaluators $1, \ldots, k$ vote as they do at profile ρ while evaluators $k+1, \ldots, n$ vote as they do at profile $\rho', k = 1, \ldots, n$. Since $v^{(0)}(x) > v^{(n)}$, there exist profiles $\rho^{(t-1)}$ and $\rho^{(t)}$ such that $v^{(t-1)}(x) > v^{(t)}(x)$. By construction $t \notin \rho^{(t-1)}(x)$ and $\rho^{(t-1)}$ and $\rho^{(t)}$ differ only at evaluator t. However, $t \notin \rho^{(t-1)}(x)$ and $v^{(t-1)}(x) > v^{(t)}$ which contradicts NM. ∎

The following condition is used in the next result.

Unanimity: For all $x \in X$ and for all $v \in V$, $\rho(x) = \emptyset$ implies $v(x) = 0$ and $\rho(x) = N$ implies $v(x) = 1$.

Lemma 9.4.3 *If a non-manipulable aggregation rule satisfies non-imposition, then it satisfies unanimity.*

Proof. Let f be an aggregation rule that is NI and NM. Let $x \in X$. Then there exists $\rho \in \mathcal{P}$ such that $v(x) = 0$. Let $\rho^* \in \mathcal{P}$ be a profile such that $\rho^*(x) = \emptyset$. By Lemma 9.4.2, f is MI since it is NM. Since $\rho^*(x) \subseteq \rho(x), v^*(x) \leq v(x)$ by MI. Hence $v^*(x) = 0$. Let $\rho^{**} \in \mathcal{P}$ be a profile such that $\rho^{**}(x) = N$. By NI there exists $\rho' \in \mathcal{P}$ with $v'(x) = 1$. Since $\rho'(x) \subseteq \rho^{**}(x)$, we have $v'(x) \leq v^{**}(x)$ by MI. Thus $v^{**}(x) = 1$. ∎

Neutrality: $\forall x, y \in X$ and $\forall \rho, \rho' \in \mathcal{P}$,

(1) if $\rho(x) = \rho'(y)$, then $v(x)$ and $v'(y)$ are both less than $1/2$ or both greater than $1/2$ or both equal to $1/2$;

(2) if $\rho(x) = N \backslash \rho'(y)$, then one of $v(x)$ and $v'(y)$ is greater than $1/2$ if and only if the other is less than $1/2$.

Lemma 9.4.4   *If a non-manipulable aggregation rule satisfies non-imposition, then it satisfies neutrality.*

Proof. Suppose f is an aggregation rule that is NM and NI. Let $x, y \in X$ and $\rho, \rho' \in \mathcal{P}$ be such that either $\rho(x) = N \backslash \rho'(y)$ or $\rho(x) = \rho'(y)$.

(1) Suppose $x \neq y$ and $\rho(x) = N \backslash \rho'(y)$. Let ρ^* be a profile such that $\rho^*(x) = \rho(x), \rho^*(y) = \rho'(y)$, and $\rho'(x) = \emptyset$ for all $z \in X \backslash \{x, y\}$. By Unanimity, $v^*(z) = 0$ for all $z \in X \backslash \{x, y\}$. Thus if either $v^*(x)$ or $v^*(y)$ is less than $1/2$, then the other is greater than $1/2$. This holds for $v(x)$ and $v(y)$ since f satisfies MI by Lemma 9.4.2 and $v(x) = v^*(x)$ and $v(y) = v^*(y)$ by MI.

By replacing $\emptyset$ with N and 0 with 1, a similar argument shows that if either $v(x)$ and $v'(y)$ is greater than $1/2$, then the other must be less than $1/2$.

(2) Suppose $x \neq y$ and $\rho(x) = \rho'(y)$. Let $z \in X \backslash \{x, y\}$. Let ρ'' be a profile such that $\rho''(z) = N \backslash \rho(x)$. Also, $\rho''(x) = N \backslash \rho'(y)$. By case (1), if either one of $v''(z)$ or $v(x)$ is less than $1/2$, then the other is greater than $1/2$. This is also the case for $v''(z)$ and $v'(y)$. Thus it follows that $v(x)$ and $v'(y)$ are both greater than or both less than or both equal to $1/2$.

(3) Suppose $x = y$ and $\rho(x) = \rho'(y)$. Then $v(x) = v'(y)$ by MI.

(4) Suppose $x = y$ and $\rho(x) = N \backslash \rho'(y)$. Let $z \in X \backslash \{x\}$ and let ρ^{**} be a profile such that $\rho^{**}(z) = \rho(x)$. Then by case (2), $v^{**}(z)$ and $v(x)$ are both greater than, both less than or both equal to $1/2$. Since $\rho^{**}(z) = N \backslash \rho'(y)$, we have by (1) that if either of $v^{**}(z)$ and $v'(y)$ is less than $1/2$, then the other is greater than $1/2$. Hence it follows that if either of $v(x)$ and $v'(y)$ is less than $1/2$, then the other is greater than $1/2$. Thus neutrality holds. ∎

Theorem 9.4.5 (*Duddy, Perote-Pena, and Piggins* [13]) *If f is a non-manipulable aggregation rule that satisfies non-imposition, then f is dictatorial.*

Proof. Let $x \in X$. Consider a sequence of profiles $\rho^{(0)}, \rho^{(1)}, \ldots, \rho^{(n)}$ such that $\rho^{(0)}(x) = \emptyset, \ldots, \rho^{(i)}(x) = \{1, \ldots, i\}$ for $i = 1, \ldots, n$. By Lemma 9.4.3, f satisfies unanimity and hence $v^{(0)}(x) = 0$ and $v^{(n)}(x) = 1$. Thus there exists $j \in N$ such that $v^{(j-1)}(x) \leq 1/2$ and $v^{(j)}(x) > 1/2$. Let ρ' be a profile such that $\rho'(y) = N \backslash \rho^{(j)}(x)$, $\rho'(x) = \{j\}$ and $\rho'(z) = \rho^{(j-1)}$ for all $z \in X \backslash \{x, y\}$. By Lemma 9.4.4, f satisfies Neutrality which in turn implies that since $v^{(j-1)} \leq 1/2$, $v'(z) \leq 1/2$ for all $z \in X \backslash \{x, y\}$. Since $v^{(j)}(x) > 1/2$, we have $v'(y) < 1/2$ by neutrality. Thus by the definition of social classification, we have that $v'(x) > 1/2$. (This is true even though j is the only individual who votes for x at profile ρ'.) Let $\rho'' \in \mathcal{P}$ be such that $\rho''(x) = N \backslash \rho'(x)$. By Neutrality, $v''(x) < 1/2$. Let $\rho \in \mathcal{P}$. If $j \in \rho(x)$, then $\rho'(x) \subseteq \rho(x)$ and so $v(x) > 1/2$ by MI. If $j \notin \rho(x)$, then $\rho(x) \subseteq \rho''(x)$ and so $v(x) < 1/2$ by MI. Thus when evaluator j votes for x, the social classification of x must be greater than $1/2$. When evaluator j does not vote for x, the social classification of x

is less than $1/2$. Since f satisfies neutrality, this must also be true of every other alternative. ∎

9.5   Application

We consider a group of individuals $N = \{1, \ldots, n\}$ that wishes to classify a finite set of alternatives X as belonging to one of two sets. The set X may be countries and the classification may classify the countries as those who are politically stable and those who are not. Individuals may disagree on the classification and so an aggregation rule is applied. The aggregation rule is a function from the set V^n into V, where

$$V = \{\nu \mid X \to [0,1] \mid \exists x \in X, \ \nu(x) = 1 \Rightarrow \exists y \in X, \ \nu(y) < 1/2$$
$$\text{and } \exists x \in X, \ \nu(x) = 0 \Rightarrow \exists y \in X, \ \nu(y) > 1/2\}.$$

We justify our definition of aggregation rule as follows. Suppose the number 1 corresponds to the classification definitely politically stable. Since politically stable and not politically stable are relative concepts, a country cannot be classified definitely stable unless another country is classified as more politically unstable than politically stable. Similar reasoning can be applied to the classification of 0.

We do not want the social classification to be imposed. The nonimposition axiom guarantees that every country is definitely stable for some profile of individual partitions and definitely unstable for some other profile of individual partitions. The axiom ensures that the social classification of any country takes at least two values. This is a defensible requirement ensuring that the social classification of any country is at least minimally responsive to individual evaluations.

It is also undesirable for the social classification to be potentially manipulable. If the social classification is manipulable, then we cannot guarantee that individual evaluators will submit partitions that are sincere. It is also desirable for the social classification to be non-dictatorial.

Impossibility theorems have been discovered by Mirkin [24], Leclerc [22], and Fishburn and Rubuinstein [16], Barthelemy et al. [4], Dimitrov et al. [12], and Chambers and Miller [6]. The formal model presented here is different to the ones presented by these authors. The model here is applicable to the possible world problem posed by List [23] and to the aggregation of so-called "dichotomous preferences" Inada [20].

Example 9.5.1 *Let* $N = \{1, 2, 3\}$ *and* $X = \{w, x, y, z\}$.
For $i = 1, 2, 3$, *define* $v_i : X \to [0, 1]$ *as follows*:

$$v_1(w) = 1, \ v_1(x) = 0, \ v_1(y) = 0, \ v_1(z) = 0,$$
$$v_2(w) = 1, \ v_2(x) = 1, \ v_2(y) = 1, \ v_2(z) = 0,$$
$$v_3(w) = 1, \ v_3(x) = 1, \ v_3(y) = 1, \ v_3(z) = 0.$$

For $C \in X$, define s_C as follows: $s_C = \sum_{i=1}^{3} v_i(C)$. Let

$$M = \max \left\{ \sum_{i=1}^{3} v_i(C) \mid C \in X \right\}$$

and

$$m = \min \left\{ \sum_{i=1}^{3} v_i(C) \mid C \in X \right\}.$$

Define $v : X \to [0,1]$ by $\forall C \in X$, $v(C) = (s_C - m)/(M - m)$. Then $M = 3$ and $m = 0$. Thus it follows that

$$v(w) = 1, \ v(x) = 2/3, \ v(y) = 2/3, \ v(z) = 0.$$

Example 9.5.2 *Let $N = \{1, 2, 3\}$ and $X = \{w, x, y, z\}$.*
For $i = 1, 2, 3$, define $v_i : X \to [0,1]$ as follows:

$$v_1(w) = 1, \ v_1(x) = 0, \ v_1(y) = 0, \ v_1(z) = 1,$$
$$v_2(w) = 1, \ v_2(x) = 1, \ v_2(y) = 1, \ v_2(z) = 0,$$
$$v_3(w) = 1, \ v_3(x) = 1, \ v_3(y) = 1, \ v_3(z) = 0.$$

Define s_C, M, m, and v as in Example 9.5.1. Then $M = 3$ and $m = 1$. Hence it follows that

$$v(w) = 1, \ v(x) = 1/2, \ v(y) = 1/2, \ v(z) = 0.$$

Thus evaluator 1 has successfully lowered the social classification of both x and y. Hence the rule is manipulable.

9.6 Exercises

1. A function $C : \mathcal{F}(X)^n \to \mathcal{F}(X)$ is called a fuzzy choice function. Let C be a fuzzy choice function.

   (1) Then C is called manipulable if there exists $x \in X$, $\sigma \in \mathcal{F}(X)^n$, $i \in N$, and $\sigma_i' \in \mathcal{F}(X)$ such that (a) $C(\sigma)(x) < \sigma_i(x)$ and $C(\sigma)(x) < C(\sigma_{M\setminus i}, \sigma_i')(x)$ or (b) $C(\sigma)(x) > \sigma_i(x)$ and $C(\sigma)(x) > C(\sigma_{M\setminus i}, \sigma_i')(x)$.

   (2) C is called strategy-proof if it's not manipulable.

   (3) C is called weakly Paretian if for all $\sigma \in \mathcal{F}(X)^n$ and all $x \in X$, $\max\{\sigma_i(x) \mid i \in N\} \geq C(\sigma)(x) \geq \min\{\sigma_i(x) \mid i \in N\}$.

   (4) C is said to satisfy the σ-only condition if for all $\sigma, \sigma' \in \mathcal{F}(X)^n$ and all $x \in X$, $\sigma_i(x) - \sigma_i'(x) \ \forall i \in N$, implies $C(\sigma)(x) = C\sigma')(x)$.

   (5) C is said to be monotonic if for all $x \in X$ and all $\sigma, \sigma' \in F(X)^n$, $\sigma_i(x) \leq \sigma_i'(x)$ implies $C(\sigma)(x) \leq (\sigma')(x)$.

   If C is a strategy-proof fuzzy choice function, prove that it satisfies the σ-only condition.

2. If C is a strategy-proof fuzzy choice function, prove that it is weakly Paretian.

3. If C is a strategy-proof fuzzy choice function, prove that it is monotonic.

4. Let M be a fuzzy choice function defined as follows: $\forall \sigma \in \mathcal{F}(X)^n \ \forall x \in X, M(\sigma)(x) = \text{median}\{p_1, \ldots, p_{n-1}, \sigma_1(x), \ldots, \sigma_n(x)\}$, where $p_i \in [0,1]$, $i = 1, \ldots, n-1$. Prove that M is strategy-proof. (M is called a fuzzy augmented median voter rule.)

5. Let $\sigma_i'(x) = 1$ and $\sigma_i''(x) = 0$ for all $x \in X$ and some $i \in N$. Prove that a fuzzy choice function C is strategy-proof if and only if for all $\sigma \in \mathcal{F}(X)^n$ and all $x \in X$ and all $i \in N, C(\sigma)(x) = \text{median}\{\sigma_i(x), C(\sigma_{N\setminus i}, \sigma_i'')(x), C(\sigma_{N\setminus i}, \sigma_i')(x)\}$.

6. Prove that a fuzzy choice function C is strategy-proof if and only if it is a fuzzy augmented median voter rule.

9.7   References

1. F. B. Abdelaziz, J. R. Figueira, and O. Meddeb, On the manipulability of the fuzzy social choice functions, *Fuzzy Sets and Systems,* 159 (2008) 177–184.

2. K. J. Arrow, *Social Choice and Individual Values,* Wiley, New York, 1951.

3. S. Barbera, An introduction to strategy-proof social choice functions, *Soc. Choice Welf.,* 18 (2001) 619–653.

4. J-P Barthelemy, B. Leclerc, and B. Monjardet, On the use of ordered sets in problems of comparison and consensus of classifications, *Journal of Classification,* 3 (1986) 187–234.

5. C. R. Barrett, P. K. Pattianaik, and M. Salles, On choosing rationality when preferences are fuzzy, *Fuzzy Sets and Systems,* 19 (1990) 1–10.

6. C. P. Chambers and A. D. Miller, Rules for aggregating information, *Social Choice and Welfare,* 36 (2011) 75–82.

7. M. Dasgupts and R. Deb, Transitivity and fuzzy preferences, *Social Choice Welfare,* 13 (1996) 305–318.

8. S. Diaz, B. De Baets, and S. Montes, Additive decomposition of fuzzy pre-orders, *Fuzzy Sets and Systems,* 158 (2007) 830–842.

9. S. Diaz, S. Montes, and B. De Baets, Transitive decompositions of fuzzy preference relations: the case of nilpotent, *Kybernetika,* 40 (2004) 71–88.

10. S. Diaz, S. Montes, and B. De Baets, Transitive bounds in additive fuzzy preference structures, *IEEE Trans. Fuzzy Systems,* 15 (2007) 275–286.

11. F. Dietrich and C. List, Strategy-proof judgement aggregation, *Economics and Philosophy,* 23 (2007) 269–300.

12. D. Dimitrov, T. Marchant, and D. Mishra, Separability and aggregation of equivalence relations, *Economic Theory,* DOI 10.1007/s00199-011-0601-2 (2011).

13. C. Duddy, J. Perote-Pena, and A. Piggins, Non-manipulable partitioning, *New Mathematics and Natural Computation,* 8 (2012) 273–282.

14. B. Dutta, Fuzzy preference and social choice, *Math. Social Sci.,* 13 (1987) 215–229.

15. B. Dutta, S, C. Panda, and P. K. Pattanaik, Exact choices and fuzzy preferences, *Math. Soc. Sci.,* 11 (1986) 53–68.

16. P. C. Fishburn and A. Rubinstein, Aggregation of equivalence relations, *Journal of Classification,* 3 (1986) 61–65.

17. J. Fodor and M. Roubens, *Fuzzy Preference Modelling and Multicriteria Decision Support,* Kluwer Academic Publishers, Dordrecht, 1994.

18. J. L. Garcia-Lapresta and B. Liamazares, Aggregation of fuzzy preferences: Some rules of the mean, *Social Choice and Welfare,* 17 9(2000) 673–690.

19. A. Gibbard, Manipulation of voting schemes: A general result, *Econometrica,* 41 (1973) 587–601.

20. K. Inada, The simple majority decision rule, *Econometrica,* 37 (1969) 490–506.

21. P. Kulshreshtha and B. Shecker, Interrelationships among fuzzy preference-based choice functions and significance of rationality conditions: a taxonomic and intuitive perspective, *Fuzzy Sets and Systems,* 109 (2000) 429–445.

22. B. Leclerc, Efficient and binary consensus functions on transitively valued relations, *Mathematical Social Sciences,* 8 (1984) 45–61.

23. C. List, Which worlds are possible? A judgment aggregation problem, *Journal of Philosophical Logic,* 37 (2008) 57–65.

24. B. G. Mirkin, *On the problem of reconciling partitions,* in H. M. Blalock (ed.) Quantitative Sociology: International Perspectives on Mathematical and Statistical Modeling, Academic Press New York, 1975.

25. B. Peleg, *Game Theoretic Analysis in Communities*, Cambridge University Press, Cambridge, 1984.

26. M. A. Satterthwaite, Strategy-proofness and Arrow's conditions: existence and corresponding theorems for voting procedures and social welfare functions, *J. Economic Theory*, 10 (1975) 187–217.

27. K. Sengupta, Fuzzy preferences and Orlovsky choice procedure, *Fuzzy Sets and Systems*, 93 (1998) 231–234.

28. K. Sengupta, Choice rules with fuzzy preferences: Some characterizations, *Social Choice and Welfare*, 16 (1999) 259–272.

29. F. Tang, Fuzzy preferences and social choice, *Bull. Economic Res.*, 46 (1994) 263–269.

30. L. A. Zadeh, Quantitative fuzzy semantics, *Information Sciences*, 3 (1971) 159–176.

Chapter 10

Similarity of Fuzzy Choice Functions

In this chapter, we focus on the work of Georgescu [19, 22]. Zadeh [47] introduced the notion of a similarity relation in order to generalize the notion of an equivalence relation. Trillas and Valverde [40] extended the notion of a similarity relation for an arbitrary t-norm. See also [27, p. 254].

Samuelson [34] introduced revealed preference theory in order to define the rational behavior of consumers in terms of a preference relation. Others including Uzawa [42], Arrow [1], Sen [35], Suzumura [38] have developed a revealed preference theory in a context of choice functions. Preferences of the individuals can be imprecise, due possibly to human subjectivity or incomplete information. Vague preferences can be mathematically modelled by fuzzy relations, [9-12, 14, 28]. However, the choice can be exact or vague even if the preference is ambiguous. When the choice is exact, it can be mathematically described by a crisp choice function. The case of vague preferences and exact choices considered in [3] has been presented in Chapter 7. In situations such as negotiations on electronic markets, the choices are fuzzy.

Various types of fuzzy choice functions have been presented in Chapters 2 and 7. See also [2, 33, 45]. Georgescu [16 , 17] presented a fuzzy generalization of classic theory on revealed preference which generalized the results of [1, 34, 35].

In this chapter, Georgescu's degree of similarity of two fuzzy choice functions is presented. It is analogous to the degree of similarity of two fuzzy sets and induces a similarity relation on the set of all choice functions defined on a collection of fuzzy choice functions. The degree of similarity of two choice functions is defined to be the pair $(*, t)$, where $*$ is a t-norm and t is an element of the interval $[0, 1]$.

There is an interesting notion involving similarity relations and that is Poincaré's paradox. Poincaré stated that in the physical world "equal" really means "indistinguishable." This consideration leads to the paradox that objects A and B may be indistinguishable and B and C may be indistinguishable and yet A and C can be distinguished. Thus Poincaré denies transitivity in the real world. One way to model indistinguishability is by the use of a t-norm in such a way that indistinguishability between objects is measured. Let E be a fuzzy relation on a set X such that E is reflexive, symmetric and max-$*$ transitive. If elements x, y in X are indistinguishable of degree t, i.e., $E(x, y) = t$ and y, z are indistinguishable of degree s, i.e., $E(y, z) = s$, then x, z are indistinguishable of degree $E(x, z) \geq t * s$. Here we use the term "similar" rather "indistinguishable."

Much of fuzzy revealed preference theory focuses on the connection of the fuzzy choice functions and fuzzy revealed preferences associated with them. These connections are expressed by several partial functions between the set of choice functions and the set of fuzzy preference relations. These functions connect the similarity of fuzzy choice functions and the similarity of fuzzy preference relations. These functions translate the $(*, t)$-equality of fuzzy choice functions into the $(*, t)$-equality of fuzzy preference relations [7, 15, 46].

The degree of similarity of two fuzzy choice functions allows for the evaluation of how rational a fuzzy choice function is. Consequently, a ranking of a collection of fuzzy choice functions with respect to their rationality can be obtained. This way of evaluating the rationality of a fuzzy choice function with the consistency condition $F\alpha$ studied in [17, 18] is also presented.

10.1   Fuzzy Choice Functions

We assume $*$ is a left-continuous t-norm throughout this chapter. The Fodor nilpotent minimum t-norm is left continuous, but not continuous. Recall that the residuum associated with a t-norm $*$ is $a \to b = \vee\{t \in [0, 1] \mid a * t \leq b\}$. The negation operation $\neg$ associated with $*$ is defined by $\neg a = a \to 0 = \vee\{t \in [0, 1] \mid a * t = 0\}$.

We next state some known and useful lemmas.

Lemma 10.1.1 (*Bělohlávek* [6], *Hájek* [25]) *The following properties hold for all $a, b, c \in [0, 1]$:*
    (1) $(a \to b) * (b \to c) \leq a \to c$.
    (2) $a * (a \to b) \leq a \wedge b$; *if $*$ is continuous, then $a * (a \to b) = a \wedge b$.*

Lemma 10.1.2 (*Bělohlávek* [6], *Hájek* [25]) *For any index set I and $a, a_i, b_j \in [0, 1]$, the following properties hold:*
    (1) $a \to (\wedge\{a_i \mid i \in I\}) = \wedge\{a \to a_i \mid i \in I\}$
    (2) $(\vee\{a_i \mid i \in I\}) \to a = \wedge\{a_i \to a \mid i \in I\}$;
    (3) $\vee\{a_i \to a \mid i \in I\} \leq (\wedge\{a_i \mid i \in I\}) \to a$;
    (4) $(\vee\{a_i \mid i \in I\}) * a = \vee\{a_i \to a \mid i \in I\}$;
    (5) $(\wedge\{a_i \mid i \in I\}) * (\wedge\{b_j \mid j \in I\}) \leq \wedge\{a_i * b_j \mid i, j \in I\}$.

Recall that the biresiduum associated with the left-continuous t-norm $*$ is defined by

$$\overleftrightarrow{\rho}(a, b) = a \leftrightarrow b = (a \to b) * (b \to a).$$

Lemma 10.1.3 (*Turunen* [41]) *The following properties hold for all* $a, b, c, d \in [0, 1]$.
(1) $\overleftrightarrow{\rho}(a, 1) = a$;
(2) $a = b \Leftrightarrow \overleftrightarrow{\rho}(a, b) = 1$;
(3) $\overleftrightarrow{\rho}(a, b) = \overleftrightarrow{\rho}(b, a)$;
(4) $\overleftrightarrow{\rho}(a, b) \leq \overleftrightarrow{\rho}(\neg a, \neg b)$;
(5) $\overleftrightarrow{\rho}(a, b) * \overleftrightarrow{\rho}(b, c) \leq \overleftrightarrow{\rho}(a, c)$;
(6) $\overleftrightarrow{\rho}(a, b) \wedge \overleftrightarrow{\rho}(c, d) \leq \overleftrightarrow{\rho}(a \vee c, b \vee d)$;
(7) $\overleftrightarrow{\rho}(a, b) * \overleftrightarrow{\rho}(c, d) \leq \overleftrightarrow{\rho}(a * c, b * d)$;
(8) $\overleftrightarrow{\rho}(a, b) * \overleftrightarrow{\rho}(c, d) \leq \overleftrightarrow{\rho}(a \to c, b \to d)$.

Lemma 10.1.4 *Let* X *be a non-empty set and let* $f, g : X \to [0, 1]$ *be functions. Then*
(1) $\overleftrightarrow{\rho}(\wedge\{f(x) \mid x \in X\}, \wedge\{g(x) \mid x \in X\}) \geq \wedge\{\overleftrightarrow{\rho}(f(x), g(x)) \mid x \in X\}$;
(2) $\overleftrightarrow{\rho}(\vee\{f(x) \mid x \in X\}, \vee\{g(x) \mid x \in X\}) \geq \wedge\{\overleftrightarrow{\rho}(f(x), g(x)) \mid x \in X\}$.

Recall that a fuzzy subset μ of X is called normal if $\mu(x) = 1$ for some $x \in X$. For $\mu, \nu \in \mathcal{FP}(X)$, let

$$I(\mu, \nu) = \wedge\{\mu(x) \to \nu(x) \mid x \in X\} \text{ and } E(\mu, \nu) = \wedge\{\mu(x) \leftrightarrow \nu(x) \mid x \in X\}.$$

Clearly, $\mu \subseteq \nu \Leftrightarrow I(\mu, \nu) = 1$ and $\mu = \nu \Leftrightarrow E(\mu, \nu) = 1$. For all $x \in X$, we have

$$I(\mu, \nu) \leq \mu(x) \to \nu(x) \text{ and } E(\mu, \nu) \leq \mu(x) \leftrightarrow \nu(x).$$

$I(\mu, \nu)$ is called the **subsethood degree** of μ and ν and $E(\mu, \nu)$ the **degree of equality** (or **degree of similarity**) of μ and ν, [6, p. 82].

Recall that a fuzzy preference relation ρ on X is called strongly total if $\rho(x, y) = 1$ or $\rho(y, x) = 1$ for any distinct $x, y \in X$. ρ is reflexive and strongly total if and only if it is strongly connected.

Let ρ_1 and ρ_2 be two fuzzy preference relations. The **degree of similarity** $E(\rho_1, \rho_2)$ of ρ_1 and ρ_2 is defined as follows:

$$E(\rho_1, \rho_2) = \wedge\{\rho_1(x, y) \leftrightarrow \rho_2(x, y) \mid x, y \in X\}.$$

A fuzzy relation E on X is said to be a **similarity relation** if it is reflexive, symmetric and max-$*$ transitive.

Let $0 \leq t \leq 1$. We say that two fuzzy subsets μ, ν of X are $(*, t)$-equal, written $\mu =_{(*, t)} \nu$, if $E(\mu, \nu) \geq t$. The $(*, t)$-equality generalizes the Cai t-equality [7]. See also [46].

We present some fuzzy choice functions introduced in [18]. We also present some fuzzy preference relations associated with fuzzy choice functions.

A fuzzy choice space is a pair $(X, \mathcal{B})$ where X is a non-empty set of alternatives and $\mathcal{B}$ is a non-empty family of non-zero fuzzy subsets of X. A fuzzy choice function on $(X, \mathcal{B})$ is a function $C : \mathcal{B} \to \mathcal{FP}^*(X)$ such that for all $\sigma \in \mathcal{B}, C(\sigma) \subseteq \sigma$.

The members of $\mathcal{B}$ can be given an interesting interpretation. They can be considered as vague criteria or vague attributes. If $x \in X$ and $\sigma \in \mathcal{B}$, then $\sigma(x)$ can be thought of as the degree to which x is available with respect to σ. It represents the degree to which x satisfies the criterion σ and $C(\sigma)(x)$ expresses the degree to which the alternative x can be chosen with respect to σ. Thus a fuzzy choice problem can be thought of as a multicriterial decision-making problem.

Results of previous chapters are proved under the assumption that the choice function C satisfies the following hypotheses:

(H1) For all $\sigma \in \mathcal{B}$, $C(\sigma)$ is a normal fuzzy subset of X;

(H2) $\mathcal{B}$ includes the characteristic functions $1_{\{x_1,\dots,x_n\}}$ for any $n \geq 1$ and $x_1, \dots, x_n \in X$.

Since $C(\sigma) \subseteq \sigma, \sigma$ is also normal in H1 for all $\sigma \in \mathcal{B}$. From H2, it follows immediately that $\mathcal{B}$ includes all non-empty finite subsets of X in the crisp case.

Let $(X, \mathcal{B})$ be a fuzzy choice space and ρ a fuzzy preference relation on X. For all $\sigma \in \mathcal{B}$, we define the fuzzy subsets $M(\sigma, \rho)$ and $G(\sigma, \rho)$ of X as in Chapter 2:

$$
\begin{aligned}
M(\sigma, \rho)(x) &= \sigma(x) * (\wedge\{(\sigma(y) * \rho(y, x)) \to \rho(x, y) \mid y \in X\}), \\
G(\sigma, \rho)(x) &= \sigma(x) * (\wedge\{\sigma(y) \to \rho(x, y) \mid y \in X\}).
\end{aligned}
$$

The functions $M(\_, \rho) : \mathcal{B} \to \mathcal{FP}(X)$ and $G(\_, \rho) : \mathcal{B} \to \mathcal{FP}(X)$ are not in general fuzzy choice functions. In the case of the Gödel t-norm $\wedge$, a sufficient condition for $M(\_, \rho)$ and $G(\_, \rho)$ to be fuzzy choice functions is for ρ to be a reflexive and max-$*$ transitive fuzzy preference relation on a finite set X and $\mathcal{B} \subseteq \mathcal{FP}(X)$, see Proposition 2.3.11.

The functions $M(\_, \rho)$ and $G(\_, \rho)$ allow the notion of rationality of fuzzy choice functions to be introduced. Let C be a fuzzy choice function on $(X, \mathcal{B})$. Then C is called M-**rational** if $C = M(\_, \rho)$ for some preference relation ρ on X and is called G-**rational** if $C = G(\_, \rho)$ for some preference relation ρ on X. In case of the Gödel t-norm $\wedge$, if ρ is reflexive and strongly total, then $M(\_, \rho) = G(\_, \rho)$ [18]. In this case, M-rationality and G-rationality coincide.

Let C be a fuzzy choice function on $(X, \mathcal{B})$. Define the fuzzy revealed preferences $\rho_C, \overline{\rho}_C$ and $\overline{\pi}_C$ on X by for all $x, y \in X$,

$$
\begin{aligned}
\rho_C(x, y) &= \vee\{C(\sigma)(x) * \sigma(y) \mid \sigma \in \mathcal{B}\}; \\
\overline{\rho}_C(x, y) &= C(1_{\{x,y\}})(x); \\
\overline{\pi}_C(x, y) &= \vee\{C(\sigma)(x) * \sigma(y) * \neg C(\sigma)(y) \mid \sigma \in \mathcal{B}\}.
\end{aligned}
$$

Let $x, y \in X$. Then $\rho_C(x, y)$ is the degree of truth of the statement that there is a criterion σ such that the alternative x is chosen with respect to criterion σ and alternative y satisfies σ. Also, $\overline{\rho}_C(x, y)$ represents the degree of truth of the statement that the alternative x is chosen from $\{x, y\}$ and $\overline{\pi}_C(x, y)$ is the degree of truth of the statement that there is a criterion σ with respect to which x is chosen and y is rejected.

The functions ρ_C, $\overline{\rho}_C$ and $\overline{\pi}_C$ are fuzzy versions of some preference relations studied in classical revealed preference theory [1, 32, 35].

Let C be a fuzzy choice function. Then C is said to be **G-normal** if $C = G(\_, \rho_C)$ and is said to be **M-normal** if $C = M(\_, \rho_C)$) [16]. G-normality is a special case of G-rationality and M-normality is a special case of M-rationality.

10.2 Similarity of Fuzzy Choice Functions

We present two of Georgescu's notions, the degree of similarity of fuzzy choice functions and the $(*, t)$-equality of fuzzy choice functions and her results that establish some correspondences between the similarity of fuzzy choice functions and the similarity of fuzzy preference relations. We determine how these correspondences translate the $(*, t)$-equality of fuzzy choice functions into the $(*, t)$-equality of fuzzy preference relations.

Recall that $*$ denotes a left-continuous t-norm.

Definition 10.2.1 *Let C_1, C_2 be two fuzzy choice functions on $(X, \mathcal{B})$. The* ***degree of similarity*** *$E(C_1, C_2)$ of C_1 and C_2 is defined to be*

$$E(C_1, C_2) = \wedge\{\overleftrightarrow{\rho}\,(C_1\,(\sigma)\,(x)\,, C_2\,(\sigma)\,(x)) \mid \sigma \in \mathcal{B}, x \in X\}.$$

*For $t \in [0, 1]$, we say that C_1 and C_2 are $(*, t)$-**equal**, written $C_1 =_{(*, t)} C_2$, if $E(C_1, C_2) \geq t$.*

Proposition 10.2.2 *Let C_1, C_2, C_3 be fuzzy choice functions on $(X, \mathcal{B})$. Then the following statements hold.*
(1) *$C_1 = C_2$ if and only if $E(C_1, C_2) = 1$;*
(2) *$E(C_1, C_2) = E(C_2, C_1)$;*
(3) *$E(C_1, C_2) * E(C_2, C_3) \leq E(C_1, C_3)$.*

Proof. Conditions (1) and (2) clearly hold.
(3) By Lemma 10.1.2(5) and Lemma 10.1.3(5), we have that

$$
\begin{aligned}
E(C_1, C_2) * E(C_2, C_3) &= (\wedge\{(C_1\,(\sigma)\,(x) \leftrightarrow C_2\,(\sigma)\,(x)) \mid \sigma \in \mathcal{B}, x \in X\}) * \\
&\quad (\wedge\{(C_2\,(\sigma)\,(x) \leftrightarrow C_3\,(\sigma)\,(x)) \mid \sigma \in \mathcal{B}, x \in X\}) \\
&\leq \wedge\{(C_1\,(\sigma)\,(x) \leftrightarrow C_2\,(\sigma)\,(x)) * \\
&\quad (C_2\,(\sigma)\,(x) \leftrightarrow C_3\,(\sigma)\,(x)) \mid \sigma \in \mathcal{B}, x \in X\} \\
&\leq \wedge\{C_1\,(\sigma)\,(x) \leftrightarrow C_3\,(\sigma)\,(x) \mid \sigma \in \mathcal{B}, x \in X\} \\
&= E(C_1, C_3).
\end{aligned}
$$

∎

Let $\mathcal{FCF}(X, \mathcal{B})$ denote the set of fuzzy choice functions on $(X, \mathcal{B})$ and by $\mathcal{FR}(X)$ the set of fuzzy preference relations on X. By Proposition 10.2.2, E is a similarity relation on $\mathcal{FCF}(X, \mathcal{B})$.

Let A, B be two non-empty sets. The notation $A \rightarrowtail B$ denotes a partial function from A to B, i.e., a function f whose domain $\mathrm{Dom}(f)$ is a subset of A and whose image $\mathrm{Im}(f)$ is a subset of B.

Consider the partial functions $\phi_i : \mathcal{FR}(X) \rightarrowtail \mathcal{FCF}(X, \mathcal{B}), i = 1, 2$, where

$$\mathrm{Dom}(\phi_1) \;=\; \{\rho \in \mathcal{FR}(X) \mid M(\_, \rho) \in \mathcal{FCF}(X, \mathcal{B})\},$$
$$\mathrm{Dom}(\phi_2) \;=\; \{\rho \in \mathcal{FR}(X) \mid G(\_, \rho) \in \mathcal{FCF}(X, \mathcal{B})\},$$

and are defined by $\phi_1(\rho) = M(\_, \rho)$ for all $\rho \in \mathrm{Dom}(\phi_1)$ and $\phi_2(\rho) = G(\_, \rho)$ for all $\rho \in \mathrm{Dom}(\phi_2)$.

Consider the partial functions, $\psi_i : \mathcal{FCF}(X, \mathcal{B}) \rightarrowtail \mathcal{FR}(X), i = 1, 2, 3$, defined by for all $C \in \mathcal{FCF}(X, \mathcal{B})$,

$$\psi_1(C) = \rho_C, \quad \psi_2(C) = \bar{\rho}_C, \quad \text{and} \quad \psi_3(C) = \bar{\pi}_C.$$

The functions ϕ_i and ψ_i provide correspondences between the fuzzy choice functions on $(X, \mathcal{B})$ and certain fuzzy preference relations on X. They provide the means by which the results for fuzzy choice functions are proved. We are interested in the way these functions preserve similarity.

The next result shows to what extent the similarity of the fuzzy preference relations is transformed by the functions ϕ_1 and ϕ_2 into the similarity of the corresponding choice functions.

Theorem 10.2.3 (*Georgescu* [19]) (1) *For all* $\rho_1, \rho_2 \in Dom(\phi_1)$,

$$E(\phi_1(\rho_1), \phi_2(\rho_2)) \geq E(\rho_1, \rho_2) * E(\rho_1, \rho_2);$$

(2) *For all* $\rho_1, \rho_2 \in Dom(\phi_2), E(\phi_2(\rho_1), \phi_2(\rho_2)) \geq E(\rho_1, \rho_2)$.

Proof. (1) By Lemma 10.1.3(2), (7), (8) and Lemma 10.1.4(1), we have that

$$
\begin{aligned}
&E(\phi_1(\rho_1), \phi_1(\rho_2)) \\
=\; &\wedge\{\overleftrightarrow{\rho}(\phi_1(\rho_1)(\sigma)(x), \phi_1(\rho_2)(\sigma)(x)) \mid \sigma \in \mathcal{B}, x \in X\} \\
=\; &\wedge\{\overleftrightarrow{\rho}(M(\sigma, \rho_1)(x), M(\sigma, \rho_2)(x)) \mid \sigma \in \mathcal{B}, x \in X\} \\
=\; &\wedge\{\overleftrightarrow{\rho}(\sigma(x) * \wedge\{(\sigma(y) * (\rho_1(y, x)) \rightarrow \rho_1(x, y)) \mid y \in X\}, \\
&\quad \sigma(x) * \wedge\{(\sigma(y) * \rho_2(y, x)) \rightarrow \rho_2(x, y) \mid y \in X\} \mid \sigma \in \mathcal{B}, x \in X\} \\
\geq\; &\wedge\{\overleftrightarrow{\rho}(\wedge(\sigma(y) * \rho_1(y, x)) \rightarrow \rho_1(x, y) \mid y \in X\}, \\
&\quad \wedge\{(\sigma(y) * \rho_2(y, x)) \rightarrow \rho_2(x, y) \mid y \in X\} \mid \sigma \in \mathcal{B}, x \in X\} \\
\geq\; &\wedge\{\overleftrightarrow{\rho}((\sigma(y) * \rho_1(y, x)) \rightarrow \rho_1(x, y), \\
&\quad (S(y) * \rho_2(y, x)) \rightarrow \rho_2(x, y)) \mid \sigma \in \mathcal{B}, x, y \in X\} \\
\geq\; &\wedge\{\overleftrightarrow{\rho}(\sigma(y) * \rho_1(y, x), \sigma(y) * \rho_2(y, x)) \\
&\quad * \overleftrightarrow{\rho}(\rho_1(x, y), \rho_2(x, y)) \mid \sigma \in \mathcal{B}, x \in X\} \\
\geq\; &\wedge \overleftrightarrow{\rho}(\rho_1(y, x), \rho_2(y, x)) * \overleftrightarrow{\rho}(\rho_1(x, y), \rho_2(x, y)) \mid x, y \in X\} \\
\geq\; &E(\rho_1, \rho_2) * E(\rho_1, \rho_2).
\end{aligned}
$$

(2) By Lemma 10.1.3(2), (7), (8) and Lemma 10.1.4(1), we have

$$
\begin{aligned}
&E(\phi_2(\rho_1), \phi_2(\rho_2)) \\
={}& \wedge\{\overleftrightarrow{\rho}(\phi_1(\rho_1)(\sigma)(x), \phi_1(\rho_2)(\sigma)(x)) \mid \sigma \in \mathcal{B}, x \in X\} \\
={}& \wedge\{\overleftrightarrow{\rho}(G(\sigma, \rho_1)(x), G(\sigma, \rho_2)(x)) \mid \sigma \in \mathcal{B}, x \in X\} \\
={}& \wedge\{\overleftrightarrow{\rho}(\sigma(x) * \wedge\{\sigma(y) \to \rho_1(x, y) \mid y \in X\}, \\
& \quad \sigma(x) * \wedge\{\sigma(y) \to \rho_2(x, y) \mid y \in X\}) \mid \sigma \in \mathcal{B}, x \in X\} \\
\geq{}& \wedge\{\overleftrightarrow{\rho}(\wedge\{\sigma(y) \to \rho_1(x, y) \mid y \in X\}, \\
& \quad \wedge\{\sigma(y) \to \rho_2(x, y)\}) \mid \sigma \in \mathcal{B}, x \in X\} \\
\geq{}& \wedge\{\overleftrightarrow{\rho}(\sigma(y) \to \rho_1(x, y), (\sigma(y) \to \rho_2(x, y)) \mid \sigma \in \mathcal{B}, x \in X\} \\
\geq{}& \wedge\{\overleftrightarrow{\rho}(\rho_1(x, y), \rho_2(x, y)) \mid x, y \in X\} = E(\rho_1, \rho_2).
\end{aligned}
$$

∎

Corollary 10.2.4 (1) If $\rho_1 =_{(*,t)} \rho_2$, then

$$\phi_1(\rho_1) =_{(*,t*t)} \phi_1(\rho_2) \text{ and } \phi_2(\rho_1) =_{(*,t)} \phi_2(\rho_2);$$

(2) Let $*$ be the Gödel t-norm. If $\rho_1 =_{(*,t)} \rho_2$, then $\phi_1(\rho_1) =_{(*,t)} \phi_1(\rho_2)$.

The next result establishes how the functions ψ_1, ψ_2 and ψ_3 transfer the similarity of choice functions into the similarity of the corresponding fuzzy preference relations.

Theorem 10.2.5 (*Georgescu* [19]) *Let* $C_1, C_2 \in \mathcal{FCF}(X, \mathcal{B})$. *Then the following statements hold.*
(1) $E(\psi_1(C_1), \psi_1(C_2)) \geq E(C_1, C_2)$;
(2) $E(\psi_2(C_1), \psi_2(C_2)) \geq E(C_1, C_2)$;
(3) $E(\psi_3(C_1), \psi_3(C_2)) \geq E(C_1, C_2) * E(C_1, C_2)$.

Proof. (1) By Lemma 10.1.3(2), (7) and Lemma 10.1.4(2), we have that

$$
\begin{aligned}
&E(\psi_1(C_1), \psi_1(C_2)) \\
={}& \wedge\{\overleftrightarrow{\rho}(\psi_1(C_1)(x, y), \psi_1(C_2)(x, y)) \mid x, y \in X\} \\
={}& \wedge\{\overleftrightarrow{\rho}(\rho_{C_1}(x, y), \rho_{C_2}(x, y)) \mid x, y \in X\} \\
={}& \wedge\overleftrightarrow{\rho}(\vee\{C_1(\sigma)(x) * \sigma(y) \mid \sigma \in \mathcal{B}\}, \\
& \quad \vee\{C_2(\sigma)(x) * \sigma(y) \mid \sigma \in \mathcal{B}\}) \mid x, y \in X\} \\
\geq{}& \wedge\{\overleftrightarrow{\rho}(C_1(\sigma)(x) * \sigma(y), C_2(\sigma)(x) * \sigma(y)) \mid x, y \in X, \sigma \in \mathcal{B}\} \\
\geq{}& \wedge\{\overleftrightarrow{\rho}(C_1(\sigma)(x), C_2(\sigma)(x)) \mid x \in X, \sigma \in \mathcal{B}\} = E(C_1, C_2).
\end{aligned}
$$

(2) By the definitions, we have that

$$
\begin{aligned}
E(\psi_2(C_1), \psi_2(C_2)) &= \wedge\{\overleftrightarrow{\rho}(\psi_2(C_1)(x, y), \psi_2(C_2)(x, y)) \mid x, y \in X\} \\
&= \wedge\{\overleftrightarrow{\rho}(\overline{\rho}_{C_1}(x, y), \overline{\rho}_{C_2}(x, y)) \mid x, y \in X\} \\
&= \wedge\{\overleftrightarrow{\rho}(C_1(1_{\{x,y\}})(x), C_2(1_{\{x,y\}})(x)) \mid x, y \in X\}.
\end{aligned}
$$

Since $\{1_{\{x,y\}} \mid x, y \in X\} \subseteq \mathcal{B}$, it follows that

$$
\begin{aligned}
&\quad \wedge\{\overleftrightarrow{\rho}(C_1(1_{\{x,y\}})(x,y), C_2(1_{\{x,y\}})(x)) \mid x, y \in X\} \\
&\leq \ \wedge\{\overleftrightarrow{\rho}(C_1(1_{\{x,y\}})(x), C_2(1_{\{x,y\}})(x)) \\
&= \ E(C_1, C_2) \mid x, y \in X\}.
\end{aligned}
$$

(3) By Lemma 10.1.3(2), (4), (7) and Lemma 10.1.4(2), it follows that

$$
\begin{aligned}
&\quad \wedge\{\overleftrightarrow{\rho}(\psi_3(C_1)(x,y), \psi_3(C_2)(x,y)) \mid x, y \in X\} \\
&\geq \ \wedge\{\overleftrightarrow{\rho}(\overline{\pi}_{C_1}(x,y), \overline{\pi}_{C_2}(x,y)) \mid x, y \in X\} \\
&\geq \ \wedge\{\overleftrightarrow{\rho}(\overline{\pi}_{C_1}(x,y), \overline{\pi}_{C_2}(x,y)) \mid x, y \in X\} \\
&\geq \ \wedge\{\overleftrightarrow{\rho}(\vee\{C_1(\sigma)(x) * \sigma(y) * \neg C_1(\sigma)(y) \mid \sigma \in \mathcal{B}\}, \\
&\qquad \vee\{C_2(\sigma)(x) * \sigma(y) * \neg C_2(\sigma)(y) \mid \sigma \in \mathcal{B}\}) \mid x, y \in X\} \\
&\geq \ \wedge\{\overleftrightarrow{\rho}(C_1(\sigma)(x), C_2(\sigma)(x)) \\
&\qquad * \overleftrightarrow{\rho}(\neg C_1(\sigma)(y), \neg C_2(\sigma)(y)) \mid x, y \in X, \sigma \in \mathcal{B}\} \\
&\geq \ \wedge\{\overleftrightarrow{\rho}(C_1(\sigma)(x), C_2(\sigma)(x)) * \\
&\qquad \overleftrightarrow{\rho}(C_1(\sigma)(y), C_2(\sigma)(y)) \mid x, y \in X, \sigma \in \mathcal{B}\} \\
&\geq \ \wedge\{\overleftrightarrow{\rho}(C_1(\sigma)(x), C_2(\sigma)(x)) \mid x, y \in X, \sigma \in \mathcal{B}\} \\
&\qquad * \wedge\{\overleftrightarrow{\rho}(C_1(\sigma)(y), C_2(\sigma)(y)) \mid x, y \in X, \sigma \in \mathcal{B}\} \\
&= \ E(C_1, C_2) * E(C_1, C_2).
\end{aligned}
$$

∎

Corollary 10.2.6 (1) If $C_1 =_{(*,t)} C_2$, then $\psi_1(C_1) =_{(*,t)} \psi_1(C_2)$, $\psi_2(C_1) =_{(*,t)} \psi_2(C_2)$ and $\psi_3(C_1) =_{(*,t*t)} \psi_3(C_2)$;

(2) If $*$ is the Gödel t-norm, then $C_1 =_{(*,t)} C_2$ implies $\psi_3(C_1) =_{(*,t)} \psi_3(C_3)$.

Let $\widehat{C}$ denote $G(\_, \rho_C)$. Then a fuzzy choice function C is G-normal if and only if $C = \widehat{C}$. That is, if and only if the rationality of C is ensured by the fuzzy revealed preference relation ρ_C.

If the t-norm $*$ is the Gödel t-norm $\wedge$, then it was shown in [16] that $C \subseteq \widehat{C}$.

The degree of similarity $E(C, \widehat{C})$ can be considered a measure of the degree of rationality of C; the larger $E(C, \widehat{C})$ is, the more rational the choice function C is.

Recall the following consistency condition for a fuzzy choice function C defined in Chapter 2.

Condition Fα. For all $\sigma, \tau \in \mathcal{B}$ and $x \in X$, $I(\sigma, \tau) \wedge \sigma(x) \wedge C(\tau)(x) \leq C(\sigma)(x)$.

The next result connects the degree of similarity $E(C, \widehat{C})$ and the consistency condition Fα.

Theorem 10.2.7 (*Georgescu* [19]) *Let C be a fuzzy choice function on* $(X, \mathcal{B})$. *For all $\sigma, \tau \in \mathcal{B}$ and $x \in X$, $E(C, \widehat{C}) \leq (I(\sigma, \tau) \wedge \sigma(x) \wedge C(\tau)(x)) \to C(\sigma)(x)$.*

Proof. In view of Lemma 2.3.1(1), it suffices to prove that $E(C, \widehat{C}) \wedge I(\sigma, \tau) \wedge \sigma(x) \wedge C(\tau)(x) \leq C(\sigma)(x)$. We have

$$
\begin{aligned}
E(C, \widehat{C}) &= \wedge\{\rho(C(\sigma)(x), \widehat{C}(\sigma)(x)) \mid x \in X, \sigma \in \mathcal{B}\} \\
&= \wedge\{(\widehat{C}(\sigma)(x) \to C(\sigma)(x)) \mid x \in X, \sigma \in \mathcal{B}\}
\end{aligned}
$$

since $\overleftrightarrow{\rho}(C(\sigma)(x), \widehat{C}(\sigma)(x)) = \widehat{C}(\sigma)(x) \to C(\sigma)(x)$.

By Lemma 10.1.1(5) and Lemma 10.1.2(5), we have that

$$
\begin{aligned}
& I(\sigma, \tau) \wedge \sigma(x) \wedge \widehat{C}(\tau)(x) \\
=\ & \wedge\{(\sigma(u) \to \tau(u)) \mid u \in X\} \wedge \sigma(x) \wedge \tau(x) \\
& \wedge(\wedge\{\tau(u) \to \rho_C(x, u) \mid u \in X\}) \\
\leq\ & \sigma(x) \wedge (\wedge\{(\sigma(u) \to \tau(u)) \wedge (\tau(u) \to \rho_C(x, u)) \mid u \in X\}) \\
\leq\ & \sigma(x) \wedge (\wedge\{\sigma(u) \to \rho_C(x, u) \mid u \in X\} \leq \widehat{C}(\sigma)(x).
\end{aligned}
$$

Thus $I(\sigma, \tau) \wedge \sigma(x) \wedge \widehat{C}(\tau)(x) \leq \widehat{C}(\sigma)(x)$.

Therefore,

$$
\begin{aligned}
& E(C, \widehat{C}) \wedge I(\sigma, \tau) \wedge \sigma(x) \wedge C(\tau)(x) \\
\leq\ & E(C, \widehat{C}) \wedge I(\sigma, \tau) \wedge \sigma(x) \wedge \widehat{C}(\tau)(x) \\
\leq\ & E(C, \widehat{C}) \wedge \widehat{C}(\sigma)(x) \\
=\ & \wedge\{\widehat{C}(\sigma')(y) \to C(\sigma')(y) \mid x \in X, \sigma' \in \mathcal{B}\} \wedge \widehat{C}(\sigma)(x) \\
\leq\ & (\widehat{C}(\sigma)(x) \to C(\sigma)(x)) \wedge \widehat{C}(\sigma)(x) \\
=\ & \widehat{C}(\sigma)(x) \wedge C(\sigma)(x) \\
=\ & C(\sigma)(x),
\end{aligned}
$$

where the next to last equality follows by Lemma 10.1.1(2). ∎

We see that the more rational the fuzzy choice function C is, the larger the degree to which C verifies the consistency condition $F\alpha$.

Theorem 10.2.8 (*Georgescu* [19]) *Suppose the fuzzy choice function C on* $(X, \mathcal{B})$. *If C satisfies hypotheses H1 and H2, then $E(C, \widehat{C}) \leq E(\rho_C, \overline{\rho}_C)$.*

Proof. By the definition of $\widehat{C}$, we have since $C \subseteq \widehat{C}$ that

$$
\begin{aligned}
E(C, \widehat{C}) &= \wedge\{\widehat{C}(\sigma)(x) \leftrightarrow C(\sigma)(x) \mid x \in X, \sigma \in \mathcal{B}\} \\
&= \wedge\{\widehat{C}(\sigma)(x) \to C(\sigma)(x) \mid x \in X, \sigma \in \mathcal{B}\}.
\end{aligned}
$$

By the definition of $\bar{\rho}_C$, we have since $\rho_C \subseteq \bar{\rho}_C$ that

$$
\begin{aligned}
E(\rho_C, \bar{\rho}_C) &= \wedge\{\rho_C(x,y) \leftrightarrow \bar{\rho}_C(x,y) \mid x,y \in X\} \\
&= \wedge\{\rho_C(x,y) \to \bar{\rho}_C(x,y) \mid x,y \in X\}.
\end{aligned}
$$

Since ρ_C is reflexive [16] and $1_{\{x,y\}}(x) = 1, 1_{\{x,y\}}(z) = 0$ for $z \notin \{x,y\}$, we have by Lemma 10.1.1(4) that

$$
\begin{aligned}
C(1_{\{x,y\}})(x) &= 1_{\{x,y\}}(x) \wedge (\wedge\{1_{\{x,y\}}(z) \to \rho_C(x,z) \mid z \in X\}) \\
&= [1_{\{x,y\}}(x) \to \rho_C(x,x)] \wedge [1_{\{x,y\}}(y) \to \rho_C(x,y)] \\
&= \rho_C(x,x) \wedge \rho_C(x,y) \\
&= \rho_C(x,y).
\end{aligned}
$$

Thus

$$
\begin{aligned}
\rho_C(x,y) &\to \bar{\rho}_C(x,y) = \widehat{C}(1_{\{x,y\}})(x) \to \bar{\rho}_C(x,y) \\
&= \widehat{C}(1_{\{x,y\}})(x) \to C(1_{\{x,y\}})(x).
\end{aligned}
$$

Hence

$$
\begin{aligned}
E(C, \widehat{C}) &= \wedge\{\widehat{C}(\sigma)(x) \to C(\sigma)(x) \mid x \in X, \sigma \in \mathcal{B}\} \\
&\leq \wedge\{\widehat{C}(1_{\{x,y\}})(x)) \to C(1_{\{x,y\}})(x) \mid x,y \in X\} \\
&= \wedge\{\rho_C(x,y) \to \bar{\rho}_C(x,y) \mid x,y \in X\} \\
&= E(\rho_C, \bar{\rho}_C).
\end{aligned}
$$

∎

Example 10.2.9 *We consider the degree of similarity of a particular collection of fuzzy choice functions. Let* $X = \{x,y\}$ *and* $\mathcal{B} = \{\mu, \nu\}$, *where* $\mu, \nu \in \mathcal{FP}(X)$ *are defined by* $\mu = s1_{\{x\}} + 1_{\{y\}}$ *and* $\nu = 1_{\{x\}} + t1_{\{y\}}$ *for* $0 \leq s, t \leq 1$.

Consider the function $C : \mathcal{B} \to \mathcal{FP}(X)$ *defined by* $C(\mu) = q1_{\{x\}} + 1_{\{y\}}$ *and* $C(\nu) = 1_{\{x\}} + r1_{\{y\}}$, *where* $0 \leq q \leq s$ *and* $0 \leq r \leq t$. *Then* C *is a fuzzy choice function on* $(X, \mathcal{B})$. *The fuzzy preference relation* ρ_C *is as follows:*
$\rho_C(x,x) = \rho_C(y,y) = 1$;

$$
\begin{aligned}
\rho_C(x,y) &= (C(\mu)(x) \wedge \mu(y)) \vee (C(\nu)(x) \wedge \nu(y)) = t \vee q; \\
\rho_C(y,x) &= (C(\mu)(y) \wedge \mu(x)) \vee (C(\nu)(y) \wedge \nu(x)) = s \vee r.
\end{aligned}
$$

The values of $\widehat{C}(\mu)$ *and* $\widehat{C}(\nu)$ *are as follows:*

$$
\begin{aligned}
\widehat{C}(\mu)(x) &= \mu(x) \wedge [\mu(x) \to \rho_C(x,x)] \wedge [\mu(y) \to \rho_C(x,y)] = s \wedge (t \vee q), \\
\widehat{C}(\mu)(y) &= \mu(y) \wedge [\mu(x) \to \rho_C(y,x)] \wedge [\mu(y) \to \rho_C(y,y)] = s \to (s \vee r) = 1.
\end{aligned}
$$

$$
\widehat{C}(\nu)(x) = 1 \text{ and } \widehat{C}(\nu)(y) = t \wedge (s \vee r).
$$

The degree of similarity $E(C, \widehat{C})$ of C and $\widehat{C}$ is as follows:

$$
\begin{aligned}
E(C, \widehat{C}) &= \wedge\{(\widehat{C}(\sigma)(z) \to C(\sigma)(z)) \mid \sigma \in \{\mu, \nu\}, z \in \{x, y\}\} \\
&= [\widehat{C}(\mu)(x) \to C(\mu)(x)] \wedge [\widehat{C}(\mu)(y) \to C(\mu)(y)] \\
&\quad \wedge [\widehat{C}(\nu)(x) \to C(\nu)(x)] \wedge [\widehat{C}(\nu)(y) \to C(\nu)(y)] \\
&= [(s \wedge (t \vee q)) \to q] \wedge [(t \wedge (s \vee r)) \to r] \\
&= \{[(s \wedge t) \vee (s \wedge q)] \to q\} \wedge \{[(t \wedge s) \vee (t \wedge r)] \to r]\} \\
&= [(s \wedge t) \to q] \wedge [(s \wedge t) \to r] = (s \wedge t) \to (q \wedge r) \\
&= \begin{cases} 1 & \text{if } s \wedge t = q \wedge r \\ q \wedge r & \text{if } s \wedge t > q \wedge r. \end{cases}
\end{aligned}
$$

In the case when the results of several independent experts are used, we can consider that a fuzzy choice problem models a multicriterial decision-making problem by interpreting the available fuzzy sets as vague attributes or criteria. Then it would be beneficial to rank the choice functions in order to be able to select the best ones.

The normality of the fuzzy choice function C expresses the rationality of C by the fuzzy revealed preference relation ρ_C. Thus the similarity degree $E(C, \widehat{C})$ can be considered a measure of the rationality of C. For two choice functions C_1 and C_2, if $E(C_1, \widehat{C_1}) \geq E(C_2, \widehat{C_2})$, then we can consider C_1 to be more rational than C_2.

In the next example, we illustrate how a ranking of a collection of fuzzy choice functions can be obtained by using the similarity degree $E(C, \widehat{C})$.

Example 10.2.10 *Let $X = \{x, y\}$ and let μ and ν be fuzzy subsets of X given by $\mu = \frac{1}{2}1_{\{x\}} + 1_{\{y\}}, \nu = 1_{\{x\}} + \frac{1}{2}1_{\{y\}}$. If we take μ and ν as available sets, we obtain the choice space $(X, \mathcal{B})$. Consider a decision-making problem for which there are 5 fuzzy choice problems. The 5 fuzzy choice problems are specified by the corresponding choice functions:*

$$
\begin{aligned}
C_1(\mu) &= \frac{1}{3}1_{\{x\}} + 1_{\{y\}}, \quad C_1(\nu) = 1_{\{x\}} + \frac{1}{5}1_{\{y\}}, \\
C_2(\mu) &= \frac{1}{8}1_{\{x\}} + 1_{\{y\}}, \quad C_2(\nu) = 1_{\{x\}} + \frac{1}{3}1_{\{y\}}, \\
C_3(\mu) &= \frac{1}{3}1_{\{x\}} + 1_{\{y\}}, \quad C_3(\nu) = 1_{\{x\}} + \frac{1}{3}1_{\{y\}}, \\
C_4(\mu) &= \frac{1}{8}1_{\{x\}} + 1_{\{y\}}, \quad C_4(\nu) = 1_{\{x\}} + \frac{1}{9}1_{\{y\}}, \\
C_5(\mu) &= \frac{1}{7}1_{\{x\}} + 1_{\{y\}}, \quad C_5(\nu) = 1_{\{x\}} + \frac{1}{8}1_{\{y\}}.
\end{aligned}
$$

Using the degree of similarity determined in Example 10.2.9, we have that

$$E(C_1, \widehat{C_1}) = \frac{1}{3} \wedge \frac{1}{5} = \frac{1}{5},$$

$$E(C_2, \widehat{C_2}) = \frac{1}{8} \wedge \frac{1}{3} = \frac{1}{8},$$

$$E(C_3, \widehat{C_3}) = \frac{1}{3} \wedge \frac{1}{3} = \frac{1}{3},$$

$$E(C_4, \widehat{C_4}) = \frac{1}{8} \wedge \frac{1}{9} = \frac{1}{9},$$

$$E(C_5, \widehat{C_5}) = \frac{1}{7} \wedge \frac{1}{8} = \frac{1}{8}.$$

We now can obtain the following ranking of the five fuzzy choice functions: C_3 followed by C_1 and then C_2, C_5 and finally C_4.

It follows that the fuzzy choice problem given by the fuzzy choice function C_3 is the most rational.

Georgescu provides the following interesting discussion: Most of the results from fuzzy revealed preference theory are achieved by following two methods: the first one from fuzzy choice functions to fuzzy preference relations, and the second one from fuzzy preference relations to fuzzy choice functions. This is expressed in this section by the functions ϕ_i and ψ_i.

Theorems 10.2.7 and 10.2.8 establish the behavior of the functions ϕ_i and ψ_j towards the similarity. Thus an evaluation of how the information is preserved by the functions ϕ_i and ψ_j can be obtained.

The degree of similarity $E(C, \widehat{C})$ of C and $\widehat{C}$ defines a measure of C's rationality. The more rational the act of choice represented by C is, the bigger $E(C, \widehat{C})$ is. When $E(C, \widehat{C}) = 1$, C is G-normal. Thus $E(C, \widehat{C})$ is an indicator that allows for a classification of the fuzzy choice functions with respect to their rationality, Example 10.2.10.

Theorem 10.2.8 connects this indicator with the consistency condition $F\alpha$.

10.3   Arrow Index of a Fuzzy Choice Function

We now consider the Arrow index as presented by Georgescu [22]. The Arrow index $\mathcal{A}(C)$ of a fuzzy choice function C is a fuzzy version of the classical Arrow axiom. It is a measure of the degree to which C satisfies the Fuzzy Arrow Axiom. We present Georgescu's result that $\mathcal{A}(C)$ characterizes the degree to which C is full rational. The Arrow index allows for the ranking of fuzzy choice functions with respect to their rationality.

The Arrow axiom (AA) was introduced in [1] in order to characterize the full rationality of choice functions. See also [39] for additional information. The properties AA and full rationality were later integrated by Sen [35] in a result regarding the characterization of the rationality of a crisp choice function by various conditions such as the axiom of revealed preferences (WARP), (SARP) and the congruence axioms (WCA), (SCA). The result is known as the Arrow-Sen Theorem.

In [20], Georgescu extended the classical result of [1] by showing that (FAA) is equivalent with the full rationality of fuzzy choice functions. A new concept defined in [20] is the Arrow index $\mathcal{A}(C)$ of a fuzzy choice function C. The $\mathcal{A}(C)$ expresses the degree to which C satisfies the Fuzzy Arrow Axiom.

As previously discussed, revealed preference is a concept introduced by Samuelson in 1938 [34] to measure the rationality of a consumer's behavior in terms of a preference relation associated with a demand function. This is a major concern of classical economic theory. The work of Uzawa [40], Arrow [1], Richter [32], Sen [35, 36, 37], Suzumura [38, 39] and others extended the area.

In Chapter 7, we presented the work by Banerjee where the choice function has the domain consisting of crisp sets and the range consisting of fuzzy sets. In Chapter 7, we also presented the work of Barrett, Pattanaik and Salles where the choice is unambiguous even though the preferences are ambiguous. See also Dasgupta and Deb [8].

In [16, 17, 18, 19, 20, 21] Georgescu worked with a general definition of the fuzzy choice function in order to develop a theory of revealed preference. Previously, we presented the fuzzy Arrow axiom (FAA) introduced by Georgescu [19] who showed that a fuzzy choice function satisfies FAA if and only if it is full rational. This result extends a classical result of [1] which established the equivalence between the Arrow Axiom and the full rationality of crisp choice functions (see also [39]).

We show that $\mathcal{A}(C)$ characterizes the degree to which the fuzzy choice function C is full rational. This result refines the above mentioned result from [19] which is obtained as a particular case.

In [20], the equivalence between FAA and the full rationality of fuzzy choice functions was established. Using the Arrow index to measure the degree of full rationality of a fuzzy choice function, the results of this section allow for the comparison of them from the point of view of their rationality. Then by using the Arrow index one can obtain a ranking of fuzzy choice functions with respect to their full rationality.

Definition 10.3.1 ([6]) *Define the fuzzy subsets of $\mathcal{FR}(X)$ as follows:* $\forall \rho \in \mathcal{FR}(X)$,

$$Ref(\rho) = \wedge\{\rho(x, x) \mid x \in X\};$$

$$Trans(\rho) = \wedge\{(\rho(x, y) * \rho(y, z)) \to \rho(x, z) \mid x, y, z \in X\};$$

$$ST(\rho) = \wedge\{\rho(x, y) * \rho(y, x) \mid x, y \in X, x \neq y\}.$$

Then Ref(ρ) is called the degree of reflexivity of ρ, Trans(ρ) is called the degree of transitivity of ρ, and $ST(\rho)$ the degree of strong totality of ρ.

Clearly, ρ is reflexive if and only if $Ref(\rho) = 1$, ρ is max-$*$ transitive if and only if $Trans(\rho) = 1$, and ρ is strongly total if and only if $ST(\rho) = 1$.

The indicators in Definition 10.3.1 allow for the comparison of two fuzzy preference relations ρ_1 and ρ_2 with respect to these properties. For example, if $Trans(\rho_1) \geq Trans(\rho_2)$, then the fuzzy preference relation ρ_1 can be considered to be more transitive than ρ_2.

Lemma 10.3.2 ([6]) *Let ρ be fuzzy preference relation on X and $x, y, z \in X$. Then $Trans(\rho) * \rho(x, y) * \rho(y, z) \leq \rho(x, z)$.*

Proof. $\rho(x, y) * \rho(y, z) \rightarrow \rho(x, z) = 1$ if $\rho(x, y) * \rho(y, z) \leq \rho(x, z)$ and the result is immediate in this case. Suppose $\rho(x, y) * \rho(y, z) > \rho(x, z)$. Then $(\rho(x, y) * \rho(y, z)) \rightarrow \rho(x, z) = \rho(x, z)$. Hence $Trans(\rho) \leq \rho(x, z)$ and the desired result follows. ∎

Often authors develop their results concerning fuzzy choice functions and fuzzy preference relations on the notion that social choice is governed by fuzzy preferences, but the act of choice is exact and so the choice functions are crisp, [3, 4, 5, 26]). However, several real world situations require the consideration of fuzzy choice functions [2, 14, 28, 45]. The first fuzzy choice function was emphasized by Orlovsky [30].

In [16, 17, 19, 20], Georgescu considered fuzzy choice functions C for which the domain and the range consist of fuzzy subsets of X. For this case, the available sets of alternatives are fuzzy subsets of X. This leads to the notion of the degree of availability of an alternative x with respect to an available set. The availability degree is useful when the decision maker possess partial information concerning the alternative x or when a criterion limits the possibility of choosing x. Hence the available sets can be considered criteria in decision making.

We assume in this section that hypotheses H_1 and H_2 hold. We fix a continuous t-norm $*$ and its residuum $\rightarrow$.

If C is a fuzzy choice function on $(X, \mathcal{B})$, then recall that

$$\rho_C(x, y) = \vee\{C(\mu)(x) \wedge \mu(y) \mid \mu \in \mathcal{B}\},$$

$$\overline{\rho}_C(x, y) = C(1_{\{x,y\}})(x)$$

for all $x, y \in X$.

Let $x, y \in X$. Then the real number $\rho_C(x, y)$ is the degree of truth of the statement that there exists a criterion μ such that alternative x is chosen with respect to μ and alternative y verifies μ. The real number $\overline{\rho}_C(x, y)$ represents the degree of truth of the statement that from the set $\{x, y\}$ at least the alternative x is chosen.

Lemma 10.3.3 ([16]) *Suppose C is a fuzzy choice function on $(X, \mathcal{B})$. Then $\bar{\rho}_C \subseteq \rho_C$ and ρ_C, $\bar{\rho}_C$ are reflexive and strongly total.*

Let $(X, \mathcal{B})$ be a fuzzy choice space and ρ be a fuzzy preference relation on X. For all $\mu \in \mathcal{B}$, recall the fuzzy subset $G(\mu, \rho)$ of X is defined as follows:

$$G(\mu, \rho)(x) = \mu(x) * \wedge \{\mu(y) \to \rho(x, y) \mid y \in X\}$$

for all $x \in X$. Then $G(\_, \rho)$ can be considered a function from $\mathcal{B}$ into $\mathcal{FP}(X)$.

Lemma 10.3.4 ([20]) *Suppose C is a fuzzy choice function on $(X, \mathcal{B})$. Then $C(\mu) \subseteq G(\mu, \rho_C)$ for all $\mu \in \mathcal{B}$.*

In general, $G(\_, Q)$ is not a fuzzy choice function. There might exist a $\mu \in \mathcal{B}$ such that $G(\mu, \rho)$ is the zero function. However, by Lemma 10.3.4, $G(\_, \rho_C)$ is a fuzzy choice function and is called the **image** of C.

A fuzzy choice function C on $(X, \mathcal{B})$ is called **rational** if $C = G(\_, \rho)$ for some fuzzy preference relation ρ on X. If ρ is reflexive, max-$*$ transitive, and total, then C is called **full rational**.

Let C_1, C_2 be two fuzzy choice functions on $(X, \mathcal{B})$. Recall from the previous section that the degree of similarity of C_1 and C_2 is defined by

$$E(C_1, C_2) = \wedge \{C_1(\mu)(x) \leftrightarrow C_2(\mu)(x) \mid x \in X, \mu \in \mathcal{B}\}$$

By Proposition 10.2.2, the function $(C_1, C_2) \to E(C_1, C_2)$ is a similarity relation on the set of fuzzy choice functions defined on $(X, \mathcal{B})$. Similarity of fuzzy functions is a concept analogous to similarity of fuzzy preference relations ([6, 14, 25, 27]).

The connection between the similarity of two fuzzy choice functions and the similarity of fuzzy preference relations associated with them is given by the next result.

Lemma 10.3.5 *Let C_1, C_2 be a fuzzy choice function on $(X, \mathcal{B})$. Then*

$$E(C_1, C_2) * C_1(\mu)(x) \leq C_2(\mu)(x)$$

for all $\mu \in \mathcal{B}$ and $x \in X$.

Proof. We have

$$
\begin{aligned}
E(C_1, C_2) * C_1(\mu)(x) &= C_1(\mu)(x) * \wedge \{C_1(\tau)(y) \leftrightarrow C_2(\tau)(y) \mid y \in X, \tau \in \mathcal{B}\} \\
&\leq C_1(\mu)(x) * [C_1(\mu)(x) \to C_2(\mu)(x)] \\
&= C_1(\mu)(x) \wedge C_2(\mu)(x) \\
&\leq C_2(\mu)(x).
\end{aligned}
$$

∎

Lemma 10.3.6 *Let C be a fuzzy choice function on $(X, \mathcal{B})$ and ρ be a fuzzy preference relation on X. Then*

$$E(C, G(\_, \rho)) * C(\mu)(x) * \mu(y) \leq \rho(x, y)$$

for all $\mu \in \mathcal{B}$ and $x, y \in X$.

Proof. By Lemma 10.3.5, we have

$$E(C, G(\_, \rho)) * C(\mu)(x) \leq G(\mu, \rho)(y) \leq \mu(y) \rightarrow \rho(x, y).$$

Hence the desired inequality follows. ∎

Arrow's axiom (AA) is a condition introduced in [1] to characterize full rationality of crisp choice functions. A crisp choice function C is said to satisfy AA if for all available sets of alternatives S_1, S_2, the following implication holds:

$$S_1 \subseteq S_2 \Rightarrow S_1 \cap C(S_2) = \emptyset \text{ or } S_1 \cap C(S_2) = C(S_1).$$

As in Georgescu [20], we say that a fuzzy choice function C on $(X, \mathcal{B})$ satisfies the **fuzzy Arrow axiom** (FAA) if for all μ_1, $\mu_2 \in \mathcal{B}$ and all $x \in X$ the following statement holds:

$$I(\mu_1, \mu_2) * \mu_1(x) * C(\mu_2)(x) \leq E(\mu_1 \cap C(\mu_2), C(\mu_1)).$$

FAA can be interpreted as follows: "For all x, μ_1, and μ_2, the maximum degree that x is chosen from μ_2, x belongs to μ_1 and μ_1 is included in μ_2, is less than or equal to the degree that the set of alternatives chosen from μ_2, and also in μ_1 is equal to the set of alternatives chosen from μ_1," [21].

In [20], Georgescu extended the classical result of [1] involving AA by showing that FAA is equivalent with the full rationality of fuzzy choice functions. A new concept defined in [20] is the Arrow index $\mathcal{A}(C)$ of a fuzzy choice function C. The Arrow index provides the degree to which C verifies the fuzzy Arrow axiom.

It is shown in this section that $\mathcal{A}(C)$ characterizes the degree to which C is full rational. As stated by Georgescu in [22]: This type of result changes the perspective of studying the rationality of fuzzy choice functions. The attention is no longer focused on rational fuzzy choice functions, but the entire class of fuzzy choice functions has been taken into account with respect to their rationality. By the Arrow index, one can appreciate how rational any fuzzy choice function is. This fact has direct consequences in concrete problems. More often, the choice of an agent does not fulfill conditions of rationality or full rationality. By using Arrow index, from a set of fuzzy choice functions we can select the ones with the maximum degree of rationality.

Definition 10.3.7 *Let C be a choice function on $(X, \mathcal{B})$. The **Arrow index** $\mathcal{A}(C)$ of C is defined by*

$$\mathcal{A}(C) \;=\; \wedge\{(I(\mu_1, \mu_2) * \mu_1(x) * C(\mu_2)(x)) \to E(\mu_1 \cap C(\mu_2), C(\mu_1)) \mid x \in X, \mu_1, \mu_2 \in \mathcal{B}\}.$$

$\mathcal{A}(C)$ is the degree to which the fuzzy choice function C satisfies the fuzzy Arrow axiom. It follows that $\mathcal{A}(C) = 1$ if and only if C verifies FAA.

Lemma 10.3.8 (*Georgescu* [22]) *Suppose $*$ is the minimum t-norm $\wedge$. If C is a fuzzy choice function $(X, \mathcal{B})$, then $\mathcal{A}(C) \wedge C(1_{\{x,y\}})(x) \wedge C(1_{\{y,z\}})(y) \le C(1_{\{x,y,z\}})(x)$.*

Proof. Assume to the contrary that

$$\mathcal{A}(C) \wedge C(1_{\{x,y\}})(x) \wedge C(1_{\{y,z\}})(y) \not\le C(1_{\{x,y,z\}})(x). \tag{10.1}$$

We show that (10.1) implies the condition

$$\mathcal{A}(C) \wedge C(1_{\{x,y\}})(x) \wedge C(1_{\{y,z\}})(y) \not\le C(1_{\{x,y,z\}})(y). \tag{10.2}$$

Assume that (10.2) does not hold, i.e.,

$$\mathcal{A}(C) \wedge C(1_{\{x,y\}})(x) \wedge C(1_{\{y,z\}})(y) \le C(1_{\{x,y,z\}})(y). \tag{10.3}$$

Now $I(1_{\{x,y\}}, 1_{\{x,y,z\}}) = 1$ and $1_{\{x,y\}}(y) = 1$. Hence

$$\begin{aligned}
&\mathcal{A}(C) \wedge C(1_{\{x,y,z\}})(y) \\
={}& \mathcal{A}(C) \wedge I(1_{\{x,y\}}, 1_{\{x,y,z\}}) \wedge 1_{\{x,y\}}(y) \wedge C(1_{\{x,y,z\}})(y) \\
\le{}& I(\chi_{\{x,y\}}, 1_{\{x,y,z\}}) \wedge 1_{\{x,y\}}(y) \wedge C(1_{\{x,y,z\}})(y) \wedge \\
& \wedge [I(1_{\{x,y\}}, 1_{\{x,y,z\}}) \wedge 1_{\{x,y\}}(y) \wedge \\
& C(1_{\{x,y,z\}})(y) \to E(1_{\{x,y\}} \cap C(1_{\{x,y,z\}}), C(1_{\{x,y\}}))].
\end{aligned}$$

Therefore, by Lemma 2.3.1(2)

$$\begin{aligned}
\mathcal{A}(C) \wedge C(1_{\{x,y,z\}})(y) \le{}& E(1_{\{x,y\}} \cap C(1_{\{x,y,z\}}), C(1_{\{x,y\}})) \\
\le{}& I(C(1_{\{x,y\}}), 1_{\{x,y\}} \cap C(1_{\{x,y,z\}})) \\
\le{}& C(1_{\{x,y\}})(x) \to (1_{\{x,y\}}(x) \wedge C(1_{\{x,y,z\}}(x)) \\
\le{}& C(1_{\{x,y\}})(x) \to C(1_{\{x,y,z\}}(x)).
\end{aligned}$$

It now follows that

$$\mathcal{A}(C) \wedge C(1_{\{x,y,z\}})(y) \wedge C(1_{\{x,y\}})(x) \le C(1_{\{x,y,z\}}(x)).$$

By (10.3),

$$\begin{aligned}
\mathcal{A}(C) \wedge C(1_{\{x,y\}})(x) \wedge C(1_{\{y,z\}})(y) \le{}& \mathcal{A}(C) \wedge C(1_{\{x,y,z\}})(y) \wedge C(1_{\{y,z\}})(x) \\
\le{}& C(1_{\{x,y,z\}})(x)
\end{aligned}$$

contradicting (10.1). Thus we obtain (10.2). Similarly, we get

$$\mathcal{A}(C) \wedge C(1_{\{x,y\}})(x) \wedge C(1_{\{y,z\}})(y) \nleq C(1_{\{x,y,z\}})(z). \qquad (10.4)$$

From (10.1), (10.2), and (10.4), we have that $C(1_{\{x,y,z\}})(x) \neq 1$, $C(1_{\{x,y,z\}})(y) \neq 1$ and $C(1_{\{x,y,z\}})(z) \neq 1$. However, since H1 holds, this contradicts that $C(1_{\{x,y,z\}})$ is a normal fuzzy subset of X. This contradiction yields the desired result. ∎

The following proposition states that max-* transitivity of the fuzzy preference relation $\bar{\rho}_C$ is ensured if the Arrow axiom holds.

Theorem 10.3.9 (*Georgescu* [22]) *Suppose* $*$ *is the minimum t-norm* $\wedge$. *If C is a fuzzy choice function* $(X, \mathcal{B})$, *then* $\mathcal{A}(C) \leq Trans(\bar{\rho}_C)$.

Proof. Let $x, y, z \in X$. We show that

$$\mathcal{A}(C) \wedge C(1_{\{x,y\}})(x) \wedge C(1_{\{y,z\}})(y) \leq C(1_{\{x,y\}})(z).$$

By Lemma 10.3.8.

$$\mathcal{A}(C) \wedge C(1_{\{x,y\}})(x) \wedge C(1_{\{y,z\}})(y) \leq C(1_{\{x,y,z\}})(x). \qquad (10.5)$$

Now

$$
\begin{aligned}
&\mathcal{A}(C) \wedge C(1_{\{x,y,z\}})(x) \\
=\ &\mathcal{A}(C) \wedge I(1_{\{x,y\}}, 1_{\{x,y,z\}}) \wedge 1_{\{x,y\}}(x) \wedge C(1_{\{x,y,z\}})(x) \\
\leq\ &I(1_{\{x,z\}}, 1_{\{x,y,z\}}) \wedge 1_{\{x,z\}}(x) \wedge C(1_{\{x,y,z\}})(y) \wedge \\
&\wedge [I(1_{\{x,z\}}, 1_{\{x,y,z\}}) \wedge 1_{\{x,z\}}(x) \wedge \\
&C(1_{\{x,y,z\}})(x) \to E(1_{\{x,z\}} \cap C(1_{\{x,y,z\}}), C(1_{\{x,z\}}))] \\
\leq\ &E(1_{\{x,z\}} \cap C(1_{\{x,y,z\}})) \\
\leq\ &(1_{\{x,y\}} \cap C(1_{\{x,y,z\}}))(x) \to C(1_{\{x,z\}})(x) \\
\leq\ &C(1_{\{x,y,z\}})(x) \to C(1_{\{x,z\}})(x).
\end{aligned}
$$

Hence we have

$$\mathcal{A}(C) \wedge C(1_{\{x,y,z\}})(x) \leq C(1_{\{x,z\}})(x).$$

By (10.5), it follows that

$$\mathcal{A}(C) \wedge C(1_{\{x,y\}})(x) \wedge C(1_{\{y,z\}})(y) \leq \mathcal{A}(C) \wedge C(1_{\{x,y,z\}})(x) \leq C(1_{\{x,z\}})(x).$$

Thus for all $x, y, z \in X$, we have

$$
\begin{aligned}
\mathcal{A}(C) &\leq (C(1_{\{x,y\}})(x) \wedge C(1_{\{y,z\}})(y)) \to C(1_{\{x,z\}})(x) \\
&= (\bar{\rho}_C(x,y) \wedge \bar{\rho}_C(y,z)) \to \bar{\rho}_C(x,z).
\end{aligned}
$$

Therefore,

$$\mathcal{A}(C) \leq \wedge\{(\bar{\rho}_C(x,y) \wedge \bar{\rho}_C(y,z)) \to \bar{\rho}_C(x,z) \mid x, y, x \in X\} \leq Trans(\bar{\rho}_C).$$

∎

Corollary 10.3.10 *Suppose $*$ is the minimum t-norm $\wedge$. If C satisfies FAA, then $\bar{\rho}_C$ is max-$*$ transitive.*

Proof. By Theorem 10.3.9, $\mathcal{A}(C) = 1$ implies $\text{Trans}(\bar{\rho}_C) = 1$. ∎

Lemma 10.3.11 (*Georgescu* [22]) *Suppose $* = \wedge$. If C is a fuzzy choice function $(X, \mathcal{B})$, then*

$$E(C, G(\_, Q)) \wedge \text{Trans}(\rho) \wedge \text{Ref}(\rho) \wedge ST(\rho) \leq \mathcal{A}(C). \qquad (10.6)$$

Proof. Let $\mu_1, \mu_2 \in \mathcal{B}$ and $x \in X$. We next prove the inequality

$$
\begin{aligned}
& E(C, G(\_, \rho)) \wedge \text{Trans}(\rho) \wedge \text{Ref}(\rho) \wedge ST(\rho) \qquad (10.7) \\
\leq\ & (I(\mu_1, \mu_2)) \wedge \mu_1(x) \wedge C(\mu_2)(x)) \rightarrow E(\mu_1 \cap C(\mu_2)), C(\mu_1)).
\end{aligned}
$$

Let $\alpha = E(C, G(\_, \rho)) \wedge \text{Trans}(\rho) \wedge \text{Ref}(\rho) \wedge ST(\rho) \wedge I(\mu_1, \mu_2) \wedge \mu_1(x) \wedge C(\mu_2)(x)$. Then the inequality in (10.7) is equivalent to the inequality

$$\alpha \leq E(\mu_1 \cap C(\mu_2), C(\mu_1)). \qquad (10.8)$$

Let $z, v \in X$. By Lemmas 10.3.6 and 10.3.2, we have that

$$
\begin{aligned}
& \alpha \wedge I(\mu_1, \mu_2) \wedge \mu_1(x) \wedge C(\mu_2)(x) \wedge C(\mu_1)(z) \wedge \mu_2(v) \\
\leq\ & \alpha \wedge [E(C, G(\_, \rho)) \wedge C(\mu_1)(z) \wedge \mu_1(x)] \\
& \wedge[E(C, G(\_, \rho)) \wedge C(\mu_1)(x) \wedge \mu_2(v)] \\
\leq\ & \alpha \wedge \rho(z, x) \wedge \rho(z, v) \\
\leq\ & \text{Trans}(\rho) \wedge \rho(z, x) \wedge \rho(x, v) \\
\leq\ & \rho(z, v).
\end{aligned}
$$

By Lemma 2.5.1, we obtain

$$\alpha \wedge C(\mu_1)(z) = \alpha \wedge I(\mu_1, \mu_2) \wedge \mu_1(x) \wedge C(\mu_2)(x) \wedge C(\mu_1)(z) \leq \mu_2(v) \rightarrow \rho(z, v)$$

for all $v \in X$. Hence

$$\alpha \wedge C(\mu_1)(z) \leq (\wedge\{\mu_2(v) \rightarrow \rho(z, v) \mid v \in X\}).$$

Also,

$$
\begin{aligned}
\alpha \wedge C(\mu_1)(z) \leq\ & I(\mu_1, \mu_2) \wedge \mu_1(x) \wedge C(\mu_2)(x) \wedge C(\mu_1)(z) \\
\leq\ & C(\mu_1)(z) \wedge I(\mu_1, \mu_2) \\
\leq\ & \mu_1(z) \wedge [\mu_1(z) \rightarrow \mu_2(z)] \\
\leq\ & \mu_2(z).
\end{aligned}
$$

Therefore,

$$\alpha \wedge C(\mu_1)(z) \leq \mu_2(z) \wedge (\wedge\{\mu_2(v) \rightarrow \rho(z, v) \mid v \in X\}) = G(\mu_2, \rho)(z).$$

Since $\alpha \wedge C(\mu_1)(z) \leq \alpha \leq E(C, G(\_, \rho))$ by Lemma 10.3.5, we have that

$$\alpha \wedge C(\mu_1)(z) \wedge E(C, G(\_, \rho)) \wedge G(\mu_2, \rho)(z) \leq C(\mu_2)(z).$$

Hence by Lemma 2.5.1(4) and (10.3), it follows that

$$
\begin{aligned}
\alpha \;\; &\leq \;\; C(\mu_1)(z) \to C(\mu_2)(z) \qquad\qquad\qquad\qquad (10.9)\\
&= \;\; [C(\mu_1)(z) \to \mu_1(z)] \wedge [C(\mu_1)(z) \to C(\mu_2)(z)]\\
&= \;\; C(\mu_1)(z) \to (\mu_1(z) \wedge C(\mu_2)(z)).
\end{aligned}
$$

Let $y, z \in X$. By Lemma 2.3.1(2), Lemma 10.3.6, and Lemma 10.3.2, we have

$$
\begin{aligned}
&\alpha \wedge C(\mu_1)(z) \wedge C(\mu_2)(z) \wedge \mu_1(y)\\
=\;\; &\alpha \wedge I(\mu_1, \mu_2) \wedge \mu_1(x) \wedge C(\mu_2)(x) \wedge \mu_1(z) \wedge C(\mu_2)(z) \wedge \mu_1(y)\\
\leq\;\; &\mu_1(y) \wedge \mu_2(y) \wedge C(\mu_2)(x) \wedge C(\mu_2)(z)\\
=\;\; &\alpha \wedge \mu_1(y) \wedge \mu_2(y) \wedge C(\mu_2)(x) \wedge C(\mu_2)(z)\\
\leq\;\; &\alpha \wedge [E(C, G(\_, \rho)) \wedge C(\mu_2)(z) \wedge \mu_2(x)] \wedge\\
&E(C, G(\_, \rho)) \wedge C(\mu_2)(x) \wedge \mu_2(y)]\\
\leq\;\; &\alpha \wedge \rho(z, x) \wedge \rho(x, y)\\
\leq\;\; &\mathrm{Trans}(\rho) \wedge \rho(z, x) \wedge \rho(x, y)\\
\leq\;\; &\rho(z, y).
\end{aligned}
$$

Thus we have that $\alpha \wedge \mu_1(z) \wedge C(\mu_2)(z) \leq \mu_1(y) \to \rho(z, y)$ for all $y \in X$. Hence

$$
\begin{aligned}
\alpha \wedge \mu_1(z) \wedge C(\mu_2)(z) \;\; &\leq \;\; \mu_1(z) \wedge (\wedge\{\mu_1(y) \to \rho(z, y) \mid y \in X\})\\
&= \;\; G(\mu_1, \rho)(z).
\end{aligned}
$$

Since $\alpha \leq E(C, G(\_, \rho))$, we obtain

$$\alpha \wedge \mu_1(z) \wedge C(\mu_2)(z) \leq E(C, G(\_, \rho)) \wedge G(\mu_1, \rho)(z) \leq C(\mu_1)(z).$$

Therefore, we have for all $z \in X$,

$$\alpha \leq (\mu_1(z) \wedge C(\mu_2)(z)) \to C(\mu_1)(z). \qquad\qquad (10.10)$$

By (10.9) and (10.10), it follows that

$$\alpha \leq \{(\mu_1(z) \wedge C(\mu_2)(z)) \leftrightarrow C(\mu_1)(z) \mid x \in X\} = E(\mu_1 \cap C(\mu_2), C(\mu_1)).$$

Thus (10.8) was proved, hence (10.7) is verified. Since (10.7) holds for all $\mu_1, \mu_2 \in B$ and $x \in X$, we obtain (10.6). ∎

Theorem 10.3.12 (*Georgescu* [22]) *Suppose $*$ is the minimum t-norm $\wedge$. If C is a fuzzy choice function $(X, \mathcal{B})$, then*

$$
\begin{aligned}
\mathcal{A}(C) \;\; &= \;\; \vee\{E(C, G(\_, \rho)) \mid \rho \in \mathcal{FR}(X)\} \wedge \mathrm{Trans}(\rho) \wedge \mathrm{Ref}(\rho) \wedge ST(\rho)]\\
&= \;\; E(C, G(\_, \bar{\rho}_C)) \wedge \mathrm{Trans}(\bar{\rho}_C).
\end{aligned}
$$

Proof. From (10.6), i.e., Lemma 10.3.11, we can derive the inequality

$$\vee\{E(C, G(\_,\rho)) \wedge \mathrm{Trans}(\rho) \wedge \mathrm{Ref}(\rho) \wedge \mathrm{ST}(\rho) \mid \rho \in \mathcal{FR}(X)\} \leq \mathcal{A}(C). \quad (10.11)$$

Now we shall prove the inequality

$$\mathcal{A}(C) \leq E(C, G(\_,\bar{\rho}_C)). \quad (10.12)$$

Let $\mu \in \mathcal{B}$ and $x \in X$. We next prove the following two inequalities:

$$\mathcal{A}(C) \leq C(\mu)(x) \to G(\mu, \bar{\rho}_C) \quad (10.13)$$

$$\mathcal{A}(C) \leq G(\mu, \bar{\rho}_C)(x) \to C(\mu)(x). \quad (10.14)$$

Let $z \in X$. Then

$$
\begin{aligned}
&\mathcal{A}(C) \wedge C(\mu)(x) \wedge C(\mu)(z) \\
={}& \mathcal{A}(C) \wedge (I(1_{\{x,y\}}, \mu) \wedge 1_{\{x,y\}}(x) \wedge C(\mu)(x) \\
\leq{}& I(1_{\{x,y\}}, \mu) \wedge 1_{\{x,y\}}(x) \wedge C(\mu)(x) \\
&\wedge[(I(1_{x,z}), \mu) \wedge 1_{\{x,y\}}(x) \wedge C(\mu)(x) \to E(1_{\{x,z\}} \cap C(\mu), C(1_{\{x,z\}}))] \\
={}& I(1_{\{x,y\}}, \mu) \wedge 1_{\{x,y\}}(x) \wedge C(\mu)(x) \wedge E(1_{\{x,z\}} \cap C(\mu), C(1_{\{x,z\}})) \\
\leq{}& E(1_{\{x,z\}} \cap C(\mu), C(1_{\{x,z\}})) \\
\leq{}& (1_{\{x,z\}}(x) \wedge C(\mu)(x) \leftrightarrow C(1_{\{x,z\}})(x)) \\
={}& C(\mu)(x) \leftrightarrow \bar{\rho}_C(x, z) \\
\leq{}& C(\mu)(x) \to \bar{\rho}_C(x, z).
\end{aligned}
$$

Hence

$$\mathcal{A}(C) \wedge C(\mu)(x) \wedge \mu(z) \leq \bar{\rho}_C(x, z).$$

Thus $\mathcal{A}(C) \wedge C(\mu)(x) \leq \mu(z) \to \bar{\rho}_C(x, z)$ for all $z \in X$. Hence

$$
\begin{aligned}
\mathcal{A}(C) \wedge C(\mu)(x) &\leq \mu(x) \wedge (\wedge\{\mu(z) \to \bar{\rho}_C(x, z) \mid z \in X\}) \\
&= G(\mu, \bar{\rho}_C)(x).
\end{aligned}
$$

Thus $\mathcal{A}(C) \wedge C(\mu)(x) \to G(\mu, \bar{\rho}_C)(x)$ and so (10.13) holds. In order to prove (10.14), we consider an element $y \in X$ such that $C(\mu)(y) = 1$. Since $C(\mu)$ is a normal fuzzy subset of X, $\mu(y) = 1$. Hence

$$G(\mu, \bar{\rho}_C)(x) \leq \mu(y) \to \bar{\rho}_C(x, y) = 1 \to \bar{\rho}_C(x, y) = \bar{\rho}_C(x, y) = C(1_{\{x,z\}})(x).$$

We note that

$$
\begin{aligned}
G(\mu, \bar{\rho}_C)(x) &\leq \mu(x) = \mu(x) \wedge \mu(y) \wedge C(\mu)(y) \\
&= I(1_{\{x,y\}}, \mu) \wedge 1_{\{x,y\}}(y) \wedge C(\mu)(y).
\end{aligned}
$$

Therefore,

$$
\begin{aligned}
&\mathcal{A}(C) \wedge G(\mu, \bar{\rho}_C) \\
\leq{}& (I(1_{\{x,y\}}, \mu) \wedge 1_{\{x,y\}}(y) \wedge C(\mu)(y)) \\
&\wedge[(I(1_{\{x,y\}}, \mu) \wedge 1_{\{x,y\}}(y) \wedge C(\mu)(y)) \to E(1_{\{x,y\}} \cap C(\mu), C(1_{\{x,y\}}))] \\
\leq{}& E(1_{\{x,y\}} \cap C(\mu), C(1_{\{x,y\}})).
\end{aligned}
$$

Since $G(\mu, \overline{\rho}_C)(x) \leq 1(\chi_{\{x,y\}})(x)$, it follows that

$$
\begin{aligned}
& \mathcal{A}(C) \wedge G(\mu, \overline{\rho}_C) \\
\leq{} & C(1_{\{x,y\}}(x)) \wedge E(1_{\{x,y\}} \cap C(\mu), C(1_{\{x,y\}})) \\
\leq{} & C(1_{\{x,y\}}(x) \wedge [(1_{\{x,y\}}(x) \vee C(\mu)(x) \to C(1_{\{x,y\}}(x))] \\
={} & \overline{\rho}_C(x,y) \wedge [C(\mu)(x) \to \overline{\rho}_C(x,y)] \\
\leq{} & \overline{\rho}_C(x,y) \wedge [\overline{\rho}_C(x,y) \to C(\mu)(x)] \\
={} & \overline{\rho}_C(x,y) \wedge C(\mu)(x) \\
\leq{} & C(\mu)(x).
\end{aligned}
$$

Hence $\mathcal{A}(C) \leq G(\mu, \overline{\rho}_C)(x) \to C(\mu)(x)$ and (10.14) holds.

The inequalities (10.13) and (10.14) hold for all $\mu \in \mathcal{B}$ and $x \in X$. Thus

$$
\mathcal{A}(C) \leq \wedge\{C_1(\mu)(x) \leftrightarrow G(\mu, \overline{\rho}_C)(x) \mid \mu \in B, x \in X\} = E(C, G(\text{-}, \overline{\rho}_C)).
$$

Thus inequality (10.12) holds. Since $\overline{\rho}_C$ is reflexive and strongly total, we have $\mathrm{Ref}(\overline{\rho}_C) = \mathrm{ST}(\overline{\rho}_C) = 1$. Thus by Theorem 10.3.9 and (10.12), we get

$$
\begin{aligned}
\mathcal{A}(C) \leq{} & E(C, G(\text{-}, \overline{\rho}_C)) \wedge \mathrm{Trans}(\overline{\rho}_C) \\
={} & E(C, G(\text{-}, \overline{\rho}_C)) \wedge \mathrm{Trans}(\overline{\rho}_C) \wedge \mathrm{Ref}(\overline{\rho}_C) \wedge \mathrm{ST}(\overline{\rho}_C) \\
\leq{} & \vee\{E(C, G(\text{-}, \rho)) \wedge \mathrm{Trans}(\rho) \wedge \mathrm{Ref}(\rho) \wedge \mathrm{ST}(\rho) \mid \rho \in \mathcal{FR}(X)\}.
\end{aligned}
$$

∎

Corollary 10.3.13 *Assume that $*$ is the minimum t-norm $\wedge$. Then the following assertions are equivalent:*
(1) *C satisfies FAA;*
(2) *$E(C, G(\text{-}, \rho)) = \mathrm{Trans}(\rho) = \mathrm{Ref}(\rho) = \mathrm{ST}(\rho) = 1$ for some $\rho \in \mathcal{FR}(X)$*
(3) *C is full rational.*

By Corollary 10.3.13, we have that the main result of [20] holds, namely, FAA is equivalent with the full rationality.

Let ρ_C^{tc} denote the transitive closure of the fuzzy revealed preference relation ρ_C. Recall the following congruence axioms for fuzzy choice functions defined in Chapter 2.

Weak fuzzy congruence axiom (WFCA)

For all $\mu \in \mathcal{B}$ and $x, y \in X$,

$$
\rho_C(x,y) \wedge C(\mu)(y) \wedge \mu(x) \leq C(\mu)(x).
$$

Strong fuzzy congruence axiom (SFCA)

For all $\mu \in \mathcal{B}$ and $x, y \in X$,

$$
\rho_C^{tc}(x,y) \wedge C(\mu)(y) \wedge \mu(x) \leq C(\mu)(x).
$$

WFCA and SFCA are fuzzy versions of the classical congruence axioms (WCA), (SCA) [30, 33, 36, 37].

We next define indicators of the axioms WFCA and SFCA.

Definition 10.3.14 ([21]) *Let C be a fuzzy choice function on a fuzzy choice space (X, B). Define the following indicators:*
(1) $WFCA(C) = \wedge\{(\mu(x) \wedge C(\mu)(y) \wedge \rho_C(x, y)) \rightarrow C(\mu)(x) \mid x, y \in X, \mu \in B\}$
(2) $SFCA(C) = \wedge\{(\mu(x) \wedge C(\mu)(y) \wedge W(x, y)) \rightarrow C(\mu)(x) \mid x, y \in X, \mu \in B\}.$

Suppose $*$ is the minimum t-norm $\wedge$. Then the following assertions are equivalent:
(1) $WFCA(C) = 1$ if and only if C satisfies the axiom WFCA;
(2) $SFCA(C) = 1$ if and only if C satisfies the axiom SFCA.

The indicator $WFCA(C)$ (resp. $SFCA(C)$) expresses the degree to which the fuzzy choice function C satisfies the axiom $WFCA(C)$ (resp. $SFCA(C)$).

Theorem 10.3.15 ([21]) *Suppose $*$ is the minimum t-norm $\wedge$. If C is a fuzzy choice function, then*

$$
\begin{aligned}
WFCA(C) &= SFCA(C) \\
&= E(C, G(\_, \overline{\rho}_C)) \wedge \ Trans(\rho_C) = E(C, G(\_, \overline{\rho}_C)) \wedge \ Trans(\overline{\rho}_C).
\end{aligned}
$$

The next result follows from Theorems 10.3.12 and 10.3.15.

Theorem 10.3.16 *Suppose $*$ is the minimum t-norm $\wedge$. If C is a fuzzy choice function, then*

$$
\begin{aligned}
A(C) &= WFCA(C) = SFCA(C) \\
&= E(C, G(\_, \overline{\rho}_C)) \wedge \ Trans(\rho_C) = E(C, G(\_, \overline{\rho}_C)) \wedge \ Trans(\overline{\rho}_C).
\end{aligned}
$$

By Theorems 10.3.15 and 10.3.16, one obtains the equality of $WFCA(C)$ and $SFCA(C)$ with the Arrow index, and an evaluation of their value in terms of the similarity and the transitivity indicator.

Suppose that $*$ is the minimum t-norm $\wedge$. The fuzzy Arrow-Sen theorem established in [18, 21] is a particular case of Theorem 10.3.15. When C is a crisp choice function, an important part of the classic Arrow-Sen theorem ([1, 35, 39]) is obtained.

The connection as developed by Georgescu between the fuzzy choice functions and the fuzzy preference relations which rationalize them is realized by conditions on fuzzy choice functions such as the axiom of revealed preference and congruence, consistency properties. The fuzzy Arrow axiom (FAA) is one such condition.

In Georgescu [20], the equivalence between FAA and the full rationality of the fuzzy choice function was shown. The notion of full rationality combines two types of properties, namely, properties of fuzzy choice functions with rationality and properties of fuzzy preference relations with reflexivity, max-$*$ transitivity, and strong connectedness.

In this section, the relation between FAA and the full rationality has been analyzed from a different point of view. The Arrow index $\mathcal{A}(C)$ of a fuzzy choice function, an indicator which expresses the degree to which C satisfies FAA, was used rather than the Arrow Axiom. Also a numerical measure of the full rationality, expressed by the degree of similarity of fuzzy choice functions and by the indicators of reflexivity, max-$*$ transitivity, and strong completeness of the fuzzy preference relation is used.

The equivalence between FAA and the full rationality is transformed here into the equality between $\mathcal{A}(C)$ and the degree of full rationality. The result of this section strengthens the one in [20] since, as noted by Georgescu, it consists of a numerical relation valid for any fuzzy choice function. It also provides a method of ranking the fuzzy choice functions with respect to their rationality.

10.4   Exercises

The following exercises are based on [23]. If ρ is a fuzzy relation on X, then the associated fuzzy strict π of ρ is defined by $\pi(x,y) = \rho(x,y) * -\rho(y,x)$ for all $x, y \in X$. Let C be a fuzzy choice function. Define the fuzzy subsets of $\mathcal{B}$: $\mathrm{Rat}_G(C) = \vee\{E(C, G(\_, \rho) \mid \rho \in \mathcal{FR}\}$, $\mathrm{Norm}_G(C) = E(C, G(\_, \rho_C)$ and $\mathrm{RAT}_M(C) = \vee\{E(C, M(\_, \rho) \mid \rho \in \mathcal{FR}\}$, $\mathrm{Norm}(C) = E(C, M(\_, \rho_C))$. $\mathrm{Rat}_G(C)$ is called the **indicator of G-rationality** of C and $\mathrm{Rat}_M(C)$ is called the **indicator of M-rationality** of C. $\mathrm{Norm}_G(C)$ is called the **indicator of G-normality** of C and $\mathrm{Norm}_M(C)$ is called the **indicator** of M-**normality** of C.

1. Let ρ_1, ρ_2 be fuzzy relations on X. Prove that for all $x, y \in X$, $E(\rho_1, \rho_2) \wedge \pi_1(x,y) \leq \pi_2(x,y)$.

2. Let C be a fuzzy choice function on $(X, \mathcal{B})$. Prove the following statements.

   (i) $\mathrm{Norm}_M(C) \leq \mathrm{Rat}_M(C) \leq \mathrm{Rat}_G(C) = \mathrm{Norm}_G(C)$;

   (ii) If C satisfies H_1, H_2, then $\mathrm{Rat}_G(C) \leq E(C, G(\_, \bar{\rho}_C))$.

   The function $A_c(\rho) = \wedge\{\wedge\{(\pi(x_1, x_2) \wedge \ldots \wedge \pi(x_{n-1} x_n) \wedge \rho(x_n, x_1)) \rightarrow \rho(x_1, x_n) \mid x_1, \ldots, x_n \in X\} \mid n \in \mathbb{N}, n \geq 2\}$ is called **degree of acyclicity** of ρ.

3. Prove that $A_c(\rho) = \wedge\{I(\pi^{n-1} \cap \rho^{-1}, \rho) \mid n \in \mathbb{N}, n \geq 2\}$, where π^{n-1} denotes the max-min composition of π with itself $n - 1$ times.

4. Prove that $\text{Trans}(\rho) \leq A_c(\rho)$.

5. Let ρ_1, ρ_2 be fuzzy preference relations. Prove that $A_c(\rho_1) \wedge E(\rho_1, \rho_2) \leq A_c(\rho_2)$.

   Define the indicators $A_c\text{Rat}_G(C) = \vee\{E(C, G(\_, \rho)) \wedge Ac(\rho)) \mid \rho \in \mathcal{FR}\}$ and $A_c\text{Rat}_M(C) = \vee\{E(C, M(\_, \rho)) \wedge Ac(\rho)) \mid \rho \in \mathcal{FR}\}$

6. If C is a fuzzy choice function on $(X, \mathcal{B})$ that satisfies H_1 and H_2, prove that $\text{Rat}_G(C) = A_c\text{Rat}(C)$.

7. If C is a fuzzy choice function on $(X, \mathcal{B})$ that satisfies H_1 and H_2, prove that $\text{Rat}_G(C) = A_c\text{Rat}G(C) = F\alpha(C) \wedge F\text{Cond}(C)$, where $F\text{Cond}(C) = \wedge\{I(G(\mu, \overline{\rho}_C), C(\mu)) \mid \mu \in \mathcal{B}\}$. ($F\text{Cond}$ is the fuzzy version of the Condorcet property.)

10.5 References

1. K. J. Arrow, Rational choice functions and orderings, *Economica,* 26 (1959) 121–127.

2. A. Banerjee, Fuzzy choice functions, revealed preference and rationality, *Fuzzy Sets and Systems,* 70 (1995) 31–43.

3. C. R. Barrett, P. K. Pattanaik, and M. Salles, On choosing rationally when preferences are fuzzy, *Fuzzy Sets and Systems,* 34 (1990) 197–212.

4. C. R. Barrett, P. K. Pattanaik, and M. Salles, Rationality and aggregation of preferences in ordinal fuzzy frameworks, *Fuzzy Sets and Systems,* 49 (1992) 9–13.

5. C. R. Barrett, P. K. Pattanaik, and M. Salles, On the structure of fuzzy social welfare functions, *Fuzzy Sets and Systems,* 19, (1996) 1–10.

6. R. Bělohlávek, *Fuzzy Relational Systems,* Foundations and Principles, Kluwer, Dordrecht, 2002.

7. K.Y. Cai, δ-Equalities of fuzzy sets, *Fuzzy Sets and Systems,* 76 (1995) 97–112.

8. M. Dasgupta and R. Deb, Fuzzy Choice functions, *Soc. Choice Welfare,* 8 (1991) 171–182.

9. B. De Baets and J. Fodor, Twenty years of fuzzy preference relations (1978-1997), *Belg. J. Oper. Res. Statist. Comput. Sci.,* 37 (1997) 61–82.

10. B. De Baets and J. Fodor (Eds.), *Principles of Fuzzy Preference Modelling and Decision Making,* Academia Press, Gent, 2003.

11. B. De Baets, B. Van De Walle, and E. Kerre, Fuzzy preference structures without incomparability, *Fuzzy Sets and Systems,* 76 (1995) 333–348.

12. B. De Baets, B. Van De Walle, and E. Kerre, Characterizable fuzzy preference structures, *Ann. Oper. Res.,* 80 (1998) 105–136.

13. D. Dubois and H. Prade, *Fuzzy Sets and Systems, Theory and Applications,* Academic Press, New York, 1980.

14. J. Fodor and M. Roubens, *Fuzzy Preference Modelling and Multicriteria Decision Support,* Kluwer, Dordrecht, 1994.

15. I. Georgescu, A generalization of the Cai δ-equality of fuzzy sets, Proceedings of the Fuzzy Information Processing Conference, Beijing, China, 1-4 March 2003, pp. 123–127.

16. I. Georgescu, On the axioms of revealed preference in fuzzy consumer theory, *J. Systems Sci. Systems Eng.,* 13 (2004) 279–296.

17. I. Georgescu, Consistency conditions in fuzzy consumers theory, *Fund. Inform.,* 61 (2004) 223–245.

18. I. Georgescu, Rational choice and revealed preference: A fuzzy approach, Ph.D. Dissertation 60, Turku Centre for Computer Science, 2005.

19. I. Georgescu, Similarity of fuzzy choice functions, *Fuzzy Sets and Systems,* 158 (2007) 1314–1326.

20. I. Georgescu, Arrow's axiom and full rationality for fuzzy choice functions, *Social Choice and Welfare,* 28 (2007) 303–319.

21. I. Georgescu, *Fuzzy choice functions: A revealed preference approach,* Berlin, Springer, 2007.

22. I. Georgescu, Arrow index of a fuzzy choice function, *Fundamenta Informaticae,* 99 (2010) 245–261.

23. I. Georgescu, Acyclic rationality indicators of fuzzy choice functions, *Fuzzy Sets and Systems,* 160 (2009) 2673–2685.

24. I. Georgescu, Ranking fuzzy choice functions by their rationality indicators, *Fuzzy Optimization and Decision Making,* 6 (2007) 367–389.

25. P. Hájek, *Methamathematics of Fuzzy Logic,* Kluwer, Dordrecht, 2000.

26. S. Klaue, K. Kurbel, and I. Loutschko, Automated negotiations on agent-based e-marketplaces: An overview. Proceedings of the 14th Bled Electronic Commerce Conference, Bled, Slovenia 508–519 2001.

27. E. P. Klement, R. Mesiar, and E. Pap, *Triangular Norms,* Kluwer, Dordrecht, 2000.

28. P. Kulshreshtha and B. Shekar, Interrelationship among fuzzy preference-based choice function and significance of rationality conditions: a taxonomic and intuitive perspective, *Fuzzy Sets and Systems,* 109 (2000) 429–445.

29. K. Kurbel, and I. Loutschko, Towards multi-agent electronic marketplaces: What is there and what is missing? *The Knowledge Engineering Review,* 18 (2003) 33–46.

30. S. Orlovsky, Decision making with a fuzzy preference relation, *Fuzzy Sets and Systems,* 1 (1978) 155–167.

31. S. Ovchinnikov, Decision-making with ternary binary relation, in Proceedings of the Conference on Information Processing Management of Uncertainty IPMU 2004, Perugia, Italy 511–516.

32. M. Richter, Revealed preference theory, *Econometrica,* 34 (1966) 635–645.

33. M. Roubens, Some properties of choice functions based on valued binary relations, *European J. Oper. Res.,* 40 (1989) 309–321.

34. P. A. Samuelson, A note of the pure theory of consumer's behavior, *Economica,* 5 (1938) 61–71.

35. A. K. Sen, Choice functions and revealed preference, *Rev. Econom. Stud.,* 38 (1971) 307–317.

36. A. K. Sen, Social choice theory: A re-examination, *Ecometrica,* 45 (1977) 53–89.

37. A. K. Sen, Choice, *Welfare and Measurement,* Cambridge MA: MIT Press and Oxford: Blackwell, 1982.

38. K. Suzumura, Rational choice and revealed preference, *Rev. Econom. Stud.,* 43 (1976) 149–159.

39. K. Suzumura, *Rational choice, collective decisions and social welfare,* Cambridge University Press, Cambridge, 1983.

40. E. Trillas and L. Valverde, *An Inquiry into Indistinguishability Operators,* in: H. J. Skala, S. Termini, E. Trillas (Eds.), Aspects of Vagueness, Reidel, Dordrecht, 1983.

41. E. Turunen, *Mathematics Behind Fuzzy Logic,* Physica-Verlag, Wurzburg, 1999.

42. H. Uzawa, A note on preference and axioms of choice, *Ann. Inst. Statist. Math.,* 8 (1956) 35–40.

43. B, Van deWalle, S. Heitch, and P. Faratin, Coping with one-to-many multicriteria negotiations in electronic marketplaces: Beyond price discovery, Munich, Germany, 2001.

44. X. Wang, B. De Baets, E. E. Kerre, A comparative study of similarity measures, *Fuzzy Sets and Systems,* 73 (1995) 259–268.

45. Ph. de Wilde, Fuzzy utility and equilibria, *IEEE Trans. Systems Man Cybernet,* 34 (2004) 1774–1785.

46. M. Ying, Perturbations of fuzzy reasoning, *IEEE Trans. Fuzzy Systems,* 7 (1999) 625–629.

47. L.A. Zadeh, Similarity relations and fuzzy orderings, *Inform. Sci.,* 3 (1971) 177–200.

Index

Milton Keynes UK
Ingram Content Group UK Ltd.
UKHW021630071024
449327UK00020BA/1259